Study Guide and Student Solutions Manual

Chemistry for Today
General, Organic, and Biochemistry

EIGHTH EDITION

Spencer L. Seager
Weber State University

Michael R. Slabaugh
Weber State University

Prepared by

Jennifer P. Harris

BROOKS/COLE
CENGAGE Learning

Australia • Brazil • Japan • Korea • Mexico • Singapore • Spain • United Kingdom • United States

BROOKS/COLE
CENGAGE Learning

For product information and technology assistance, contact us at **Cengage Learning Customer & Sales Support, 1-800-354-9706**

For permission to use material from this text or product, submit all requests online at **www.cengage.com/permissions**
Further permissions questions can be emailed to **permissionrequest@cengage.com**

ISBN-13: 978-1-133-60699-4
ISBN-10: 1-133-60699-7

Brooks/Cole
20 Davis Drive
Belmont, CA 94002-3098
USA

Cengage Learning is a leading provider of customized learning solutions with office locations around the globe, including Singapore, the United Kingdom, Australia, Mexico, Brazil, and Japan. Locate your local office at: **www.cengage.com/global**

Cengage Learning products are represented in Canada by Nelson Education, Ltd.

To learn more about Brooks/Cole, visit **www.cengage.com/brookscole**

Purchase any of our products at your local college store or at our preferred online store **www.cengagebrain.com**

Printed in the United States of America
1 2 3 4 5 6 7 16 15 14 13 12

Table of Contents

General Chemistry

Chapter 1	Matter, Measurements, and Calculations	1
Chapter 2	Atoms and Molecules	25
Chapter 3	Electronic Structure and the Periodic Law	41
Chapter 4	Forces Between Particles	53
Chapter 5	Chemical Reactions	71
Chapter 6	The States of Matter	89
Chapter 7	Solutions and Colloids	107
Chapter 8	Reaction Rates and Equilibrium	129
Chapter 9	Acids, Bases, and Salts	145
Chapter 10	Radioactivity and Nuclear Processes	167

Organic Chemistry

Chapter 11	Organic Compounds: Alkanes	179
Chapter 12	Unsaturated Hydrocarbons	199
Chapter 13	Alcohols, Phenols, and Ethers	217
Chapter 14	Aldehydes and Ketones	233
Chapter 15	Carboxylic Acids and Esters	247
Chapter 16	Amines and Amides	263

Biochemistry

Chapter 17	Carbohydrates	277
Chapter 18	Lipids	291
Chapter 19	Proteins	305
Chapter 20	Enzymes	319
Chapter 21	Nucleic Acids and Protein Synthesis	331
Chapter 22	Nutrition and Energy for Life	343
Chapter 23	Carbohydrate Metabolism	355
Chapter 24	Lipid and Amino Acid Metabolism	369
Chapter 25	Body Fluids	381
	Answers to the Chapter Self-Tests	391

Preface

This book is designed to serve as both a reference and a study tool. Each chapter includes a Chapter Outline, Learning Objectives/Assessment statements, Solutions for the Even-numbered End of Chapter (EOC) Exercises, Additional Activities, Answers to the Additional Activities, and Self-Test Questions. The Additional Activities include specific questions for each section as well as a summary activity entitled, "Tying It All Together with a Laboratory Application." The Answers to the Self-Test Questions are at the end of the book.

The EOC Exercises marked with the ☑ symbol are questions identified as "quick checks." Your ability to answer the "quick check" questions correctly indicates how well you have met the learning objectives for the chapter. It is important to remember that your instructor may have additional learning objectives that are not identified in this text.

I hope this book is useful to you and that your experience studying chemistry is positive. The subject matter is fascinating (ask any chemist) and applies to daily life. As with any meaningful scholastic experience, you will need to study. Ideally, using this book will focus that process. If you have questions, comments, or concerns about the book, please, e-mail me at jpharris@gmail.com.

Thank you to all of the people who helped me with this and the previous editions of the book, especially my students, Sylvia Krick, Lynn Carlson, and Ruth Leslie. I also want to thank a few of the many notable chemists in my life: Dr. Fred R. Harris, my husband and best friend; Dr. Barinder Deu, my first high school chemistry teacher; Dr. Debra Feakes, my mentor at Southwest Texas State University (now, Texas State University-San Marcos); and my colleagues both at Portland Community College in Oregon and Citrus Community College in California. Additionally, I want to thank my mom and dad, COL (Ret.) and Mrs. David L. Pointer, and my sister, Colleen E. Parks, MS, RD, LD (dietitian extraordinaire), for their unfailing support.

Chapter 1: Matter, Measurements, and Calculations

CHAPTER OUTLINE

1.1 What Is Matter?
1.2 Properties and Changes
1.3 A Model of Matter
1.4 Classifying Matter
1.5 Measurement Units
1.6 The Metric System
1.7 Large and Small Numbers
1.8 Significant Figures
1.9 Using Units in Calculations
1.10 Calculating Percentages
1.11 Density

LEARNING OBJECTIVES/ASSESSMENT

When you have completed your study of this chapter, you should be able to:

1. Explain what matter is. (Section 1.1; Exercise 1.2)
2. Explain differences between the terms *physical* and *chemical* as applied to:
   a. Properties of matter (Section 1.2; Exercises 1.10 b & c)
   b. Changes in matter (Section 1.2; Exercises 1.8 a & b)
3. Describe matter in terms of the accepted scientific model. (Section 1.3; Exercise 1.12)
4. On the basis of observation or information given to you, classify matter into the correct category of each of the following pairs:
   a. Heterogeneous or homogeneous (Section 1.4; Exercise 1.22)
   b. Solution or pure substance (Section 1.4; Exercise 1.24)
   c. Element or compound (Section 1.4; Exercise 1.18)
5. Recognize the use of measurement units in everyday activities. (Section 1.5; Exercise 1.28)
6. Recognize units of the metric system, and convert measurements done using the metric system into related units. (Section 1.6; Exercises 1.30 and 1.40)
7. Express numbers using scientific notation, and do calculations with numbers expressed in scientific notation. (Section 1.7; Exercises 1.48 and 1.60)
8. Express the results of measurements and calculations using the correct number of significant figures. (Section 1.8; Exercises 1.64 and 1.66)
9. Use the factor-unit method to solve numerical problems. (Section 1.9; Exercise 1.82)
10. Do calculations involving percentages. (Section 1.10; Exercise 1.92)
11. Do calculations involving densities. (Section 1.11; Exercise 1.98)

SOLUTIONS FOR THE END OF CHAPTER EXERCISES

WHAT IS MATTER? (SECTION 1.1)

☑1.2 All matter occupies space and has mass. Mass is a measurement of the amount of matter in an object. The mass of an object is constant regardless of where the mass is measured. Weight is a measurement of the gravitational force acting on an object. The weight of an object will change with gravity; therefore, the weight of an object will be different at different altitudes and on different planets.

1.4 The distance you can throw a bowling ball will change more than the distance you can roll a bowling ball on a flat, smooth surface. When throwing a ball, gravity pulls the ball toward the ground and air resistance slows its decent. The gravitational force on the moon is approximately 1/6th the gravitational force that is present on the earth; therefore, when throwing a ball on the moon, you should be able to throw it further than you can on earth. The moon does not have air resistance. When rolling a ball, friction helps to slow down the ball. If the flat, smooth surface is the same on the earth and the moon, the amount of friction should remain constant.

1.6 The attractive force of gravity for objects near the earth's surface increases as you get closer to the center of the earth (Exercise 1.5). If the earth bulges at the equator, the people at the equator are further from the center of the earth than people at the North Pole. If two people with the same mass were weighed at the equator and at the North Pole, the person at the equator would weigh less than the person at the North Pole because the gravitational force at the North Pole is stronger than the gravitational force at the equator.

PROPERTIES AND CHANGES (SECTION 1.2)

1.8 ☑a. The two pieces of the stick still have the same chemical composition as the original stick. This was a change that did not involve composition; therefore, it is a **physical change**.

☑b. As the candle burns, it produces carbon dioxide, water, soot, and other products. This is a change that involves composition; therefore, it is a **chemical change**.

c. The pieces of rock salt have the same chemical composition as the original larger piece of rock salt. This was a change that did not involve composition; therefore, it is a **physical change**.

d. Many tree leaves are green in the spring and summer because of the green chlorophyll that is used in photosynthesis to produce energy for the tree. During these seasons, the tree stores the extra energy so that in autumn when the days grow shorter, the chlorophyll is no longer needed. As the leaves in the cell stop producing chlorophyll, the other colors present in the leaves become more visible. This change involves composition; therefore, it is a **chemical change**.

1.10 a. The phase of matter at room temperature is a **physical property** because the composition does not change while making this observation.

☑b. The reaction between two substances is a **chemical property** because the composition of the products differs from the reactants. The products for the reaction between sodium metal and water are sodium hydroxide and hydrogen gas. (Note: Predicting the products for this type of chemical reaction is covered in Section 9.6.)

☑c. Freezing point is a **physical property** because the composition does not change while making this observation.

d. The inability of a material to form new products by rusting is a **chemical property** because rust would have a different chemical composition than gold. Attempting to change the chemical composition of a material is a test of a chemical property regardless of whether the attempt is successful.

e. The color of a substance is a **physical property** because the composition does not change while making this observation.

A MODEL OF MATTER (SECTION 1.3)

☑1.12 a. Yes, the succinic acid molecules have been changed by the process. The molecules of succinic acid released at least one atom each in the form of a gas. Without those atoms, the molecules cannot be succinic acid molecules. Also, if they were still succinic acid molecules, the melting point of the remaining solid would still be 182°C.

b. No, the white solid that remains after heating is not succinic acid. The melting point of succinic acid is 182°C, but the melting point of this new solid is not 182°C.

c. The succinic acid molecules contain more atoms than the molecules of the white solid produced by this process. Some of the atoms that were originally part of the succinic acid molecules are given off as a gas. That leaves fewer atoms to be a part of the molecules of new white solid.

d. Succinic acid is **heteroatomic**. Some of the atoms were able to leave the succinic acid molecule in the form of a gas. Other atoms remained as part of a new molecule. If all of the atoms were the same type, they would have all turned into a gas or they would have all remained as a solid.

1.14 Sulfur dioxide is **heteroatomic**. If oxygen gas and solid sulfur react to form one product, then sulfur dioxide must contain oxygen atoms and sulfur atoms.

1.16 Water is **heteroatomic**. If breaking water apart into its components produces both hydrogen gas and oxygen gas, then water must contain two types of atoms.

CLASSIFYING MATTER (SECTION 1.4)

☑1.18
a. Substance A is a **compound** because it is composed of molecules that contain more than one type of atom.
b. Substance D is an **element** because it is composed of molecules that contain only one type of atom.
c. Substance E is a **compound** because it is a pure substance that can break down into at least two different materials. Substances G and J **cannot be classified** because no tests were performed on them.

1.20
a. Substance R might appear to be an element based on the tests performed. It has not decomposed into any simpler substances based on these tests; however, this is not an exhaustive list of tests that could be performed on substance R. Substance R **cannot be classified** as an element or a compound based on the information given.
b. Substance T is a **compound**. It is composed of at least two different elements because it produced two different substances on heating.
c. The solid left in part b **cannot be classified** as an element or a compound. No tests have been performed on it.

☑1.22

a.	A gold chain	It is **homogeneous** because it has the same composition throughout.
b.	Liquid eye drops	It is **homogeneous** because it has the same composition throughout.
c.	Chunky peanut butter	It is **heterogeneous** because it does not have the same composition throughout (peanut chunk vs. smooth regions).
d.	A slice of watermelon	It is **heterogeneous** because it does not have the same composition throughout (rind, meat, and seeds).
e.	Cooking oil	It is **homogeneous** because it has the same composition throughout.
f.	Italian salad dressing	It is **heterogeneous** because it does not have the same composition throughout (oil, vinegar, and seasonings).
g.	Window glass	It is **homogeneous** because it has the same composition throughout.

☑1.24 a. A pure gold chain — The chain is only made of gold; therefore, it is a **pure substance**.

b. Liquid eye drops — This is a **solution** because it contains saline (water and sodium chloride) as well as other substances.

e. Cooking oil — It can be a **solution** or a **pure substance** depending on the oil. Some oils only contain a single compound (pure substances); however, most oils are a mixture of several different compounds (solution).

g. Window glass — It is a **solution** because it contains a mixture of silicon dioxide and other components like iron.

MEASUREMENT UNITS (SECTION 1.5)

1.26 Modern society is complex and interdependent. Accomplishing projects like building a bridge, constructing a house, or machining an engine may require many different people to participate. Some people design the project, others supply the necessary materials, and yet another group does the construction. In order for the project to be successful, all of these people need a common language of measurement. Measurement is also important for giving directions, keeping track of the time people work, and keeping indoor environments at a comfortable temperature and pressure.

☑1.28 In cooking, measuring cups (units = cups) and measuring spoons (units = teaspoon, Tablespoon) are used for volume measurements. In sewing and woodworking, a tape measure (units = yards, feet, inches) is used for length measurements. In machining, dial calipers (units = inches or centimeters) are used for length measurements. In athletic events, a timer or stopwatch (units = hours, minutes, seconds) are used for time measurements. In driving, a speedometer (units = miles per hour), odometer (units = miles), and a tachometer (rotations per minute) are used for speed, distance, and rotation rate measurements, respectively. (Many other examples are possible.)

THE METRIC SYSTEM (SECTION 1.6)

☑1.30 The metric units are (a) degrees Celsius, (b) liters, (d) milligrams, and (f) seconds. The English units are (c) feet and (e) quarts.

1.32 Meters are a metric unit that could replace the English unit feet in the measurement of the ceiling height. Liters are a metric unit that could replace the English unit quarts in the measurement of the volume of a cooking pot.

1.34 a. $1.00\ \cancel{\text{L}}\left(\dfrac{1\ \mu\text{L}}{10^{-6}\ \cancel{\text{L}}}\right) = 1.00 \times 10^{6}\ \mu\text{L}$

b. $75\ \cancel{\text{kilowatts}}\left(\dfrac{1000\ \text{watts}}{1\ \cancel{\text{kilowatt}}}\right) = 7.5 \times 10^{4}\ \text{watts}$

c. $15\ \cancel{\text{megahertz}}\left(\dfrac{10^{6}\ \text{hertz}}{1\ \cancel{\text{megahertz}}}\right) = 1.5 \times 10^{7}\ \text{hertz}$

d. $200\ \cancel{\text{picometers}}\left(\dfrac{10^{-12}\ \text{meters}}{1\ \cancel{\text{picometer}}}\right) = 2.00 \times 10^{-10}\ \text{meters}$

1.36 $1\,\cancel{\text{cup}}\left(\dfrac{240\,\cancel{\text{mL}}}{1\,\cancel{\text{cup}}}\right)\left(\dfrac{1\,\text{L}}{1000\,\cancel{\text{mL}}}\right)=0.240\,\text{L}$ or $1\,\cancel{\text{cup}}\left(\dfrac{240\,\cancel{\text{mL}}}{1\,\cancel{\text{cup}}}\right)\left(\dfrac{1\,\text{cm}^3}{1\,\cancel{\text{mL}}}\right)=240\,\text{cm}^3$

1.38 $4.0\,\cancel{\text{kg}}\left(\dfrac{2.20\,\text{lbs.}}{1\,\cancel{\text{kg}}}\right)=8.8\,\text{lbs.}$

☑1.40 a. $1\,\text{m}=1.094\,\text{yd, so: }1\,\text{m}-1\,\text{yd}=1.094\,\text{yd}-1\,\text{yd}=0.094\,\cancel{\text{yd}}\left(\dfrac{3\,\cancel{\text{ft}}}{1\,\cancel{\text{yd}}}\right)\left(\dfrac{12\,\text{in}}{1\,\cancel{\text{ft}}}\right)=3.4\,\text{in}$

b. The size of 1°C is the same as 1 K; therefore, a change of 65°C is also a change of 65 K.

c. $5\,\cancel{\text{lbs.}}\left(\dfrac{1\,\text{kg}}{2.20\,\cancel{\text{lbs.}}}\right)=2.27\,\text{kg}=2\,\text{kg}$ with significant figures

1.42 a.

$1.0\,\cancel{\text{cm}^3}\left(\dfrac{1\,\cancel{\text{dm}^3}}{1000\,\cancel{\text{cm}^3}}\right)\left(\dfrac{1\,\text{kg}}{1.0\,\cancel{\text{dm}^3}}\right)=1.0\times10^{-3}\,\text{kg}$ or

$1.0\,\cancel{\text{cm}^3}\left(\dfrac{1\,\cancel{\text{dm}^3}}{1000\,\cancel{\text{cm}^3}}\right)\left(\dfrac{1\,\cancel{\text{kg}}}{1.0\,\cancel{\text{dm}^3}}\right)\left(\dfrac{1000\,\text{g}}{1\,\cancel{\text{kg}}}\right)=1.0\,\text{g}$

b. $2.0\,\cancel{\text{L}}\left(\dfrac{1.057\,\cancel{\text{qt}}}{1\,\cancel{\text{L}}}\right)\left(\dfrac{32\,\text{fl oz}}{1\,\cancel{\text{qt}}}\right)=68\,\text{fl oz}$

c. $5\,\cancel{\text{grain}}\left(\dfrac{1\,\text{mg}}{0.015\,\cancel{\text{grain}}}\right)=333\,\text{mg}=3\times10^2\,\text{mg}$ with significant figures

1.44 $°\text{F}=\frac{9}{5}(°\text{C})+32$ $\quad °\text{F}=\frac{9}{5}(36.1°\text{C})+32=97.0°\text{F}$ $\quad °\text{F}=\frac{9}{5}(37.2°\text{C})+32=99.0°\text{F}$

LARGE AND SMALL NUMBERS (SECTION 1.7)

1.46
- a. 02.7×10^{-3} Improper form because no leading zero is necessary. (2.7×10^{-3})
- b. 4.1×10^{2} Correct.
- c. 71.9×10^{-6} Improper form because only one digit should be to the left of the decimal point. (7.19×10^{-5})
- d. 10^{3} Improper form because a nonexponential term should be written before the exponential term. (1×10^{3})
- e. 0.0405×10^{-2} Improper form because one nonzero digit should be to the left of the decimal point. (4.05×10^{-4})
- f. 0.119 Improper form because one nonzero digit should be to the left of the decimal point and an exponential term should be to the right of the nonexponential term. (1.19×10^{-1})

☑1.48
- a. 14 thousand = 14,000 = 1.4×10^{4}
- b. 365 3.65×10^{2}
- c. 0.00204 2.04×10^{-3}
- d. 461.8 4.618×10^{2}

e.	0.00100		1.00×10^{-3}
f.	9.11 hundred	$= 9.11 \times 100 =$	9.11×10^{2}

1.50 a. 186 thousand mi/s $186 \times 1000 = 1.86 \times 10^{5}$ mi/s

b. 1100 million km/h $1100 \times 1{,}000{,}000 = 1.1 \times 10^{9}$ km/h

1.52 0.000 000 000 000 000 000 000 105 g

The decimal point has been moved 22 places to the left. This places 21 zeros to the right of the decimal point and before the numbers 105 g.

1.54

a.	$(8.2 \times 10^{-3})(1.1 \times 10^{-2})$	$= 9.02 \times 10^{-5}$	$= 9.0 \times 10^{-5}$ with significant figures
b.	$(2.7 \times 10^{2})(5.1 \times 10^{4})$	$= 1.377 \times 10^{7}$	$= 1.4 \times 10^{7}$ with significant figures
c.	$(3.3 \times 10^{-4})(2.3 \times 10^{2})$	$= 7.59 \times 10^{-2}$	$= 7.6 \times 10^{-2}$ with significant figures
d.	$(9.2 \times 10^{-4})(2.1 \times 10^{4})$	$= 1.932 \times 10^{1}$	$= 1.9 \times 10^{1}$ with significant figures
e.	$(4.3 \times 10^{6})(6.1 \times 10^{5})$	$= 2.623 \times 10^{12}$	$= 2.6 \times 10^{12}$ with significant figures

1.56

a.	(144)(0.0876)	$= (1.44 \times 10^{2})(8.76 \times 10^{-2})$	$= 1.26144 \times 10^{1}$	$= 1.26 \times 10^{1}$ with SF
b.	(751)(106)	$= (7.51 \times 10^{2})(1.06 \times 10^{2})$	$= 7.9606 \times 10^{4}$	$= 7.96 \times 10^{4}$ with SF
c.	(0.0422)(0.00119)	$= (4.22 \times 10^{-2})(1.19 \times 10^{-3})$	$= 5.0218 \times 10^{-5}$	$= 5.02 \times 10^{-5}$ with SF
d.	(128,000)(0.0000316)	$= (1.28 \times 10^{5})(3.16 \times 10^{-5})$	$= 4.0448 \times 10^{0}$	$= 4.04 \times 10^{0}$ with SF

1.58

a. $\dfrac{3.1\times10^{-3}}{1.2\times10^{2}} = 2.583 \times 10^{-5}$ $= 2.6 \times 10^{-5}$ with SF

b. $\dfrac{7.9\times10^{4}}{3.6\times10^{2}} = 2.194 \times 10^{2}$ $= 2.2 \times 10^{2}$ with SF

c. $\dfrac{4.7\times10^{-1}}{7.4\times10^{2}} = 6.35135 \times 10^{-4}$ $= 6.4 \times 10^{-4}$ with SF

d. $\dfrac{0.00229}{3.16} = 7.2468354 \times 10^{-4}$ $= 7.25 \times 10^{-4}$ with SF

e. $\dfrac{119}{3.8\times10^{3}} = 3.131578947 \times 10^{-2}$ $= 3.1 \times 10^{-2}$ with SF

☑1.60

a. $\dfrac{(5.3)(0.22)}{(6.1)(1.1)} = 1.7377 \times 10^{-1}$ $= 1.7 \times 10^{-1}$ with SF

b. $\dfrac{(3.8\times10^{-4})(1.7\times10^{-2})}{6.3\times10^{3}} = 1.025 \times 10^{-9}$ $= 1.0 \times 10^{-9}$ with SF

c. $\dfrac{4.8\times10^{6}}{(7.4\times10^{3})(2.5\times10^{-4})} = 2.59459 \times 10^{6}$ $= 2.6 \times 10^{6}$ with SF

d. $\dfrac{5.6}{(0.022)(109)} = 2.335279 \times 10^{0}$ $= 2.3 \times 10^{0}$ with SF

e. $\dfrac{(4.6\times10^{-3})(2.3\times10^{2})}{(7.4\times10^{-4})(9.4\times10^{-5})} = 1.520989 \times 10^{7}$ $= 1.5 \times 10^{7}$ with SF

SIGNIFICANT FIGURES (SECTION 1.8)

1.62
a. A ruler with smallest scale marking of 0.1 cm — 0.01 cm
b. A measuring telescope with smallest scale marking of 0.1 mm — 0.01 mm
c. A protractor with smallest scale marking of 1° — 0.1°
d. A tire pressure with smallest scale marking of 1 lb/in² — 0.1 lb/in²

☑1.64
a. Exactly 6 mL of water measured with a graduated cylinder that has smallest markings of 0.1 mL. — 6.00 mL
b. A temperature that appears to be exactly 37 degrees using a thermometer with smallest markings of 1°C. — 37.0°C
c. A time of exactly nine seconds measured with a stopwatch that has smallest markings of 0.1 second. — 9.00 seconds
d. Fifteen and one-half degrees measured with a protractor that has 1 degree scale markings. — 15.5°

☑1.66
a. Measured = 5.06 lbs. Exact = 16 potatoes
$$\frac{5.06\,\text{lb.}}{16\,\text{potatoes}} = 0.31625\frac{\text{lb.}}{\text{potato}} = 0.316\frac{\text{lb.}}{\text{potato}}\ \text{with SF}$$
b. Measured = percentages Exact = 5 players
$$\frac{71.2\% + 66.9\% + 74.1\% + 80.9\% + 63.6\%}{5\,\text{players}} = 71.34\%\ \text{with SF}$$

1.68

a.	0.0400	3 SF (0.0<u>400</u>)	d.	4.4×10^{-3}	2 SF
b.	309	3 SF	e.	1.002	4 SF
c.	4.006	4 SF	f.	255.02	5 SF

1.70

a.	$(3.71)(1.4)$	5.194	= 5.2 with significant figures
b.	$(0.0851)(1.2262)$	0.10434962	= 0.104 with significant figures
c.	$\frac{(0.1432)(2.81)}{(0.7762)}$	0.518412780211	= 0.518 with significant figures
d.	$(3.3\times10^{4})(3.09\times10^{-3})$	101.97	= 1.0×10^{2} with significant figures
e.	$\frac{(760)(2.00)}{6.02\times10^{20}}$	$2.52491694352 \times 10^{-18}$	= 2.52×10^{-18} with significant figures (assuming 0 in 760 is significant)

1.72

a.	$0.208 + 4.9 + 1.11$	= 6.218	= 6.2 with significant figures
b.	$228 + 0.999 + 1.02$	= 230.019	= 2.30×10^{2} with significant figures
c.	$8.543 - 7.954$	= 0.589	= 0.589 with significant figures
d.	$(3.2\times10^{-2}) + (5.5\times10^{-1})$ (Hint: Write in decimal form first, then add.)	= 0.582	= 0.58 with significant figures
e.	$336.86 - 309.11$	= 27.75	= 27.75 with significant figures
f.	$21.66 - 0.02387$	= 21.63613	= 21.64 with significant figures

1.74
a. $\frac{(0.0267 + 0.00119)(4.626)}{28.7794}$ = 0.004483037867 = 0.00448 with significant figures

b. $\dfrac{212.6-21.88}{86.37}$ $= 2.20817413454$ $= 2.208$ with significant figures

c. $\dfrac{27.99-18.07}{4.63-0.88}$ $= 2.6453333$ $= 2.65$ with significant figures

d. $\dfrac{18.87}{2.46}-\dfrac{18.07}{0.88}$ $= -12.8633592018$ $= -13$ with significant figures
(HINT: Do divisions first, then subtract.)

e. $\dfrac{(8.46-2.09)(0.51+0.22)}{(3.74+0.07)(0.16+0.2)}$ $= 3.3902741324$ $= 3$ with significant figures

f. $\dfrac{12.06-11.84}{0.271}$ $= 0.811808118081$ $= 0.81$ with significant figures

1.76 a. **Area** $(A = l \times w)$ / **Perimeter** $(P = 2(l) + 2(w))$

Black $A = 12.00\text{ cm} \times 10.40\text{ cm} = 124.8\text{ cm}^2$ $\quad P = 2(12.00\text{ cm}) + 2(10.40\text{ cm}) = 44.80\text{ cm}$

Red $A = 20.20\text{ cm} \times 2.42\text{ cm} = 48.884\text{ cm}^2 = 48.9\text{ cm}^2$ $\quad P = 2(20.20\text{ cm}) + 2(2.42\text{ cm}) = 45.24\text{ cm}$

Green $A = 3.18\text{ cm} \times 2.55\text{ cm} = 8.109\text{ cm}^2 = 8.11\text{ cm}^2$ $\quad P = 2(3.18\text{ cm}) + 2(2.55\text{ cm}) = 11.46\text{ cm}$

Orange $A = 13.22\text{ cm} \times 0.68\text{ cm} = 8.9896\text{ cm}^2 = 9.0\text{ cm}^2$ $\quad P = 2(13.22\text{ cm}) + 2(0.68\text{ cm}) = 27.80\text{ cm}$

b. **Length** / **Width**

Black $12.00\ \cancel{\text{cm}}\left(\dfrac{1\text{ m}}{100\ \cancel{\text{cm}}}\right) = 0.1200\text{ m}$ $\quad 10.40\ \cancel{\text{cm}}\left(\dfrac{1\text{ m}}{100\ \cancel{\text{cm}}}\right) = 0.1040\text{ m}$

Red $20.20\ \cancel{\text{cm}}\left(\dfrac{1\text{ m}}{100\ \cancel{\text{cm}}}\right) = 0.2020\text{ m}$ $\quad 2.42\ \cancel{\text{cm}}\left(\dfrac{1\text{ m}}{100\ \cancel{\text{cm}}}\right) = 0.0242\text{ m}$

Green $3.18\ \cancel{\text{cm}}\left(\dfrac{1\text{ m}}{100\ \cancel{\text{cm}}}\right) = 0.0318\text{ m}$ $\quad 2.55\ \cancel{\text{cm}}\left(\dfrac{1\text{ m}}{100\ \cancel{\text{cm}}}\right) = 0.0255\text{ m}$

Orange $13.22\ \cancel{\text{cm}}\left(\dfrac{1\text{ m}}{100\ \cancel{\text{cm}}}\right) = 0.1322\text{ m}$ $\quad 0.68\ \cancel{\text{cm}}\left(\dfrac{1\text{ m}}{100\ \cancel{\text{cm}}}\right) = 0.0068\text{ m}$

Area $(A = l \times w)$ / **Perimeter** $(P = 2(l) + 2(w))$

Black $A = 0.1200\text{ m} \times 0.1040\text{ m} = 0.01248\text{ m}^2$ $\quad P = 2(0.1200\text{ m}) + 2(0.1040\text{ m}) = 0.4480\text{ m}$

Red $A = 0.2020\text{ m} \times 0.0242\text{ m} = 0.0048884\text{ m}^2 = 0.00489\text{ m}^2$ $\quad P = 2(0.2020\text{ m}) + 2(0.0242\text{ m}) = 0.4524\text{ m}$

Green $A = 0.0318\text{ m} \times 0.0255\text{ m} = 8.109 \times 10^{-4}\text{ m}^2 = 8.11 \times 10^{-4}\text{ m}^2$ $\quad P = 2(0.0318\text{ m}) + 2(0.0255\text{ m}) = 0.1146\text{ m}$

Orange $A = 0.1322\text{ m} \times 0.0068\text{ m} = 8.9896 \times 10^{-4}\text{ m}^2 = 9.0 \times 10^{-4}\text{ m}^2$ $\quad P = 2(0.1322\text{ m}) + 2(0.0068\text{ m}) = 0.2780\text{ m}$

c. No, the number of significant figures in the answers remains constant. The numbers of places past the decimal are different; however, that could be fixed by rewriting all of the answers in scientific notation.

USING UNITS IN CALCULATIONS (SECTION 1.9)

1.78 a. 20 mg to grains $\frac{0.015 \text{ grains}}{1 \text{ mg}}$ c. 4 qt to liters $\frac{1 \text{ L}}{1.057 \text{ qt}}$

b. 350 mL to fl oz $\frac{0.0338 \text{ fl oz}}{1 \text{ mL}}$ d. 5 yd to meters $\frac{1 \text{ m}}{1.094 \text{ yd}}$

1.80 $$26 \cancel{\text{miles}}\left(\frac{1 \text{ km}}{0.621 \cancel{\text{miles}}}\right) = 41.8679549114 \text{ km} = 42 \text{ km}$$

☑1.82 $$250 \cancel{\text{mL}}\left(\frac{0.0338 \cancel{\text{fl oz}}}{1 \cancel{\text{mL}}}\right)\left(\frac{1 \text{ cup}}{8 \cancel{\text{fl oz}}}\right) = 1.05625 \text{ cups} = 1.1 \text{ cups}$$

(Note: Cups are not measured in 0.1 increments.)

1.84 $$18.0 \cancel{\text{kg}}\left(\frac{2.20 \text{ lbs.}}{1 \cancel{\text{kg}}}\right) = 39.6 \text{ lbs}$$ The bag is not overweight.

1.86 $$131 \frac{\cancel{\text{mg}}}{\cancel{\text{dL}}}\left(\frac{1 \text{ g}}{1000 \cancel{\text{mg}}}\right)\left(\frac{10 \cancel{\text{dL}}}{1 \text{ L}}\right) = 1.31 \frac{\text{g}}{\text{L}}$$

CALCULATING PERCENTAGES (SECTION 1.10)

1.88 $$\frac{\$25.73}{\$467.80} \times 100 = 5.500\%$$

1.90 $$\frac{1.0 \frac{\text{mg}}{\text{day}}}{1.4 \frac{\text{mg}}{\text{day}}} \times 100 = 71\%$$

☑1.92 Total = 987.1 mg + 213.3 mg + 99.7 mg + 14.4 mg + 0.1 mg = 1314.6 mg

$$\text{IgG} = \frac{987.1 \text{ mg}}{1314.6 \text{ mg}} \times 100 = 75.09\%; \text{IgA} = \frac{213.3 \text{ mg}}{1314.6 \text{ mg}} \times 100 = 16.23\%; \text{IgM} = \frac{99.7 \text{ mg}}{1314.6 \text{ mg}} \times 100 = 7.58\%;$$

$$\text{IgD} = \frac{14.4 \text{ mg}}{1314.6 \text{ mg}} \times 100 = 1.10\%; \text{IgE} = \frac{0.1 \text{ mg}}{1314.6 \text{ mg}} \times 100 = 0.008\%$$

DENSITY (SECTION 1.11)

1.94 a. $\frac{39.6 \text{ g}}{50.0 \text{ mL}} = 0.792 \frac{\text{g}}{\text{mL}}$ c. $\frac{39.54 \text{ g}}{20.0 \text{ L}} = 1.98 \frac{\text{g}}{\text{L}}$

b. $\frac{243 \text{ g}}{236 \text{ mL}} = 1.03 \frac{\text{g}}{\text{mL}}$ d. $\frac{222.5 \text{ g}}{25.0 \text{ cm}^3} = 8.90 \frac{\text{g}}{\text{cm}^3}$

1.96 $$\text{Volume} = (3.98 \text{ cm})^3 = 63.0 \text{ cm}^3$$

$$\text{Density} = \frac{\text{mass}}{\text{volume}} = \frac{718.3 \text{ g}}{(3.98 \text{ cm})^3} = 11.4 \frac{\text{g}}{\text{cm}^3}$$

☑1.98 $280\ \text{g}\left(\frac{1\ \text{mL}}{0.736\ \text{g}}\right) = 380\ \text{mL}$

ADDITIONAL EXERCISES

1.100 a. $4.5\ \text{km}\left(\frac{1000\ \text{m}}{1\ \text{km}}\right)\left(\frac{1000\ \text{mm}}{1\ \text{m}}\right) = 4.5\times10^{6}\ \text{mm}$

b. $6.0\times10^{6}\ \text{mg}\left(\frac{1\ \text{g}}{1000\ \text{mg}}\right) = 6.0\times10^{3}\ \text{g}$

c. $9.86\times10^{15}\ \text{m}\left(\frac{1\ \text{km}}{1000\ \text{m}}\right) = 9.86\times10^{12}\ \text{km}$

d. $1.91\times10^{-4}\ \text{kg}\left(\frac{1000\ \text{g}}{1\ \text{kg}}\right)\left(\frac{1000\ \text{mg}}{1\ \text{g}}\right) = 1.91\times10^{2}\ \text{mg}$

e. $5.0\ \text{ng}\left(\frac{1\ \text{g}}{10^{9}\ \text{ng}}\right)\left(\frac{1000\ \text{mg}}{1\ \text{g}}\right) = 5.0\times10^{-6}\ \text{mg}$

1.102 $170\ \text{lbs. body weight}\left(\frac{14\ \text{lbs. fat}}{100\ \text{lbs. body weight}}\right)\left(\frac{4500\ \text{kcal}}{1\ \text{lb. fat}}\right)\left(\frac{1\ \text{day}}{2000\ \text{kcal}}\right) = 53.55\ \text{days}$

= 54 days with significant figures

1.104 $175\ \text{lbs.}\left(\frac{1\ \text{kg}}{2.2\ \text{lbs.}}\right)\left(\frac{12\ \text{mg}}{1\ \text{kg}}\right) = 954.54\ \text{mg}$

$= 9.5\times10^{2}$ mg with significant figures

ALLIED HEALTH EXAM CONNECTION

1.106 The physical properties include (b) boiling point and (d) osmolarity because these properties can be observed without changing (or attempting to change) the chemical composition of the materials.

1.108 Only (b) rice and beans is a mixture.

1.110 $°\text{F} = \frac{9}{5}(°\text{C}) + 32$ $\quad °\text{F} = \frac{9}{5}(25°\text{C}) + 32 = 77°\text{F}$

(d) 25°C is approximately 77°F.

1.112 The freezing point of water is (d) 273 K.

1.114 A calorie is a form of (b) heat.

1.116 $4.50\times10^{2}\ \text{nm}\left(\frac{10^{-9}\ \text{m}}{1\ \text{nm}}\right)\left(\frac{1\ \text{pm}}{10^{-12}\ \text{m}}\right) = 4.50\times10^{5}\ \text{pm}$

4.50×10^{2} nm is (d) 4.50×10^{5} pm.

1.118 $$4.50\times10^{2}\ \text{nm}\left(\frac{10^{-9}\ \text{m}}{1\ \text{nm}}\right)=4.50\times10^{-7}\ \text{m}$$

4.50×10^2 nm is (c) 4.50×10^{-7} m.

1.120 (b) $1{,}000{,}000 = 10^6$

1.122 (d) $0.05620 = 5.620\times10^{-2}$

1.124 $$\text{(c) } 36\ \text{calories}\left(\frac{1\ \text{g CHO}}{4\ \text{calories}}\right)=9\ \text{g CHO}$$

1.126 $$10\%\ \text{solution}=\frac{10\ \text{g glucose}}{100\ \text{mL solution}}\times100$$

A 10% glucose solution contains (d) 10 grams of glucose per 100 milliliters of solution.

CHEMISTRY FOR THOUGHT

1.128

a. To separate wood sawdust and sand, I would add water. The sawdust will float, while the sand will sink. The top layer of water and sawdust can be poured off into a filter. The water will run through the filter leaving the sawdust on the filter. The sawdust can then be allowed to dry. The remainder of the water and sand can be poured off into a filter and the sand can be allowed to dry.

b. To separate sugar and sand, I would add water to dissolve the sugar. I would then filter the mixture to isolate the sand. I would evaporate the water to isolate the sugar.

c. To separate iron filings and sand, I would use a magnet. The iron filings will be attracted to the magnet, while the sand will not be attracted to the magnet.

d. To separate sand soaked with oil, I would pour the mixture through a filter. The oil will go through the filter and leave the sand behind on the filter.

1.130 $$44.5\ \text{kg}\left(\frac{2.2\ \text{lbs.}}{1\ \text{kg}}\right)=97.9\ \text{lbs.}$$

$$44.5\div2.2=20.2$$

This student should have used the relationship 2.2 lbs. = 1 kg to multiply 44.5 kg by 2.2 lbs./kg to find a weight of 97.9 lbs. The mistake she made appears to be that she divided 44.5 kg by 2.2 rather than multiplying by it. Consequently, she found a weight of only 20.2 lbs. Since she knows 2.2 lbs. = 1 kg, she was expecting the pound value to be larger than the kilogram value and she determined she had made a calculation error.

1.132 Hang gliding confirms that air is an example of matter because air occupies space and has mass. If air did not occupy space or have mass, the hang glider would fall to the ground rather than gliding through the air.

1.134 $$\text{density} = \frac{240.8 \text{ g}}{60.1 \text{ mL} - 32.6 \text{ mL}} = \frac{240.8 \text{ g}}{27.5 \text{ mL}} = 8.76 \tfrac{\text{g}}{\text{mL}}$$

The density of the object is only 8.76 g/mL; therefore, it does not have the same density as silver and is not silver.

1.136 When two teaspoons of sugar are dissolved in a small glass of water, the volume of the resulting solution is not significantly larger than the original volume of the water because as they dissolve, the sugar molecules are separated from one another and surrounded by water molecules. The sugar molecules fit in between the water molecules and do not significantly increase the volume of the solution.

ADDITIONAL ACTIVITIES

Section 1.1 Review: Is air matter? Describe how you might test your answer.

Which of these devices will give the same readings for an object at the top of a mountain, at the bottom of a valley, and on another planet? Explain.

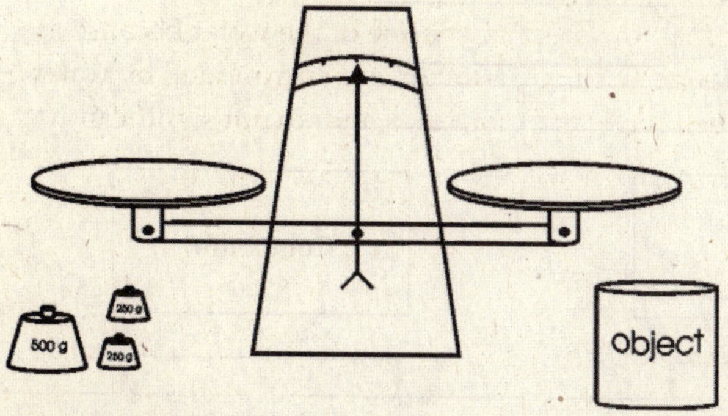

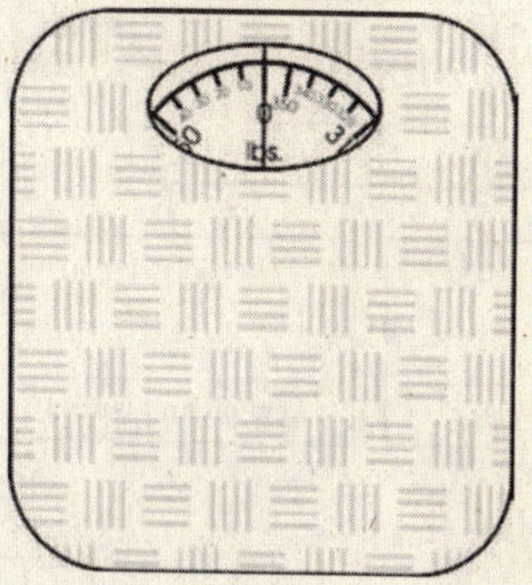

Section 1.2 Review: Identify the following as chemical or physical AND as properties or changes:

		Chemical or Physical	Property or Change
a.	the boiling point of water		
b.	mixing baking soda and vinegar to produce carbon dioxide gas		
c.	freezing vinegar		
d.	milk reacts with lemon juice		

Section 1.3 Review: Prefix review. Scientists make use of many prefixes. Complete the table below.

Prefix	Meaning	Meaning when combined with "–atomic molecule"	
mono-		monoatomic molecule*	
di-		diatomic molecule	
tri-		triatomic molecule	
poly-		polyatomic molecule	
homo-		homoatomic molecule	
hetero-		heteroatomic molecule	

* This term is not actually used. Can you think of a reason why that might be?

Draw examples of a:

a. heteroatomic diatomic molecule
b. homoatomic triatomic molecule
c. polyatomic molecule

Section 1.4 Review: Complete the following organizational chart to visualize the relationships between the definitions related to the classification of matter.

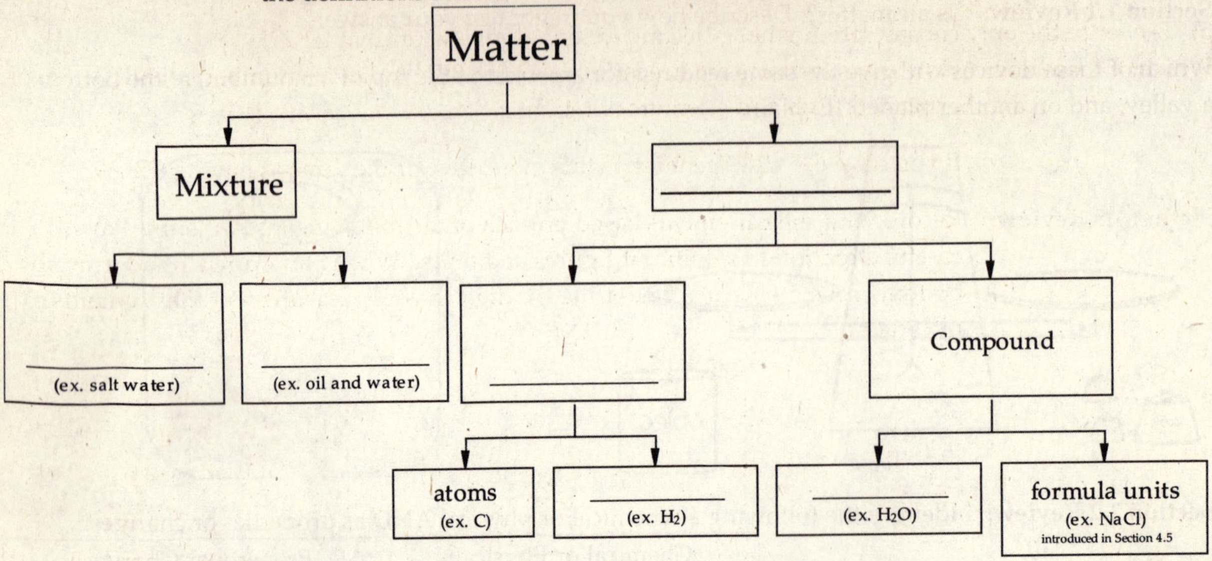

Section 1.5 Review: Identify the measurement units in the following paragraphs.

Sally Smith drove 4.3 miles to the grocery store to pick up a few items. The trip was fairly uneventful, but she was stopped by 4 red lights, and while driving through a school zone, she was very careful not to exceed 25 mph.

Once at the store Ms. Smith picked up a dozen eggs, a ½ gallon of milk, a 2-L bottle of sparkling water, a case of soda, a box of animal crackers, a pint of ice cream, 2 tomatoes, and a bunch of bananas. At the checkout, the grocery clerk placed the tomatoes and the bananas on the scale. They weighed 1.28 lbs. and 2.41 lbs., respectively. The tomatoes cost a $1.50/lb. and the bananas were on sale for 4 lbs./$1. Her total bill was $25.38.

On the way home, Sally stopped at the gas station because she had less than ¼ tank of gas. The mid-grade gas was $2.89/gallon and she filled her tank for $28.73. She reset her trip meter, which had read 218.4 miles before it returned to zero.

Section 1.6 Review: Two nutrition labels are shown below. Answer the following questions based on those labels.

Caffeine-Free Lemon-Lime Beverage

Nutrition Facts	
Serving Size	1 can (355 mL)
Amount per serving	
Calories 160	
	% Daily Value*
Total Fat 0 g	0%
Saturated Fat 0 g	0%
Polyunsaturated Fat 0 g	
Monounsaturated Fat 0 g	
Cholesterol 0 mg	0%
Sodium 75 mg	3%
Total Carbohydrates 40 g	13%
Dietary Fiber 0 g	0%
Sugars 40 g	
Protein 0 g	

Fat Free Milk

Nutrition Facts	
Serving Size	1 cup (240 mL)
Amount per serving	
Calories 90	
	% Daily Value*
Total Fat 0 g	0%
Saturated Fat 0 g	0%
Polyunsaturated Fat 0 g	
Monounsaturated Fat 0.1 g	
Cholesterol less than 5 mg	1%
Sodium 130 mg	5%
Total Carbohydrates 13 g	4%
Dietary Fiber 0 g	0%
Sugars 12 g	
Protein 9 g	

(1) How many nutritional calories are in 355 mL of fat free milk?

(2) How many kilocalories are in 355 mL of both beverages?

(3) How many Joules are in 355 mL of both beverages?

(4) Which beverage provides more energy per 355 mL serving?

(5) Is energy the only consideration when selecting foods? (Preview of Chapter 22)

Section 1.7 Review: How many pennies are needed to have \$58.39? Write this number in both expanded form and scientific notation.

If you have 6.4×10^{-2} pennies, is this more, less, or the same as having \$58.39?

Section 1.8 Review: Record your measurements based on each of the following instruments. Pay careful attention to significant figures and units. When appropriate, read from the bottom of the meniscus. Underline the digit in your measurement that contains uncertainty.

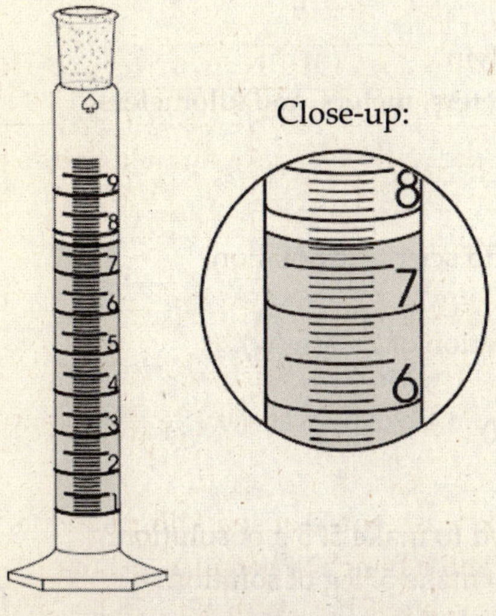

10-mL graduated cylinder

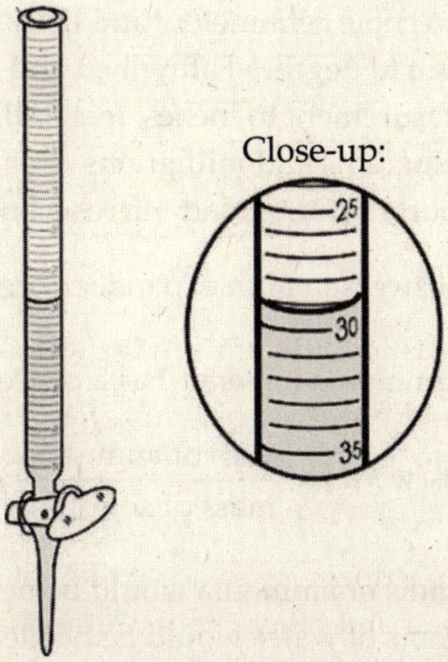

50-mL buret

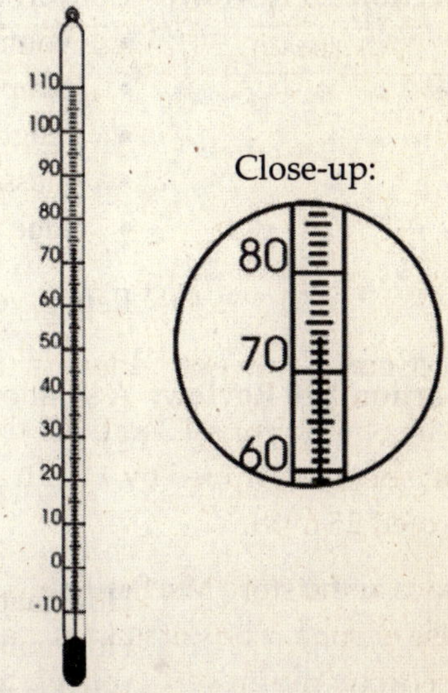

-10°C – 110°C thermometer

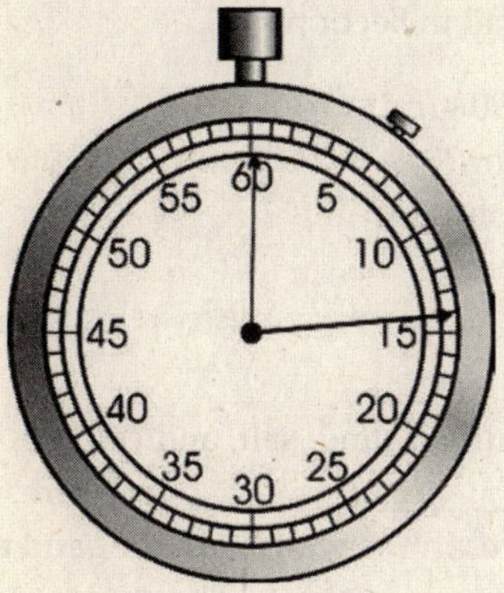

60-minute stopwatch
(when reset, both hands point to 60)

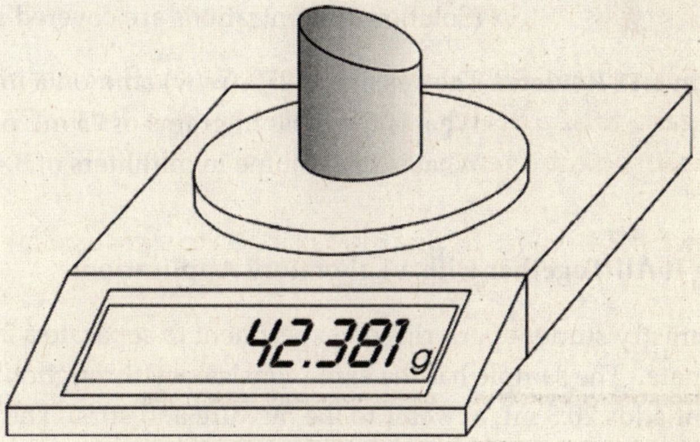

top-loading balance

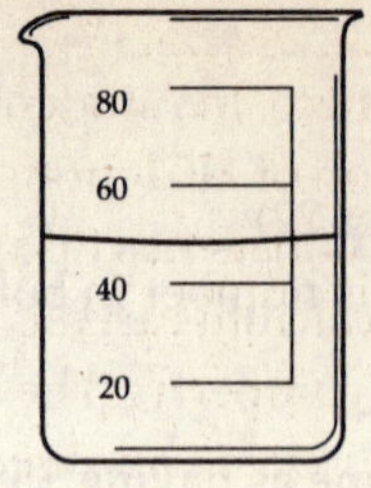

100-mL beaker

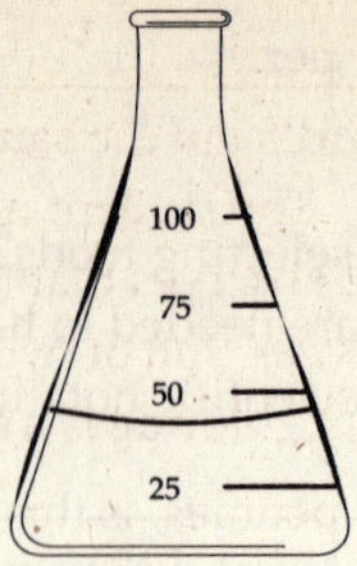

125-mL Erlenmeyer flask

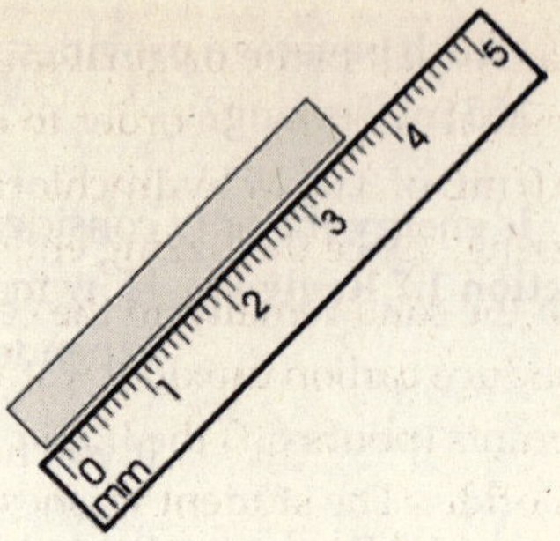

5-cm ruler*

*Many rulers have the centimeter scale marked with mm. On a ruler that has both inches and centimeters, you can double check whether the markings are millimeters or centimeters by remembering: 2.54 cm = 1 inch.

Section 1.9 Review: Convert the units in the Section 1.8 review as shown below:

- volumes to cubic centimeters and liters
- temperature to degrees Fahrenheit and Kelvin
- length measurement to inches, feet, millimeters, meters, and kilometers
- mass to kilograms and milligrams
- time to hours, minutes, and microseconds.

Report your answers in both expanded form and scientific notation.

Section 1.10 Review: A solution of ammonia in water has a concentration of 27%(w/w).

$$\%(w/w) = \frac{\text{mass of ammonia}}{\text{mass of solution}} \times 100$$

How many grams of ammonia would be needed to make 575 g of solution?
How many grams of water would be needed to make 575 g of solution?
What is the weight percentage of water in this solution?

(Solution concentrations are covered in more detail in Section 7.4.)

Section 1.11 Review: The density of 27%(w/w) ammonia in water is 0.90 g/mL.
What is the mass in grams of 75 mL of this solution?
What is the volume in milliliters of 8.4 g of this solution?

Tying It All Together with a Laboratory Application:

A chemistry student performs an experiment to separate a 2.503 g mixture of sand, salt, and calcium carbonate. The sample has the same consistency throughout; it is a (1) ________________ mixture. The student adds 20.8 mL of water to the mixture and stirs. The salt dissolves in the water, but the sand and calcium carbonate do not. Salt dissolving in water is a (2) ________________ process. The student allows the mixture to stand and the sand and calcium carbonate fall to the bottom of the container. This is a (3) ________________ mixture. The student filters the mixture. The salt water goes through the filter paper, but the sand and calcium carbonate do not. The salt water is a (4) ________________ mixture. The student pours the salt water into a beaker with a mass of 25.842 g, then places the beaker on a hotplate. The water boils and leaves salt crystals in the beaker. The student used a (5) ______________ change to isolate the salt from the water. The mass of the salt crystals and the beaker are 26.048 g. The mass of salt in the original mixture is (6) ________________, which is (7) __________ % of the original mixture. In order to isolate the sand from the sand and calcium carbonate, the student adds 20.0 mL of 3.00 M hydrochloric acid (HCl). The student notices that the addition of HCl produces a slight fizzing. Once the fizzing ends, the student observes that the white calcium carbonate has disappeared, but the sand remains in the container. The reaction of hydrochloric acid with calcium carbonate to produce carbon dioxide, water, and calcium chloride is a (8) ______________ property. The student decants (pours off) the liquid from the (9) ________________ mixture of sand, water, and calcium chloride. The student washes the sand with distilled water and decants it into the same container as the water and calcium chloride. The student dries the sand in a container on the hotplate. The mass of the sand is 1.942 x 10^{-3} kg, which is (10) ______________ g, or (11) ____________% of the original mixture. The student adds 20.00 mL of 1.25 M potassium carbonate to the clear calcium chloride and water mixture. A white precipitate of calcium carbonate forms immediately. This was a (12) ________________ change. The student filters the mixture and the calcium carbonate remains on the 425 mg piece of filter paper. After allowing the calcium carbonate and filter paper to dry, the student finds their combined mass is 765 mg. The mass of the calcium carbonate is (13) ____________ mg or (14) ____________ g, which is (15) ______________% of the original mixture. The total mass of the recovered substances, sand, salt, and calcium carbonate, is (16) ________________________ which is (17) ____________ (greater than, less than, or the same as) the mass of original mixture. Which of the following might have occurred (18)? ________________

A. All of the components were completely recovered.
B. Some of the material was lost during the many transferring steps of this procedure.
C. One of the components was not dried thoroughly.

SOLUTIONS FOR THE ADDITIONAL ACTIVITIES

Section 1.1 Review:
Air is matter because it has mass and occupies space. The mass of air can be verified by finding the mass of an empty balloon, blowing up the balloon, and taking the mass of the filled balloon. The difference in the mass of the full balloon and the empty balloon is the mass of the air in the balloon. The volume of the air is visible because the balloon is filled.

The lab balance will give the same readings for an object regardless of where the measurement is taken because it measures mass. The bathroom scale will not be consistent because it measures weight. Weight is the measurement of the gravitational force on an object, not the mass of the object.

Section 1.2 Review:

		Chemical or Physical	Property or Change
a.	the boiling point of water	physical	property
b.	mixing baking soda and vinegar to produce carbon dioxide gas	chemical	change
c.	freezing vinegar	physical	change
d.	milk reacts with lemon juice	chemical	property

Section 1.3 Review:

Prefix	Meaning	Meaning when combined with "–atomic molecule"	
mono-	1	monoatomic molecule*	a molecule containing 1 atom
di-	2	diatomic molecule	a molecule containing 2 atoms

tri-	3	triatomic molecule	a molecule containing 3 atoms
poly-	many	polyatomic molecule	a molecule containing many atoms
homo-	same	homoatomic molecule	a molecule containing only one type of atom
hetero-	different	heteroatomic molecule	a molecule containing at least 2 types of atoms

* A molecule must contain at least two atoms, otherwise it is called an atom.

Draw examples of a:

a.	heteroatomic diatomic molecule	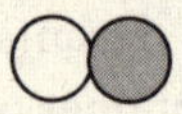	This could be any molecule containing 2 different types of atoms.
b.	homoatomic triatomic molecule	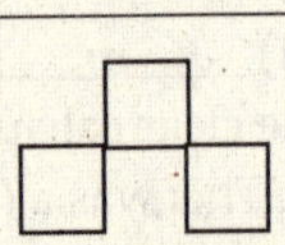	This could be any molecule containing 3 of the same type of atoms.
c.	polyatomic molecule		This could be any molecule containing 4 or more atoms.

Section 1.4 Review:

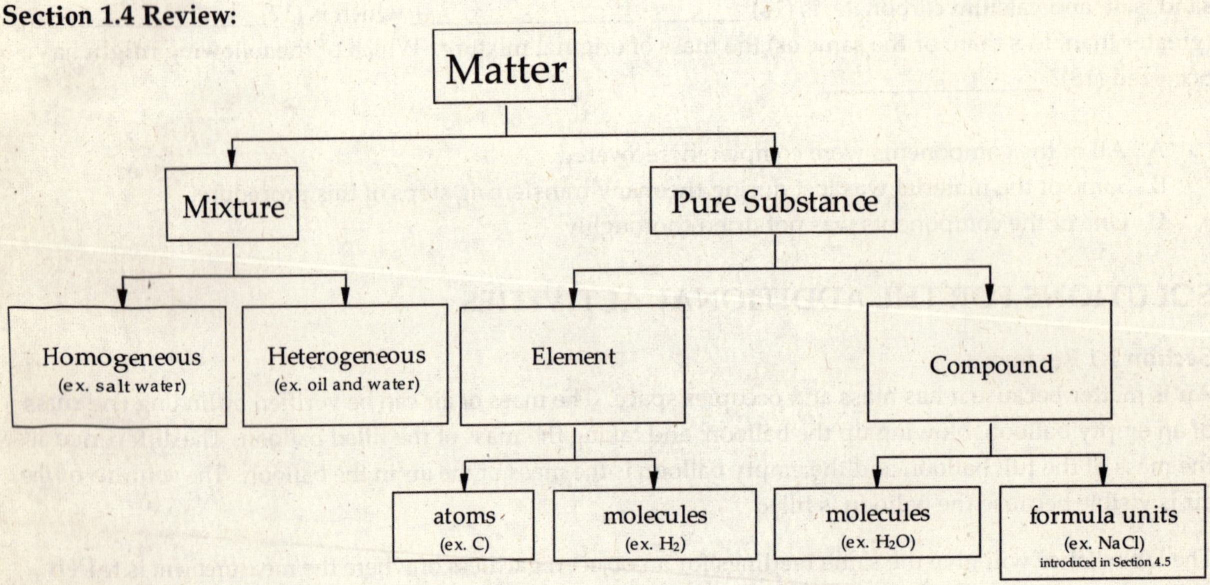

Section 1.5 Review:

The measurement units were miles (twice), mph, gallon, L, pint, and lbs. (twice).

The non-measurement units were red lights, dozen, case, box, tomatoes, bunch, \$/lb., lbs./\$, \$ (twice), tank, and \$/gallon.

Section 1.6 Review:

(1) $$355\ \cancel{\text{mL}}\ \text{milk} \times \frac{90\ \text{Cal}}{240\ \cancel{\text{mL}}} = 133.125\ \text{Cal} = 130\ \text{Calories}$$

(2) $$355\ \cancel{\text{mL}}\ \text{soda} \times \frac{160\ \cancel{\text{Calories}}}{355\ \cancel{\text{mL}}} \times \frac{1\ \text{kilocalorie}}{1\ \cancel{\text{Calorie}}} = 160\ \text{kcal}$$

$$355\ \cancel{\text{mL}}\ \text{milk} \times \frac{90\ \cancel{\text{Cal}}}{240\ \cancel{\text{mL}}} \times \frac{1\ \text{kilocalorie}}{1\ \cancel{\text{Calorie}}} = 130\ \text{kcal}$$

(3)

$$355\ \cancel{\text{mL}}\ \text{soda}\times\frac{160\ \cancel{\text{Cal}}}{355\ \cancel{\text{mL}}}\times\frac{1\ \cancel{\text{kilocalorie}}}{1\ \cancel{\text{Calorie}}}\times\frac{1000\ \cancel{\text{cal}}}{1\ \cancel{\text{kcal}}}\times\frac{4.184\ \text{J}}{1\ \cancel{\text{cal}}}=669440\ \text{J}=6.7\times10^{5}\ \text{J}$$

$$355\ \cancel{\text{mL}}\ \text{milk}\times\frac{90\ \cancel{\text{Cal}}}{240\ \cancel{\text{mL}}}\times\frac{1\ \cancel{\text{kilocalorie}}}{1\ \cancel{\text{Calorie}}}\times\frac{1000\ \cancel{\text{cal}}}{1\ \cancel{\text{kcal}}}\times\frac{4.184\ \text{J}}{1\ \cancel{\text{cal}}}=556995\ \text{J}=5.6\times10^{5}\ \text{J}$$

(4) The soda provides more energy per 355 mL serving.

(5) Energy is not the only consideration when selecting food. Nutritional value is also important. The milk has more nutritional value than the soda.

Section 1.7 Review:

$$\$58.39\times\frac{100\ \text{pennies}}{\$1.00}=5839\ \text{pennies}=5.839\times10^{3}\ \text{pennies}$$

6.4×10^{-2} = 0.064 pennies; This value is less than one penny and much less than $58.39.

Section 1.8 and 1.9 Review:

Instrument	Section 1.8	Section 1.9
graduated cylinder	$7.7\underline{4}$ mL	$7.74\ \cancel{\text{mL}}\left(\dfrac{1\ \text{cm}^3}{1\ \cancel{\text{mL}}}\right) = 7.74\ \text{cm}^3 = 7.74\times10^0\ \text{cm}^3$ $7.74\ \cancel{\text{mL}}\left(\dfrac{1\ \text{L}}{1000\ \cancel{\text{mL}}}\right) = 0.00774\ \text{L} = 7.74\times10^{-3}\ \text{L}$
buret	$29.\underline{2}$ mL	$29.2\ \cancel{\text{mL}}\left(\dfrac{1\ \text{cm}^3}{1\ \cancel{\text{mL}}}\right) = 29.2\ \text{cm}^3 = 2.92\times10^1\ \text{cm}^3$ $29.2\ \cancel{\text{mL}}\left(\dfrac{1\ \text{L}}{1000\ \cancel{\text{mL}}}\right) = 0.0292\ \text{L} = 2.92\times10^{-2}\ \text{L}$
thermometer	$73.\underline{3}$°C	$°\text{F} = \frac{9}{5}(73.3°\text{C}) + 32 = 164°\text{F}$ $73.3°\text{C} + 273 = 346.3\ \text{K}$
stopwatch	$14.\underline{2}$ sec	$14.2\ \cancel{\text{sec.}}\left(\dfrac{1\ \cancel{\text{min.}}}{60\ \cancel{\text{sec.}}}\right)\left(\dfrac{1\ \text{hour}}{60\ \cancel{\text{min.}}}\right) = 0.00394\ \text{hr.} = 3.94\times10^{-3}\ \text{hr.}$ $14.2\ \cancel{\text{sec.}}\left(\dfrac{1\ \text{min.}}{60\ \cancel{\text{sec.}}}\right) = 0.237\ \text{min.} = 2.37\times10^{-1}\ \text{min.}$ $14.2\ \cancel{\text{sec.}}\left(\dfrac{10^6\ \mu\text{s}}{1\ \cancel{\text{sec.}}}\right) = 14200000\ \mu\text{s} = 1.42\times10^7\ \mu\text{s}$
balance	$42.38\underline{1}$ g	$42.381\ \cancel{\text{g}}\left(\dfrac{1\ \text{kg}}{1000\ \cancel{\text{g}}}\right) = 0.042381\ \text{kg} = 4.2381\times10^{-2}\ \text{kg}$ $42.381\ \cancel{\text{g}}\left(\dfrac{1000\ \text{mg}}{1\ \cancel{\text{g}}}\right) = 42381\ \text{mg} = 4.2381\times10^4\ \text{mg}$
beaker	$4\underline{7}$ mL	$47\ \cancel{\text{mL}}\left(\dfrac{1\ \text{cm}^3}{1\ \cancel{\text{mL}}}\right) = 47\ \text{cm}^3 = 4.7\times10^1\ \text{cm}^3$ $47\ \cancel{\text{mL}}\left(\dfrac{1\ \text{L}}{1000\ \cancel{\text{mL}}}\right) = 0.047\ \text{L} = 4.7\times10^{-2}\ \text{L}$
Erlenmeyer flask	$4\underline{0}$ mL	$40\ \cancel{\text{mL}}\left(\dfrac{1\ \text{cm}^3}{1\ \cancel{\text{mL}}}\right) = 40\ \text{cm}^3 = 4.0\times10^1\ \text{cm}^3$ $40\ \cancel{\text{mL}}\left(\dfrac{1\ \text{L}}{1000\ \cancel{\text{mL}}}\right) = 0.040\ \text{L} = 4.0\times10^{-2}\ \text{L}$
ruler	$3.6\underline{6}$ cm	$3.66\ \cancel{\text{cm}}\left(\dfrac{0.394\ \text{in.}}{1\ \cancel{\text{cm}}}\right) = 1.44\ \text{in.} = 1.44\times10^0\ \text{in.}$ $3.66\ \cancel{\text{cm}}\left(\dfrac{0.394\ \cancel{\text{in.}}}{1\ \cancel{\text{cm}}}\right)\left(\dfrac{1\ \text{ft.}}{12\ \cancel{\text{in.}}}\right) = 0.120\ \text{ft.} = 1.20\times10^{-1}\ \text{ft.}$ $3.66\ \cancel{\text{cm}}\left(\dfrac{10\ \text{mm}}{1\ \cancel{\text{cm}}}\right) = 36.6\ \text{mm} = 3.66\times10^1\ \text{mm}$ $3.66\ \cancel{\text{cm}}\left(\dfrac{1\ \text{m}}{100\ \cancel{\text{cm}}}\right) = 0.0366\ \text{m} = 3.66\times10^{-2}\ \text{m}$

		$3.66\ \cancel{\text{cm}}\left(\dfrac{1\ \cancel{\text{m}}}{100\ \cancel{\text{cm}}}\right)\left(\dfrac{1\ \text{km}}{1000\ \cancel{\text{m}}}\right) = 0.0000366\ \text{km} = 3.66\times10^{-5}\ \text{km}$

Section 1.10 Review: $575\ \text{g}\times 27\% = 155.25\ \text{g} = 160\ \text{g ammonia};\ 575\ \text{g} - 155.25\ \text{g} = 419.75\ \text{g} = 420\ \text{g water}$

$$\frac{419.75\ \text{g}}{575\ \text{g}}\times 100 = 73\%(\text{w/w}) \quad \text{or} \quad 100\% - 27\% = 73\%(\text{w/w})$$

Section 1.11 Review: $75\ \text{mL}\times\dfrac{0.90\ \text{g}}{1\ \text{mL}} = 67.5\ \text{g} = 68\ \text{g}$ $\qquad 8.4\ \text{g}\times\dfrac{1\ \text{mL}}{0.90\ \text{g}} = 9.\bar{3}\ \text{mL} = 9.3\ \text{mL}$

Tying It All Together with a Laboratory Application:

(1) homogeneous
(2) physical
(3) heterogeneous
(4) homogeneous
(5) physical
(6) 0.206 g
(7) 8.23%
(8) chemical
(9) heterogeneous
(10) 1.942 g
(11) 77.59%
(12) chemical
(13) 340 mg
(14) 0.340 g
(15) 13.6%
(16) 2.488 g
(17) less than
(18) B

SELF-TEST QUESTIONS

Multiple Choice

1. Which of the following involves a chemical change?
   a. stretching a rubber band
   b. breaking a stick
   c. lighting a candle
   d. melting an ice cube

2. A solid substance is subjected to a number of test and observations. Which of the following would be classified as a chemical property of the substance?
   a. It is gray in color.
   b. It has a density of 2.04 g/mL.
   c. It dissolves in acid and a gas is liberated.
   d. It is not attracted to either pole of a magnet.

3. Which of the following terms could not be properly used in the description of a compound?
   a. solution
   b. polyatomic
   c. pure substance
   d. heteroatomic

4. Which of the following is an example of heterogeneous matter?
   a. water containing sand
   b. a sample of salt water
   c. a pure sample of iron
   d. a sample of pure table salt

5. When a substance undergoes a physical change, which of the following is always true?
   a. It melts.
   b. A new substance is produced.
   c. Heat is given off.
   d. The molecular composition is unchanged.

6. Which of the following is not a chemical change?
   a. burning magnesium
   b. pulverizing sulfur
   c. exploding nitroglycerine
   d. rusting iron

7. Which of the following is the basic unit of length in the metric system?
   a. centimeter
   b. meter
   c. millimeter
   d. kilometer

8. Which of the following is a derived unit?

a. calorie b. cubic decimeter c. Joule d. kilogram

9. In the number 3.91×10^{-3}, the original decimal position is located ______ from its current position.
   a. 3 places to the right
   b. 3 places to the left
   c. 2 places to the right
   d. 2 places to the left

10. How many significant figures are included in the number 0.02102?
    a. two b. three c. four d. five

11. Twenty-one (21) students in a class of 116 got a B grade on an exam. What percent of the students in the class got B's?
    a. 21.0 b. 22.1 c. 15.3 d. 18.1

12. What single factor derived from Table 1.3 would allow you to calculate the number of quarts in a 2.0 L bottle of soft drink?
    a. $\frac{1.057 \text{ quarts}}{1 \text{ L}}$ b. $\frac{0.0338 \text{ fl oz}}{1 \text{ mL}}$ c. $\frac{1 \text{ L}}{1.057 \text{ quarts}}$ d. $\frac{1 \text{ mL}}{0.0338 \text{ fl oz}}$

13. On a hot day, a Fahrenheit thermometer reads 97.3°F. What would this reading be on a Celsius thermometer?
    a. 118°C b. 22.1°C c. 36.3°C d. 143°C

14. The density of a 1 mL sample of a patient's blood is 1.08 g/mL. The density of a pint of blood taken at the same time from the same patient would be ______________ 1.08 g/mL.
    a. greater than
    b. less than
    c. equal to
    d. more than one possible answer

15. A 125 mL urine specimen weighs 136.0 g. The density of the specimen is:
    a. $1.09 \frac{\text{g}}{\text{mL}}$ b. $0.919 \frac{\text{g}}{\text{mL}}$ c. $261 \frac{\text{g}}{\text{mL}}$ d. $11.0 \frac{\text{g}}{\text{mL}}$

Matching

Match the type of measurement on the right to the measurement units given on the left.

16. Kelvin
17. milliliter
18. gram
19. centimeter
20. cubic decimeter
21. kilometer
22. pounds per cubic foot

a. mass
b. volume
c. length
d. temperature
e. density

True-False

23. The mass of an object is the same as its weight.
24. A physical property can be observed without attempting any composition changes.
25. The cooking of food involves chemical changes.
26. The smallest piece of water that has the properties of water is called an atom.
27. Carbon monoxide molecules are diatomic and heteroatomic.
28. The prefix milli- means one thousand times.
29. One meter is shorter than one yard.

30. A pure substance containing sulfur and oxygen atoms must be classified as a compound.
31. The calorie and Joule are both units of energy.
32. In scientific notation, the exponent on the 10 cannot be larger than 15.
33. The correctly rounded sum resulting from adding 13.0, 1.094, and 0.132 will contain five significant figures.
34. If an object floats in water, it must have a higher density than water.
35. Most gases are less dense than liquids.

ANSWERS TO THE SELF-TEST QUESTIONS

1. C	8. B	15. A	22. E	29. F
2. C	9. B	16. D	23. F	30. T
3. A	10. C	17. B	24. T	31. T
4. A	11. D	18. A	25. T	32. F
5. D	12. A	19. C	26. F	33. F
6. B	13. C	20. B	27. T	34. F
7. B	14. C	21. C	28. F	35. T

Chapter 2: Atoms and Molecules

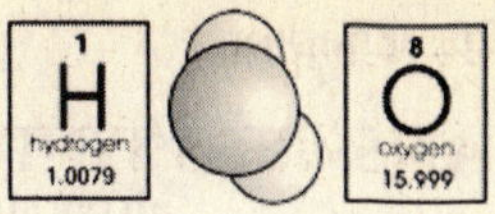

CHAPTER OUTLINE

2.1 Symbols and Formulas
2.2 Inside the Atom
2.3 Isotopes
2.4 Relative Masses of Atoms and Molecules
2.5 Isotopes and Atomic Weights
2.6 Avogadro's Number: The Mole
2.7 The Mole and Chemical Formulas

LEARNING OBJECTIVES/ASSESSMENT

When you have completed your study of this chapter, you should be able to:

1. Use symbols for chemical elements to write formulas for chemical compounds. (Section 2.1; Exercise 2.4)
2. Identify the characteristics of protons, neutrons, and electrons. (Section 2.2; Exercises 2.10 and 2.12)
3. Use the concepts of atomic number and mass number to determine the number of subatomic particles in isotopes and to write correct symbols for isotopes. (Section 2.3; Exercises 2.16 and 2.22)
4. Use atomic weights of the elements to calculate molecular weights of compounds. (Section 2.4; Exercise 2.32)
5. Use isotope percent abundances and masses to calculate atomic weights of elements. (Section 2.5; Exercise 2.38)
6. Use the mole concept to obtain relationships between number of moles, number of grams, and number of atoms for elements, and use those relationships to obtain factors for use in factor-unit calculations. (Section 2.6; Exercises 2.44 a & b and 2.46 a & b)
7. Use the mole concept and molecular formulas to obtain relationships between number of moles, number of grams, and number of atoms or molecules for compounds, and use those relationships to obtain factors for use in factor-unit calculations. (Section 2.7; Exercise 2.50 b and 2.52 b)

SOLUTIONS FOR THE END OF CHAPTER EXERCISES

SYMBOLS AND FORMULAS (SECTION 2.1)

2.2 a. A triatomic molecule of a compound*

b. A molecule of a compound containing two atoms of one element and two atoms of a second element*

c. A molecule of a compound containing two atoms of one element, one atom of a second element, and four atoms of a third element*

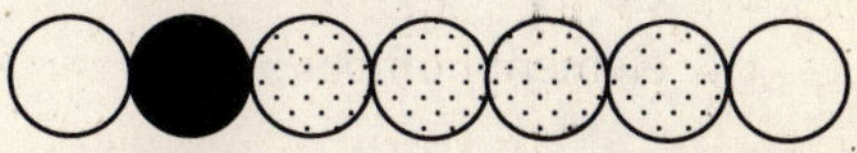

d. A molecule containing two atoms of one element, six atoms of a second element, and one atom of a third element*

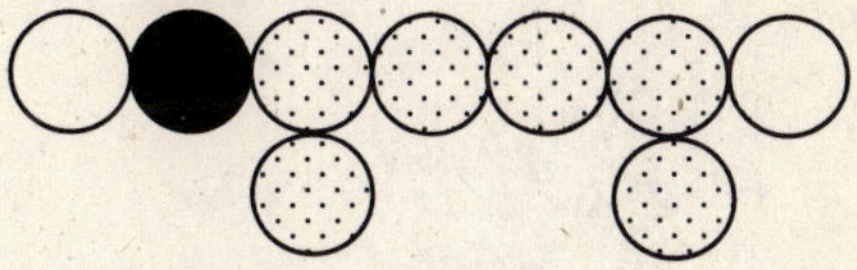

*Note: Each of these structures could be drawn in many different ways.

☑2.4 a. A molecule of water (two hydrogen atoms and one oxygen atom) — H_2O; like Exercise 2.2 a*

b. A molecule of hydrogen peroxide (two hydrogen atoms and two oxygen atoms) — H_2O_2; like Exercise 2.2 b*

*The number and variety of atoms are alike. The actual structures of the molecules are different.

c. A molecule of sulfuric acid (two hydrogen atoms, one sulfur atom, and four oxygen atoms) — H_2SO_4; like Exercise 2.2 c*

d. A molecule of ethyl alcohol (two carbon atoms, six hydrogen atoms, and one oxygen atom) — C_2H_6O; like Exercise 2.2 d*

*The number and variety of atoms are alike. The actual structures of the molecules are different.

2.6
a. methane (CH_4) — 1 carbon atom; 4 hydrogen atoms
b. perchloric acid ($HClO_4$) — 1 hydrogen atom; 1 chlorine atom; 4 oxygen atoms
c. methylamine (CH_5N) — 1 carbon atom; 5 hydrogen atoms; 1 nitrogen atom
d. propane (C_3H_8) — 3 carbon atoms; 8 hydrogen atoms

2.8
a. HSH (hydrogen sulfide) — More than one H is part of the compound; a subscript should be used: H_2S
b. $HCLO_2$ (chlorous acid) — The elemental symbol for chlorine is Cl (the second letter of a symbol must be lowercase): $HClO_2$
c. $2HN_2$ (hydrazine – two hydrogen atoms and four nitrogen atoms) — The subscripts should reflect the actual number of each type of atom in the compound: H_2N_4
d. C2H6 (ethane) — The numbers should be subscripted: C_2H_6

INSIDE THE ATOM (SECTION 2.2)

☑2.10

	Charge	Mass (u)
a. 4 protons and 5 neutrons	4	9
b. 9 protons and 10 neutrons	9	19
c. 20 protons and 23 neutrons	20	43
d. 47 protons and 60 neutrons	47	107

☑2.12 The number of protons and electrons are equal in a neutral atom.
a. 4 electrons b. 9 electrons c. 20 electrons d. 47 electrons

ISOTOPES (SECTION 2.3)

2.14

	Electrons	Protons
a. potassium	19	19
b. Cd	48	48
c. element number 51	51	51

☑2.16

	Protons	Neutrons	Electrons
a. $^{34}_{16}S$	16	18	16
b. $^{91}_{40}Zr$	40	51	40
c. $^{131}_{54}Xe$	54	77	54

2.18
a. silicon-28 — $^{28}_{14}Si$
b. argon-40 — $^{40}_{18}Ar$
c. strontium-88 — $^{88}_{38}Sr$

2.20

	Mass Number	Atomic Number	Symbol
a. 4 protons and 5 neutrons	9	4	$^{9}_{4}Be$
b. 9 protons and 10 neutrons	19	9	$^{19}_{9}F$

c.	20 protons and 23 neutrons	43	20	${}^{43}_{20}Ca$
d.	47 protons and 60 neutrons	107	47	${}^{107}_{47}Ag$

☑2.22 a. contains 17 electrons and 20 neutrons ${}^{37}_{17}Cl$

b. a copper atom with a mass number of 65 ${}^{65}_{29}Cu$

c. a zinc atom that contains 36 neutrons ${}^{66}_{30}Zn$

RELATIVE MASSES OF ATOMS AND MOLECULES (SECTION 2.4)

2.24 $12\ u\left(\frac{1\text{ atom He}}{4\text{ u He}}\right) = 3\text{ atoms He}$

2.26 $77.1\% \times 52.00\ u = 0.771 \times 52.00u = 40.1\ u$; Ca; calcium

2.28 $\frac{1}{2} \times 28.09\ u = 14.05\ u$; N; nitrogen

2.30 a. sulfur trioxide (SO_3) $(1\times32.07\ u)+(3\times16.00\ u)=80.07\ u$

b. glycerin ($C_3H_8O_3$) $(3\times12.01\ u)+(8\times1.008\ u)+(3\times16.00)=92.09\ u$

c. sulfuric acid (H_2SO_4) $(2\times1.008\ u)+(1\times32.07\ u)+(4\times16.00\ u)=98.09\ u$

d. nitrogen (N_2) $2\times14.01\ u = 28.02\ u$

e. propane (C_3H_8) $(3\times12.01\ u)+(8\times1.008\ u)=44.09\ u$

☑2.32 The gas is most likely to be ethylene based on the following calculations:

acetylene: $(2\times12.01\ u)+(2\times1.008\ u)=26.04\ u$

ethylene: $(2\times12.01\ u)+(4\times1.008\ u)=28.05\ u$

ethane: $(2\times12.01\ u)+(6\times1.008\ u)=30.07\ u$

The experimental value for the molecular weight of a flammable gas known to contain only carbon and hydrogen is 28.05 u, which is identical to the theoretical value of 28.05 u, which was calculated for ethylene.

2.34 The y in the formula for serine stands for 3, the number of carbon atoms in the chemical formula.

$$(y\times12.01\ u)+(7\times1.008\ u)+(1\times14.01\ u)+(3\times16.00\ u)=105.10\ u$$

$$y\times12.01\ u + 69.07\ u = 105.10\ u$$

$$y\times12.01\ u = 36.03\ u$$

$$y = 3$$

ISOTOPES AND ATOMIC WEIGHTS (SECTION 2.5)

2.36 a. The number of neutrons in the nucleus $26.982 - 13 = 13.982 \approx 14$ neutrons

b. The mass (in u) of the nucleus (to three significant figures) 27.0 u

☑2.38 $19.78\% \times 10.0129\text{ u} + 80.22\% \times 11.0093\text{ u} =$

$0.1978 \times 10.0129\text{ u} + 0.8022 \times 11.0093\text{ u} = 10.81221208\text{ u}$; 10.812 u with SF

or

$$\frac{(19.78 \times 10.0129\text{ u}) + (80.22 \times 11.0093\text{ u})}{100} = 10.81221208\text{ u; } 10.812\text{ u with SF}$$

The atomic weight listed for boron in the periodic table is 10.81 u. The two values are close to one another.

2.40 $69.09\% \times 62.9298\text{ u} + 30.91\% \times 64.9278\text{ u} =$

$0.6909 \times 62.9298\text{ u} + 0.3091 \times 64.9278\text{ u} = 63.5473818\text{ u}$; 63.55 u with SF

or

$$\frac{(69.09 \times 62.9298\text{ u}) + (30.91 \times 64.9278\text{ u})}{100} = 63.5473818\text{ u; } 63.55\text{ u with SF}$$

The atomic weight listed for copper in the periodic table is 63.55 u. The two values are the same.

AVOGADRO'S NUMBER: THE MOLE (SECTION 2.6)

2.42

$$1.60\ \cancel{\text{g O}}\left(\frac{6.02 \times 10^{23}\text{ atoms O}}{16.00\ \cancel{\text{g O}}}\right) = 6.02 \times 10^{22}\text{ atoms O}$$

$$6.02 \times 10^{22}\ \cancel{\text{atoms F}}\left(\frac{19.0\text{ g F}}{6.02 \times 10^{23}\ \cancel{\text{atoms F}}}\right) = 1.90\text{ g F}$$

2.44 ☑a. silicon

$1\text{ mol Si atoms} = 6.02 \times 10^{23}\text{ Si atoms}$

$6.02 \times 10^{23}\text{ Si atoms} = 28.1\text{ g Si}$

$1\text{ mol Si atoms} = 28.1\text{ g Si}$

☑b. calcium

$1\text{ mol Ca atoms} = 6.02 \times 10^{23}\text{ Ca atoms}$

$6.02 \times 10^{23}\text{ Ca atoms} = 40.1\text{ g Ca}$

$1\text{ mol Ca atoms} = 40.1\text{ g Ca}$

c. argon

$1\text{ mol Ar atoms} = 6.02 \times 10^{23}\text{ Ar atoms}$

$6.02 \times 10^{23}\text{ Ar atoms} = 39.9\text{ g Ar}$

$1\text{ mol Ar atoms} = 39.9\text{ g Ar}$

2.46 ☑a. The number of grams of silicon in 1.25 mol of silicon

$1\text{ mol Si atoms} = 28.1\text{ g Si};\ \dfrac{28.1\text{ g Si}}{1\text{ mol Si atoms}}$

$$1.25\ \cancel{\text{mol Si}}\left(\frac{28.1\text{ g Si}}{1\ \cancel{\text{mol Si}}}\right) = 35.1\text{ g Si}$$

☑b. The mass in grams of one calcium atom

$6.02 \times 10^{23}\text{ Ca atoms} = 40.1\text{ g Ca};\ \dfrac{40.1\text{ g Ca}}{6.02 \times 10^{23}\text{ Ca atoms}}$

$$1\ \cancel{\text{atom Ca}}\left(\frac{40.1\text{ g Ca}}{6.02 \times 10^{23}\ \cancel{\text{Ca atoms}}}\right) = 6.66 \times 10^{-23}\text{ g Ca}$$

(Note: One atom is assumed to be an exact number.)

c. The number of argon atoms in a 20.5-g sample of argon

$$6.02\times10^{23}\ \text{Ar atoms} = 39.9\ \text{g Ar};\ \frac{6.02\times10^{23}\ \text{Ar atoms}}{39.9\ \text{g Ar}}$$

$$20.5\ \cancel{\text{g Ar}}\left(\frac{6.02\times10^{23}\ \text{Ar atoms}}{39.9\ \cancel{\text{g Ar}}}\right) = 3.09\times10^{23}\ \text{Ar atoms}$$

THE MOLE AND CHEMICAL FORMULAS (SECTION 2.7)

2.48 $(1\times10.8\ \text{u})+(3\times19.0\ \text{u})=67.8\ \text{u};\ 1\ \text{mole}\ BF_3 = 67.8\ \text{g}\ BF_3$

$(2\times1.01\ \text{u})+(1\times32.1\ \text{u})=34.1\ \text{u};\ 1\ \text{mole}\ H_2S = 34.1\ \text{g}\ H_2S$

$$0.34\ \cancel{\text{g}\ H_2S}\left(\frac{6.02\times10^{23}\ \text{molecules}\ H_2S}{34.1\ \cancel{\text{g}\ H_2S}}\right) = 6.0\times10^{21}\ \text{molecules}\ H_2S$$

$$6.0\times10^{21}\ \cancel{\text{molecules}\ BF_3}\left(\frac{67.8\ \text{g}\ BF_3}{6.02\times10^{23}\ \cancel{\text{molecules}\ BF_3}}\right) = 0.68\ \text{g}\ BF_3$$

2.50 a. benzene (C_6H_6)

1. 2 C_6H_6 molecules contain 12 C atoms and 12 H atoms.
2. 10 C_6H_6 molecules contain 60 C atoms and 60 H atoms.
3. 100 C_6H_6 molecules contain 600 C atoms and 600 H atoms.
4. 6.02×10^{23} C_6H_6 molecules contains 36.12×10^{23} C atoms and 36.12×10^{23} H atoms.
5. 1 mol of C_6H_6 molecules contain 6 moles of C atoms and 6 moles of H atoms.
6. 78.1 g of benzene contains 72.0 g of C and 6.1 g of H.

☑b. nitrogen dioxide (NO_2)

1. 2 NO_2 molecules contain 2 N atoms and 4 O atoms.
2. 10 NO_2 molecules contain 10 N atoms and 20 O atoms.
3. 100 NO_2 molecules contain 100 N atoms and 200 O atoms.
4. 6.02×10^{23} NO_2 molecules contain 6.02×10^{23} N atoms and 12.04×10^{23} O atoms.
5. 1 mol of NO_2 molecules contains 1 mole of N atoms and 2 moles of O atoms.
6. 46.0 g of nitrogen dioxide contains 14.0 g of N and 32.0 g of O.

c. hydrogen chloride (HCl)

1. 2 HCl molecules contain 2 H atoms and 2 Cl atoms.
2. 10 HCl molecules contain 10 H atoms and 10 Cl atoms.
3. 100 HCl molecules contain 100 H atoms and 100 Cl atoms.
4. 6.02×10^{23} HCl molecules contain 6.02×10^{23} H atoms and 6.02×10^{23} Cl atoms.
5. 1 mol of HCl molecules contains 1 mole of H atoms and 1 mole Cl atoms.
6. 36.5 g of hydrogen chloride contains 1.01 g of H and 35.5 g of Cl.

2.52 a. **Statement 5.** 1 mol of C_6H_6 molecules contains 6 moles of C atoms and 6 moles of H atoms.

$$\text{Factor}: \left(\frac{6 \text{ moles H atoms}}{1 \text{ mole } C_6H_6}\right)$$

$$0.75 \cancel{\text{mol } C_6H_6}\left(\frac{6 \text{ moles H atoms}}{1 \cancel{\text{mole } C_6H_6}}\right) = 4.5 \text{ moles H atoms}$$

☑b. **Statement 4.** 6.02×10^{23} NO_2 molecules contain 6.02×10^{23} N atoms and 12.04×10^{23} O atoms.

$$\text{Factor}: \left(\frac{12.04\times10^{23} \text{ O atoms}}{1 \text{ mole } NO_2}\right)$$

$$0.50 \cancel{\text{mole } NO_2}\left(\frac{12.04\times10^{23} \text{ O atoms}}{1 \cancel{\text{mole } NO_2}}\right) = 6.0\times10^{23} \text{ O atoms}$$

c. **Statement 6.** 36.5 g of hydrogen chloride contains 1.01 g of H and 35.5 g of Cl.

$$\text{Factor}: \left(\frac{35.5 \text{ g Cl}}{36.5 \text{ g HCl}}\right)$$

$$\frac{35.5 \text{ g Cl}}{36.5 \text{ g HCl}}\times100 = 97.3\% \text{ Cl in HCl}$$

2.54

$$0.75 \cancel{\text{mole } H_2O}\left(\frac{1 \cancel{\text{mole O atoms}}}{1 \cancel{\text{mole } H_2O}}\right)\left(\frac{6.02\times10^{23} \text{ O atoms}}{1 \cancel{\text{mole O atoms}}}\right) = 4.515\times10^{23} \text{ O atoms}$$

$$4.515\times10^{23} \cancel{\text{O atoms}}\left(\frac{1 \text{ mole } \cancel{\text{O atoms}}}{6.02\times10^{23} \cancel{\text{O atoms}}}\right)\left(\frac{1 \cancel{\text{mole } C_2H_6O}}{1 \text{ mole } \cancel{\text{O atoms}}}\right)\left(\frac{46.1 \text{ g } C_2H_6O}{1 \cancel{\text{mole } C_2H_6O}}\right)$$

$$= 34.575 \text{ g } C_2H_6O \approx 35 \text{ g with SF}$$

2.56

$$\frac{4.04 \text{ g H}}{16.0 \text{ g } CH_4}\times100 = 25.3\% \text{ H in } CH_4 \qquad \frac{6.06 \text{ g H}}{30.1 \text{ g } C_2H_6}\times100 = 20.1\% \text{ H in } C_2H_6$$

2.58 **Statement 4.** 6.02×10^{23} H_3PO_4 molecules contain 18.06×10^{23} H atoms, 6.02×10^{23} P atoms, and 24.08×10^{23} O atoms.

Statement 5. 1 mol H_3PO_4 molecules contains 3 moles of H atoms, 1 mole of P atoms, and 4 moles of O atoms.

Statement 6. 98.0 g of phosphoric acid contains 3.03 g of H, 31.0 g of P, and 64.0 g of O.

a. **Statement 6.** 98.0 g of phosphoric acid contains 3.03 g of H, 31.0 g of P, and 64.0 g of O.

$$\text{Factor}: \left(\frac{3.03 \text{ g H}}{98.0 \text{ g } H_3PO_4}\right)$$

$$46.8 \cancel{\text{g } H_3PO_4}\left(\frac{3.03 \text{ g H}}{98.0 \cancel{\text{g } H_3PO_4}}\right) = 1.45 \text{ g H}$$

b. **Statement 5.** 1 mol H_3PO_4 molecules contains 3 moles of H atoms, 1 mole of P atoms, and 4 moles of O atoms.

$$\text{Factor}: \left(\frac{4 \text{ moles of O atoms}}{1 \text{ mole } H_3PO_4}\right)$$

$$1.25 \cancel{\text{moles } H_3PO_4}\left(\frac{4 \text{ moles of O atoms}}{1 \cancel{\text{mole } H_3PO_4}}\right) = 5.00 \text{ moles of O atoms}$$

c. **Statement 4.** 6.02×10^{23} H_3PO_4 molecules contain 18.06×10^{23} H atoms, 6.02×10^{23} P atoms, and 24.08×10^{23} O atoms.

$$\text{Factor}: \left(\frac{6.02\times10^{23} \text{ P atoms}}{6.02\times10^{23} \text{ } H_3PO_4 \text{ molecules}}\right)$$

$$8.42\times10^{21} \cancel{\text{molecules } H_3PO_4}\left(\frac{6.02\times10^{23} \text{ P atoms}}{6.02\times10^{23} \cancel{H_3PO_4 \text{ molecules}}}\right) = 8.42\times10^{21} \text{ P atoms}$$

2.60 Magnetite (Fe_3O_4) contains the higher mass percentage of iron as shown in the calculation below:

$$\frac{167 \text{ g Fe}}{231 \text{ g } Fe_3O_4}\times100 = 72.3\% \text{ Fe in } Fe_3O_4 \qquad \frac{112 \text{ g Fe}}{160 \text{ g } Fe_2O_3}\times100 = 70.0\% \text{ Fe in } Fe_2O_3$$

ADDITIONAL EXERCISES

2.62 U-238 contains 3 more neutrons in its nucleus than U-235. U-238 and U-235 have the same volume because the extra neutrons in U-238 do not change the size of the electron cloud. U-238 is 3u heavier than U-235 because of the 3 extra neutrons. Density is a ratio of mass to volume; therefore, U-238 is more dense than U-235 because it has a larger mass divided by the same volume.

2.64

$$\frac{1.99\times10^{-23} \text{ g}}{1 \cancel{\text{C}-12 \text{ atom}}}\left(\frac{1 \cancel{\text{C}-12 \text{ atom}}}{12 \cancel{\text{protons + neutrons}}}\right)\left(\frac{14 \cancel{\text{protons + neutrons}}}{1 \text{C}-14 \text{ atom}}\right) = \frac{2.32\times10^{-23} \text{ g}}{1 \text{C}-14 \text{ atom}}$$

2.66 In Figure 2.2, the electrons are much closer to the nucleus than they would be in a properly scaled drawing. Consequently, the volume of the atom represented in Figure 2.2 is much less than it should be. Density is calculated as a ratio of mass to volume. The mass of this atom has not changed; however, the volume has decreased. Therefore, the atom in Figure 2.2 is much more dense than an atom that is 99.999% empty.

ALLIED HEALTH EXAM CONNECTION

2.68 (b) Water is a chemical compound. (a) Blood and (d) air are mixtures, while (c) oxygen is an element.

2.70 $^{34}_{17}Cl$ has (a) 17 protons, 17 neutrons (34-17=17), and 17 electrons (electrons = protons in neutral atom).

2.72 Copper has (b) 29 protons because the atomic number is the number of protons.

2.74 The negative charged particle found within the atom is the (b) electron.

2.76 The major portion of an atom's mass consists of (a) neutrons and protons.

2.78 (d) $^{33}_{16}S^{2-}$ has 16 protons, 17 neutrons, and 18 electrons.

2.80 The mass number of an atom with 60 protons, 60 electrons, and 75 neutrons is (b) 135.

2.82 (c) 1.0 mol NO_2 has the greatest number of atoms (1.8 x 10^{24} atoms). 1.0 mol N has 6.0 x 10^{23} atoms, 1.0 g N has 4.3 x 10^{22} atoms, and 0.5 mol NH_3 has 1.2 x 10^{24} atoms.

2.84 The molar mass of calcium oxide, CaO, is (a) 56 g (40 g Ca + 16 g O).

2.86 (b) 2.0 moles Al are contained in a 54.0 g sample of Al.

$$54.0\ \cancel{\text{g Al}}\left(\frac{1\text{ mole Al}}{27.0\ \cancel{\text{g Al}}}\right) = 2.00\text{ mole Al}$$

CHEMISTRY FOR THOUGHT

2.88 Aluminum exists as one isotope; therefore, all atoms have the same number of protons and neutrons as well as the same mass. Nickel exists as several isotopes; therefore, the individual atoms do not have the weighted average atomic mass of 58.69 u.

2.90
$$\frac{\text{dry bean mass}}{\text{jelly bean mass}} = \frac{1}{1.60}$$

$$472\ \cancel{\text{g jelly beans}}\left(\frac{1\text{ g dry beans}}{1.60\ \cancel{\text{g jelly beans}}}\right) = 295\text{ g dry beans}$$

$$472\ \cancel{\text{g jelly beans}}\left(\frac{1\text{ jelly bean}}{1.18\ \cancel{\text{g jelly bean}}}\right) = 400\text{ jelly beans}$$

Each jar contains 400 beans.

2.92 If the atomic mass unit were redefined as being equal to 1/24th the mass of a carbon-12 atom, then the atomic weight of a carbon-12 atom would be 24 u. Changing the definition for an atomic mass unit does not change the relative mass ratio of carbon to magnesium. Magnesium atoms are approximately 2.024 times as heavy as carbon-12 atoms; therefore, the atomic weight of magnesium would be approximately 48.6 u.

2.94 The value of Avogadro's number would not change even if the atomic mass unit were redefined. Avogadro's number is the number of particles in one mole and has a constant value of 6.022 x 10^{23}.

ADDITIONAL ACTIVITIES

Section 2.1 Review: The crossword game pieces are chemical symbols combined to spell the name of different substances. The molecular formulas for these named substances (or one of their components) are written to the right of the pieces. Write out the element names for the all of the elemental symbols.

Ca	F	Fe	I	Ne	$C_8H_{10}N_4O_2$

As	P	Ir	In	$C_9H_8O_4$

S	U	Cr	O	Se	$C_{12}H_{22}O_{11}$

W	At	Er	H_2O

C	Ho	Co	La	Te	$C_7H_8N_4O_2$ (theobromine – one of the components of chocolate)

Section 2.2 Review: Assume the following picture represents an atom.

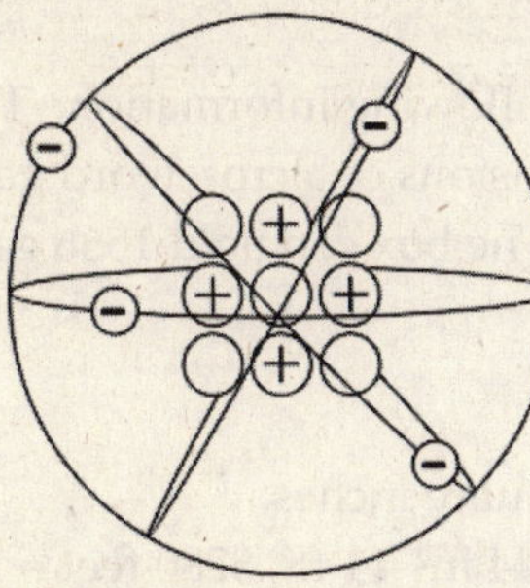

(1) Circle the nucleus.
(2) Which particle(s) is(are) found in the nucleus?
(3) Which particle(s) is(are) found outside the nucleus?
(4) What is the charge on the nucleus?
(5) What is the charge on the whole atom?
(6) What is the mass of this atom in amu?
(7) What is the mass of this atom in grams?

Section 2.3 Review:

$_{20}$Ca

(1) Add the atomic numbers to each of the game pieces from the Section 2.1 Review.
(2) What is the atomic number for the atom represented in the Section 2.2 Review?
(3) What is the mass number for the atom represented in the Section 2.2 Review?

(4) What are the name and symbol for the isotope represented in the Section 2.2 Review?
(5) Write the symbol for carbon-14.
(6) What is the atomic number for carbon-14?
(7) What is the mass number for carbon-14?
(8) How many protons are in carbon-14?
(9) How many neutrons are in carbon-14?
(10) How many electrons are in carbon-14?
(11) Sketch an atom of carbon-14 (like the drawing in the Section 2.2 Review).

Section 2.4 Review:

(1) Calculate the molecular weights for the crossword game "words" and their molecular formulas from the Section 2.1 Review.

J jupiterium

A scientist claims to have discovered a new element, jupiterium. Its atomic mass is 5.994 times the mass of sulfur.

(2) What is the atomic weight of this element?
(3) Add the atomic weight of this element to the box from the periodic table.
(4) Does any known element have this atomic weight? If so, what are the name and symbol for this element?
(5) What information in addition to the atomic mass could confirm the real identity of this element?

Section 2.5 Review:

Gallium has two naturally occurring isotopes as shown in the table below:

Isotope symbol	Natural Abundance	Isotope Mass
^{69}Ga	60.11%	68.925580 amu
^{71}Ga	39.89%	70.9247005 amu

(1) In a sample of 10,000 gallium atoms, how many are gallium-69?
(2) If you were able to pull out individual gallium atoms from sample of 10,000 gallium atom, what is the likelihood of the atom being a gallium-71 isotope?
(3) Calculate the average atomic weight of gallium. Does this match the periodic table?
(4) Will any individual gallium atom have this average atomic weight?
(5) How many atoms are in 5.00 g of gallium?
(6) How many gallium-69 atoms are in a 5.00 g sample of gallium?

Section 2.6 Review:

(1) Write Avogadro's number in expanded form.

This number is enormous! To gain perspective on this number, consider the following information. The surface area of the earth is approximately 196,935,000 square miles. The dimensions of a crossword game box are as follows: 7.5 inches wide, 15.75 inches long, and 1.515 inches deep. The box can hold 1566 game pieces (although each game is sold with only 100 pieces).

(2) Write the surface area of the earth in scientific notation.
(3) Calculate the surface area (length x width) of the top of the game box in square inches.
(4) Convert the surface area of the box into square miles. (12 inches = 1 foot, 144 in^2 = 1 ft^2, 5280 feet = 1 mile, 2.78784×10^7 ft^2 = 1 mi^2)
(5) How many boxes could be placed on the earth's surface?
(6) How many game pieces could all of these boxes contain?

(7) Divide Avogadro's number by the number of game pieces in the boxes that would cover the earth's surface. This will tell you how many game boxes deep each stack would need to be in order to cover the earth with one mole of game pieces.
(8) Convert the depth of the stack into miles by multiplying the number of boxes in a stack by the depth of each box in miles.

As a point of comparison of the size of atoms to the game pieces:
(9) What is the mass of one mole of silicon atoms?
(10) If the density of silicon is 2.33 g/cm^3, what is the volume (in cm^3) of 1 mole of silicon atoms?
(11) Convert this volume to Tablespoons. (15 mL = 1 Tablespoon)
(12) Could you hold one mole of silicon atoms in your hand?
(13) Assuming the atoms occupy all of the space in the 1 mole sample of silicon*, what is the volume (in cm^3) of a silicon atom?
(14) How many silicon atoms will fit into the volume of a game piece (0.715 in. x 0.795 in. x 0.175 in.)?

*Note: This assumption is not valid. The solid will contain empty space and the atoms are actually smaller than this calculation implies.

Section 2.7 Review:
(1) Calculate the molecular weight for all of the molecular formulas given in the right column of the table in the Section 2.1 Review.
(2) How many molecules are in 1 mole of each of those compounds?
(3) What is the mass in grams for 1 mole of each of those compounds?
(4) How many moles of carbon are in 1 mole of each of those compounds?
(5) How many atoms of hydrogen are in 1 mole of each of those compounds?
(6) How many grams of oxygen are in 1 mole of each of those compounds?
(7) What is the mass percentage of nitrogen in each of those compounds?

Tying It All Together with a Laboratory Application:
In 1911 and 1914, Ernest Rutherford published papers in *Philosophical Magazine* about the structure of the atom. At this time, the charge and mass of an electron were known; however, the proton and the neutron had not been discovered. The experimental setup is shown below:

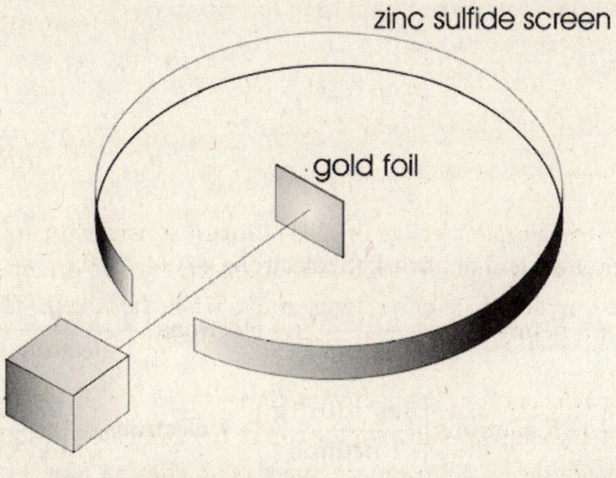

The gold foil is 4 x 10^{-5} cm thick. If the gold foil was 4 x 10^{-5} cm wide by 4 x 10^{-5} cm high, the volume of the gold foil would be (1) ___________ cm^3. The density of gold is 19.3 g/cm^3; therefore the mass of this gold foil would be (2) ___________ g or (3) ___________ u. The number of moles of gold atoms in this sample of gold foil would be (4) ___________ moles. The number of atoms in this sample of gold foil would be (5) ___________ atoms. The gold foil would be approximately (6) ___________ atoms thick. The actual width and the height of the gold foil were greater than 4×10^{-5} cm for ease of experimental setup.

The alpha particles are produced by radioactive decay, which is covered in Chapter 10. One of the accepted chemical symbols for an alpha particle is ${}^{4}_{2}\alpha$. The mass number of an alpha particle is (7) ___________. The atomic number of an alpha particle is (8) ___________. The element with the same atomic number as an alpha particle is (9) ___________. The relative mass of gold to an alpha particle is (10) ___________. An alpha particle contains (11) ___________ protons and (12) ___________ neutrons. The alpha particle has a 2+ charge, which means it contains (13) ___________ electrons. The alpha particles travel at 2.09×10^{9} cm/sec. Light travels at 3.00×10^{8} m/s. The speed of an alpha particle is (14) ___________ percentage the speed of light.

The alpha particles were aimed at the gold foil which was surrounded by a circular zinc sulfide (ZnS) screen. ZnS is a phosphor. It emits light after being struck by an energetic material (like an alpha particle). In this experiment, the angle was measured between the initial path of the alpha particle to the gold foil and the location on the screen that the alpha particle struck after interacting with the gold foil. Most of the alpha particles in this experiment went straight through the gold foil (~98%) or were deflected only a small amount (~2%). Approximately 1 in 20,000 alpha particles, though, were turned back through an average angle of 90°. Rutherford's explanation for this phenomenon was that an atom is mainly empty space with an exceedingly small, dense, positively charged (15) ___________ at the center of the atom and (16) ___________ distributed around the outside of the nucleus that maintain a neutral atom. If an alpha particle passed close to the nucleus of one of the gold atoms it would be (pick one: attracted or repelled) (17) ___________ because the alpha particle is positive and the nucleus is positive.

SOLUTIONS FOR THE ADDITIONAL ACTIVITIES

Section 2.1 Review:

Caffeine – calcium, fluorine, iron, iodine, neon, carbon, hydrogen, nitrogen, oxygen
Aspirin – arsenic, phosphorus, iridium, indium, carbon, hydrogen, oxygen
Sucrose – sulfur, uranium, chromium, oxygen, selenium, carbon, hydrogen, oxygen
Water – tungsten, astatine, erbium, hydrogen, oxygen
Chocolate – carbon, holmium, cobalt, lanthanum, tellurium, carbon, hydrogen, nitrogen, oxygen

Section 2.2 Review:

(1) 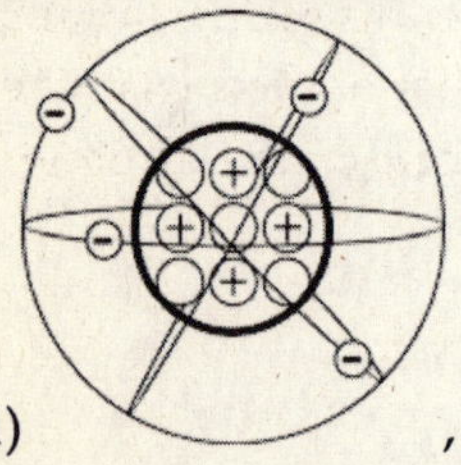 , (2) protons and neutrons, (3) electrons, (4) +4, (5) 0,

$$(6)\ 4\ \text{protons}\left(\frac{1\ \text{u}}{1\ \text{proton}}\right)+5\ \text{neutrons}\left(\frac{1\ \text{u}}{1\ \text{neutron}}\right)+4\ \text{electrons}\left(\frac{\frac{1}{1836}\ \text{u}}{1\ \text{electron}}\right)=9.0021786\ \text{u}=9\ \text{u}\ ,$$

$$(7)\ 4\ \text{protons}\left(\frac{1.67\times10^{-24}\ \text{g}}{1\ \text{proton}}\right)+5\ \text{neutrons}\left(\frac{1.67\times10^{-24}\ \text{g}}{1\ \text{neutron}}\right)+4\ \text{electrons}\left(\frac{9.07\times10^{-28}\ \text{g}}{1\ \text{electron}}\right)=1.50\times10^{-23}\ \text{g}$$

Section 2.3 Review:

(1)

$_{20}$Ca	$_{9}$F	$_{26}$Fe	$_{53}$I	$_{10}$Ne
$_{33}$As	$_{15}$P	$_{77}$Ir	$_{49}$In	
$_{16}$S	$_{92}$U	$_{24}$Cr	$_{8}$O	$_{34}$Se
$_{74}$W	$_{85}$At	$_{68}$Er		
$_{6}$C	$_{67}$Ho	$_{27}$Co	$_{57}$La	$_{52}$Te

(2) 4, (3) 9, (4) beryllium-9, $^{9}_{4}\text{Be}$,

(5) $^{14}_{6}\text{C}$, (6) 6, (7) 14, (8) 6, (9) 8, (10) 6,

(11)

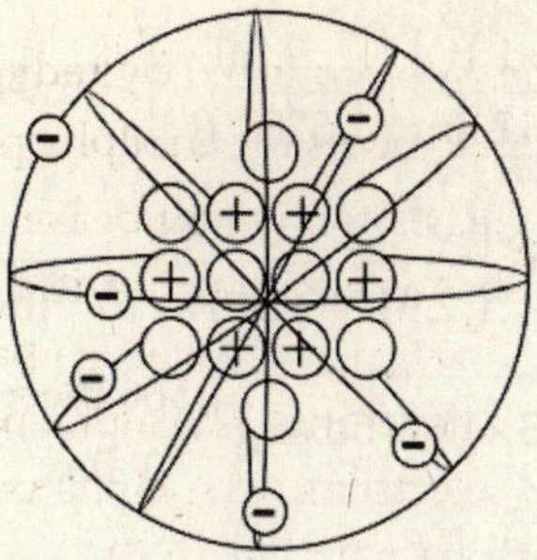

Section 2.4 Review:

(1) Caffeine – $40.08\text{ u} + 19.00\text{ u} + 55.85\text{ u} + 126.90\text{ u} + 20.18\text{ u} = 262.01\text{ u}$;

$$C_8H_{10}N_4O_2 - \ 8\,C\left(\frac{12.01\text{ u}}{1\,C}\right) + 10\,H\left(\frac{1.0079\text{ u}}{1\,H}\right) + 4\,N\left(\frac{14.01\text{ u}}{1\,N}\right) + 2\,O\left(\frac{16.00\text{ u}}{1\,O}\right) = 194.20\text{ u};$$

Aspirin – $74.92\text{ u} + 30.97\text{ u} + 192.22\text{ u} + 114.82\text{ u} = 412.93\text{ u}$;

$$C_9H_8O_4 - \ 9\,C\left(\frac{12.01\text{ u}}{1\,C}\right) + 8\,H\left(\frac{1.0079\text{ u}}{1\,H}\right) + 4\,O\left(\frac{16.00\text{ u}}{1\,O}\right) = 180.15\text{ u};$$

Sucrose – $32.07\text{ u} + 238.03\text{ u} + 52.00\text{ u} + 16.00\text{ u} + 78.96\text{ u} = 417.06\text{ u}$;

$$C_{12}H_{22}O_{11} - \ 12\,C\left(\frac{12.01\text{ u}}{1\,C}\right) + 22\,H\left(\frac{1.0079\text{ u}}{1\,H}\right) + 11\,O\left(\frac{16.00\text{ u}}{1\,O}\right) = 342.29\text{ u};$$

Water – $183.9\text{ u} + 210\text{ u} + 167.26\text{ u} = 561\text{ u}$;

$$H_2O - \ 2\,H\left(\frac{1.0079\text{ u}}{1\,H}\right) + 1\,O\left(\frac{16.00\text{ u}}{1\,O}\right) = 18.02\text{ u};$$

Chocolate – $12.01\text{ u} + 164.93\text{ u} + 58.93\text{ u} + 138.91\text{ u} + 127.60\text{ u} = 502.38\text{ u}$;

$$C_7H_8N_4O_2 - \ 7\,C\left(\frac{12.01\text{ u}}{1\,C}\right) + 8\,H\left(\frac{1.0079\text{ u}}{1\,H}\right) + 4\,N\left(\frac{14.01\text{ u}}{1\,N}\right) + 2\,O\left(\frac{16.00\text{ u}}{1\,O}\right) = 180.17\text{ u};$$

(2) $5.994 \times 32.07\text{ u} = 192.2\text{ u}$; (3)

J
jupiterium
192.2

; (4) iridium, Ir; (5) If the atomic number of jupiterium is 77, then this element is really iridium. The atomic number is the number of protons in the nucleus of an atom. The atomic number is unique for each element.

Section 2.5 Review:

(1) $60.11\% \times 10{,}000 = 0.6011 \times 10{,}000 = 6{,}011$; (2) $\dfrac{3989}{10000} \times 100 = 39.89\%$;

(3) $60.11\%(68.925580\text{ u}) + 39.89\%(70.9247005\text{ u}) = 0.6011(68.925580\text{ u}) + 0.3989(70.9247005\text{ u}) = 69.72\text{ u}$,

yes; (4) no; (5) $5.00\,\cancel{\text{g}}\left(\dfrac{1\,\cancel{\text{u}}}{1.661 \times 10^{-24}\,\cancel{\text{g}}}\right)\left(\dfrac{1\text{ atom}}{69.72\,\cancel{\text{u}}}\right) = 4.32 \times 10^{22}\text{ atoms}$;

(6) $$5.00\,\cancel{\text{g}}\left(\frac{1\,\cancel{\text{u}}}{1.661 \times 10^{-24}\,\cancel{\text{g}}}\right)\left(\frac{1\text{ atom}}{69.72\,\cancel{\text{u}}}\right)60.11\% = 5.00\,\cancel{\text{g}}\left(\frac{1\,\cancel{\text{u}}}{1.661 \times 10^{-24}\,\cancel{\text{g}}}\right)\left(\frac{1\text{ atom}}{69.72\,\cancel{\text{u}}}\right)0.6011 =$$

$$= 2.60 \times 10^{22}\text{ gallium} - 69\text{ atoms}$$

Section 2.6 Review:

(1) 602 200 000 000 000 000 000 000; (2) 1.96935 x 10^8 square miles;

(3) $7.5 \text{ inches} \times 15.75 \text{ inches} = 118.125 \text{ in}^2 = 120 \text{ in}^2$;

(4) $118.125 \text{ in}^2 \left(\frac{1 \text{ ft}}{12 \text{ in}}\right)^2 \left(\frac{1 \text{ mi}}{5280 \text{ ft}}\right)^2 = 2.9424662104 \times 10^{-8} \text{mi}^2 = 2.9 \times 10^{-8} \text{mi}^2$;

(5) $196935000 \text{ mi}^2 \left(\frac{1 \text{ box}}{2.9424662104 \times 10^{-8} \text{ mi}^2}\right) = 6.69285510583 \times 10^{15} \text{ boxes} = 6.7 \times 10^{15} \text{ boxes}$;

(6) $6.69285510583 \times 10^{15} \text{ boxes} \left(\frac{1566 \text{ pieces}}{1 \text{ box}}\right) = 1.04810110957 \times 10^{19} \text{ pieces} = 1.0 \times 10^{19} \text{ pieces}$;

(7) $6.022 \times 10^{23} \text{ pieces} \left(\frac{\text{boxes to cover earth's surface}}{1.04810110957 \times 10^{19} \text{ pieces}}\right) = 57456.2888 \text{ layers of boxes} = 5.7 \times 10^4 \text{ boxes deep}$;

(8) $57456.2888 \text{ layers of boxes} \left(\frac{1.515 \text{ inches}}{1 \text{ box}}\right)\left(\frac{1 \text{ foot}}{12 \text{ inches}}\right)\left(\frac{1 \text{ mile}}{5280 \text{ feet}}\right) = 1.37383644996 \text{ miles} = 1.4 \text{ miles}$;

(9) 28.09 g; (10) $28.09 \text{ g} \left(\frac{1 \text{ cm}^3}{2.33 \text{ g}}\right) = 12.0557939914 \text{ cm}^3 = 12.1 \text{ cm}^3$;

(11) $12.0557939914 \text{ cm}^3 \left(\frac{1 \text{ mL}}{1 \text{ cm}^3}\right)\left(\frac{1 \text{ Tablespoon}}{15 \text{ mL}}\right) = 0.803719599427 \text{ Tablespoons} = 0.80 \text{ Tablespoons}$;

(12) yes; (13) $\frac{12.0557939914 \text{ cm}^3}{1 \text{ mole of atoms}}\left(\frac{1 \text{ mole}}{6.022 \times 10^{23} \text{ atoms}}\right) = 2.00195848413 \times 10^{-23} \frac{\text{cm}^3}{\text{atoms}} = 2.0 \times 10^{-23} \frac{\text{cm}^3}{\text{atoms}}$;

(14) $0.715 \text{ in} \times 0.795 \text{ in} \times 0.175 \text{ in} \left(\frac{2.54 \text{ cm}}{1 \text{ in}}\right)^3 \left(\frac{2.33 \text{ g}}{1 \text{ cm}^3}\right)\left(\frac{1 \text{ mole}}{28.09 \text{ g}}\right)\left(\frac{6.022 \times 10^{23} \text{ atoms}}{1 \text{ mole}}\right) = 8.14 \times 10^{22} \text{ atoms}$

Section 2.7 Review:

(1) $C_8H_{10}N_4O_2$ = 194.20 u; $C_9H_8O_4$ = 180.15 u; $C_{12}H_{22}O_{11}$ = 342.29 u; H_2O = 18.02 u; $C_7H_8N_4O_{12}$ = 180.17 u; (2) 1 mole of any compound contains 6.022 x 10^{23} molecules; (3) $C_8H_{10}N_4O_2$ = 194.20 g; $C_9H_8O_4$ = 180.15 g; $C_{12}H_{22}O_{11}$ = 342.29 g; H_2O = 18.02 g; $C_7H_8N_4O_{12}$ = 180.17 g; (4) $C_8H_{10}N_4O_2$ = 8 moles C; $C_9H_8O_4$ = 9 moles C; $C_{12}H_{22}O_{11}$ = 12 moles C; H_2O = 0 moles C; $C_7H_8N_4O_{12}$ = 7 moles C; (5) $C_8H_{10}N_4O_2$ =

$1 \text{ mole } C_8H_{10}N_4O_2 \left(\frac{6.022 \times 10^{23} \text{ molecules}}{1 \text{ mole } C_8H_{10}N_4O_2}\right)\left(\frac{10 \text{ atoms H}}{1 \text{ molecule } C_8H_{10}N_4O_2}\right) = 6.022 \times 10^{24} \text{ atoms H}$; $C_9H_8O_4$ = 4.818 x 10^{24} atoms H; $C_{12}H_{22}O_{11}$ = 1.32 x10^{25} atoms H; H_2O = 1.204 x 10^{24} atoms H; $C_7H_8N_4O_{12}$ = 4.818 x 10^{24} atoms H; (6) $C_8H_{10}N_4O_2$ = $1 \text{ mole } C_8H_{10}N_4O_2 \left(\frac{2 \text{ moles O}}{1 \text{ mole } C_8H_{10}N_4O_2}\right)\left(\frac{16.00 \text{ g O}}{1 \text{ mole O}}\right) = 32.00 \text{ g O}$; $C_9H_8O_4$ = 64.00 g O; $C_{12}H_{22}O_{11}$ = 176.00 g O; H_2O = 16.00 g O; $C_7H_8N_4O_{12}$ = 192.00 g O; (7) $C_8H_{10}N_4O_2$ =

$\frac{56.04 \text{ g N}}{194.20 \text{ g } C_8H_{10}N_4O_2} \times 100 = 28.86\% \text{ N}$; $C_9H_8O_4$ = 0% N; $C_{12}H_{22}O_{11}$ = 0% N; H_2O = 0% N; $C_7H_8N_4O_{12}$ = 31.10% N

Tying It All Together with a Laboratory Application:

(1) 6.4×10^{-14} cm^3 = 6×10^{-14} cm^3
(2) 1.2352×10^{-12} g = 1×10^{-12} g
(3) 7.4365×10^{11} u = 7×10^{11} u
(4) 6.2710×10^{-15} moles = 6×10^{-15} moles
(5) 3.7764×10^{9} atoms = 4×10^{9} atoms
(6) 1557.25 atoms = 2000 atoms = 2×10^{3} atoms
(7) 4
(8) 2
(9) helium
(10) 49
(11) 2
(12) 2
(13) 0
(14) 6.97%
(15) nucleus
(16) electrons
(17) repelled

SELF-TEST QUESTIONS

Multiple Choice

1. Which of the following is an incorrect symbol for an element?
   a. Ce b. Au c. K d. CR

2. Which of the following is an incorrect formula for a compound?
   a. CO_2 b. CO_1 c. N_2O d. NO_2

3. Two objects have masses of 3.2 g and 0.80 g. What is the relative mass of the 3.2 g object compared to the other?
   a. 4.0 to 1 b. 2.0 to 1 c. 0.50 to 1 d. 0.25 to 1

4. Suppose the atomic weights of the elements were assigned in such a way that the atomic weight of helium, He, was 1.00 u. What would be the atomic weight of oxygen, O, in u, on this scale?
   a. 16.0 b. 8.00 c. 4.00 d. 0.250

5. What is the molecular weight of phosphoric acid, H_3PO_4, in u?
   a. 48.0 b. 50.0 c. 96.0 d. 98.0

6. How many neutrons are there in the nucleus of a potassium-39 atom?
   a. 1 b. 19 c. 20 d. 39

7. What is the mass in grams of 1.00 mole of chlorine molecules, Cl_2?
   a. 6.02×10^{23} b. 71.0 c. 35.5 d. 1.18×10^{-22}

8. Calculate the weight percent of sulfur, S, in SO_2.
   a. 50.1 b. 33.3 c. 66.7 d. 25.0

Matching

Match the molecules represented on the left with the terms on the right to the correct classification given.

9. 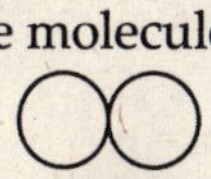

10. 

11. 

12. 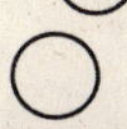

13. 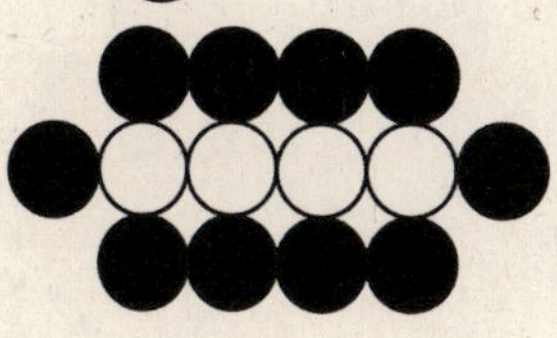

a. homoatomic and diatomic
b. homoatomic and triatomic
c. homoatomic and polyatomic
d. heteroatomic and diatomic
e. heteroatomic and triatomic
f. heteroatomic and polyatomic
g. none of the above

Match the number given as responses to the following:

14.	The number of moles of oxygen atoms in 2 moles of NO_2.	a.	1
15.	The number of moles of NH_3 that contain 3 moles of nitrogen atoms.	b.	2
16.	The number of moles of nitrogen atoms in one-half mole of N_2O_5.	c.	3
17.	The number of moles of electrons in one mole of helium atoms.	d.	4
18.	The number of moles of neutrons in one mole of $^{3}_{1}H$.		

True-False

19. In some instances, two different elements are represented by the same symbol.
20. The mass of a single atom of silicon, Si, is 28.1 g.
21. All isotopes of a specific element have the same atomic number.
22. All atoms of a specific element have the same number of protons and electrons.
23. One mole of water molecules, H_2O, contain two moles of hydrogen atoms, H.
24. 1.00 mol of sulfur, S, contains the same number of atoms as 14.0 g of nitrogen, N.
25. 6.02×10^{23} molecules of methane, CH_4, contains 6.02×10^{23} atoms of hydrogen.

Chapter 3: Electronic Structure and the Periodic Law

CHAPTER OUTLINE

3.1 The Periodic Law and Table
3.2 Electronic Arrangements in Atoms
3.3 The Shell Model and Chemical Properties
3.4 Electronic Configurations
3.5 Another Look at the Periodic Table
3.6 Property Trends within the Periodic Table

LEARNING OBJECTIVES/ASSESSMENT

When you have completed your study of this chapter, you should be able to:

1. Locate elements in the periodic table on the basis of group and period designations. (Section 3.1; Exercise 3.4)
2. Determine the number of electrons in designated atomic orbitals, subshells, or shells. (Section 3.2; Exercise 3.12)
3. Determine the number of valence shell electrons and the electronic structure for atoms, and relate this information to the location of elements in the periodic table. (Section 3.3; Exercises 3.18 and 3.22)
4. Determine the following for elements: the electronic configuration of atoms, the number of unpaired electrons in atoms, and the identity of atoms based on provided electronic configurations. (Section 3.4; Exercises 3.24 and 3.28)
5. Determine the shell and subshell locations of the distinguishing electrons in elements, and based on their location in the periodic table, classify elements into the categories given in Figures 3.10 (representative element, transition element, inner-transition element, noble gas) and 3.12. (metal, metalloid, nonmetal). (Section 3.5; Exercises 3.34 and 3.36)
6. Recognize property trends of elements within the periodic table, and use the trends to predict selected properties of the elements. (Section 3.6; Exercises 3.40 and 3.42)

SOLUTIONS FOR THE END OF CHAPTER EXERCISES

THE PERIODIC LAW AND TABLE (SECTION 3.1)

3.2			Group	Period
	a.	Sb	VA (15)	5
	b.	cadmium	II B (12)	5
	c.	element number 31	III A (13)	4
	d.	Zr	IV B (4)	5

☑3.4			Symbol	Name
	a.	The noble gas belonging to period 4	Kr	krypton
	b.	The fourth element (reading down) in group IVA (14)	Sn	tin
	c.	Belongs to group VIB (6) and period 5	Mo	molybdenum
	d.	The sixth element (reading left to right) in period 6	Nd	neodymium

3.6			
	a.	How many elements are located in group VIIB (7) of the periodic table?	4
	b.	How many total elements are found in periods 1 and 2 of the periodic table?	10
	c.	How many elements are found in period 5 of the periodic table?	18

3.8			
	a.	This is a horizontal arrangement of elements in the periodic table.	period
	b.	Element 11 begins this arrangement in the periodic table.	period
	c.	The element nitrogen is the first member of this arrangement.	group
	d.	Elements 9, 17, 35, and 53 belong to this arrangement.	group

ELECTRONIC ARRANGEMENTS IN ATOMS (SECTION 3.2)

3.10 Protons are subatomic particles with a positive charge that are located in the nucleus.

☑3.12
a. A $2p$ orbital — 2 electrons
b. A $2p$ subshell — 6 electrons
c. The second shell — 8 electrons

3.14 Four (4) orbitals are found in the second shell: one $2s$ orbital and three $2p$ orbitals.

3.16 Seven (7) orbitals are found in a $4f$ subshell. The maximum number of electrons that can be located in this subshell is 14 because each of the seven orbitals can hold two electrons.

THE SHELL MODEL AND CHEMICAL PROPERTIES (SECTION 3.3)

☑3.18
a. element number 54 — 8 electrons
b. The first element (reading down) in group V A (15) — 5 electrons
c. Sn — 4 electrons
d. The fourth element (reading left to right) in period 3 — 4 electrons

3.20 Cesium is the period 6 element with chemical properties most like sodium. Cesium has 1 valence-shell electron. Sodium also has only 1 valence-shell electron.

☑3.22 I would expect to find silver and gold in addition to the copper because these elements are all in the same group on the periodic table. Elements that are in the same group have similar chemical properties; therefore, if copper is part of this ore, then the other elements that are most similar to it are also likely to be part of the ore.

ELECTRONIC CONFIGURATIONS (SECTION 3.4)

☑3.24

		Electron Configuration	Unpaired Electrons
a.	element number 37	$1s^22s^22p^63s^23p^64s^23d^{10}4p^65s^1$	1
b.	Si	$1s^22s^22p^63s^23p^2$	2
c.	titanium	$1s^22s^22p^63s^23p^64s^23d^2$	2
d.	Ar	$1s^22s^22p^63s^23p^6$	0

3.26

		Electron Configuration	Solutions
a.	s electrons in potassium	$1s^22s^22p^63s^23p^64s^1$	7
b.	unpaired electrons in Al	$1s^22s^22p^63s^23p^1$	1
c.	filled subshells in magnesium	$1s^22s^22p^63s^2$	4

☑3.28

		Symbol	Name
a.	Contains only two $2p$ electrons	C	carbon
b.	Contains an unpaired $3s$ electron	Na	sodium
c.	Contains two unpaired $3p$ electrons	Si or S	silicon or sulfur
d.	Contains three $4d$ electrons	Nb	niobium
e.	Contains three unpaired $3d$ electrons	V or Co	vanadium or cobalt

3.30

a.	arsenic	$[Ar]4s^23d^{10}4p^3$	c.	silicon	$[Ne]3s^23p^2$
b.	An element that contains 25 electrons	$[Ar]4s^23d^5$	d.	element number 53	$[Kr]5s^24d^{10}5p^5$

3.32

a.	sodium	$[Ne]3s^1$	e.	phosphorus	$[Ne]3s^23p^3$
b.	magnesium	$[Ne]3s^2$	f.	sulfur	$[Ne]3s^23p^4$
c.	aluminum	$[Ne]3s^23p^1$	g.	chlorine	$[Ne]3s^23p^5$
d.	silicon	$[Ne]3s^23p^2$	h.	argon	$[Ne]3s^23p^6$

ANOTHER LOOK AT THE PERIODIC TABLE (SECTION 3.5)

☑3.34

a.	lead	*p* area
b.	element 27	*d* area
c.	Tb	*f* area
d.	Rb	*s* area

☑3.36

a.	iron	transition
b.	element 15	representative
c.	U	inner-transition
d.	xenon	noble gas
e.	tin	representative

3.38

a.	argon	nonmetal
b.	element 3	metal
c.	Ge	metalloid
d.	boron	metalloid
e.	Pm	metal

PROPERTY TRENDS WITHIN THE PERIODIC TABLE (SECTION 3.6)

☑3.40

a.	K or Ti	K more metallic
b.	As or Bi	Bi
c.	Mg or Sr	Sr
d.	Sn or Ge	Sn

☑3.42

a.	Ga or Se	Ga larger radius
b.	N or Sb	Sb
c.	O or C	C
d.	Te or S	Te

3.44

a.	Si or Cl	Si loses e^- more easily
b.	Mg or Ba	Ba
c.	F or Br	Br
d.	Ca or S	Ca

ADDITIONAL EXERCISES

3.46 Chemical properties are dependent on the number of valence electrons an atom contains, not the number of neutrons an atom contains; therefore, the chemical properties of isotopes of the same element are the same because all isotopes of the same element contain the same number of electrons, including valence electrons.

3.48 The atom with the electron configuration of $1s^2\,2s^2\,2p^4$ is oxygen, which has an atomic weight of 16.00 u; therefore, the mass of 3.0×10^{20} oxygen atoms is 8.0 mg.

$$3.0\times10^{20}\ \cancel{\text{atoms O}}\left(\frac{16.00\ \cancel{\text{u}}}{1\ \cancel{\text{atom O}}}\right)\left(\frac{1.661\times10^{-24}\ \cancel{\text{g}}}{1\ \cancel{\text{u}}}\right)\left(\frac{1000\ \text{mg}}{1\ \cancel{\text{g}}}\right)=8.0\ \text{mg}$$

$$\text{or}\quad 3.0\times10^{20}\ \cancel{\text{atoms O}}\left(\frac{16.00\ \cancel{\text{g O}}}{6.022\times10^{23}\ \cancel{\text{atoms O}}}\right)\left(\frac{1000\ \text{mg}}{1\ \cancel{\text{g}}}\right)=8.0\ \text{mg}$$

3.50 $$\text{Molar Mass} = \frac{\text{grams}}{\text{mole}} = \frac{10.02\ \text{g}}{0.250\ \text{moles}} = 40.08\,\frac{\text{grams}}{\text{mole}}$$

The element is calcium, which is a representative element that conducts electricity.

ALLIED HEALTH EXAM CONNECTION

3.52 The horizontal row of the periodic table are called (d) periods.

3.54 (d) Mg and Ca have similar chemical properties because they are part of the same group IIA (2).

3.56 Nonmetals are located on the (a) upper right of the periodic table.

3.58 (b) Sodium is an alkali metal; it belongs to group IA.

3.60 The maximum number of electrons that each *p* orbital can hold is (b) 2.

3.62 The element with the smallest atomic radius is (b) Mg. The order of increasing atomic radius is magnesium, strontium, barium, radium.

3.64 (d) Sr has the largest first ionization energy.

3.66 (d) Valence describes the electrons in the outermost principal energy level of an atom.

3.68 (a) Na/K have the same number of electrons in their outermost energy level. They both have 1 valence electron.

3.70 (b) 2 valence electrons are needed to complete the outer valence shell of sulfur.

CHEMISTRY FOR THOUGHT

3.72 Calcium bromide is $CaBr_2$ and strontium bromide is $SrBr_2$, because calcium and strontium have the same number of valence electrons as magnesium, which forms $MgBr_2$ when it reacts with bromine.

3.74 Gold, silver, copper, nickel, platinum, and zinc are used in coins and medals because they have limited chemical reactivity, they are malleable, and they have low enough melting points to be melted and poured into uniform shapes.

Gold is useful in electrical connectors for critical electronic parts such as computers in space craft because it not only conducts electricity, but also has limited chemical reactivity, which means it can be trusted to continue to conduct electricity without undergoing a chemical change.

3.76 The number of floors in Hotel Five is 5 and the number of rooms on the top floor of Hotel Five is 9. Each hotel has the same number of floors as the hotel number. The number of rooms on each floor is an increasing odd integer (1, 3, 5, 7).

ADDITIONAL ACTIVITIES

Section 3.1 Review: Does this periodic table of shapes have the same orientation of periods and groups as the periodic table of elements? Explain why or why not.

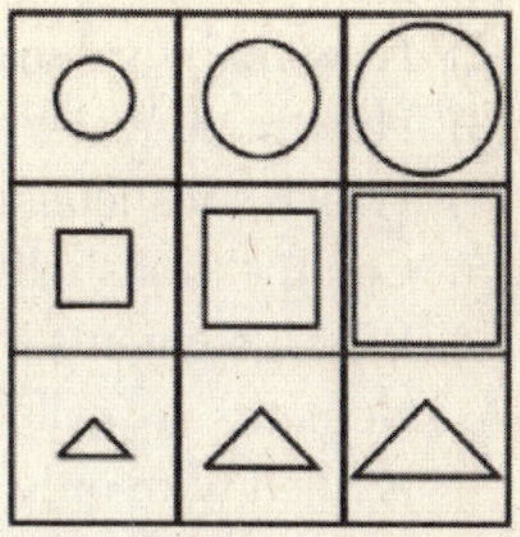

Section 3.2 Review: The periodic table can be used to help remember the relationships between shells, subshells, orbitals, and electrons.

(1) Number the periods on the left side of the periodic table. The period numbers are the same as the shell numbers. How many shells are known?

(2) The blocks on the periodic table represent the subshells. Color the blocks of the periodic table. Use the following instructions and/or Figure 3.9 as a guide.

(a) The s block consists of the first 2 columns of the periodic table and the box that represents He.

(b) The p block consists of the last 6 columns of the periodic table (excluding the box that represents He).

(c) The d block consists of 10 columns in the middle of the periodic table. The first of these ten columns is next to the s block. The last nine columns are immediately to the right of the p block.

(d) The f block consists of 14 columns in the middle of the periodic table. The f block splits the d block into 2 groups and is usually placed below the periodic table.

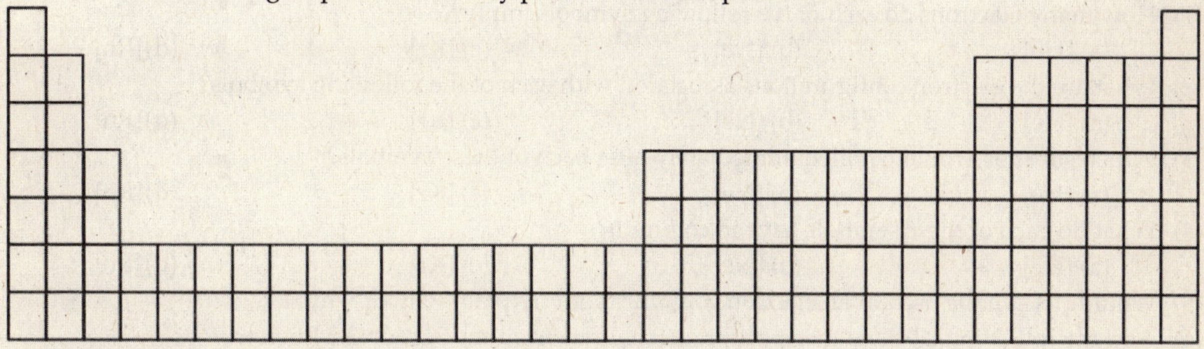

(3) The shell number not only describes the location and energy of the electrons around a nucleus, but it also gives the number of subshells within the shell. How many subshells should each of the labeled periods have? Name these subshells. (Only the first four subshells are given names in the textbook. After f the lettering becomes alphabetical (except letter *j* is excluded).) Based on your colored periodic table, why does the textbook exclude the higher energy subshells?

(4) Trace across each period and count the number of subshells in each period. Does this match your prediction from (3)? Propose a reason for any discrepancies.

(5) The maximum number of electrons in a subshell can be obtained by counting the number of boxes in one row of the subshell. What is the maximum number of electrons in the…

(a) s subshell? (b) p subshell? (c) d subshell? (d) f subshell?

(6) The number of orbitals in a subshell can be obtained by dividing the maximum number of electrons in a subshell by two (because two electrons can be placed in each orbital). How many orbitals are in the…

(a) s subshell? (b) p subshell? (c) d subshell? (d) f subshell?

Section 3.3 Review:

(1) How many valence electrons are in cesium (#55)?
(2) How many valence electrons are in selenium (#34)?
(3) How many valence electrons are in xenon (#54)?
(4) Which subshells must be filled in order to have 8 electrons in the valence shell?
(5) Can the first shell have 8 electrons in its valence shell? Explain why or why not.
(6) Which elements in the periodic table contain valence shells with filled s and p orbitals? (These elements are called **noble gases**. Helium is normally included as a noble gas because its valence shell is filled with only 2 electrons.)
(7) Add a "Noble Gases" label to the appropriate column of the periodic table in the Section 3.2 Review.

Section 3.4 Review:

(1) Rank the three subshells in the third shell in order of increasing energy.
(2) Based on Figure 3.7 in the textbook, how does the energy level of the 4s subshell compare to the energy levels of each of the subshells in the third shell?
(3) Referring back to the coloring exercise from the Section 3.2 Review, label the start of each subshell on the periodic table (ex. 1s).

For questions 4-7, you may refer to the Section 3.2 Review or Figure 3.8 in the textbook.
(4) Which subshell is completely filled immediately before the 4f subshell is filled?
(5) Which subshell is completely filled immediately before the 5f subshell is filled?
(6) Which subshell is completely filled immediately before the 5d subshell is filled?
(7) Which subshell is completely filled immediately before the 6d subshell is filled?

Section 3.5 Review:

(1) How many electrons do each of the following symbols imply?
(a) [He] (b) [Ne] (c) [Kr] (d)[Rn]
(2) What are the electron configurations associated with each of the following symbols?
(a) [He] (b) [Ne] (c) [Kr] (d)[Rn]
(3) Which subshell would be filled immediately after each of these symbols?
(a) [He] (b) [Ne] (c) [Kr] (d)[Rn]
(4) What do each of these symbols have in common?
(a) [He] (b) [Ne] (c) [Kr] (d)[Rn]
(5) Would [Ir] ever be used in an electron configuration? Explain why or why not.
(6) Add the following labels to the appropriate columns from the Section 3.2 Review.
(a) representative elements (b) inner-transition elements (c) transition elements
(7) In which subshell(s) can the distinguishing electron be found for each of the following classes?
(a) metal (b) nonmetal (c) metalloid
(8) How many electrons are in the valence shell of each of the following elements?
(a) Sr (b) W (c) U (d) As
(9) In which subshell is the distinguishing electron for each of the following elements found?
(a) Sr (b) W (c) U (d) As

Section 3.6 Review:

(1) On the periodic table below, draw and label the trend for increasing metallic character in a period and a group.
(2) On the periodic table below, draw and label the trend for increasing atomic size in a period and a group.
(3) On the periodic table below, draw and label the trend for increasing ionization energy in a period and a group.

(4) What is the relationship between the trends for metallic character and atomic size?
(5) What is the relationship between the trends for metallic character and ionization energy?
(6) What is the relationship between the trends for atomic size and ionization energy?

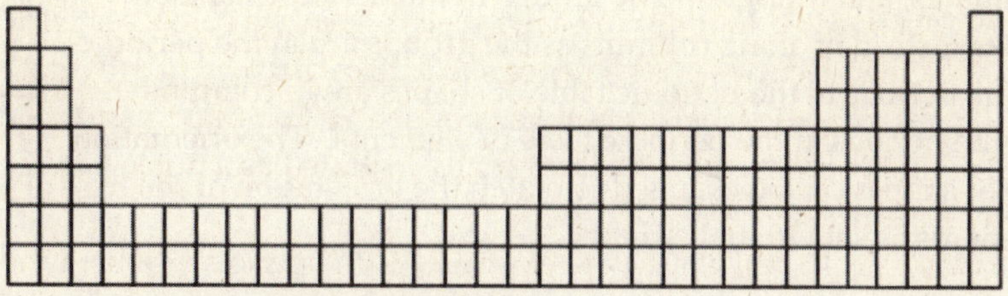

Tying It All Together with a Laboratory Application:
When sodium metal is placed in water, it reacts to produce sodium hydroxide and hydrogen gas. When potassium is placed in water, it will have a (pick one: different or similar) (1) _______ chemical reaction because sodium and potassium are part of the same (2) _______. The number of electrons in the (3) _______ shell of sodium is 1. The number of electrons in the valence shell of potassium is (4) _______. The distinguishing electron for sodium is in the (5) _______ subshell. The distinguishing electron for potassium is in shell (6) _______. Sodium is a (pick one: representative element, noble gas, inner-transition element, transition element) (7) _______. Potassium is a (pick one: metal, nonmetal, metalloid) (8) _______. Potassium atoms are (pick one: larger or smaller) (9) _______ than sodium atoms. Potassium atoms are (10) _______ metallic than sodium atoms. In terms of chemical behavior, potassium atoms will likely react (11) _______ vigorously with water because they lose electrons (12) _______ easily than sodium atoms.

Magnesium is in group (13) _______ and period (14) _______. The chemical behavior of magnesium in water (15) _______ be predicted based on the knowledge of the chemical behavior of sodium because elements in the same (16) _______ do not necessarily share the same chemical properties. Based on periodic trends, though, magnesium atoms are (pick one: larger or smaller) (17) _______ than sodium atoms. Magnesium atoms are (18) _______ metallic than sodium atoms. Magnesium atoms lose electrons (19) _______ easily than sodium atoms. (In Section 9.6, we will find that magnesium atoms will not react with cold water, but they will react with steam.)

In terms of physical properties, magnesium will have a (20) _______ thermal conductivity, a (21) _______ electrical conductivity, and a (22) _______ luster. Magnesium can be drawn into wires because it is (23) _______ and hammered into sheets because it is (24) _______. Magnesium has these properties because it is a (25) _______.

SOLUTIONS FOR THE ADDITIONAL ACTIVITIES

Section 3.1 Review: The periodic table of shapes does not have the same orientation as the periodic table of elements. The groups in the periodic table of shapes are circles, squares, and triangles. The groups in the periodic table of shapes are along rows instead of along columns as the groups are in the periodic table of elements. The periods in the periodic table of shapes are in columns instead of rows like the periods in the periodic table of elements. The orientation of the periodic table of shapes can be changed to match the orientation of the periodic table of elements as shown to the right.

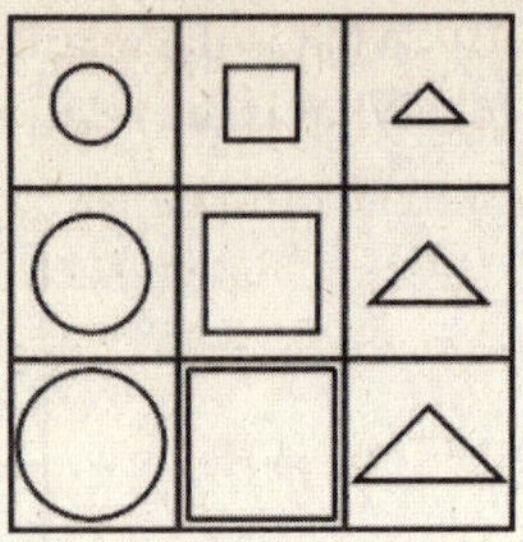

Corrected periodic table of shapes.

Section 3.2 Review:

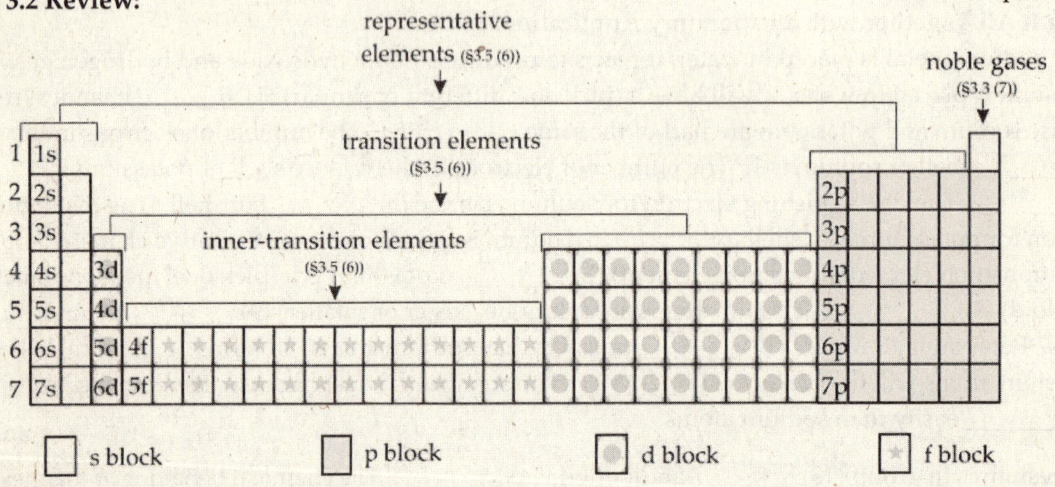

(1) 7 known shells

(3)

shell number	1	2	3	4	5	6	7
number of subshells	1	2	3	4	5	6	7
name of subshells	s	s, p	s, p, d	s, p, d, f	s, p, d, f, g	s, p, d, f, g, h	s, p, d, f, g, h, i

The higher energy subshells (g, h, and i) are not actually occupied in the elements on the periodic table.

(4)

period number	1	2	3	4	5	6	7
number of subshells	1	2	2	3	3	4	4

Periods 1 and 2 match the prediction from (3); however, periods 3, 4, 5, 6, and 7 do not. Periods 5, 6, and 7 could not match the proposed values from (3) because the higher energy subshells (g, h, and i) are not represented on the periodic table. Period 3 does not contain a d subshell and period 4 has a d subshell before the p subshell. Perhaps, the 3d subshell is in period 4 (which means that the 4d subshell is in period 5, the 5d subshell is in period 6, and the 6d subshell is in period 7). Periods 4 and 5 do not contain f subshells. Perhaps the 4f subshell is in period 6 and the 5f subshell is in period 7.

(5) (a) 2; (b) 6; (c) 10; (d) 14; (6) (a) 1; (b) 3; (c) 5; (d) 7

Section 3.3 Review:

(1) 1; (2) 6; (3) 8; (4) s and p; (5) No, the first shell can have a maximum of 2 electrons in its valence shell because it only contains an *s* subshell, which contains only one *s* orbital, which contains a maximum of 2 electrons. (6) neon, argon, krypton, xenon, radon

Section 3.4 Review:

(1) 3s < 3p < 3d; (2) The 4s subshell is higher in energy than the 3s and 3p subshells, but lower in energy than the 3d subshell. 3s < 3p < 4s < 3d; (4) 6s; (5) 7s; (6) 4f; (7) 5f

Section 3.5 Review:

(1) (a) 2; (b) 10; (c) 36; (d) 86; (2) (a) $1s^2$; (b) $1s^2 2s^2 2p^6$; (c) $1s^2 2s^2 2p^6 3s^2 3p^6 4s^2 3d^{10} 4p^6$; (d) $1s^2 2s^2 2p^6 3s^2 3p^6 4s^2 3d^{10} 4p^6 5s^2 4d^{10} 5p^6 6s^2 4f^{14} 5d^{10} 6p^6$; (3) (a) 2s; (b) 3s; (c) 5s; (d) 7s; (4) All of these elements are noble gases. (5) [Ir] would never be used in an abbreviated electron configuration because iridium is not a noble gas. The abbreviated electron configurations must start with a noble gas in brackets. (7) (a) s, p, d, f; (b) s, p; (c) p; (8) (a) 2; (b) 2; (c) 2; (d) 5; (9) (a) 5s; (b) 5d; (c) 5f; (d) 4p

Section 3.6 Review:

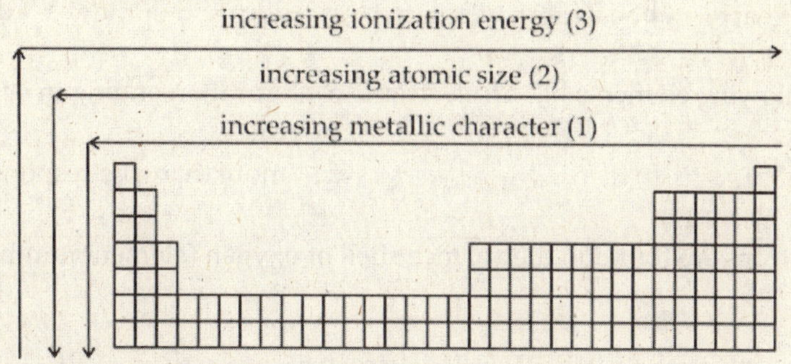

(4) the trends are the same; (5) the trends are opposite; (6) the trends are opposite

Tying It All Together with a Laboratory Application:

(1) similar
(2) group or family
(3) valence
(4) 1
(5) 3s
(6) 4
(7) representative element
(8) metal
(9) larger
(10) more
(11) more
(12) more
(13) II A or (2)
(14) 3
(15) cannot
(16) period
(17) smaller
(18) less
(19) less
(20) high
(21) high
(22) metallic
(23) ductile
(24) malleable
(25) metal

SELF-TEST QUESTIONS

Multiple Choice

1. Elements in the same group
   a. have similar chemical properties.
   b. have consecutive atomic numbers.
   c. are called isotopes.
   d. constitute a period of elements.

2. Which element of those with the following atomic numbers should have properties similar to those of oxygen (element number 8)?
   a. 15 b. 4 c. 2 d. 34

3. Which of the following elements is found in period 3 of the periodic table?
   a. Al
   b. Ga
   c. B
   d. more than one response is correct

4. The electronic configuration for an element containing 15 protons would be
   a. $1s^2 2s^2 2p^6 3s^2 3p^6$ b. $1s^2 2s^2 2p^6 3s^2 4p^3$ c. $1s^2 2s^2 2p^6 3s^2 3p^6 4s^2$ d. $1s^2 2s^2 2p^6 3s^2 3p^3$

5. Which of the following is a true statement for an electronic configuration of $1s^2 2s^2 2p^6 3s^2 3p^6$?
   a. there are 6 electrons in the 3p orbital
   b. there are 6 electrons in the 3p shell
   c. there are 6 electrons in the 3p subshell
   d. more than one response is correct

6. The maximum number of electrons that may occupy a 4*d* orbital is
   a. 10 b. 4 c. 2 d. 8

7. How many unpaired electrons are found in titanium (element number 22)?
   a. 1 b. 2 c. 3 d. 4

8. Which element has the electronic configuration $1s^22s^22p^63s^23p^6$?
   a. Ne b. Ar c. K d. Kr

9. The element with unpaired electrons in a *d* subshell is element number
   a. 33 b. 27 c. 20 d. 53

10. Which of the following elements has an electronic configuration ending in $4d^7$?
   a. Co
   b. Ir
   c. Rh
   d. more than one response is correct

11. How many electrons are found in the valence shell of oxygen (element number 8)?
   a. 3 b. 4 c. 5 d. 6

12. The distinguishing electronic configuration of np^4 is characteristic of which group in the periodic table?
   a. II A (2) b. IV A (14) c. VI A (16) d. noble gases

13. The element with the electronic configuration $1s^22s^22p^3$ will be found in
   a. period 1, group VI A (16)
   b. period V A (15), group 2
   c. period 2, group V A (15)
   d. period III A (13), group V A (15)

14. How many electrons would be contained in the valence shell of an atom in group VI A (16)?
   a. 6 b. 2 c. 4 d. 16

15. Which of the following atoms is the most metallic?
   a. Sr b. Ge c. Br d. Rb

16. Which of the following atoms has the largest radius?
   a. Si b. C c. Sn d. Ge

17. Which of the following atoms would have the lowest ionization potential?
   a. Ca b. Na c. K d. Rb

True-False

18. Elements 20 and 21 are in the same period of the periodic table.
19. Hund's rule states that electrons within a subshell remain unpaired if possible.
20. A $2p$ and a $3p$ subshell contain the same number of orbitals.
21. The maximum number of electrons an orbital may contain does not vary with the type of orbital.
22. The distinguishing electron in Br is found in a *p* orbital.

Matching

From the list on the right, choose a response that is consistent with the description on the left as far as electronic configurations are concerned. You may use a response more than once.

23. a noble gas
24. Mg
25. a transition element
26. an element just completing the filling of the n=2 shell

a. $1s^22s^22p^63s^2$
b. $1s^22s^22p^6$
c. $1s^22s^22p^63s^23p^64s^23d^3$
d. $1s^22s^22p^4$

Match each of the categories on the right with an element on the left.

27. neon (Ne)
28. phosphorus (P)
29. calcium (Ca)
30. element number 47
31. element number 82
32. top element of group V A (15)
33. the element containing 66 protons
34. the element with $5s^2$ as its distinguishing electron
35. the element in period 3 with 8 valence electrons

a. a representative metal
b. a noble gas nonmetal
c. a nonmetal but not a noble gas
d. a transition metal
e. an inner-transition metal

Chapter 4: Forces Between Particles

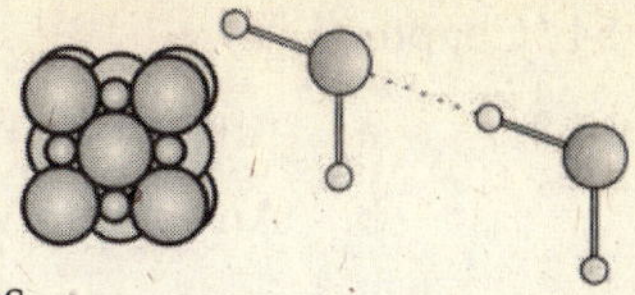

CHAPTER OUTLINE

4.1 Noble Gas Configurations
4.2 Ionic Bonding
4.3 Ionic Compounds
4.4 Naming Binary Ionic Compounds
4.5 The Smallest Unit of Ionic Compounds
4.6 Covalent Bonding
4.7 Polyatomic Ions
4.8 Shapes of Molecules and Polyatomic Ions
4.9 The Polarity of Covalent Molecules
4.10 More about Naming Compounds
4.11 Other Interparticle Forces

LEARNING OBJECTIVES/ASSESSMENT

When you have completed your study of this chapter, you should be able to:

1. Draw correct Lewis structures for atoms of representative elements. (Section 4.1; Exercise 4.2)
2. Use electronic configurations to determine the number of electrons gained or lost by atoms as they achieve noble gas configurations. (Section 4.2; Exercise 4.12)
3. Use the octet rule to correctly predict the ions formed during the formation of ionic compounds, and write correct formulas for binary ionic compounds containing a representative metal and a representative nonmetal. (Section 4.3; Exercises 4.20 and 4.22)
4. Correctly name binary ionic compounds. (Section 4.4; Exercise 4.30)
5. Determine formula weights for ionic compounds. (Section 4.5; Exercise 4.38)
6. Draw correct Lewis structures for covalent molecules. (Section 4.6; Exercise 4.48)
7. Draw correct Lewis structures for polyatomic ions. (Section 4.7; Exercise 4.50)
8. Use VSEPR theory to predict the shapes of molecules and polyatomic ions (Section 4.8; Exercises 4.52 and 4.54)
9. Use electronegativities to classify covalent bonds of molecules, and determine whether covalent molecules are polar or nonpolar. (Section 4.9; Exercises 4.58 and 4.64)
10. Write correct formulas for ionic compounds containing representative metals and polyatomic ions, and correctly name binary covalent compounds and compounds containing polyatomic ions. (Section 4.10; Exercises 4.66, 4.70, and 4.72)
11. Relate melting and boiling points of pure substances to the strength and type of interparticle forces present in the substances. (Section 4.11; Exercises 4.78 and 4.80)

SOLUTIONS FOR THE END OF CHAPTER EXERCISES

NOBLE GAS CONFIGURATIONS (SECTION 4.1)

☑4.2
- a. iodine ·Ï: (7 dots)
- b. strontium ·Sr·
- c. tin ·Ṡn·
- d. sulfur ·S̤̈: (6 dots)

4.4
- a. tin [Kr]$5s^2 4d^{10} 5p^2$
- b. Cs [Xe]$6s^1$
- c. element number 49 [Kr]$5s^2 4d^{10} 5p^1$
- d. strontium [Kr]$5s^2$

4.6
- a. Tin ·Ṡn·
- b. Cs Ċs
- c. element number 49 ·İn·
- d. strontium ·Sr·

4.8 a. Any group IIA (2) element $\dot{E}\cdot$

b. Any group VA (15) element $\cdot\dot{\underset{\cdot}{E}}\cdot$

IONIC BONDING (SECTION 4.2)

4.10

		Added electrons	Removed electrons
a.	tin	4	14 (4 not including *d* electrons)
b.	Cs	31 (7 not including *d* or *f* electrons)	1
c.	element number 49	5	13 (3 not including *d* electrons)
d.	strontium	16 (6 not including *d* electrons)	2

☑4.12

		Number of electrons lost/gained	Equation
a.	Cs	1 electron lost	$Cs \rightarrow Cs^{+} + e^{-}$
b.	oxygen	2 electrons gained	$O + 2e^{-} \rightarrow O^{2-}$
c.	element number 7	3 electrons gained	$N + 3e^{-} \rightarrow N^{3-}$
d.	iodine	1 electron gained	$I + e^{-} \rightarrow I^{-}$

4.14

		Equation	Ion Symbol
a.	A phosphorus atom that has gained three electrons	$P + 3e^{-} \rightarrow P^{3-}$	P^{3-}
b.	A strontium atom that has lost two electrons	$Sr \rightarrow Sr^{2+} + 2e^{-}$	Sr^{2+}
c.	An iodine atom that has gained one electron	$I + e^{-} \rightarrow I^{-}$	I^{-}

4.16 a. E^{2-} sulfur c. E^{+} sodium

b. E^{3+} aluminum d. E^{-} chlorine

4.18 a. Ca^{2+} argon c. As^{3-} krypton

b. S^{2-} argon d. Cs^{+} xenon

IONIC COMPOUNDS (SECTION 4.3)

☑4.20

		Cation formation	Anion formation	Formula
a.	Ca and Cl	$Ca \rightarrow Ca^{2+} + 2e^{-}$	$Cl + e^{-} \rightarrow Cl^{-}$	$CaCl_2$
b.	lithium and bromine	$Li \rightarrow Li^{+} + e^{-}$	$Br + e^{-} \rightarrow Br^{-}$	LiBr
c.	elements number 12 and 16	$Mg \rightarrow Mg^{2+} + 2e^{-}$	$S + 2e^{-} \rightarrow S^{2-}$	MgS

☑4.22 a. Se^{2-} BaSe c. I^{-} BaI_2

b. P^{3-} Ba_3P_2 d. As^{3-} Ba_3As_2

4.24 a. CaS binary d. $NaHCO_3$ not binary

b. Na_3P binary e. P_2O_5 binary

c. $MgSO_4$ not binary

NAMING BINARY IONIC COMPOUNDS (SECTION 4.4)

4.26 a. Li^{+} lithium ion; b. Mg^{2+} magnesium ion; c. Ba^{2+} barium ion; d. Cs^{+} cesium ion

4.28 a. Br^{-} bromide ion; b. O^{2-} oxide ion; c. P^{3-} phosphide ion; d. Te^{2-} telluride ion

☑4.30 a. SrS strontium sulfide; b. CaF_2 calcium fluoride; c. $BaCl_2$ barium chloride; d. Li_2O lithium oxide; e. MgO magnesium oxide

4.32
a. PbO and PbO_2 lead (II) oxide and lead (IV) oxide
b. CuCl and $CuCl_2$ copper (I) chloride and copper (II) chloride
c. Au_2S and Au_2S_3 gold (I) sulfide and gold (III) sulfide
d. CoO and Co_2O_3 cobalt (II) oxide and cobalt (III) oxide

4.34
a. PbO and PbO_2 plumbous oxide and plumbic oxide
b. CuCl and $CuCl_2$ cuprous chloride and cupric chloride
c. Au_2S and Au_2S_3 aurous sulfide and auric sulfide
d. CoO and Co_2O_3 cobaltous oxide and cobaltic oxide

4.36 a. lead(IV) oxide PbO_2; b. cobalt(II) chloride $CoCl_2$; c. copper(II) sulfide CuS; d. nickel(III) nitride NiN; e. platinum(II) phosphide Pt_3P_2

THE SMALLEST UNIT OF IONIC COMPOUNDS (SECTION 4.5)

(Note: Based on Section 2.6, calculations involving atomic weights, molecular weights, and Avogadro's number will use three significant figures.)

☑4.38
a. NaBr $(1\times 23.0\text{ u})+(1\times 79.9\text{ u})=102.9\text{ u}$
b. CaF_2 $(1\times 40.1\text{ u})+(2\times 19.0\text{ u})=78.1\text{ u}$
c. Cu_2S $(2\times 63.5\text{ u})+(1\times 32.1\text{ u})=159.1\text{ u}$
d. Li_3N $(3\times 6.94\text{ u})+(1\times 14.0\text{ u})=34.8\text{ u}$

4.40 a. NaBr Na^{+}, Br^{-}; b. CaF_2 Ca^{2+}, F^{-}; c. Cu_2S Cu^{+}, S^{2-}; d. Li_3N Li^{+}, N^{3-}

4.42
a. NaBr $1\times 23.0\text{ g}=23.0\text{ g }Na^{+}$ $1\times 79.9\text{ g}=79.9\text{ g }Br^{-}$
b. CaF_2 $1\times 40.1\text{ g}=40.1\text{ g }Ca^{2+}$ $2\times 19.0\text{ g}=38.0\text{ g }F^{-}$
c. Cu_2S $2\times 63.5\text{ g}=127\text{ g }Cu^{2+}$ $1\times 32.1\text{ g}=32.1\text{ g }S^{2-}$
d. Li_3N $3\times 6.94\text{ g}=20.8\text{ g }Li^{+}$ $1\times 14.0\text{ g}=14.0\text{ g }N^{3-}$

4.44 a. NaBr

$$1.00 \text{ mole NaBr} \times \frac{6.02\times10^{23} \text{ fomula units}}{1 \text{ mole NaBr}} \times \frac{1 \text{ Na}^+ \text{ ion}}{1 \text{ formula unit NaBr}} = 6.02\times10^{23} \text{ Na}^+ \text{ ions}$$

$$1.00 \text{ mole NaBr} \times \frac{6.02\times10^{23} \text{ fomula units}}{1 \text{ mole NaBr}} \times \frac{1 \text{ Br}^- \text{ ion}}{1 \text{ formula unit NaBr}} = 6.02\times10^{23} \text{ Br}^- \text{ ions}$$

b. CaF_2

$$1.00 \text{ mole CaF}_2 \times \frac{6.02\times10^{23} \text{ fomula units}}{1 \text{ mole CaF}_2} \times \frac{1 \text{ Ca}^{2+} \text{ ions}}{1 \text{ formula unit CaF}_2} = 6.02\times10^{23} \text{ Ca}^{2+} \text{ ion}$$

$$1.00 \text{ mole CaF}_2 \times \frac{6.02\times10^{23} \text{ fomula units}}{1 \text{ mole CaF}_2} \times \frac{2 \text{ F}^- \text{ ions}}{1 \text{ formula unit CaF}_2} = 1.20\times10^{24} \text{ F}^- \text{ ions}$$

c. Cu_2S

$$1.00 \text{ mole Cu}_2\text{S} \times \frac{6.02\times10^{23} \text{ fomula units}}{1 \text{ mole Cu}_2\text{S}} \times \frac{2 \text{ Cu}^+ \text{ ions}}{1 \text{ formula unit Cu}_2\text{S}} = 1.20\times10^{24} \text{ Cu}^+ \text{ ions}$$

$$1.00 \text{ mole Cu}_2\text{S} \times \frac{6.02\times10^{23} \text{ fomula units}}{1 \text{ mole Cu}_2\text{S}} \times \frac{1 \text{ S}^{2-} \text{ ion}}{1 \text{ formula unit Cu}_2\text{S}} = 6.02\times10^{23} \text{ S}^{2-} \text{ ions}$$

d. Li_3N

$$1.00 \text{ mole Li}_3\text{N} \times \frac{6.02\times10^{23} \text{ fomula units}}{1 \text{ mole Li}_3\text{N}} \times \frac{3 \text{ Li}^+ \text{ ions}}{1 \text{ formula unit Li}_3\text{N}} = 1.81\times10^{24} \text{ Li}^+ \text{ ions}$$

$$1.00 \text{ mole Li}_3\text{N} \times \frac{6.02\times10^{23} \text{ fomula units}}{1 \text{ mole Li}_3\text{N}} \times \frac{1 \text{ N}^{3-} \text{ ion}}{1 \text{ formula unit Li}_3\text{N}} = 6.02\times10^{23} \text{ N}^{3-} \text{ ions}$$

COVALENT BONDING (SECTION 4.6)

4.46

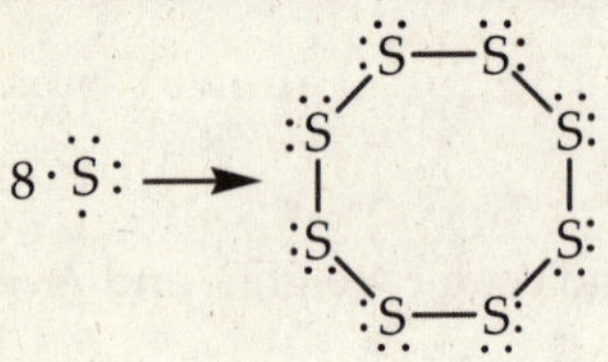

☑4.48 a. CH_4 (each H atom is bonded to the C atom)

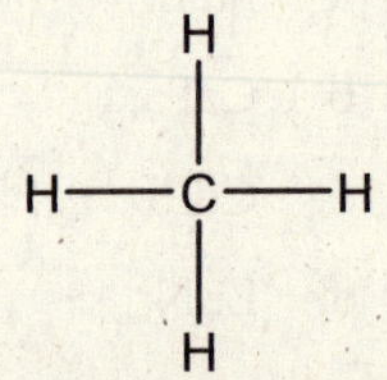

b. CO_2 (each O atom is bonded to the C atom)

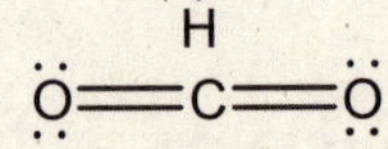

c. H_2Se (each H atom is bonded to the Se atom)

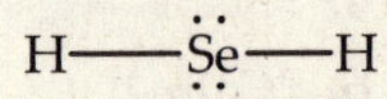

d. NH_3 (each H atom is bonded to the N atom)

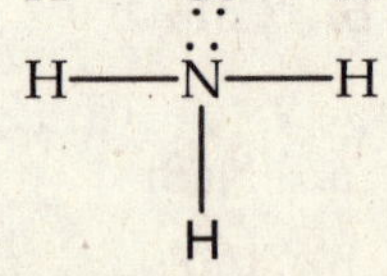

POLYATOMIC IONS (SECTION 4.7)

☑4.50 a. NH_4^+ (each H atom is bonded to the N atom)

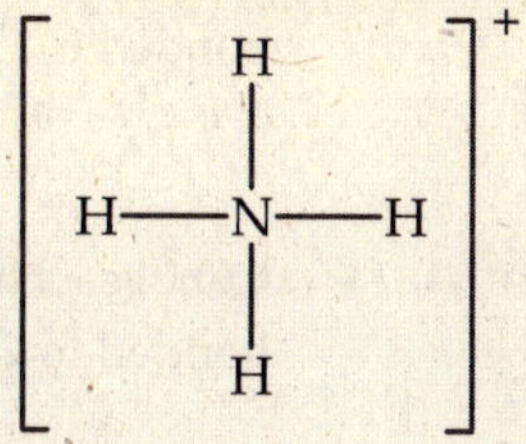

b. PO_4^{3-} (each O atom is bonded to the P atom)

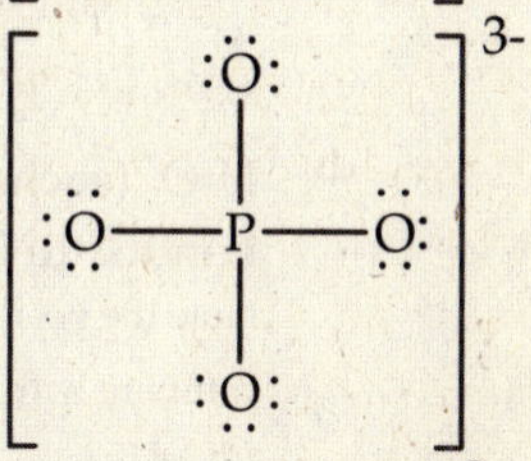

c. SO_3^{2-} (each O atom is bonded to the S atom)

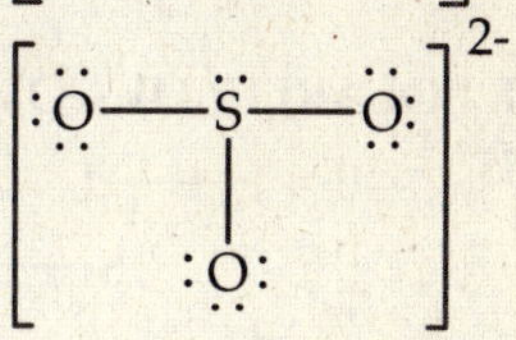

SHAPES OF MOLECULES AND POLYATOMIC IONS (SECTION 4.8)

☑4.52

	Lewis structure	3D Structure	VSEPR
a. H_2S (each H atom is bonded to the S atom)	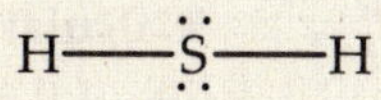	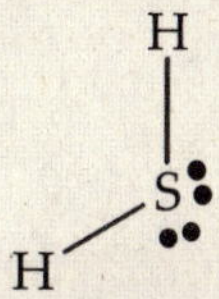	bent or angular
b. PCl_3 (each Cl atom is bonded to the P atom)	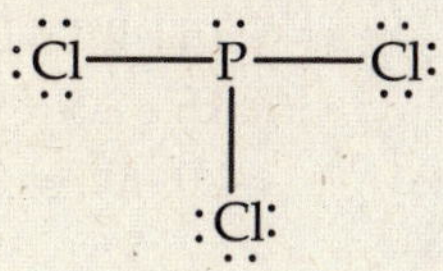	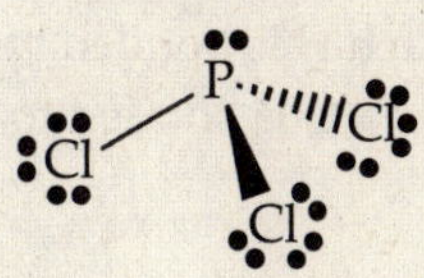	triangular pyramid with P at the top
c. OF_2 (each F atom is bonded to the O atom)	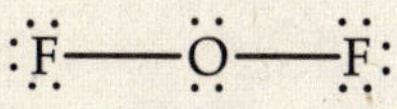	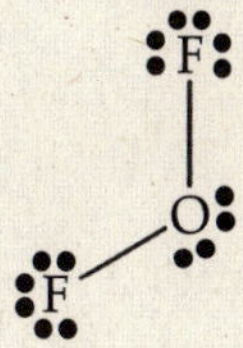	bent or angular
d. SnF_4 (each F atom is bonded to the Sn atom)	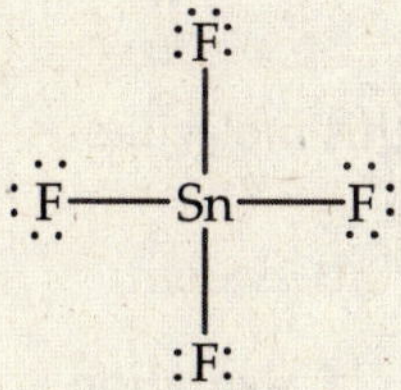	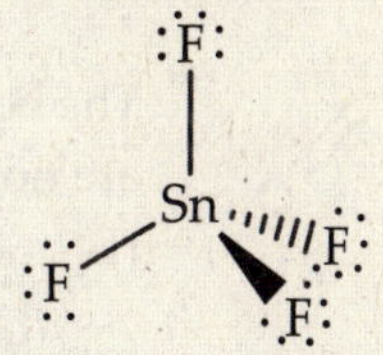	tetrahedral with Sn in the center

☑4.54

	Lewis structure	3D Structure	VSEPR
a. NO_2^- (each O is bonded to N)	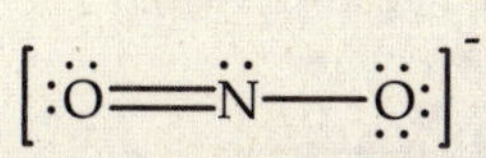	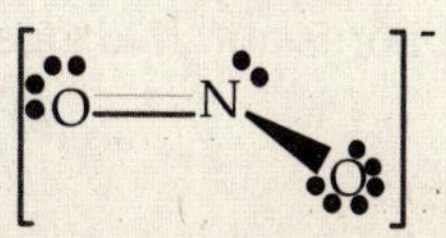	bent or angular

b. ClO_3^- (each O is bonded to Cl)

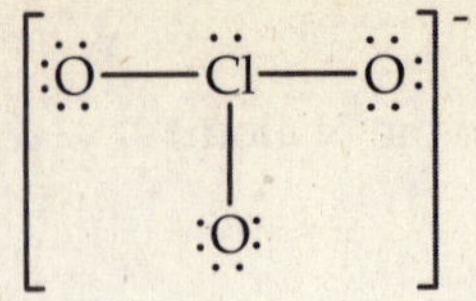

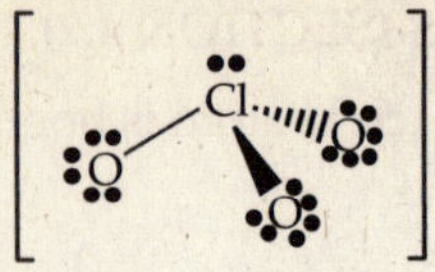

triangular pyramid with Cl at the top

c. CO_3^{2-} (each O is bonded to C)

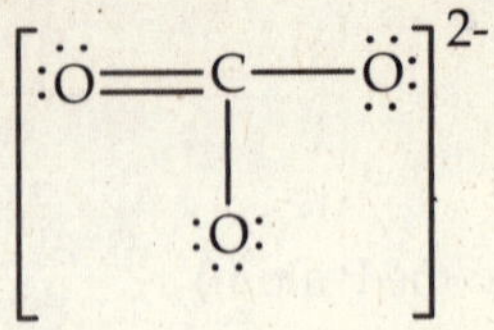

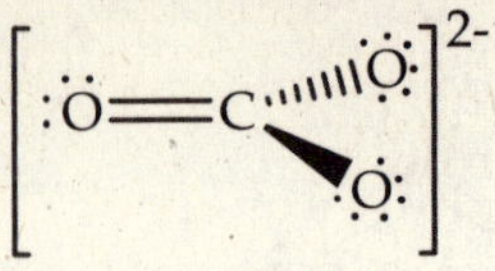

flat triangle with C in the middle

d. H_3O^+ (each H is bonded to O)
Note the positive charge; compare with NH_4^+.

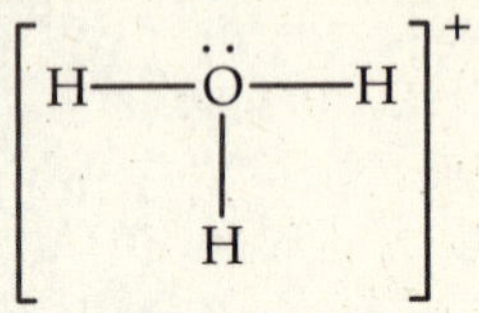

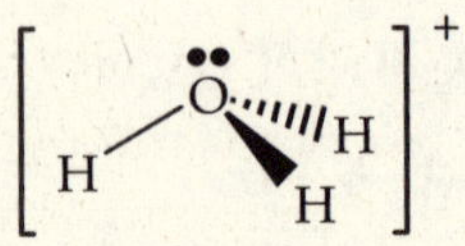

triangular pyramid with O at the top

THE POLARITY OF COVALENT MOLECULES (SECTION 4.9)

4.56 a. H—I $\qquad \Delta EN = 2.5 - 2.1 = 0.4$

b.

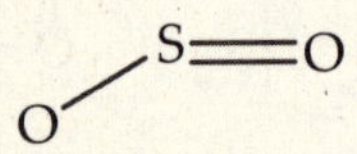

$\Delta EN = 3.5 - 2.5 = 1.0$

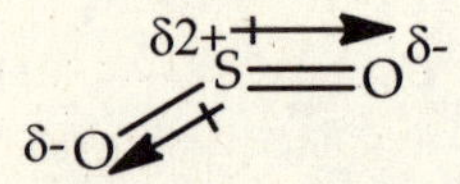

c.

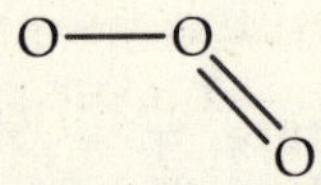

$\Delta EN = 3.5 - 3.5 = 0$ $\qquad$ The bonds in this molecule are nonpolar.

☑4.58

		Calculation	Classification
a.	LiBr	$2.8 - 1.0 = 1.8$	polar covalent
b.	HCl	$3.0 - 2.1 = 0.9$	polar covalent
c.	PH_3 (each H is bonded to P)	$2.1 - 2.1 = 0.0$	nonpolar covalent
d.	SO_2 (each O is bonded to S)	$3.5 - 2.5 = 1.0$	polar covalent
e.	CsF	$4.0 - 0.7 = 3.3$	ionic

4.60 a. H—I

This is a polar molecule because the charge distribution from bond polarization is nonsymmetric.

b.

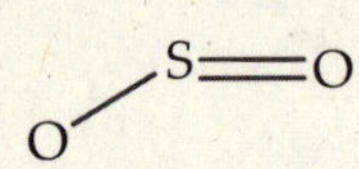

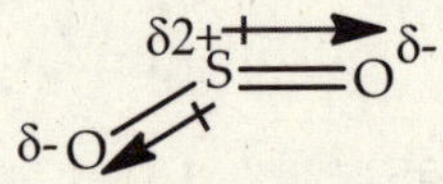

This is a polar molecule because the charge distribution from bond polarization is nonsymmetric.

c.

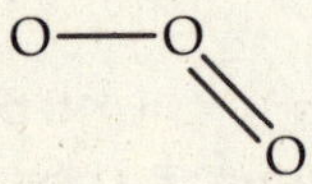

The bonds in this molecule are nonpolar.

This is a nonpolar molecule because the molecule does not have a charge distribution (the bonds are nonpolar).

4.62

		Calculation	Classification
a.	magnesium and chlorine	$3.0 - 1.2 = 1.8$	polar covalent
b.	carbon and hydrogen	$2.5 - 2.1 = 0.4$	polar covalent
c.	phosphorus and hydrogen	$2.1 - 2.1 = 0.0$	nonpolar covalent

☑4.64 a. C≡O

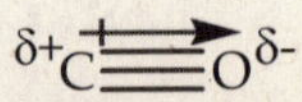

polar molecule (nonsymmetric charge distribution)

b. 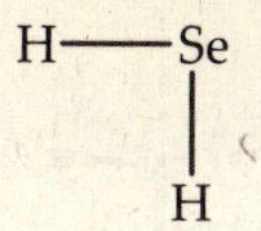 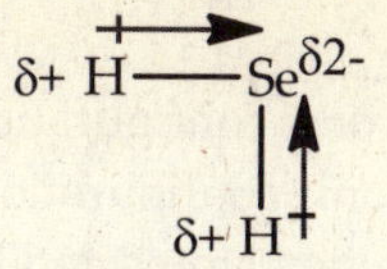

polar molecule
(nonsymmetric charge distribution)

c. 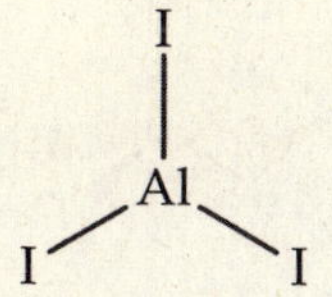 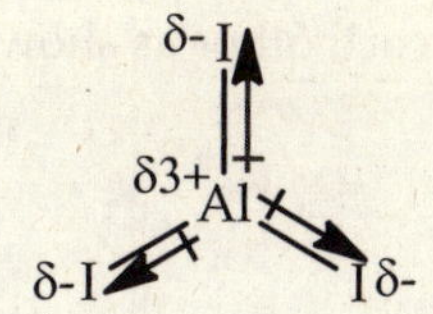

nonpolar molecule
(symmetric charge distribution)

MORE ABOUT NAMING COMPOUNDS (SECTION 4.10)

☑4.66 a. PCl_3 phosphorus trichloride
b. N_2O_5 dinitrogen pentoxide
c. CCl_4 carbon tetrachloride
d. BF_3 boron trifluoride
e. CS_2 carbon disulfide

4.68 a. chlorine dioxide ClO_2
b. dinitrogen monoxide N_2O
c. sulfur dioxide SO_2
d. carbon tetrachloride CCl_4

☑4.70			
a.	calcium and the nitrite ion	$Ca(NO_2)_2$	calcium nitrite
b.	magnesium and the hypochlorite ion	$Mg(ClO)_2$	magnesium hypochlorite
c.	Cs and $Cr_2O_7^{2-}$	$Cs_2Cr_2O_7$	cesium dichromate
d.	K and SO_3^{2-}	K_2SO_3	potassium sulfite

☑4.72 a. barium hydroxide $Ba(OH)_2$
b. magnesium sulfite $MgSO_3$
c. calcium carbonate $CaCO_3$
d. ammonium sulfate $(NH_4)_2SO_4$
e. lithium hydrogen carbonate $LiHCO_3$

4.74		
a.	Any group I A (1) element and SO_3^{2-}	M_2SO_3
b.	Any group I A (1) element and $C_2H_3O_2^-$	$MC_2H_3O_2$
c.	Any metal that forms M^{2+} ions and $Cr_2O_7^{2-}$	MCr_2O_7
d.	Any metal that forms M^{3+} ions and PO_4^{3-}	MPO_4
e.	Any metal that forms M^{3+} ions and NO_3^-	$M(NO_3)_3$

OTHER INTERPARTICLE FORCES (SECTION 4.11)

4.76 The alcohol has higher melting and boiling points than the ether. The forces that hold the alcohol molecules together must be stronger and harder to break than the forces that hold the ether molecules together.

☑4.78 The noble gases are nonpolar; therefore, they only experience dispersion forces. These forces increase with the size of the particles; therefore, the order of increasing boiling point for the noble gases is: helium, neon, argon, krypton, xenon, and radon.

☑4.80 Dispersion forces and dipolar forces are unlikely to be the predominant ones in the lattice of solid sucrose because these forces are relatively weak. Ionic bonds, metallic bonds, and covalent bonds are unlikely to be the predominant ones in the lattice of solid sucrose because they are very strong.

ADDITIONAL EXERCISES

4.82 Hydrogen is the element with an electronic configuration of $1s^1$. Nitrogen is the element with the electronic configuration of $1s^2\ 2s^2\ 2p^3$. The molecule that contains 3 hydrogen atoms and 1 nitrogen atom is NH_3. They hydrogen bond to each other as shown here:

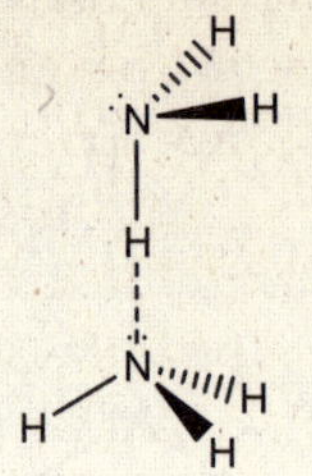

4.84 When magnesium metal reacts with oxygen, magnesium oxide (MgO) is formed. The mass of 0.200 moles of MgO is calculated as follows:

$$0.200 \text{ moles MgO}\left(\frac{40.31 \text{ g MgO}}{1 \text{ mole MgO}}\right) = 8.06 \text{ g MgO}$$

4.86 If one atom of oxygen reacted with two atoms of nitrogen to form a molecule, the formula of the molecule would be N_2O. The elecronegativity difference between nitrogen and oxygen is 0.5; therefore, the bond between nitrogen and oxygen is covalent. The name of N_2O is dinitrogen monoxide.

ALLIED HEALTH EXAM CONNECTION

4.88 Inert gases have (a) 8 electrons in their outermost energy levels (except for helium, which has 2 electrons in it outermost energy level).

4.90 Li_2SO_3 is (a) lithium sulfite.

4.92 When calcium reacts with chlorine to form calcium chloride, it (c) loses two electrons.

4.94 A covalent bond is believed to be caused by a (b) sharing of electrons.

4.96 The formula for bismuth (III) hydroxide is (b) $Bi(OH)_3$.

4.98 When bromine and magnesium are reacted to form $MgBr_2$, an (c) ionic bond is formed. (Note: Answer is based on the generalization that a metal and nonmetal react to form an ionic bond.)

4.100 In forming an ionic bond with an atom of chlorine, a sodium atom will (c) give up 1 electron to the chlorine atom.

4.102 (a) CO contains a bond with the least ionic character because it contains a bond between two nonmetals, while all the other compounds contain a bond between a metal and a nonmetal.

4.104 (a) F_2 is nonpolar and contains a nonpolar covalent bond.

4.106 The type of bond that forms between two molecules of water is a (b) hydrogen bond.

CHEMISTRY FOR THOUGHT

4.108 Yes, a reaction similar to the one that occurs between sodium metal and chlorine gas would be expected to occur between potassium metal and fluorine gas because potassium is in the same group as sodium and fluorine is in the same group as chlorine. In fact, any combination of a metal from group IA (1) (lithium, sodium, potassium, rubidium, cesium, francium) and nonmetal from group VIIA (17) (fluorine, chlorine, bromine, iodine, astatine) should produce a similar reaction.

4.110 The central atom in the balloon models will always be at the point the balloons are tied together. A group of 5 balloons would have a triangular bipyramidal shape with 3 balloons forming a triangular shape in the middle and 2 balloons forming an axis perpendicular to the triangular shape. A group of 6 balloons would have an octahedral shape. If the center of the 6 balloons were placed in the middle of a cube, then the top of each balloon would be in the center of a cube face because the bond angles are all 90°.

5-balloon structure

6-balloon structure

4.112 When a metal changes to form a positively charged metal ion, it loses electrons. The remaining electrons are then pulled closer to the positively charged nucleus, which makes the size of the metal ion smaller than the size of the original metal atom.

4.114 Neon atoms do not combine to form Ne_2 molecules because each neon atom has a completed valence shell and does not need to form bonds with another atom in order to satisfy its valence.

4.116 When CH_4 is burned in air at a high temperature, CO_2 and H_2O are the primary products; however, air is mostly nitrogen and as the combustion occurs at high temperatures, free radical oxygen atoms produced during the reaction with methane can react with high temperature nitrogen molecules to produce NO as a side product.

ADDITIONAL ACTIVITIES

Section 4.1 Review:

(1) Complete the electron-dot formulas for each of the elements in the abbreviated periodic table below.

(1) I A	(2) II A	(13) III A	(14) IV A	(15) V A	(16) VI A	(17) VII A	(18) VIII A
H							He
Li	Be	B	C	N	O	F	Ne
Na	Mg	Al	Si	P	S	Cl	Ar

(2) What patterns do you notice after completing (1)?

(3) What is the relationship between the group numbers in parenthesis and the dots on the elements in the *s* block?
(4) What is the relationship between the group numbers in parenthesis and the dots on the elements in the *p* block?
(5) How many dots would be placed on an element from the *d* block?
(6) How many dots would be placed on an element from the *f* block?
(7) Is there a relationship between the number of dots and the group numbers (Roman numeral or parenthetical) for the elements in the *d* or *f* blocks? Explain your answer.

Section 4.2 Review:

(1) Referring to the abbreviated periodic table in the Section 4.1 Review, how many electrons must an element from groups I A (1) through VII A (17) gain in order to have the same valence shell electron configuration as the noble gas from the same period?
(2) For hydrogen and the elements in period 2 from groups IA (1) through VII A (17), write the symbols for the ions formed by gaining enough electrons to have the same electron configuration as the noble gas from the same period.
(3) Which ions from (2) are the most likely to exist? Why?
(4) Referring to the abbreviated periodic table in the Section 4.1 Review, how many electrons must an element from groups I A (1) through VII A (17) lose in order to have the same valence shell electron configuration as the noble gas from the previous period?
(5) For hydrogen and the elements in period 3 from groups IA (1) through VII A (17), write the symbols for the ions formed by losing enough electrons to have the same electron configuration as the noble gas from the previous period.
(6) Which ions from (5) are the most likely to exist? Why?
(7) Rb^+ is isoelectronic with krypton. Is the Rb^+ ion the same as a krypton atom? Explain your answer.

Section 4.3 Review:

(1) For the elements in groups I A (1) through IIV A (17), write the charge of the mostly likely ion for each group above the group numbers on the abbreviated periodic table from the Section 4.1 Review.
(2) Write formulas for the compounds formed between hydrogen and each of the elements in groups I A (1) through IIV A (17) from period 2.
(3) The elements in the *d* and *f* blocks usually form ions with a 2+ charge; however, many exceptions exist. All of the metal atoms in the *p* block except aluminum can also have charges other than the ones recorded in (1). Identify the numbers and charges of the ions in the following compounds.

$IrCl_4$	Pr_2O_3	BiF_5	PbO_2	$SnBr_2$	InN	CuI	U_3N_2	VSe_2	YbTe	Au_2S	GaP	WC

Section 4.4 Review:

(1) Name the compounds from the Section 4.3 Review using a roman numeral in parenthesis to indicate the charge on the positive ion.
(2) Name the following compounds:

Li_2S	BeO	NaF	$MgCl_2$	K_3N	Ca_3P_2	RbI	SrBr	Cs_2Te	BaSe	Fr_4C	Ra_3As_2	AlF_3

(3) Do the compounds from (2) need roman numerals like the compounds from (1)? Explain why or why not.

Section 4.5 Review:

(1) Calculate the formula weight for the following compounds: $IrCl_4$, Pr_2O_3, Li_2S, BeO.
(2) Calculate the mass of 1 mole of the following compounds: BiF_5, PbO_2, NaF, $MgCl_2$.
(3) Calculate the number of formula units in 1 mole of the following compounds: $SnBr_2$, InN, K_3N, Ca_3P_2.
(4) Calculate the number and identify the types of ions in 1 mole of the following compounds: U_3N_2, RbI.

Section 4.6 Review:

(1) Based on the electron-dot symbols in the abbreviated periodic table from the Section 4.1 Review, what is the minimum number of electrons each nonmetal would need to share with other atoms in order to have a completed valence shell?

(2) What is wrong with each of these Lewis structures? Propose a better structure, if possible.

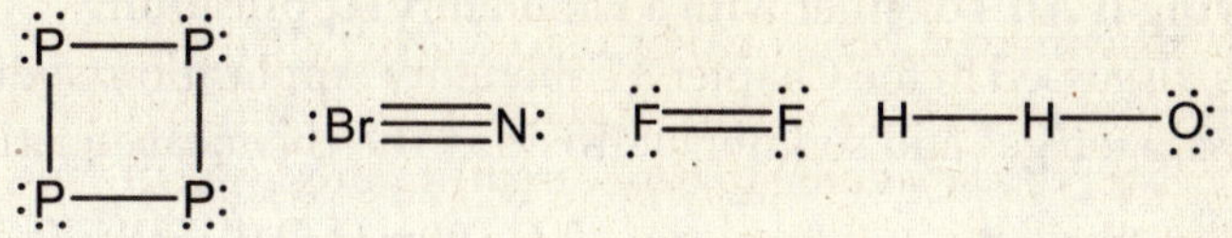

Section 4.7 Review:

What charge should each of the following polyatomic ions have?

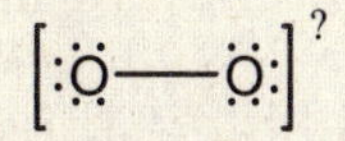

peroxide

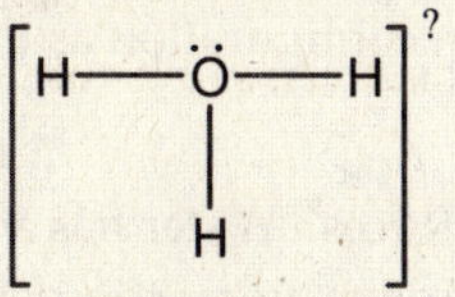

hydronium

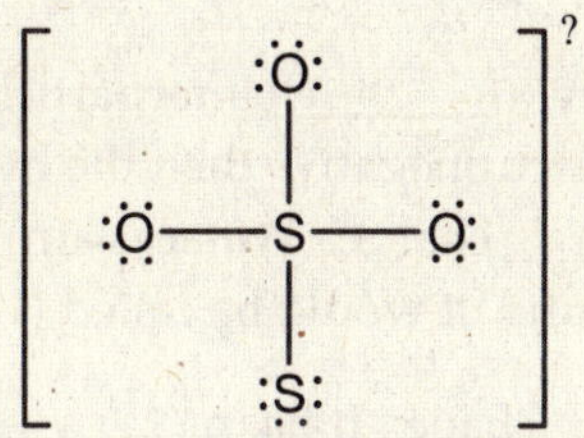

thiosulfate

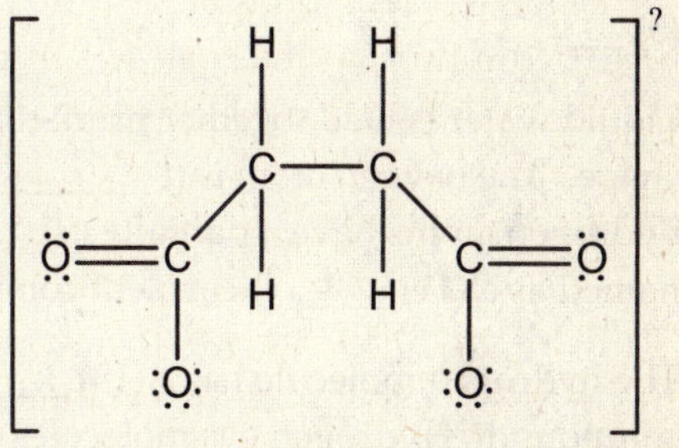

succinate

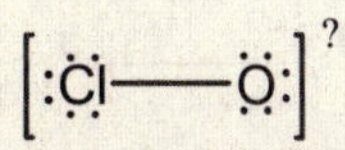

hypochlorite

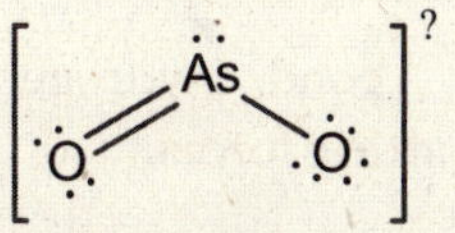

arsenite

Section 4.8 Review:

Predict the shapes of each of the "corrected" Lewis structures from the Section 4.6 Review as well as the polyatomic ions in the Section 4.7 Review. (For the succinate ion, predict the shape around each carbon atom.)

Section 4.9 Review:

Perform the following tasks for each of the "corrected" Lewis structures from the Section 4.6 Review as well as the polyatomic ions in the Section 4.7 Review.

(1) Draw the charge distribution on the Lewis Structure.
(2) Using Table 4.4 from the textbook, classify each bond as polar covalent, nonpolar covalent, or ionic.
(3) Classify the molecule or ion as polar or nonpolar.

Section 4.10 Review:

(1) Identify the following chemical formulas as covalent or ionic compounds or polyatomic ions.
(2) Name each of the chemical formulas.

Na_2O	N_2O	MnO_2	MnO_4^-	SCl_2	$MgCl_2$	P_2F_4	KF	NH_3	NH_4Br

Section 4.11 Review:

(1) What is the difference between the prefixes *inter-* and *intra-*?
(2) Identify the types of "particles" in each of the following types of materials.

network solids	ionic compounds	polar covalent compounds
metals	nonpolar covalent compounds	polar covalent compounds containing one of the following bonds: N—H, O—H, or F—H

(3) Match the materials from (2) to the major interparticle force that hold these materials together.

ionic bond	metallic bond	covalent bond
dipolar force	hydrogen bonding	dispersion force

(4) Match the following chemical substances with a material from (2) and an interparticle force from (3).

NH_3 (ammonia)	brass (an alloy of copper and zinc)	diamond (a form of pure C atoms)	Ne	KCl	SO_2

Tying It All Together with a Laboratory Application:
As discussed in the Chapter 3 Laboratory Application, sodium metal reacts with water to produce hydrogen gas and sodium hydroxide. The chemical equation for this reaction is:

$$2\ Na\ (s) + H_2O\ (l) \rightarrow H_2\ (g) + 2\ NaOH\ (aq)$$

Rewrite this equation using Lewis Structures (1).
Sodium metal is held together by (2) _______. During the reaction the sodium atoms (3) _______ electrons to form (4) _______ ions. The Na^+ ion has the same electron configuration as the noble gas (5) _______.

Liquid water is held together primarily by (6) _______ interparticle forces. Water has a (7) _______ shape. The oxygen atom is (8) _______ electronegative than the hydrogen atom; consequently, the hydrogen atoms have a partially (9) _______ charge. Water is a(n) (10) _______ compound. If water were named as a (11) _______ covalent compound, it would be called (12) _______.

The hydrogen molecule has a (13) _______ shape. It has a (14) _______ bond. Hydrogen (15) _______ a compound. Hydrogen gas molecules experience (16) _______ interparticle forces.

Sodium hydroxide is a(n) (17) _______ compound. It is soluble in water; however, if all the water evaporated, it would form a(n) (18) _______. The sodium ion and the hydroxide ion would occupy different (19) _______. These ions would be held together by a(n) (20) _______ bond. The smallest unit of sodium hydroxide is called a(n) (21) _______ and has a formula weight of (22) _______.

The hydroxide ion is a(n) (23) _______ (simple or polyatomic) ion. It contains a (24) _______ bond. It has a (25) _______ shape.

SOLUTIONS FOR THE ADDITIONAL ACTIVITIES

Section 4.1 Review:

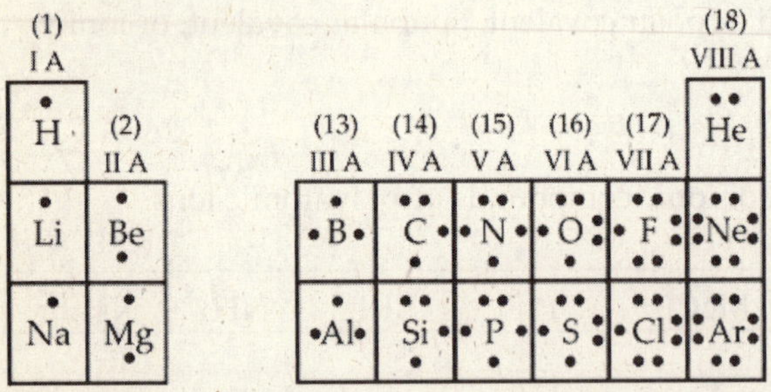

(2) The electron-dot symbols for elements in the same group have the same number of electrons. The number of electrons increases one per box from left to right across a period. (3) In the *s* block, the number in parenthesis is the same as the number of electrons in the electron-dot symbol. (4) In the *p* block, the number in parenthesis is 10 higher than the number of electrons in the electron-dot symbol. (5) 2; (6) 2; (7) No obvious connection exists between the Roman numerals or parenthetical numbers for the *d* block and the 2 valence-shell electrons these elements have. The *f* block elements do not have Roman numerals or parenthetical numbers, thus no connection can be established for these elements either.

Section 4.2 Review:

	Group	I A (1)	II A (2)	III A (13)	IV A (14)	V A (15)	VI A (16)	VII A (17)
(1)	e^- to gain	7	6	5	4	3	2	1
(2)	symbols	H^-; Li^{7-}	Be^{6-}	B^{5-}	C^{4-}	N^{3-}	O^{2-}	F^-

(3) The ions from (2) that are most likely to form are H^-, C^{4-}, N^{3-}, O^{2-}, and F^-. These ions formed when the neutral atoms gained no more than half the total number of valence-shell electrons for the noble gas from the same period (≤1 for He, ≤4 Ne).

	Group	I A (1)	II A (2)	III A (13)	IV A (14)	V A (15)	VI A (16)	VII A (17)
(4)	e^- to lose	1	2	3	4	5	6	7
(5)	symbols	H^+ *; Na^+	Mg^{2+}	Al^{3+}	Si^{4+}	P^{5+}	S^{6+}	Cl^{7+}

*Hydrogen is in period 1; therefore, a noble gas from the previous period would have to be in period 0 which does not exist. The electron in the valence-shell of hydrogen was removed to form this ion.

(6) The ions from (5) that are most likely to form are H^+, Na^+, Mg^{2+}, Al^{3+}, and Si^{4+}. These ions formed when the neutral atoms lost no more than half the total number of valence-shell electrons for the noble gas from the previous row (1 for H*, ≤4 Ne).

(7) No, Rb^+ (the rubidium ion) is not the same as a krypton atom. The rubidium ion has one more proton than the krypton atom. Rb^+ has the same electron configuration as a krypton atom as well as the same total number of electrons as a krypton atom. Consequently, the rubidium ion has a positive charge because it has one more proton than electron. The rubidium ion has chemical properties that more similar to a krypton atom than a rubidium atom because many chemical properties result from the electron configuration; however, the rubidium ion is not a krypton atom. The rubidium ion is smaller than a krypton atom because the additional proton pulls the electrons closer to the nucleus of the rubidium ion. The positive charge on the rubidium ion also causes it to be attracted to negatively charged ions. Krypton is electrically neutral and is not attracted to negative ions.

Section 4.3 Review:

(1)

1+* (1) I A	2+ (2) II A	3+ (13) III A	4+/4- (14) IV A	3- (15) V A	2- (16) VI A	1- (17) VII A	no ions (18) VIII A
H							He
Li	Be	B	C	N	O	F	Ne
Na	Mg	Al	Si	P	S	Cl	Ar

*Hydrogen can make both 1+ and 1- ions.

(2)

H^-			
LiH	BeH_2	BH_3	CH_4*

H^+			
H_4C*	H_3N	H_2O	HF

*$CH_4 = H_4C$

(3)	$IrCl_4$	Pr_2O_3	BiF_5	PbO_2	$SnBr_2$	InN	CuI	U_3N_2	VSe_2	YbTe	Au_2S	GaP	WC
-	4 Cl^-	3 O^{2-}	5 F^-	2 O^{2-}	2 Br^-	N^{3-}	I^-	2 N^{3-}	2 Se^{2-}	Te^{2-}	S^{2-}	P^{3-}	C^{4-}
+	Ir^{4+}	2 Pr^{3+}	Bi^{5+}	Pb^{4+}	Sn^{2+}	In^{3+}	Cu^+	3 U^{2+}	V^{4+}	Yb^{2+}	2 Au^+	Ga^{3+}	W^{4+}

Section 4.4 Review:

(1)

$IrCl_4$ iridium (IV) chloride	Pr_2O_3 praseodymium (III) oxide	BiF_5 bismuth (V) fluoride	PbO_2 lead (IV) oxide
$SnBr_2$ tin (II) bromide	InN indium (III) nitride	CuI copper (I) iodide	U_3N_2 uranium (II) nitride
VSe_2 vanadium (IV) selenide	YbTe ytterbium (II) telluride	Au_2S gold (I) sulfide	GaP gallium (III) phosphide
WC tungsten (IV) carbide			

(2)

Li_2S lithium sulfide	BeO beryllium oxide	NaF sodium fluoride	$MgCl_2$ magnesium chloride

K_3N potassium nitride	Ca_3P_2 calcium phosphide	RbI rubidium iodide	SrBr strontium bromide
Cs_2Te cesium telluride	BaSe barium selenide	Fr_4C francium carbide	Ra_3As_2 radium arsenide
AlF_3 aluminum fluoride			

(3) The compounds in (2) do not need roman numerals because the metals ions from group I A (1) always have a 1+ charge, the metals ions from group II A (2) always have a 2+ charge, and the aluminum ion always has a 3+ charge.

Section 4.5 Review:

(1) $IrCl_4$ = 334.02 u; Pr_2O_3 = 329.82 u; Li_2S = 45.95 u; BeO = 25.01 u; (2) BiF_5 = 303.98 g; PbO_2 = 239.2 g; NaF = 41.99 g; $MgCl_2$ = 95.21 g; (3) 6.022 x 10^{23} formula units $SnBr_2$; 6.022 x 10^{23} formula units InN; 6.022 x 10^{23} formula units K_3N; 6.022 x 10^{23} formula units Ca_3P_2; (4) 1.8066 x 10^{24} U^{2+} ions, 1.2044 x 10^{24} N^{3-} ions; 6.022 x 10^{23} Rb^{+} ions, 6.022 x 10^{23} I^{-} ions.

Section 4.6 Review:

(1) H = 1; C = 4; N = 3; P = 3; O = 2; S = 2; F = 1; Cl = 1; He = 0; Ne = 0; Ar = 0; (2) The given structure for P_4 is incorrect because the picture shows 24 electrons (16 electrons in lone pairs and 8 electrons in bonds), but the structure should only contain 20 electrons (4 P atoms, each with 5 valence-shell electrons).

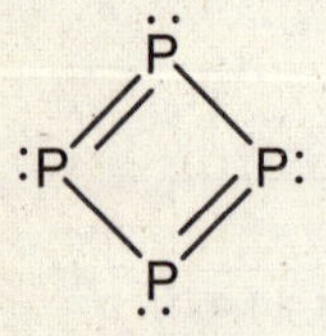
Structure A

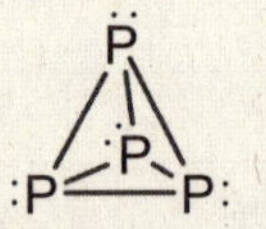
Structure B

Structure A is a viable alternative because this structure has only 20 electrons (8 electrons in lone pairs and 12 electrons in bonds). Structure B is not only a viable alternative to the given structure, but is also the actual structure. Like structure A, structure B contains 20 electrons (8 electrons in lone pairs and 12 electrons in bonds). Each phosphorus atom is bonded to 3 other phosphorus atoms in order to form a tetrahedron. This structure can be constructed with 4 paper clips (P atoms) and 6 straws (bonds).

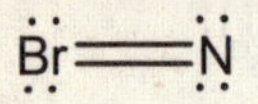
Structure C

The given structure for BrN is incorrect because the structure only contains 10 electrons (4 electrons in lone pairs and 6 electrons in bonds) when it should contain 12 electrons (Br has 7 valence-shell electrons and N has 5 valence-shell electrons). A possible alternative structure is structure C.

Structure D

The structure for F_2 is incorrect because the structure only contains 12 electrons (8 electrons in lone pairs and 4 electrons in bonds) when it should contain 14 electrons (each F atom has 7 valence-shell electrons). A possible alternative structure is structure D.

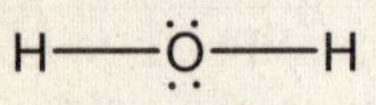
Structure E

The structure for H_2O is incorrect because the structure contains 10 electrons (6 electrons in lone pairs and 4 electrons in bonds) when it should contain 8 electrons (H has 1 valence-shell electrons and O has 6 valence-shell electrons). Also, the hydrogen atom in the middle of this structure has 2 bonds, but hydrogen is only capable of making 1 bond. A possible alternative structure is structure E.

Section 4.7 Review:

peroxide = 2-; hydronium = 1+; hypochlorite = 1-; arsenite = 1-; thiosulfate = 2-; succinate = 2-

Section 4.8 Review:

Structure A = bent or angular (based on each P atom); Structure B = pyramid with triangle base (based on each P atom); Structure C = linear; Structure D = linear; Structure E = bent or angular; peroxide = linear; hydronium = pyramid with triangle base; hypochlorite = linear; arsenite = bent or angular; thiosulfate =

tetrahedron; succinate = triangle (both C atoms double bonded to an O atom and single bonded to an O atom and another C atom), tetrahedron (both C atoms bonded to 2 H and 2 other C atoms)

Section 4.9 Review:

Structures A & B — (1) no charge distribution because only P atoms, (2) nonpolar covalent bonds, (3) nonpolar molecule

Structure C — (1) 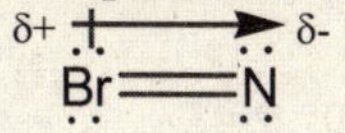 , (2) polar covalent bond, (3) polar molecule

Structure D — (1) no charge distribution because only F atoms, (2) nonpolar covalent bond, (3) nonpolar molecule

Structure E — (1) 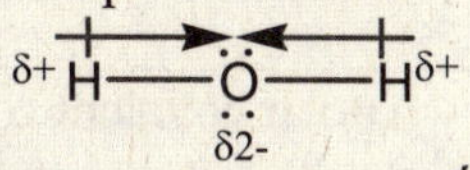 , (2) polar covalent bonds, (3) polar molecule

peroxide ion — (1) no charge distribution because only O atoms, (2) nonpolar covalent bond, (3) nonpolar ion

hydronium ion — (1) 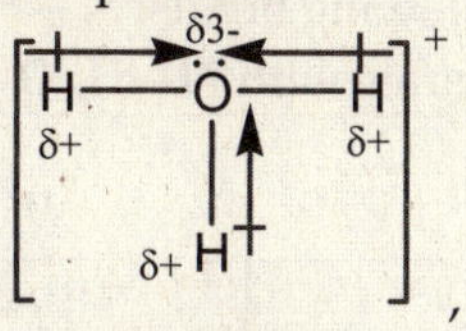 , (2) polar covalent bonds, (3) polar ion

hypochlorite ion — (1) 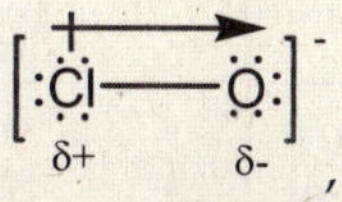 , (2) polar covalent bond, (3) polar ion

arsenite ion — (1) 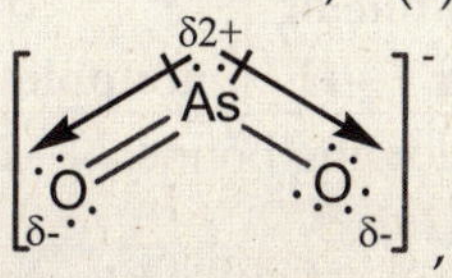 , (2) polar covalent bonds, (3) polar ion

thiosulfate ion — (1) 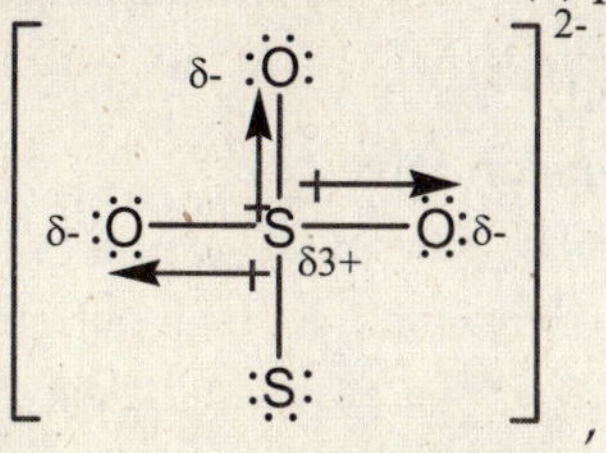 , (2) S-O bonds = polar covalent, S-S bond = nonpolar covalent, (3) polar ion

succinate ion — (1) 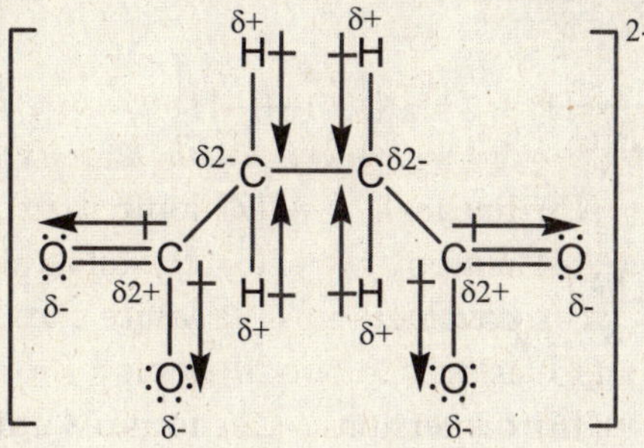 (2) C-C bonds = nonpolar covalent, all other bonds = polar covalent, (3) polar ion

Section 4.10 Review:

Na_2O	ionic compound	sodium oxide	$MgCl_2$	ionic compound	magnesium chloride
N_2O	covalent compound	dinitrogen monoxide	P_2F_4	covalent compound	diphosphorus tetrafluoride
MnO_2	ionic compound	manganese (IV) oxide	KF	ionic compound	potassium fluoride

MnO_4^-	polyatomic ion	permanganate ion	NH_3	covalent compound	nitrogen trihydride*
SCl_2	covalent compound	sulfur dichloride	NH_4Br	ionic compound	ammonium bromide

*Note: NH_3 has the common name ammonia.

Section 4.11 Review:

(1) *inter-* means between two different parts, *intra-* means within one part;

(2)

network solids **atoms**	ionic compounds **ions**	polar covalent compounds **polar molecules**
metals **metal atoms**	nonpolar covalent compounds **nonpolar covalent molecules**	polar covalent compounds containing one of the following bonds: N—H, O—H, or F—H **molecules with N—H, O—H, or F—H bonds**

(3)

ionic bond **ionic compounds**	metallic bond **metals**	covalent bond **network solids**
dipolar force **polar covalent compounds**	hydrogen bonding **polar covalent compounds containing one of the following bonds: N—H, O—H, or F—H**	dispersion force **nonpolar covalent compounds**

(4) NH_3 – polar covalent compounds containing one of the following bonds: N—H, O—H, or F—H and hydrogen bonding;
brass – metal and metallic bond;
diamond – covalent network solid and covalent bond;
Ne – nonpolar covalent molecules (atoms) and dispersion force;
KCl – ionic compound and ionic bond;
SO_2 – polar covalent compound and dipolar force

Tying It All Together with a Laboratory Application:

(1)

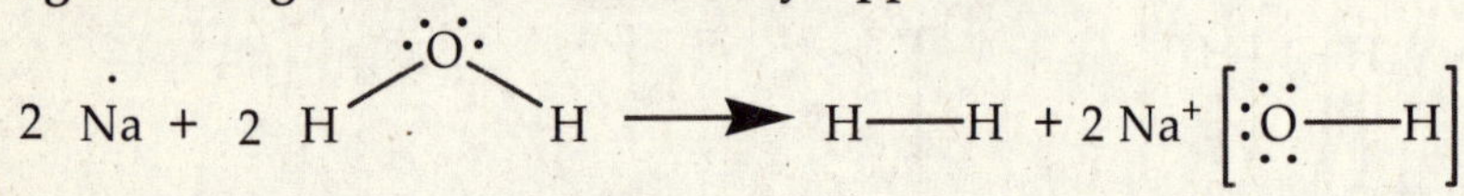

(2) metallic bonding
(3) lose
(4) sodium
(5) neon
(6) hydrogen bonding
(7) bent
(8) more
(9) positive
(10) polar covalent
(11) binary
(12) dihydrogen monoxide
(13) linear
(14) nonpolar covalent
(15) is not
(16) dispersion
(17) ionic
(18) crystal lattice
(19) lattice sites
(20) ionic
(21) formula unit
(22) 40.00 u
(23) polyatomic
(24) polar covalent
(25) linear

SELF-TEST QUESTIONS

Multiple Choice

1. Which of the following elements has the lowest electronegativity?
   a. As b. P c. Br d. Cl

2. In describing the strength of interparticle forces we discover that the weakest forces or bonds are:
   a. ionic. b. dipolar. c. covalent. d. dispersion.

3. The formula for the compound formed between the elements Ba and O would be:
   a. BaO   b. Ba_2O c. BaO_2 d. Ba_2O_3

4. The formula for the ionic compound containing Al^{3+} and SO_4^{2-} ions would be:
   a. $Al(SO_4)_2$ b. $AlSO_4$ c. $Al_3(SO_4)_2$ d. $Al_2(SO_4)_3$

5. The expected fomula of the molecule formed when nonmetals C and H combine in compliance with the octet rule is:
   a. CH_4 b. CH_2 c. C_4H d. CH_3

6. The name of the covalent compound PCl_3 is:
   a. trichlorophosphide.
   b. phosphorus trichlorine.
   c. phosphorus trichloride.
   d. phophorus chloride.

7. The compound $MgSO_4$ is correctly named:
   a. magnesium sulfur tetroxide.
   b. magnesium sulfoxide.
   c. magnesium sulfide.
   d. magnesium sulfate.

8. This bond is found in molecules such as HCl and H_2O.
   a. nonpolar covalent bond
   b. polar covalent bond
   c. ionic bond
   d. metallic bond

9. If the electronegativity difference between two elements A and B is 1.0, what type of bond is A-B?
   a. nonpolar covalent
   b. polar covalent
   c. ionic
   d. metallic

10. Which of the following is a correct electron dot formula for sulfur (element 16)?
   a. 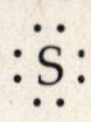   b. 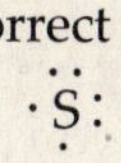   c. 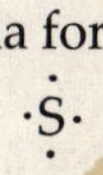   d. 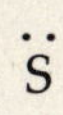

11. In the structures below, each bonding electron pair is denoted by a dash. A correct structure of SO_3^{2-} is
   a. 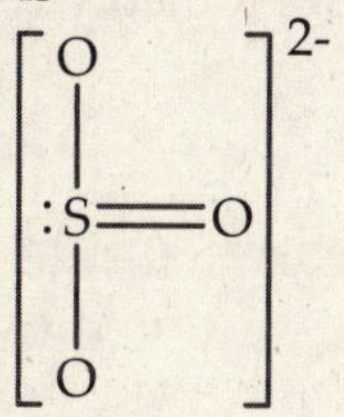   b. 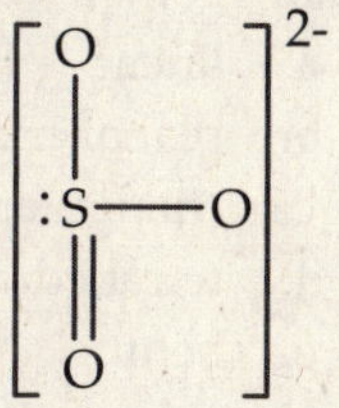   c. 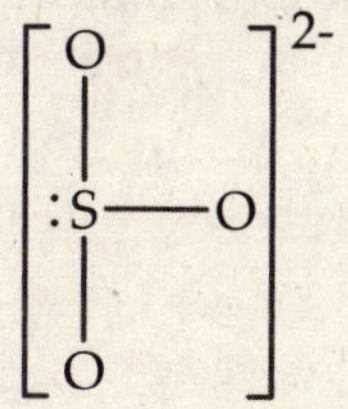   d. 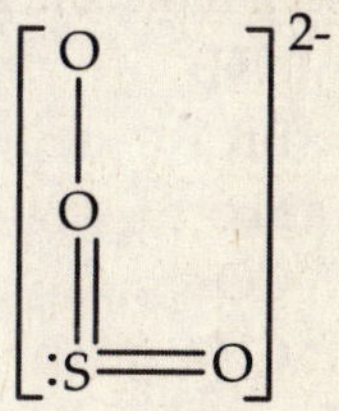

12. A covalent molecule forms between elements A and B. B is more electronegative. Which of the following molecules would be polar?
   a. 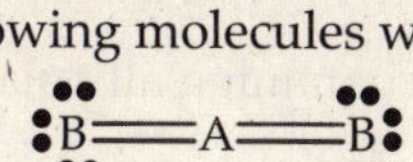   b. 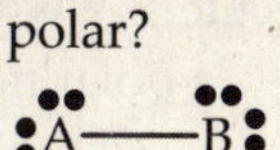   c. 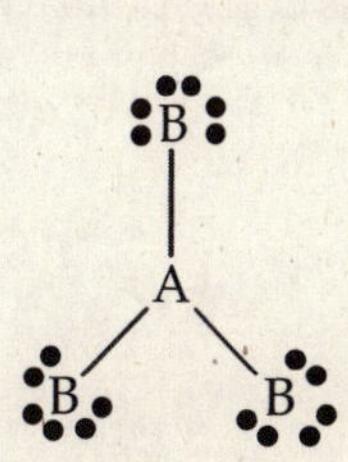
   (planar)
   d. 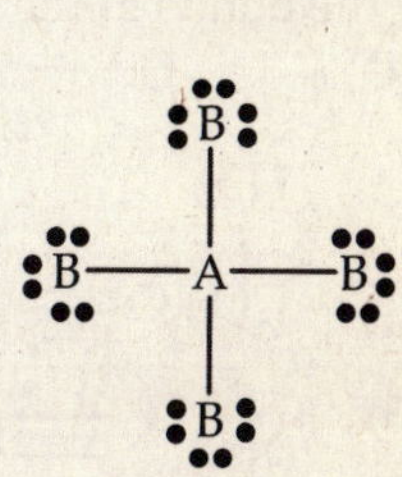

13. Which of the following contains polar bonds, but is a nonpolar molecule?

a. 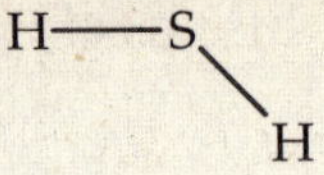 b. O═C═O c. H—Cl d. F—F

14. According to the VSEPR theory, the shape of OF_2 will be

a. linear b. bent c. triangle d. tetrahedral

15. The name for OF_2 is

a. monoxygen difluoride
b. oxygen difluorine
c. oxygen fluorine
d. oxygen difluoride

True-False

16. No more than one pair of electrons can be shared to form covalent bonds between atoms.
17. Dispersion forces between particles are correctly classified as very strong.
18. A compound between elements with atomic numbers 7 and 9 will contain covalent bonds.
19. All covalent bonds are polar.
20. The interparticle forces in a solid noble gas would have to be polar in nature.
21. Neon (Ne) has a higher boiling point than krypton (Kr).
22. Nonpolar molecules only contain nonpolar bonds.

Matching

An ionic compound is formed from each of the pairs given on the left. For each pair, choose the correct formula for the resulting compound from the responses on the right.

23. X^{+} and Y^{2-} a. X_3Y_2
24. X^{+} and Y^{-} b. XY
25. X^{3+} and Y^{3-} c. XY_2
26. X is a group II A (2) ion and Y is a group VI A (16) ion d. X_2Y

For each molecule given on the left, predict the molecular geometry based on VSEPR theory.

27. NH_3 a. linear
28. BrCl b. planar triangle
29. H_2S c. triangular pyramid
30. CO_2 d. tetrahedral
31. CH_4 e. bent

For each of the molecules on the left, choose the statement from the right that correctly gives the polarity of the bonds in the molecule and the polarity of the molecule as a whole.

32. H_2S, 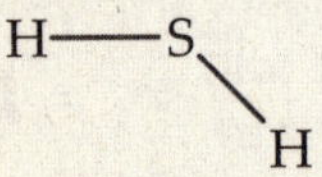 a. a polar molecule containing all polar bonds

33. CO_2, O═C═O b. a nonpolar molecule containing all nonpolar bonds

34. N_2O, N≡N—O c. a nonpolar molecule containing all polar bonds

35. O_3, 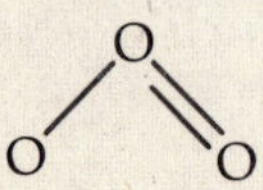 d. a polar molecule containing polar bonds and nonpolar bonds

Chapter 5: Chemical Reactions

CHAPTER OUTLINE

5.1 Chemical Equations
5.2 Types of Reactions
5.3 Redox Reactions
5.4 Decomposition Reactions
5.5 Combination Reactions
5.6 Replacement Reactions
5.7 Ionic Equations
5.8 Energy and Reactions
5.9 The Mole and Chemical Equations
5.10 The Limiting Reactant
5.11 Reaction Yields

LEARNING OBJECTIVES/ASSESSMENT

When you have completed your study of this chapter, you should be able to:

1. Identify the reactants and products in written reaction equations, and balance the equations by inspection. (Section 5.1; Exercises 5.2 and 5.6)
2. Assign oxidation numbers to elements in chemical formulas, and identify the oxidizing and reducing agents in redox reactions. (Section 5.3; Exercises 5.10 and 5.15)
3. Classify reactions into the categories of redox or nonredox, then into the categories of decomposition, combination, single replacement, or double replacement. (Sections 5.4, 5.5, and 5.6; Exercise 5.20)
4. Write molecular equations in total ionic and net ionic forms. (Section 5.7; Exercise 5.30 a, b, & c)
5. Classify reactions as exothermic or endothermic. (Section 5.8; Exercise 5.34)
6. Use the mole concept to do calculations based on chemical reaction equations. (Section 5.9; Exercise 5.42)
7. Use the mole concept to do calculations based on the limiting-reactant principle. (Section 5.10; Exercise 5.52)
8. Use the mole concept to do percentage-yield calculations. (Section 5.11; Exercise 5.56)

SOLUTIONS FOR THE END OF CHAPTER EXERCISES

CHEMICAL EQUATIONS (SECTION 5.1)

☑5.2

	Equation	Reactants	Products
a.	H_2 (g) + Cl_2 (g) → 2 HCl (g)	H_2, Cl_2	HCl
b.	2 $KClO_3$ (s) → 2 KCl (s) + 3 O_2 (g)	$KClO_3$	KCl, O_2
c.	magnesium oxide + carbon → magnesium + carbon monoxide	magnesium oxide, carbon	magnesium, carbon monoxide
d.	ethane + oxygen → carbon dioxide + water	ethane, oxygen	carbon dioxide, water

5.4

a. ZnS (s) + O_2 (g) → ZnO (s) + SO_2 (g) is not consistent with the law of conservation of matter because the reactant (left) side of the equation has two moles of oxygen atoms, while the product (right) side of the equation has three moles of oxygen atoms.

b. Cl_2 (aq) + 2 I^- (aq) →I_2 (aq) + 2Cl^- (aq) is consistent with the law of conservation of matter.

c. 1.50 g oxygen + 1.50 g carbon → 2.80 g carbon monoxide is not consistent with the law of conservation of matter because the mass of the reactants is 3.00 g, while the mass of the products is only 2.80 g.

Notice, the number of moles of oxygen and carbon are not equal on both sides of the equation either.

$$\text{Reactants}: 1.50\text{ g } O_2\left(\frac{1\text{ mole } O_2}{32.0\text{ g } O_2}\right) = 0.0469\text{ moles } O_2;$$

$$1.50\text{ g C}\left(\frac{1\text{ mole C}}{12.0\text{ g C}}\right) = 0.125\text{ moles C}$$

$$\text{Products}: 2.80\text{ g CO}\left(\frac{1\text{ mole CO}}{28.0\text{ g CO}}\right)\left(\frac{1\text{ mole O}}{1\text{ mole CO}}\right) = 0.100\text{ mole O};$$

$$2.80\text{ g CO}\left(\frac{1\text{ mole CO}}{28.0\text{ g CO}}\right)\left(\frac{1\text{ mole C}}{1\text{ mole CO}}\right) = 0.100\text{ mole C}$$

d. $2\ C_2H_6\ (g) + 7\ O_2\ (g) \rightarrow 4\ CO_2\ (g) + 6\ H_2O\ (g)$ is consistent with the law of conservation of matter.

☑5.6 a. $Ag\ (s) + Cu(NO_3)_2\ (aq) \rightarrow Cu\ (s) + AgNO_3\ (aq)$

Elements	Ag	Cu	N	O
Reactant	1	1	2	6
Product	1	1	1	3

This equation is not balanced because the number of moles of nitrogen and oxygen are not balanced.

b. $2\ N_2O\ (g) + 3\ O_2\ (g) \rightarrow 4\ NO_2\ (g)$

Elements	N	O
Reactant	4	8
Product	4	8

This equation is balanced.

c. $Mg\ (s) + O_2\ (g) \rightarrow 2\ MgO\ (s)$

Elements	Mg	O
Reactant	1	2
Product	2	2

This equation is not balanced because the number of moles of magnesium is not balanced.

d. $H_2SO_4\ (aq) + Ca(OH)_2\ (aq) \rightarrow CaSO_4\ (s) + 2\ H_2O\ (l)$

Elements	H	S	O	Ca
Reactant	4	1	6	1
Product	4	1	6	1

This equation is balanced.

5.8 a. $KClO_3\ (s) \rightarrow KCl\ (s) + O_2\ (g)$ — $2\ KClO_3\ (s) \rightarrow 2\ KCl\ (s) + 3\ O_2\ (g)$

b. $C_2H_6\ (g) + O_2\ (g) \rightarrow CO_2(g) + H_2O\ (l)$ — $2\ C_2H_6\ (g) + 7\ O_2\ (g) \rightarrow 4\ CO_2(g) + 6\ H_2O\ (l)$

c. nitrogen + oxygen → dinitrogen pentoxide — $2\ N_2\ (g) + 5\ O_2\ (g) \rightarrow 2\ N_2O_5\ (g)$

d. $MgCl_2\ (s) + H_2O\ (g) \rightarrow MgO\ (s) + HCl\ (g)$ — $MgCl_2\ (s) + H_2O\ (g) \rightarrow MgO\ (s) + 2\ HCl\ (g)$

e. $CaH_2\ (s) + H_2O\ (l) \rightarrow Ca(OH)_2\ (s) + H_2\ (g)$ — $CaH_2\ (s) + 2\ H_2O\ (l) \rightarrow Ca(OH)_2\ (s) + 2\ H_2\ (g)$

f. $Al\ (s) + Fe_2O_3\ (s) \rightarrow Al_2O_3\ (s) + Fe\ (s)$ — $2\ Al\ (s) + Fe_2O_3\ (s) \rightarrow Al_2O_3\ (s) + 2\ Fe\ (s)$

g. aluminum + bromine → aluminum bromide — $2\ Al\ (s) + 3\ Br_2\ (l) \rightarrow 2\ AlBr_3\ (s)$

h. $Hg_2(NO_3)_2\ (aq) + NaCl\ (aq) \rightarrow Hg_2Cl_2\ (s) + NaNO_3\ (aq)$ — $Hg_2(NO_3)_2\ (aq) + 2\ NaCl\ (aq) \rightarrow Hg_2Cl_2\ (s) + 2\ NaNO_3\ (aq)$

REDOX REACTIONS (SECTION 5.3)

☑5.10

	Calculations	O.N.
a. HC̲lO	$H = +1 \quad Cl = ? \quad O = -2 \quad HClO = 0$ $(+1) + ? + (-2) = 0 \Rightarrow ? = +1$	+1
b. NaN̲O₂	$Na = +1 \quad N = ? \quad O = -2 \quad NaNO_2 = 0$ $(+1) + ? + 2(-2) = 0 \Rightarrow ? = +3$	+3
c. N̲₂		0

d. $Ca(\underline{Cl}O)_2$ — $Ca = +2 \quad Cl = ? \quad O = -2 \quad N_2O = 0$

$$(+2) + 2(? + (-2)) = 0 \quad \Rightarrow \quad ? = +1$$

+1

e. $\underline{Al}^{3+}$ — +3

f. $\underline{N}_2O_5$ — $N = ? \quad O = -2 \quad N_2O = 0$

$$2(?) + 5(-2) = 0 \quad \Rightarrow \quad ? = +5$$

+5

5.12

		Oxidation Numbers	Highest O.N.
a.	$Na_2Cr_2O_7$	Na = +1, Cr = +6, O = -2	chromium
b.	$K_2S_2O_3$	K = +1, S = +2, O = -2	sulfur
c.	HNO_3	H = +1, N = +5, O = -2	nitrogen
d.	P_2O_5	P = +5, O = -2	phosphorus
e.	$Mg(ClO_4)_2$	Mg = +2, Cl = +7, O = -2	chlorine
f.	$HClO_2$	H = +1, Cl = +3, O = -2	chlorine

5.14

		O.N.	Change	Classification
a.	$4\ \underline{Al}\ (s) + 3\ O_2\ (aq) \rightarrow 2\ Al_2O_3\ (s)$	$0 \rightarrow +3$	lost e^-	oxidized
b.	$SO_2\ (g) + \underline{H}_2O\ (l) \rightarrow H_2SO_3\ (aq)$	$+1 \rightarrow +1$	no change	neither
c.	$2\ K\underline{Cl}O_3\ (s) \rightarrow 2\ KCl\ (s) + 3\ O_2\ (g)$	$+5 \rightarrow -1$	gained e^-	reduced
d.	$2\ \underline{C}O\ (g) + O_2\ (g) \rightarrow 2\ CO_2\ (g)$	$+2 \rightarrow +4$	lost e^-	oxidized
e.	$2\ \underline{Na}\ (s) + 2\ H_2O\ (l) \rightarrow 2\ NaOH\ (aq) + H_2\ (g)$	$0 \rightarrow +1$	lost e^-	oxidized

5.16 a. $2\ Cu\ (s) + O_2\ (g) \rightarrow 2\ CuO\ (s)$

Elements	Cu	O
Reactant	0	0
Product	+2	-2

The oxidizing agent is O_2.
The reducing agent is Cu.

b. $Cl_2\ (aq) + 2\ KI\ (aq) \rightarrow 2\ KCl\ (aq) + I_2\ (aq)$

Elements	Cl	K	I
Reactant	0	+1	-1
Product	-1	+1	0

The oxidizing agent is Cl_2.
The reducing agent is KI.

c. $3\ MnO_2\ (s) + 4\ Al\ (s) \rightarrow 2\ Al_2O_3\ (s) + 3\ Mn\ (s)$

Elements	Mn	O	Al
Reactant	+4	-2	0
Product	0	-2	+3

The oxidizing agent is MnO_2.
The reducing agent is Al.

d. $2\ H^+\ (aq) + 3\ SO_3^{2-}\ (aq) + 2\ NO_3^-\ (aq) \rightarrow 2\ NO\ (g) + H_2O\ (l) + 3\ SO_4^{2-}\ (aq)$

Elements	H	S	O	N
Reactant	+1	+4	-2	+5
Product	+1	+6	-2	+2

The oxidizing agent is NO_3^-.
The reducing agent is SO_3^{2-}.

e. $Mg\ (s) + 2\ HCl\ (aq) \rightarrow MgCl_2\ (aq) + H_2\ (g)$

Elements	Mg	H	Cl
Reactant	0	+1	-1
Product	+2	0	-1

The oxidizing agent is HCl.
The reducing agent is Mg.

f. $4\ NO_2\ (g) + O_2\ (g) \rightarrow 2\ N_2O_5\ (g)$

Elements	N	O	O
Reactant	+4	-2	0
Product	+5	-2	-2

The oxidizing agent is O_2.
The reducing agent is NO_2.

5.18 $6\ NaOH\ (aq) + 2\ Al\ (s) \rightarrow 3\ H_2\ (g) + 2\ Na_3AlO_3\ (aq) + heat$

The oxidizing agent is NaOH because the oxidation number for the hydrogen changes from +1 to 0. The reducing agent is Al because the oxidation number for the aluminum changes from 0 to +3.

DECOMPOSITION, COMBINATION, AND REPACEMENT REACTIONS (SECTION 5.4-5.6)

☑5.20
a. $K_2CO_3\ (s) \rightarrow K_2O\ (s) + CO_2\ (g)$ — nonredox: decomposition

b. $Ca\ (s) + 2\ H_2O\ (l) \rightarrow Ca(OH)_2\ (s) + H_2\ (g)$ — redox: single-replacement
   0 / +1 -2 / +2 -2 +1 / 0

c. $BaCl_2\ (aq) + H_2SO_4\ (aq) \rightarrow BaSO_4\ (s) + 2\ HCl\ (aq)$ — nonredox: double-replacement

d. $SO_2\ (g) + H_2O\ (l) \rightarrow H_2SO_3\ (aq)$ — nonredox: combination

e. $2\ NO\ (g) + O_2\ (g) \rightarrow 2\ NO_2\ (g)$ — redox: combination
   +2 -2 / 0 / +4 -2

f. $2\ Zn\ (s) + O_2\ (g) \rightarrow 2\ ZnO\ (s)$ — redox: combination
   0 / 0 / +2 -2

5.22 $2\ NaHCO_3\ (s) \xrightarrow{Heat} Na_2CO_3\ (s) + H_2O\ (g) + CO_2\ (g)$
   +1 +1 +4 -2 / +1 +4 -2 / +1 -2 / +4 -2

This reaction is a nonredox decomposition reaction.

5.24 $CH_4\ (g) + 2\ O_2\ (g) \rightarrow CO_2\ (g) + 2\ H_2O\ (g)$; This reaction is a redox reaction.
   -4 +1 / 0 / +4 -2 / +1 -2

5.26 $Cl_2\ (aq) + H_2O\ (l) \rightarrow HOCl\ (aq) + HCl\ (aq)$; This is a redox reaction.
   0 / +1 -2 / +1 -2 +1 / +1 -1

IONIC EQUATIONS (SECTION 5.7)

5.28

	Compound	Ions		Compound	Ions
a.	$LiNO_3$	Li^+, NO_3^-	d.	KOH	K^+, OH^-
b.	Na_2HPO_4	2 Na^+, HPO_4^{2-}	e.	$MgBr_2$	Mg^{2+}, 2 Br^-
c.	$Ca(ClO_3)_2$	Ca^{2+}, 2 ClO_3^-	f.	$(NH_4)_2SO_4$	2 NH_4^+, SO_4^{2-}

5.30 ☑ a. $\underline{SO_2}\ (aq) + \underline{H_2O}\ (l) \rightarrow H_2SO_3\ (aq)$

Total ionic equation: $SO_2\ (aq) + H_2O\ (l) \rightarrow 2\ H^+(aq) + SO_3^{2-}\ (aq)$

Spectator ions: none

Net ionic equation: $SO_2\ (aq) + H_2O\ (l) \rightarrow 2\ H^+(aq) + SO_3^{2-}\ (aq)$

☑ b. $CuSO_4\ (aq) + Zn\ (s) \rightarrow Cu\ (s) + ZnSO_4\ (aq)$

Total ionic equation: $Cu^{2+}(aq) + SO_4^{2-}(aq) + Zn\ (s) \rightarrow Cu\ (s) + Zn^{2+}(aq) + SO_4^{2-}(aq)$

Spectator ions: $SO_4^{2-}(aq)$

Net ionic equation: $Cu^{2+}(aq) + Zn\ (s) \rightarrow Cu\ (s) + Zn^{2+}(aq)$

☑ c. $2\ KBr\ (aq) + 2\ H_2SO_4\ (aq) \rightarrow \underline{Br_2}\ (aq) + \underline{SO_2}\ (aq) + K_2SO_4\ (aq) + 2\ \underline{H_2O}\ (l)$

Total ionic equation:

$$2\ K^+(aq) + 2\ Br^-(aq) + 4\ H^+(aq) + 2\ SO_4^{2-}(aq) \rightarrow Br_2\ (aq) + SO_2\ (aq) + 2\ K^+\ (aq) + SO_4^{2-}\ (aq) + 2\ H_2O\ (l)$$

Spectator ions: $2\ K^+\ (aq), SO_4^{2-}\ (aq)$

Net ionic equation: $2\ Br^-(aq) + 4\ H^+(aq) + SO_4^{2-}(aq) \rightarrow Br_2\ (aq) + SO_2\ (aq) + 2\ H_2O\ (l)$

d. $AgNO_3\ (aq) + NaOH\ (aq) \rightarrow AgOH\ (s) + NaNO_3\ (aq)$

Total ionic equation:

$$Ag^+(aq) + NO_3^-(aq) + Na^+(aq) + OH^-(aq) \rightarrow AgOH\ (s) + Na^+(aq) + NO_3^-(aq)$$

Spectator ions: $Na^+(aq), NO_3^-(aq)$

Net ionic equation: $Ag^+(aq) + OH^-(aq) \rightarrow AgOH\ (s)$

e. $BaCO_3\ (s) + 2\ HNO_3\ (aq) \rightarrow Ba(NO_3)_2\ (aq) + \underline{CO_2}\ (g) + \underline{H_2O}\ (l)$

Total ionic equation:

$$BaCO_3\ (s) + 2\ H^+(aq) + 2\ NO_3^-(aq) \rightarrow Ba^{2+}(aq) + 2\ NO_3^-(aq) + CO_2\ (g) + H_2O\ (l)$$

Spectator ions: $2\ NO_3^-(aq)$

Net ionic equation: $BaCO_3\ (s) + 2\ H^+(aq) \rightarrow Ba^{2+}(aq) + CO_2\ (g) + H_2O\ (l)$

f. $\underline{N_2O_5}\ (aq) + \underline{H_2O}\ (l) \rightarrow 2\ HNO_3\ (aq)$

Total ionic equation: $N_2O_5\ (aq) + H_2O\ (l) \rightarrow 2\ H^+(aq) + 2\ NO_3^-(aq)$

Spectator ions: none

Net ionic equation: $N_2O_5\ (aq) + H_2O\ (l) \rightarrow 2\ H^+(aq) + 2\ NO_3^-(aq)$

5.32 a. $HNO_3\ (aq) + KOH\ (aq) \rightarrow KNO_3\ (aq) + H_2O\ (l)$

Total ionic equation: $H^+(aq) + NO_3^-(aq) + K^+\ (aq) + OH^-(aq) \rightarrow K^+(aq) + NO_3^-(aq) + H_2O\ (l)$

Spectator ions: $K^+(aq), NO_3^-(aq)$

Net ionic equation: $H^+\ (aq) + OH^-(aq) \rightarrow H_2O\ (l)$

b. $H_3PO_4\ (aq) + 3\ NH_4OH\ (aq) \rightarrow (NH_4)_3PO_4\ (aq) + 3\ H_2O\ (l)$

Total ionic equation:

$$3\ H^+(aq) + PO_4^{3-}\ (aq) + 3\ NH_4^+\ (aq) + 3\ OH^-(aq) \rightarrow 3\ NH_4^+\ (aq) + PO_4^{3-}\ (aq) + 3\ H_2O\ (l)$$

Spectator ions: $3\ NH_4^+\ (aq), PO_4^{3-}\ (aq)$

Net ionic equation: $3\ H^+\ (aq) + 3\ OH^-(aq) \rightarrow 3\ H_2O\ (l)$; $H^+\ (aq) + OH^-(aq) \rightarrow H_2O\ (l)$

c. $HI\ (aq) + NaOH\ (aq) \rightarrow NaI\ (aq) + H_2O\ (l)$

Total ionic equation: $H^+(aq) + I^-(aq) + Na^+(aq) + OH^-(aq) \rightarrow Na^+(aq) + I^-(aq) + H_2O\ (l)$

Spectator ions: $Na^+(aq),\ I^-(aq)$

Net ionic equation: $H^+\ (aq) + OH^-(aq) \rightarrow H_2O\ (l)$

The reactants and products are identical for all three net ionic equations.

ENERGY AND REACTIONS (SECTION 5.8)

☑5.34 The emergency hot pack becoming warm is an exothermic process because, as water is mixed with the solid, heat is released.

5.36 By insulating the ice, the heat in the food will not be able to travel into the ice as easily. It would be better if the ice were not wrapped in a thick insulating blanket. Wrapping the blanket around the ice and the food would be a better arrangement for keeping the food cold.

THE MOLE AND CHEMICAL EQUATIONS (SECTION 5.9)

5.38 a. $S (s) + O_2 (g) \rightarrow SO_2 (g)$

Statement 1: 1 S atom + 1 O_2 molecule $\rightarrow$ 1 SO_2 molecule

Statement 2: 1 mole S + 1 mole O_2 $\rightarrow$ 1 mole SO_2

Statement 3: 6.02×10^{23} S atoms + 6.02×10^{23} O_2 molecules $\rightarrow$ 6.02×10^{23} SO_2 molecules

Statement 4: 32.1 g S + 32.0 g O_2 $\rightarrow$ 64.1 g SO_2

b. $Sr (s) + 2 H_2O (l) \rightarrow Sr(OH)_2 (s) + H_2 (g)$

Statement 1: 1 Sr atom + 2 H_2O molecules $\rightarrow$ 1 $Sr(OH)_2$ formula unit + 1 H_2 molecule

Statement 2: 1 mole Sr + 2 moles H_2O $\rightarrow$ 1 mole $Sr(OH)_2$ + 1 mole H_2

Statement 3: 6.02×10^{23} Sr atoms + 12.0×10^{23} H_2O molecules $\rightarrow$
6.02×10^{23} $Sr(OH)_2$ formula units + 6.02×10^{23} H_2 molecules

Statement 4: 87.6 g Sr + 36.0 g H_2O $\rightarrow$ 121.6 g $Sr(OH)_2$ + 2.0 g H_2

c. $2 H_2S (g) + 3 O_2 (g) \rightarrow 2 H_2O (g) + 2 SO_2 (g)$

Statement 1: 2 H_2S molecules + 3 O_2 molecules $\rightarrow$ 2 H_2O molecules + 2 SO_2 molecules

Statement 2: 2 moles H_2S + 3 moles O_2 $\rightarrow$ 2 moles H_2O + 2 moles SO_2

Statement 3: 12.0×10^{23} H_2S molecules + 18.1×10^{23} O_2 molecules $\rightarrow$
12.0×10^{23} H_2O molecules + 12.0×10^{23} SO_2 molecules

Statement 4: 68.2 g H_2S + 96.0 g O_2 $\rightarrow$ 36.0 g H_2O + 128.2 g SO_2

d. $4 NH_3 (g) + 5 O_2 (g) \rightarrow 4 NO (g) + 6 H_2O (g)$

Statement 1: 4 NH_3 molecules + 5 O_2 molecules $\rightarrow$ 4 NO molecules + 6 H_2O molecules

Statement 2: 4 moles NH_3 + 5 moles O_2 $\rightarrow$ 4 moles NO + 6 moles H_2O

Statement 3: 24.1×10^{23} NH_3 molecules + 30.1×10^{23} O_2 molecules $\rightarrow$
24.1×10^{23} NO molecules + 36.1×10^{23} H_2O molecules

Statement 4: 68.0 g NH_3 + 160.0 g O_2 $\rightarrow$ 120.0 g NO + 108.0 g H_2O

e. $CaO (s) + 3 C (s) \rightarrow CaC_2 (s) + CO (g)$

Statement 1: 1 CaO formula unit + 3 C atoms $\rightarrow$ 1 CaC_2 formula unit + 1 CO molecule

Statement 2: 1 mole CaO + 3 moles C $\rightarrow$ 1 mole CaC_2 + 1 mole CO

Statement 3: 6.02×10^{23} CaO formula units + 18.1×10^{23} C atoms $\rightarrow$
6.02×10^{23} CaC_2 formula units + 6.02×10^{23} CO molecules

Statement 4: 56.1 g CaO + 36.0 g C $\rightarrow$ 64.1 g CaC_2 + 28.0 g CO

5.40 $2\,SO_2 + O_2 \rightarrow 2\,SO_3\,(g)$

Statement 1: $2\,SO_2$ molecules + 1 O_2 molecule $\rightarrow$ 2 SO_3 molecules

Statement 2: 2 moles SO_2 + 1 mole O_2 $\rightarrow$ 2 moles SO_3

Statement 3: 12.0×10^{23} SO_2 molecules + 6.02×10^{23} O_2 molecules $\rightarrow$ 12.0×10^{23} SO_3 molecules

Statement 4 : 128 g SO_2 + 32.0 g O_2 $\rightarrow$ 160 g SO_3

Factors :

$$\frac{12.0\times10^{23}\ SO_2\ \text{molecules}}{6.02\times10^{23}\ O_2\ \text{molecules}},\ \frac{6.02\times10^{23}\ O_2\ \text{molecules}}{12.0\times10^{23}\ SO_2\ \text{molecules}},\ \frac{12.0\times10^{23}\ SO_2\ \text{molecules}}{12.0\times10^{23}\ SO_3\ \text{molecules}};$$

$$\frac{12.0\times10^{23}\ SO_3\ \text{molecules}}{12.0\times10^{23}\ SO_2\ \text{molecules}},\ \frac{6.02\times10^{23}\ O_2\ \text{molecules}}{12.0\times10^{23}\ SO_3\ \text{molecules}},\ \frac{12.0\times10^{23}\ SO_3\ \text{molecules}}{6.02\times10^{23}\ O_2\ \text{molecules}};$$

$$\frac{2\ \text{moles}\ SO_2}{1\ \text{mole}\ O_2};\ \frac{1\ \text{mole}\ O_2}{2\ \text{moles}\ SO_2};\ \frac{2\ \text{moles}\ SO_2}{2\ \text{moles}\ SO_3},\ \frac{2\ \text{moles}\ SO_3}{2\ \text{moles}\ SO_2};\ \frac{1\ \text{mole}\ O_2}{2\ \text{moles}\ SO_3};\ \frac{2\ \text{moles}\ SO_3}{1\ \text{mole}\ O_2};$$

$$\frac{128\ g\ SO_2}{32.0\ g\ O_2};\ \frac{32.0\ g\ O_2}{128\ g\ SO_2};\ \frac{128\ g\ SO_2}{160\ g\ SO_3},\ \frac{160\ g\ SO_3}{128\ g\ SO_2};\ \frac{32.0\ g\ O_2}{160\ g\ SO_3};\ \frac{160\ g\ SO_3}{32.0\ g\ O_2}$$

This list does not include all possible factors.

☑5.42 $CaCO_3\,(s) \rightarrow CaO\,(s) + CO_2\,(g)$

$$500\ g\ CaO\left(\frac{100\ g\ CaCO_3}{56.1\ g\ CaO}\right) = 891.265597148\ g\ CaCO_3$$

$$\approx 891\ g\ CaCO_3$$

5.44 $2\,Al\,(s) + 3\,Br_2\,(l) \rightarrow 2\,AlBr_3\,(s)$

$$50.1\ g\ Al\left(\frac{479\ g\ Br_2}{54.0\ g\ Al}\right) = 444.4055556\ g\ Br_2$$

$$\approx 444\ g\ Br_2\ \text{with SF}$$

5.46

$$50.1\ g\ Al\left(\frac{533\ g\ AlBr_3}{54.0\ g\ Al}\right) = 494.50555556\ g\ AlBr_3$$

$$\approx 495\ g\ AlBr_3\ \text{with SF}$$

or

$$50.1\ g\ Al + 445\ g\ Br_2 = 495.1\ g\ Al\ Br_3 \approx 495\ g\ AlBr_3\ (SF)$$

5.48 $TiCl_4\,(s) + 2\,Mg\,(s) \rightarrow Ti\,(s) + 2\,MgCl_2\,(s)$

$$1.00\ kg\ Ti\left(\frac{1000\ g\ Ti}{1\ kg\ Ti}\right)\left(\frac{48.6\ g\ Mg}{47.9\ g\ Ti}\right) = 1014.61377871\ g\ Mg$$

$$\approx 1.01\times10^3\ g\ Mg\ \text{with SF}$$

5.50 $C_6H_{12}O_2\,(aq) + 8\,O_2\,(aq) \rightarrow 6\,CO_2\,(aq) + 6\,H_2O\,(l)$

$$1.00\ \text{mol caproic acid}\left(\frac{256\ g\ O_2}{1\ \text{mol caproic acid}}\right) = 256\ g\ O_2$$

THE LIMITING REACTANT (SECTION 5.10)

☑5.52 a. $N_2\ (g) + 2\ O_2\ (g) \rightarrow 2\ NO_2\ (g)$

O_2 is the limiting reactant as shown by the calculations below:

$$1.25 \text{ moles } N_2\left(\frac{2 \text{ moles } NO_2}{1 \text{ mole } N_2}\right) = 2.50 \text{ moles } NO_2$$

$$50.0 \text{ g } O_2\left(\frac{2 \text{ moles } NO_2}{64.00 \text{ g } O_2}\right) = 1.56 \text{ moles } NO_2$$

b. Only 71.9 g NO_2 can be produced because O_2 is the limiting reactant. Enough N_2 is present to make 115 g of NO_2, but not enough oxygen is present to react with the excess amount of N_2.

$$50.0 \text{ g } O_2\left(\frac{92.0 \text{ g } NO_2}{64.0 \text{ g } O_2}\right) = 71.9 \text{ g } NO_2$$

$$1.25 \text{ moles } N_2\left(\frac{92.0 \text{ g } NO_2}{1 \text{ mole } N_2}\right) = 115 \text{ g } NO_2$$

5.54 $NH_3\ (aq) + CO_2\ (aq) + H_2O\ (l) \rightarrow NH_4HCO_3\ (aq)$

144 g NH_4HCO_3 is the maximum mass in grams that can be produced by these reactants. CO_2 is the limiting reactant because the amount given can produce less product than the amounts of either the NH_3 or H_2O given. The NH_3 and H_2O are both excess reactants.

$$50.0 \text{ g } NH_3\left(\frac{79.1 \text{ g } NH_4HCO_3}{17.0 \text{ g } NH_3}\right) = 233 \text{ g } NH_4HCO_3$$

$$80.0 \text{ g } CO_2\left(\frac{79.1 \text{ g } NH_4HCO_3}{44.0 \text{ g } CO_2}\right) = 144 \text{ g } NH_4HCO_3$$

$$2.00 \text{ moles } H_2O\left(\frac{79.1 \text{ g } NH_4HCO_3}{1 \text{ moles } H_2O}\right) = 158 \text{ g } NH_4HCO_3$$

REACTION YIELDS (SECTION 5.11)

☑5.56 $\frac{11.74 \text{ g}}{16.09 \text{ g}} \times 100 = 72.96\%$ yield

5.58 The theoretical yield is the sum of the masses of reactant *A* with reactant *B* because the amounts given are said to react exactly, which means both reactants will be completely used to make the product.

$$\frac{9.04 \text{ g}}{7.59 \text{ g} + 4.88 \text{ g}} \times 100 = 72.5\% \text{ yield}$$

5.60 $2\ HgO\ (s) \rightarrow 2\ Hg\ (l) + O_2\ (g)$

$$7.22 \text{ g HgO}\left(\frac{402 \text{ g Hg}}{434 \text{ g HgO}}\right) = 6.6876 \text{ g Hg}$$

$$\frac{5.95 \text{ g}}{6.6876 \text{ g Hg}} \times 100 = 89.0\% \text{ yield}$$

or

$$\frac{5.95 \text{ g}}{7.22 \text{ g HgO}\left(\frac{402 \text{ g Hg}}{434 \text{ g HgO}}\right)} \times 100 = 89.0\% \text{ yield}$$

ADDITIONAL EXERCISES

5.62 barium chloride (aq) + sodium sulfate (aq) → sodium chloride (aq) + barium sulfate (s)

Molecular equation:

$$BaCl_2\ (aq) + Na_2SO_4\ (aq) \rightarrow 2\ NaCl\ (aq) + BaSO_4\ (s)$$

Total Ionic equation:

$$Ba^{2+}(aq) + 2\ Cl^-(aq) + 2\ Na^+(aq) + SO_4^{2-}(aq) \rightarrow 2\ Na^+(aq) + 2\ Cl^-(aq) + BaSO_4(s)$$

Spectator Ions: 2 Na^+ (aq) and 2 Cl^- (aq)

Net Ionic equation:

$$Ba^{2+}(aq) + SO_4^{2-}(aq) \rightarrow BaSO_4(s)$$

5.64

$$6.983\ g\ \frac{\text{naturally occuring}}{\text{elemental iron}}\left(\frac{60\ g\ ^{60}Fe}{55.9\ g\ \frac{\text{naturally occuring}}{\text{elemental iron}}}\right) = 7.5\ g\ ^{60}Fe$$

ALLIED HEALTH EXAM CONNECTION

5.66 (c) is the balanced equation.

$$Mg\ (s) + 2\ H_2O\ (g) \rightarrow Mg(OH)_2\ (s) + H_2\ (g)$$

5.68 (c) is the balanced equation. $C_{12}H_{22}O_{11} \rightarrow 12\ C + 11\ H_2O$

5.70 The coefficient before ammonia, NH_3, is (b) 2 in the balanced equation, as shown below:

$$N_2\ (g) + 3\ H_2\ (g) \rightleftharpoons 2\ NH_3\ (g)$$

5.72 The oxidation number for nitrogen in HNO_3 is (b) +5.

$H = +1 \quad N = ? \quad O = -2 \quad HNO_3 = 0$

$(+1) + ? + 3(-2) = 0$

$? = +5$

5.74 The oxidation number for sulfur in $Na_2S_2O_3$ is (c) +2.

$Na = +1 \quad S = ? \quad O = -2 \quad Na_2S_2O_3 = 0$

$2(+1) + 2(?) + 3(-2) = 0$

$2(?) = +4$

$? = +2$

5.76 (a) C_2H_4 (g) is oxidized and $3O_2$ (g) is reduced, in the equation, $C_2H_4\ (g) + 3\ O_2\ (g) \rightarrow 2\ CO_2\ (g) + 2\ H_2O\ (g)$. The oxygen changes from an oxidation number of 0 to -2 by gaining electrons (reduction). The carbon changes from an oxidation number of -2 to +4 by losing electrons (oxidation).

5.78 The answer is (b) oxidizing agent: 4 NO_3^- (aq), reducing agent: Sn(s).

$$8\ H^+\ (aq) + 6\ Cl^-\ (aq) + Sn\ (s) + 4\ NO_3^-\ (aq) \rightarrow SnCl_6^{2-}\ (aq) + 4\ NO_2\ (g) + 4\ H_2O\ (l)$$

The tin changes from an oxidation number of 0 to +4 by losing electrons (oxidation); therefore,

Sn is the reducing agent. The nitrogen changes from an oxidation number of +5 to +4 by gaining electrons (reduction); therefore, NO_3^- is the oxidizing agent.

5.80 This equation, $Cl_2 + 2e^- \rightarrow 2Cl^-$, is (a) a reduction reaction because Cl_2 is gaining electrons.

5.82 This equation, $Mg(OH)_2 + 2\,HCl \rightarrow MgCl_2 + 2\,H_2O$, is (d) double replacement reaction because the reactants are trading partners.

5.84 The net ionic equation is (b) $Ca^{2+}\,(aq) + 2\,Cl^-\,(aq) \rightarrow CaCl_2\,(s)$.

molecular equation: $Ca(NO_3)_2\,(aq) + 2\,KCl\,(aq) \rightarrow CaCl_2\,(s) + 2\,KNO_3\,(aq)$
total ionic equation:
$Ca^{2+}\,(aq) + 2\,NO_3^-\,(aq) + 2\,K^+\,(aq) + 2\,Cl^-\,(aq) \rightarrow CaCl_2\,(s) + 2\,K^+\,(aq) + 2\,NO_3^-\,(aq)$
spectator ions: $2\,NO_3^-\,(aq)$, $2\,K^+\,(aq)$
net ionic equation: $Ca^{2+}\,(aq) + 2\,Cl^-\,(aq) \rightarrow CaCl_2\,(s)$

5.86 (c) 53.5 g NH_4Cl can be produced.

$$NH_3 + HCl \rightarrow NH_4Cl$$

$$17\text{ g }NH_3\left(\frac{1\text{ mole }NH_3}{17.0\text{ g }NH_3}\right) = 1.0\text{ mole }NH_3 \qquad 36.5\text{ g HCl}\left(\frac{1\text{ mole HCl}}{36.5\text{ g HCl}}\right) = 1.0\text{ mole HCl}$$

$$17\text{ g }NH_3 + 36.5\text{ g HCl} = 53.5\text{ g }NH_4Cl$$

5.88 (d) 0.25 mole of $CaCl_2$ is needed to form 0.5 mole of NaCl.

$$CaCl_2 + Na_2CO_3 \rightarrow CaCO_3 + 2NaCl$$

$$0.5\text{ mole NaCl}\left(\frac{1\text{ mole }CaCl_2}{2\text{ moles NaCl}}\right) = 0.25\text{ mole }CaCl_2$$

$$0.5\text{ mole NaCl}\left(\frac{1\text{ mole }Na_2CO_3}{2\text{ moles NaCl}}\right) = 0.25\text{ mole }Na_2CO_3$$

$$0.5\text{ mole NaCl}\left(\frac{1\text{ mole }CaCO_3}{2\text{ moles NaCl}}\right) = 0.25\text{ mole }CaCO_3$$

CHEMISTRY FOR THOUGHT

5.90 When a yield of more than 100% occurs for a compound prepared by precipitation from water solutions, it is likely that the "dry" compound is still moist and contains extra water mass.

5.92 The bubbles formed during the reaction of hydrogen peroxide indicate that at least one of the products is a gas. Another material that might provide the enzyme catalyst needed for hydrogen peroxide decomposition is blood.

5.94 $2\,Cr_2O_7^{2-}\,(aq) + 3\,CH_3CH_2OH\,(aq) + 16\,H^+\,(aq) \rightarrow 4Cr^{3+}\,(aq) + 3\,CH_3COOH\,(aq) + 11\,H_2O\,(l)$
The initial color of the solution is orange-colored, but as it reacts with alcohol, the solution becomes pale violet. The darker violet the solution becomes, the more alcohol that was present in the breath.

5.96 Zinc changing from zinc metal into zinc ions is an oxidation reaction because the zinc atoms are losing electrons to become cations. This reaction is the source of electrons as shown in the following equation:

$$Zn\ (s) \rightarrow Zn^{2+}\ (aq) + 2\ e^-$$

ADDITIONAL ACTIVITIES

Section 5.1 Review:

(1) Punctuate the following chemical reactions.
(2) Add the appropriate symbols for the states of matter.
(3) Do the following chemical reactions violate the law of conservation of matter? If so, correct the mistakes.

Reaction A: Solid white phosphorus reacts with oxygen gas to produce solid tetraphophorus decaoxide. P_4 ( ) O_2 ( ) P_4O_{10} ( )

Reaction B: Aqueous sodium hydroxide neutralizes aqueous sulfuric acid and forms aqueous sodium sulfate and water. $NaOH$ ( ) H_2SO_4 ( ) Na_2SO_4 ( ) H_2O ( )

Section 5.2 Review:

Complete the diagram by adding the reaction types (combination, decomposition, single replacement, and double replacement) to the appropriate locations. The circle on the left represents reaction types that can be classified only as nonredox reactions. The circle on the right represents reactions types that can be classified only as redox reactions. The area where they overlap represents reaction types that can be redox or nonredox reactions. The area outside of the circles represents reaction types that cannot be classified as redox or nonredox reactions.

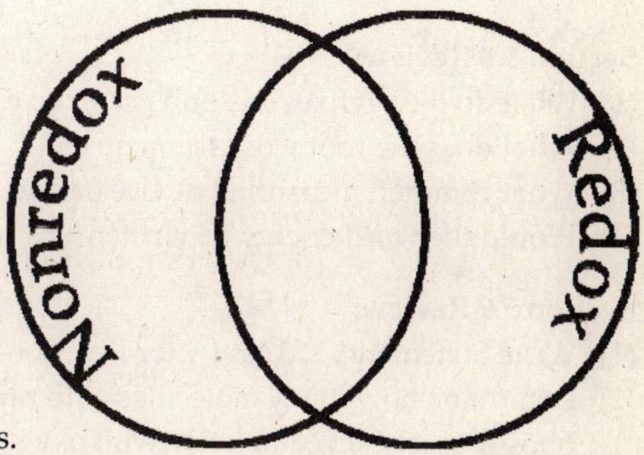

Section 5.3 Review:

(1) Calculate the oxidation numbers for all of the elements in the chemical formulas in the reactions from the Section 5.1 Review.
(2) Which of the reactions in Section 5.1 Review are redox reactions?
(3) Which element(s) from (2) were oxidized? Which element(s) from (2) were reduced?
(4) Identify the oxidizing agent and the reducing agent in the redox reaction(s) from (2).

Section 5.4 Review:

(1) What does it mean for something to "decompose"?
(2) Is the label "decomposition" an appropriate description for this type of reaction?
(3) Are the reactions in the Section 5.1 Review decomposition reactions?

Section 5.5 Review:

(1) What does it mean to "combine"?
(2) Is the label "combination" an appropriate description for this type of reaction?
(3) Do the terms "addition reaction" and "synthesis reaction" also have a descriptive meaning for this type of reaction?
(4) Are the reactions in the Section 5.1 Review combination reactions?

Section 5.6 Review:

(1) What does it mean to "replace"?
(2) What does it mean to "substitute"?
(3) Do the terms "single-replacement reaction" and "substitution reaction" describe the process that occurs during these reactions?
(4) Are the reactions in the Section 5.1 Review single-replacement reactions?
(5) What does "metathesis" mean?
(6) Do the terms "double-replacement reaction" and "metathesis reaction" describe the process that occurs during these reactions?
(7) Are the reactions in the Section 5.1 Review double-replacement reactions?

Section 5.7 Review:

(1) Write the total ionic equation for Reaction B from the Section 5.1 Review.
(2) What are the overall charges for the reactant and product sides of the reaction?
(3) Does the law of conservation of matter hold in the total ionic equation?
(4) Identify the spectator ions in the total ionic equation.
(5) Write the net ionic equation for Reaction B.
(6) Can the net ionic equation for Reaction B be simplified? If so, simplify the equation.
(7) What are the overall charges for the reactant and product sides of the reaction?
(8) Does the law of conservation of matter hold in the net ionic equation?

Section 5.8 Review:

(1) What do the prefixes *exo-* and *endo-* mean?
(2) What does the root word *therm* mean?
(3) What common measuring device (found in a laboratory) uses the root word *therm*?
(4) Would the word energy be written as a reactant or a product in an endothermic reaction?

Section 5.9 Review:

(1) Write Statements 2, 3, and 4 for Reaction A from the Section 5.1 Review.
(2) Determine how many molecules of tetraphosphorus decaoxide can be formed if 5.9 g of white phosphorus react completely with oxygen as shown in Reaction A from the Section 5.1 Review.
(3) Determine how many molecules of tetraphosphous decaoxide can be formed if 5.9 g of oxygen react completely with white phosphorus as shown in Reaction A from the Section 5.1 Review.

Section 5.10 Review:

(1) If 5.9 g of white phosphorus react with 5.9 g of oxygen, how many molecules of tetraphosphorus decaoxide can be formed?
(2) What are the identities of the limiting reactant and the excess reactant in (1)?

Section 5.11 Review:

(1) If only 10.0 g of P_4O_{10} are formed from the reaction described in (1) of the Section 5.10 Review, what is the percent yield?
(2) Other than poor lab technique, what could account for this apparent violation of the law of conservation of matter?

Tying It All Together with a Laboratory Application:

The "copper cycle" is a series of chemical reactions that begins with copper metal, forms several intermediate compounds, and finally returns copper metal.

In a fume hood, 0.48 g of copper metal reacts with 5.0×10^{-2} moles of nitric acid to form copper (II) nitrate, water, and nitrogen dioxide. The equation is:

(1) ______ Cu (s) + ______ HNO_3 (aq) → ______ $Cu(NO_3)_2$ (aq) + ______ H_2O (l) + ______ NO_2 (g)

The (2) ______ (percent or theoretical) yield of brown nitrogen dioxide gas is (3) ______ g. (4) ______ is the excess reactant. This is a (5) ______ reaction. Copper is the (6) ______ agent. (7) ______ is reduced.

Once the copper has completely reacted, a blue solution remains. Sodium hydroxide is added to the solution. The following reaction occurs:

(8) ______ $Cu(NO_3)_2$ (aq) + ______ NaOH (aq) → ______ $Cu(OH)_2$ (s) + ______ $NaNO_3$ (aq)

This reaction can be classified both (9) ______ and ______. Write the (10) total ionic equation and (11) net ionic equation for this reaction. (12) Identify the spectator ions in the equation.

The sodium nitrate solution is decanted from the copper (II) hydroxide solid which is then transferred to an evaporating dish. The copper (II) hydroxide is gently heated above a beaker of boiling water and the following reaction occurs:

(13) ______ $Cu(OH)_2$ (s) → ______ CuO (s) + ______ H_2O (l)

This reaction can be classified as both (14) ______ and ______. This reaction occurs because the copper hydroxide is heated. This is an (15) ______ (endothermic or exothermic) reaction.

Sulfuric acid (H_2SO_4) is then added to the solid copper (II) oxide and aqueous copper (II) sulfate and liquid water are formed. The reactants for this reaction are (16) ______ and the products for this reaction are (17) ______. (18) Write the balanced molecular equation for this reaction. This reaction can be classified as both (19) ______ and ______.

A coil of zinc wire is added to the blue copper sulfate solution. The following reaction occurs:

(20) ______ $CuSO_4$ (aq) + ______ Zn (s) → ______ Cu (s) + ______ $ZnSO_4$ (aq)

This reaction can be classified as both (21) ______ and ______. (22) Which type of reaction was not represented in the copper cycle?

The mass of copper that theoretically could have been recovered from the copper cycle is (23) ______ because the copper compounds were always the (24) ______ reactants in the reactions. The actual amount of copper recovered was 0.42 g. The percent yield for this reaction cycle is (25) ______. (26) Why was the percent yield not 100%?

SOLUTIONS FOR THE ADDITIONAL ACTIVITIES

Section 5.1 Review:

Reaction A: P_4 (s) + 5 O_2 (g) → P_4O_{10} (s); the coefficient of 5 before the O_2 balances the reaction

Reaction B: 2 NaOH (aq) + H_2SO_4 (aq) → Na_2SO_4 (aq) + 2 H_2O (l); the coefficients of 2 before the NaOH and the H_2O balance the reaction

Section 5.2 Review:

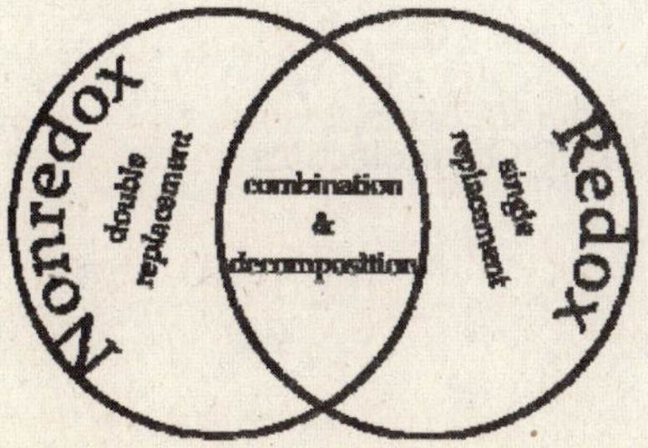

Section 5.3 Review:

(1) Reaction A: P_4 (s) + 5 O_2 (g) → P_4 O_{10} (s)

0 0 +5 -2

Reaction B: 2 Na O H (aq) + H_2 S O_4 (aq) →

+1 -2 +1 +1 +6 -2

Na_2 S O_4 (aq) + 2 H_2 O (l)

+1 +6 -2 +1 -2

(2) Reaction A is a redox reaction. Reaction B is a nonredox reaction.
(3) Phosphorus is oxidized. Oxygen is reduced.
(4) Oxygen is the oxidizing agent. Phosphorus is the reducing agent.

Section 5.4 Review:

(1) When something decomposes, it breaks into smaller pieces. (2) During a decomposition reaction, a single reactant breaks apart into two or more products which are smaller pieces than the original reactant. This label is appropriate. (3) Neither of the reactions from the Section 5.1 Review is a decomposition reaction.

Section 5.5 Review:

(1) When two or more materials are joined or brought into a union, they are said to combine.
(2) Combination reactions occur when two or more reactants join to form one product. The label is appropriate. (3) The term addition reaction implies that two or more reactants are joined together. The term synthesis reaction also implies elements are put together to form a whole (the product). Both of these terms have descriptive meaning for this type of reaction. (4) Reaction A from the Section 5.1 Review is a combination reaction.

Section 5.6 Review:

(1) Replace means to put something in the position of a previous object. (2) Substitute means to let something use the place of another. (3) Both single-replacement reaction and substitution reaction imply that one element will take the place of another element during the reaction. These names reflect the process that occurs. (4) Neither of the reactions in the Section 5.1 Review is a single-replacement reaction. (5) Metathesis means transposition or interchange. (6) Double-replacement reaction and metathesis reaction both imply that two elements trade their positions as the reaction progresses. These names reflect the process that occurs. (7) Reaction B from the Section 5.1 Review is a double-replacement reaction.

Section 5.7 Review:

(1) $2\ Na^+(aq) + 2\ OH^-(aq) + 2\ H^+(aq) + SO_4^{2-}(aq) \rightarrow 2\ Na^+(aq) + SO_4^{2-}(aq) + 2\ H_2O\ (l)$
(2) Reactant side = 2(+1) + 2(-1) + 2 (+1) + 1(-2) = 0; Product side = 2(+1) + 1(-2) + 2(0) = 0
(3) The law of conservation of matter holds for the total ionic equation.
(Reactant side: 2 Na, 4 H, 6 O, and 1 S; Product side: 2 Na, 1 S, 6 O, and 4 H)
(4) Spectator ions = $2\ Na^+(aq)$ and $SO_4^{2-}(aq)$
(5) Net ionic equation = $2\ OH^-(aq) + 2\ H^+(aq) \rightarrow 2\ H_2O\ (l)$
(6) Simplified net ionic equation = $OH^-(aq) + H^+(aq) \rightarrow H_2O\ (l)$
(7) Reactant side = (-1) + (+1) = 0, Product side = 0
(8) The law of conservation of matter holds for the net ionic equation. (Reactant side: 1 O and 2 H, Product side: 2 H and 1 O)

Section 5.8 Review:

(1) *exo-* means outside, outer, outer part; *endo-* means in, within, inner; (2) *therm-* means heat;
(3) thermometer; (4) Energy is a reactant in an endothermic reaction.

Section 5.9 Review:

(1) $P_4\ (s) + 5\ O_2\ (g) \rightarrow P_4O_{10}\ (s)$
Statement 2: $1 \text{ mole } P_4 + 5 \text{ moles } O_2 \rightarrow 1 \text{ mole } P_4O_{10}$
Statement 3: $6.02\times10^{23}\ P_4 \text{ molecules} + 3.01\times10^{24}\ O_2 \text{ molecules} \rightarrow 6.02\times10^{23}\ P_4O_{10} \text{ molecules}$;
Statement 4: $123.88 \text{ g } P_4 + 160.00 \text{ g } O_2 \rightarrow 283.88 \text{ g } P_4O_{10}$

(2) $5.9 \text{ g } P_4\left(\frac{6.02\times10^{23} \text{ molecules } P_4O_{10}}{123.88 \text{ g } P_4}\right) = 2.9\times10^{22} \text{ molecules } P_4O_{10}$;

(3) $5.9\text{ g O}_2\left(\dfrac{6.02\times10^{23}\text{ molecules P}_4\text{O}_{10}}{160.00\text{ g O}_2}\right)=2.2\times10^{22}\text{ molecules P}_4\text{O}_{10}$

Section 5.10 Review:

(1) 2.2 x 10^{22} molecules P_4O_{10} can be formed

(2) limiting reactant = oxygen, excess reactant = white phosphorus

Section 5.11 Review:

(1) $\dfrac{10.0\text{ g}}{5.9\text{ g O}_2\left(\dfrac{283.90\text{ g P}_4\text{O}_{10}}{160.00\text{ g O}_2}\right)}\times100=96\%\text{ yield}$

(2) Side reactions may have occurred. The mass of the products from the side reactions added to the mass of the P_4O_{10} should equal the theoretical mass.

Tying It All Together with a Laboratory Application:

(1) 1, 4, 1, 2, 2
(2) theoretical
(3) 0.70 g
(4) HNO_3
(5) redox
(6) reducing
(7) HNO_3
(8) 1, 2, 1, 2
(9) nonredox, double replacement
(10) Cu^{2+} (aq) + 2 NO_3^- (aq) + 2 Na^+ (aq) + 2 OH^- (aq) → $Cu(OH)_2$ (s) + 2 Na^+ (aq) + 2 NO_3^- (aq)
(11) Cu^{2+} (aq) + 2 OH^- (aq) → $Cu(OH)_2$ (s)
(12) 2 Na^+ (aq), 2 NO_3^- (aq)
(13) 1, 1, 1
(14) nonredox, decomposition
(15) endothermic
(16) sulfuric acid, copper (II) oxide
(17) copper (II) sulfate, water
(18) H_2SO_4 (aq) + CuO (s) → $CuSO_4$ (aq) + H_2O (l)
(19) nonredox, double replacement
(20) 1, 1, 1, 1
(21) redox, single replacement
(22) combination
(23) 0.48 g
(24) limiting
(25) 88%
(26) This cycle of reactions requires excellent lab technique. Some of the copper may not have completely reacted in the first reaction. Some of the copper hydroxide may not have been transferred to the evaporating dish. Some of the copper may not have been scraped off the zinc wire. Also, side reactions may have occurred.

SELF-TEST QUESTIONS

Multiple Choice

1. Which of the following is a reactant in the reaction: $2H^+ + CaCO_3 \rightarrow H_2O + Ca^{2+} + CO_2$

   a. H_2O b. H^+ c. Ca^{2+} d. CO_2

2. What is the coefficient to the left of H_2 when the following equation is balanced?

   $$Na + H_2O \rightarrow NaOH + H_2$$

   a. 1 b. 2 c. 3 d. 4

3. A decomposition reaction can also be classified as:

   a. a combination reaction.
   b. a single replacement reaction.
   c. a double replacement reaction.
   d. a redox or nonredox reaction.

4. The oxidation number of a monatomic ion is always:

   a. +1. b. 0. c. -1. d. equal to the charge on the ion.

5. In a redox reaction, the reducing agent:
   a. is reduced.
   b. is oxidized.
   c. gains electrons.
   d. contains an element whose oxidation number decreases.

6. Identify the spectator ion in the following reaction: $Cl_2 + 2\,K^+ + 2\,Br^- \rightarrow 2\,K^+ + 2\,Cl^- + Br_2$
   a. Br^-
   b. Cl_2
   c. K^+
   d. Cl^-

7. Which of the following statements is consistent with the balanced equation: $2\,SO_2 + O_2 \rightarrow 2\,SO_3$
   a. one mol SO_2 reacts with one mol O_2
   b. two mol SO_2 produces two mol SO_3
   c. 64.07 g SO_2 reacts with 32.0 g O_2
   d. 32.0 g O_2 produces 80.07 g SO_3

8. According to the reaction $N_2 + 3\,H_2 \rightarrow 2\,NH_3$, how many grams of H_2 are needed to produce 4.0 moles of NH_3?
   a. 3.0
   b. 6.0
   c. 9.0
   d. 12.0

9. For the reaction $2\,H_2 + O_2 \rightarrow 2\,H_2O$, how many moles of H_2O could be obtained by reacting 64.0 grams of O_2 and 8.0 grams of H_2?
   a. 2.0
   b. 4.0
   c. 8.0
   d. 72

10. Nitrogen and hydrogen react as follows to form ammonia: $N_2 + 3\,H_2 \rightarrow 2\,NH_3$. In a reaction mixture consisting of 1.50 g H_2 and 6.00 g N_2, it is true that:
    a. H_2 is the limiting reactant.
    b. N_2 is the limiting reactant.
    c. N_2 is present in excess.
    d. an exact reacting ratio is present.

True-False

11. Oxidation numbers never change in combination reactions.
12. The sum of the oxidation numbers of elements in a compound always equals zero.
13. The loss of hydrogen corresponds to an oxidation process.
14. The number of reactants is greater than the number of products in all decomposition reactions.
15. A reaction that releases heat is classified as endothermic.
16. A balanced equation represents a statement of the law of conservation of matter.
17. The percentage yield of a reaction cannot exceed 100%, if the measured product is pure.

Matching

Match the oxidation number of the underlined element to the correct value given as a response.

18. $H_2\underline{P}O_4$
19. $\underline{Br}_2$
20. $\underline{C}O_2$
21. $\underline{Cs}_2O$
22. $\underline{Cr}_2O_3$
23. $\underline{S}O_3^{2-}$
24. $\underline{Ba}^{2+}$
25. $\underline{Na}$

a. 0
b. +1
c. +2
d. +3
e. +4
f. none of the above

Match each of the reactions below with the correct category from the responses.

26. $3\ Fe + 2\ O_2 \rightarrow Fe_3O_4$
27. $CuO + H_2 \rightarrow Cu + H_2O$
28. $CuCO_3 \rightarrow CuO + CO_2$
29. $KOH + HBr \rightarrow H_2O + KBr$
30. $2\ Ag_2O \rightarrow 4\ Ag + O_2$

a. decomposition
b. single replacement
c. double replacement
d. combination

Match each of the reactions below with the correct category from the responses. For redox reactions, also identify the reducing agent.

31. $3\ Fe + 2\ O_2 \rightarrow Fe_3O_4$
32. $CuO + H_2 \rightarrow Cu + H_2O$
33. $CuCO_3 \rightarrow CuO + CO_2$
34. $KOH + HBr \rightarrow H_2O + KBr$
35. $2\ Ag_2O \rightarrow 4\ Ag + O_2$

a. redox reaction
b. nonredox reaction

Chapter 6: The States of Matter

CHAPTER OUTLINE

6.1 Observed Properties of Matter
6.2 The Kinetic Molecular Theory of Matter
6.3 The Solid State
6.4 The Liquid State
6.5 The Gaseous State
6.6 The Gas Laws
6.7 Pressure, Temperature, and Volume Relationships
6.8 The Ideal Gas Law
6.9 Dalton's Law
6.10 Graham's Law
6.11 Changes in State
6.12 Evaporation and Vapor Pressure
6.13 Boiling and the Boiling Point
6.14 Sublimation and Melting
6.15 Energy and the States of Matter

LEARNING OBJECTIVES/ASSESSMENT

When you have completed your study of this chapter, you should be able to:

1. Do calculations based on the property of density. (Section 6.1; Exercise 6.2)
2. Demonstrate an understanding of the kinetic molecular theory of matter. (Section 6.2, 6.3, and 6.4; Exercise 6.8)
3. Use the kinetic molecular theory to explain and compare the properties of matter in different states. (Sections 6.5; Exercises 6.12 and 6.16)
4. Do calculations to convert pressure and temperature values into various units. (Section 6.6; Exercises 6.20 and 6.22)
5. Do calculations based on Boyle's law, Charles's law, and the combined gas law. (Section 6.7; Exercises 6.24, 6.32, and 6.34)
6. Do calculations based on the ideal gas law. (Section 6.8; Exercise 6.46)
7. Do calculations based on Dalton's law. (Section 6.9; Exercise 6.58)
8. Do calculations based on Graham's law. (Section 6.10; Exercise 6.60)
9. Classify changes of state as exothermic or endothermic. (Section 6.11; Exercise 6.64)
10. Demonstrate an understanding of the concepts of vapor pressure and evaporation. (Section 6.12; Exercise 6.68)
11. Demonstrate an understanding of the process of boiling and the concept of boiling point. (Section 6.13; Exercise 6.70)
12. Demonstrate an understanding of the processes of sublimation and melting. (Section 6.14; Exercise 6.74)
13. Do calculations based on energy changes that accompany heating, cooling, or changing the state of a substance. (Section 6.15; Exercises 6.76 and 6.78)

SOLUTIONS FOR THE END OF CHAPTER EXERCISES

OBSERVED PROPERTIES OF MATTER (SECTION 6.1)

☑6.2 a. Sea water ($d = 1.03$ g/mL)

$$125\text{ g}\left(\frac{1\text{ mL}}{1.03\text{ g}}\right) = 121\text{ mL}$$

b. Methyl alcohol ($d = 0.792$ g/mL)

$$125\text{ g}\left(\frac{1\text{ mL}}{0.792\text{ g}}\right) = 158\text{ mL}$$

c. Concentrated sulfuric acid ($d = 1.84$ g/mL)

$$125\text{ g}\left(\frac{1\text{ mL}}{1.84\text{ g}}\right) = 67.9\text{ mL}$$

6.4 The volume of water increases by 2 mL on heating as shown by the calculation below:

$$500\text{ mL}\left(\frac{1.00\text{ g}}{1\text{ mL}}\right) = 500\text{ g};\ 500\text{ g}\left(\frac{1\text{ mL}}{0.996\text{ g}}\right) = 502\text{ mL};\ 502\text{ mL} - 500\text{ mL} = 2\text{ mL}$$

6.6 a. The density of the gas will decrease as the gas is heated. The mass of the gas will not change, but the volume of the gas will increase. Volume and density have an inverse relationship; therefore, the density will decrease.

b.

$$\text{in 1 step: density}_2 = \frac{\text{volume}_1 \times \text{density}_1}{\text{volume}_2} \Rightarrow \frac{1.50\text{ L}\left(\frac{1.98\text{ g}}{1\text{ L}}\right)}{1.78\text{ L}} = \frac{2.97\text{ g}}{1.78\text{ L}} = 1.67\,\tfrac{\text{g}}{\text{L}}$$

$$\text{or in 2 steps: mass} = \text{volume}_1 \times \text{density}_1 \Rightarrow 1.50\text{ L}\left(\frac{1.98\text{ g}}{1\text{ L}}\right) = 2.97\text{ g}$$

$$\text{density}_2 = \frac{\text{mass}}{\text{volume}_2} \Rightarrow \frac{2.97\text{ g}}{1.78\text{ L}} = 1.67\,\tfrac{\text{g}}{\text{L}}$$

The mass of the sample is the same before and after heating.

THE KINETIC MOLECULAR THEORY OF MATTER (SECTION 6.2)

☑6.8 As the ball is moving upward, it has kinetic energy that is transferred into potential energy as the ball slows and reaches the highest point in its path. At its highest point, the ball stops and all of the energy is potential energy. As the ball falls from its highest point, the potential energy is transferred into kinetic energy. When the ball is caught, the kinetic energy is transferred into the hand of the catcher.

6.10 The kinetic energy of helium and methane molecules is the same as shown in the calculations below. The kinetic energy of all gases is the same at a given temperature.

$$KE = \tfrac{1}{2}mv^2$$

$$KE_{He} = \tfrac{1}{2}(4.00\text{ u})\left(1.26\times10^5\,\tfrac{\text{cm}}{\text{s}}\right)^2 = 3.1752\times10^{10}\,\tfrac{\text{u}\cdot\text{cm}^2}{\text{s}^2} = 3.18\times10^{10}\,\tfrac{\text{u}\cdot\text{cm}^2}{\text{s}^2}$$

$$KE_{CH_4} = \tfrac{1}{2}(16.0\text{ u})\left(6.30\times10^4\,\tfrac{\text{cm}}{\text{s}}\right)^2 = 3.1752\times10^{10}\,\tfrac{\text{u}\cdot\text{cm}^2}{\text{s}^2} = 3.18\times10^{10}\,\tfrac{\text{u}\cdot\text{cm}^2}{\text{s}^2}$$

THE SOLID, LIQUID, AND GASEOUS STATES (SECTION 6.3-6.5)

☑6.12 a. The molecules of a liquid possess kinetic and potential energy. The kinetic energy is not great enough to overcome the attractive forces between the molecules; therefore, the molecules are able to flow together into the shape of the container, but the volume of the liquid remains constant.

b. Solid and liquids are composed of molecules with considerable attractive forces between the molecules. These attractive forces bring the molecules close together and cause them to be difficult to compress further.

c. Gases are composed of molecules with high kinetic energy and low potential energy. The molecules are small compared to the amount of space occupied by the gas. The molecules strike each other and the walls of the container; however, all of the collisions are elastic and no net energy is lost from the system. Since the molecules are in constant motion, the same average number of gas molecules will strike the walls of the container at any given time and the elastic collisions result in uniform pressure on the walls of its container.

6.14 Kinetic energy is the energy of motion. The particles in a solid have the lowest kinetic energy because the particles are moving the least in this phase of matter. The particles in a liquid have higher kinetic energy than a solid, as well as lower kinetic energy than a gas. The particles in a gas have the highest kinetic energy because the particles are moving the most in this phase of matter.

The potential energy of the three phases of matter is easiest to compare during a phase transition. When a solid melts into a liquid, the temperature does not change, even though energy is added to the solid; therefore, both the liquid and the solid particles have the same average kinetic energy at the melting point, and the liquid must have higher potential energy than the solid. Similarly, when a liquid evaporates into a gas, the temperature does not change, even though energy is added to the liquid; therefore, both the liquid and the gas particles have the same average kinetic energy at the melting point, and the gas must have higher potential energy than the liquid. (See Section 6.15 and Figure 6.14 for addition explanation.)

☑6.16
a. Temperature changes influence the volume of this state substantially. — gaseous
b. In this state, constituent particles are less free to move about than in other states. — solid
c. Pressure changes influence the volume of this state more than that of the other two states. — gaseous
d. This state is characterized by an indefinite shape and a low density. — gaseous

THE GAS LAWS (SECTION 6.6)

6.18
a. atm
$$28.6\text{ in. Hg}\left(\frac{1\text{ atm}}{29.9\text{ in. Hg}}\right)=0.957\text{ atm}$$
b. torr
$$28.6\text{ in. Hg}\left(\frac{1\text{ atm}}{29.9\text{ in. Hg}}\right)\left(\frac{760\text{ torr}}{1\text{ atm}}\right)=727\text{ torr}$$
c. psi
$$28.6\text{ in. Hg}\left(\frac{1\text{ atm}}{29.9\text{ in. Hg}}\right)\left(\frac{14.7\text{ psi}}{1\text{ atm}}\right)=14.1\text{ psi}$$
d. bars
$$28.6\text{ in. Hg}\left(\frac{1.01\text{ bars}}{29.9\text{ in. Hg}}\right)=0.966\text{ bars}$$

☑6.20
a. atm $$210\text{ psi}\left(\frac{1\text{ atm}}{14.7\text{ psi}}\right)=14.3\text{ atm}$$
b. bars $$210\text{ psi}\left(\frac{1\text{ atm}}{14.7\text{ psi}}\right)\left(\frac{1.01\text{ bars}}{1\text{ atm}}\right)=14.4\text{ bars}$$
c. mm Hg $$210\text{ psi}\left(\frac{1\text{ atm}}{14.7\text{ psi}}\right)\left(\frac{760\text{ mm Hg}}{1\text{ atm}}\right)=1.09\times10^{4}\text{ mm Hg}$$
d. in. Hg $$210\text{ psi}\left(\frac{1\text{ atm}}{14.7\text{ psi}}\right)\left(\frac{29.9\text{ in. Hg}}{1\text{ atm}}\right)=427\text{ in. Hg}$$

☑6.22
a. The melting point of potassium metal, 63.7°C, to kelvins. — $63.7°\text{C}+273=337\text{ K}$
b. The freezing point of liquid hydrogen, 14.1 K, to degrees Celsius. — $14.1\text{ K}-273=-259°\text{C}$
c. The boiling point of liquid helium, -268.9°C to kelvins. — $-268.9°\text{C}+273=4\text{ K}$

PRESSURE, TEMPERATURE, AND VOLUME RELATIONSHIPS (SECTION 6.7)

☑6.24 Gas A: $\dfrac{(1.50\text{ atm})(2.00\text{ L})}{300\text{ K}} = \dfrac{(P_f)(3.00\text{ L})}{450\text{ K}}$; Gas B: $\dfrac{(2.35\text{ atm})(1.97\text{ L})}{293\text{ K}} = \dfrac{(1.09\text{ atm})(V_f)}{310\text{ K}}$;

$P_f = 1.50\text{ atm}$ $\quad$ $V_f = 4.49\text{ L}$

Gas C: $\dfrac{(9.86\text{ atm})(11.7\text{ L})}{500\text{ K}} = \dfrac{(5.14\text{ atm})(9.90\text{ L})}{T_f}$

$T_f = 221\text{ K}$

6.26 $\dfrac{(650\text{ torr})(200\text{ mL})}{45.0°\text{C}+273} = \dfrac{(760\text{ torr})(V_f)}{0.0°\text{C}+273} \Rightarrow V_f = 147\text{ mL}$

6.28 $\dfrac{(1.00\text{ atm})(2.50\text{ L})}{0.00°\text{C}+273} = \dfrac{(P_f)(1.00\text{ L})}{25.0°\text{C}+273} \Rightarrow P_f = 2.73\text{ atm}$

6.30 $(1.00\text{ atm})\left(\dfrac{14.7\text{ psi}}{1\text{ atm}}\right)(V_i) = (32.0\text{ psi})(12.9\text{ L}) \Rightarrow V_i = 28.1\text{ L}$

☑6.32 $(650\text{ torr})(800\text{ mL}) = (760\text{ torr})(V_f) \Rightarrow V_f = 684\text{ mL}$

☑6.34 $\dfrac{3.8\text{ L}}{20°\text{C}+273} = \dfrac{V_f}{85°\text{C}+273} \Rightarrow V_f = 4.6\text{ L}$

6.36 $\dfrac{V_i}{120°\text{C}+273} = \dfrac{2.0\text{ L}}{40°\text{C}+273} \Rightarrow V_i = 2.5\text{ L}$

6.38 $(1.00\text{ atm})(2500\text{ L}) = (P_f)(25.0\text{ L}) \Rightarrow P_f = 100\text{ atm}$

6.40 $\dfrac{(0.98\text{ atm})(8000\text{ ft}^3)}{23°\text{C}+273} = \dfrac{(400\text{ torr})\left(\frac{1\text{ atm}}{760\text{ torr}}\right)(V_f)}{5.3°\text{C}+273} \Rightarrow V_f = 1.4\times10^4\text{ ft}^3$

6.42 $(2.50\text{ atm})(200\text{ mL}) = (60.0\text{ kPa})\left(\dfrac{1\text{ atm}}{101\text{ kPa}}\right)(V_f) \Rightarrow V_f = 842\text{ mL}$

6.44 $(760\text{ torr})(2.00\text{ L}) = (4.00\text{ atm})\left(\dfrac{760\text{ torr}}{1\text{ atm}}\right)(V_f) \Rightarrow V_f = 0.500\text{ L}$

$\text{density} = \dfrac{\text{mass}}{\text{volume}} \Rightarrow \dfrac{2.50\text{ g}}{0.500\text{ L}} = 5.00\,\frac{\text{g}}{\text{L}}$

THE IDEAL GAS LAW (SECTION 6.8)

☑6.46 a. $PV = nRT \Rightarrow (735\text{ torr})\left(\frac{1\text{ atm}}{760\text{ torr}}\right)(400\text{ mL})\left(\frac{1\text{ L}}{1000\text{ mL}}\right) = n\left(0.0821\,\frac{\text{L}\cdot\text{atm}}{\text{mol}\cdot\text{K}}\right)((90.0+273)\text{K})$

$n = \dfrac{(735\text{ torr})\left(\frac{1\text{ atm}}{760\text{ torr}}\right)(400\text{ mL})\left(\frac{1\text{ L}}{1000\text{ mL}}\right)}{\left(0.0821\,\frac{\text{L}\cdot\text{atm}}{\text{mol}\cdot\text{K}}\right)((90.0+273)\text{K})} = 0.0130\text{ moles}$

b. $PV = nRT \Rightarrow P(2.60\text{ L}) = (0.750\text{ mol})(0.0821\,\frac{\text{L}\cdot\text{atm}}{\text{mol}\cdot\text{K}})((50.0+273)\text{K})$

$$P = \frac{(0.750\text{ mol})(0.0821\,\frac{\text{L}\cdot\text{atm}}{\text{mol}\cdot\text{K}})((50.0+273)\text{K})}{(2.60\text{ L})} = 7.65\text{ atm}$$

c. $PV = nRT \Rightarrow (4.32\text{ atm})V = (1.50\text{ mol})(0.0821\,\frac{\text{L}\cdot\text{atm}}{\text{mol}\cdot\text{K}})((20.0+273)\text{K})$

$$V = \frac{(1.50\text{ mol})(0.0821\,\frac{\text{L}\cdot\text{atm}}{\text{mol}\cdot\text{K}})((20.0+273)\text{K})}{(4.32\text{ atm})} = 8.35\text{ L}$$

6.48 $PV = nRT \Rightarrow P(1.25\text{ L}) = (10.0\text{ g})(\frac{1\text{ mole}}{64.07\text{ g SO}_2})(0.0821\,\frac{\text{L}\cdot\text{atm}}{\text{mol}\cdot\text{K}})((38.0+273)\text{K})$

$$P = \frac{(10.0\text{ g})(\frac{1\text{ mole}}{64.07\text{ g SO}_2})(0.0821\,\frac{\text{L}\cdot\text{atm}}{\text{mol}\cdot\text{K}})((38.0+273)\text{K})}{(1.25\text{ L})} = 3.19\text{ atm}$$

6.50 $PV = nRT \Rightarrow (400\text{ psi})(\frac{1\text{ atm}}{14.7\text{ psi}})(1.50\text{ L}) = n(0.0821\,\frac{\text{L}\cdot\text{atm}}{\text{mol}\cdot\text{K}})((30.0+273)\text{K})$

$$n = \frac{(400\text{ psi})(\frac{1\text{ atm}}{14.7\text{ psi}})(1.50\text{ L})}{(0.0821\,\frac{\text{L}\cdot\text{atm}}{\text{mol}\cdot\text{K}})((30.0+273)\text{K})} = 1.6407715991\text{ moles}$$

$$\text{mass} = 1.6407715991\text{ moles}\left(\frac{16.04\text{ g CH}_4}{1\text{ mole CH}_4}\right) = 26.3\text{ g}$$

6.52 $PV = nRT \Rightarrow (1\text{ atm})(60\text{ L}) = n(0.0821\,\frac{\text{L}\cdot\text{atm}}{\text{mol}\cdot\text{K}})(273\text{ K})$

$$n = \frac{(1\text{ atm})(60\text{ L})}{(0.0821\,\frac{\text{L}\cdot\text{atm}}{\text{mol}\cdot\text{K}})(273\text{ K})} = 2.7\text{ moles}$$

6.54 $PV = nRT \Rightarrow (1\text{ atm})(3.96\text{ L}) = n(0.0821\,\frac{\text{L}\cdot\text{atm}}{\text{mol}\cdot\text{K}})(273\text{ K})$

$$n = \frac{(1\text{ atm})(3.96\text{ L})}{(0.0821\,\frac{\text{L}\cdot\text{atm}}{\text{mol}\cdot\text{K}})(273\text{ K})} = 0.1766808\text{ moles} \Rightarrow MW = \frac{8.12\text{ g}}{0.1766808\text{ moles}} = 46.0\,\frac{\text{g}}{\text{mol}}$$

6.56 The gas is CO_2 (MW = 44.0 g/mole) according to the calculation below:

$$PV = nRT$$

$$(640\text{ torr})(\frac{1\text{ atm}}{760\text{ torr}})(114.0\text{ mL})(\frac{1\text{L}}{1000\text{ mL}}) = n(0.0821\,\frac{\text{L}\cdot\text{atm}}{\text{mol}\cdot\text{K}})((20+273)\text{ K})$$

$$n = \frac{(640\text{ torr})(\frac{1\text{ atm}}{760\text{ torr}})(114.0\text{ mL})(\frac{1\text{L}}{1000\text{ mL}})}{(0.0821\,\frac{\text{L}\cdot\text{atm}}{\text{mol}\cdot\text{K}})((20+273)\text{ K})} = 0.00399\text{ moles}$$

$$MW = \frac{0.176\text{ g}}{0.00399\text{ moles}} = 44.1\,\frac{\text{g}}{\text{mol}}$$

DALTON'S LAW (SECTION 6.9)

☑6.58 $P_{N_2} + P_{O_2} + P_{CO_2} = P_T \Rightarrow 810\text{ torr} + 920\text{ torr} + P_{CO_2} = 2100\text{ torr} \Rightarrow P_{CO_2} = 370\text{ torr}$

GRAHAM'S LAW (SECTION 6.10)

☑6.60 The molecular mass of oxygen is 16 times the molecular mass of hydrogen according to the following calculation:

$$\frac{\text{effusion rate A}}{\text{effusion rate B}} = \sqrt{\frac{\text{molecular mass of B}}{\text{molecular mass of A}}}$$

$$4 = \sqrt{\frac{MM_{O_2}}{MM_{H_2}}} \Rightarrow (4)^2 = \left(\sqrt{\frac{MM_{O_2}}{MM_{H_2}}}\right)^2 \Rightarrow 16 = \frac{MM_{O_2}}{MM_{H_2}}$$

The molecular mass of oxygen according to the periodic table is 32.00 u and the molecular mass of hydrogen according to the periodic table is 2.01 u. The molecular mass of oxygen is approximately 16 times greater than the molecular mass of hydrogen.

6.62 The helium balloon is the one that appeared to be going "flat." The helium gas was able to escape more easily than the nitrogen gas through the porous rubber surface of the balloon because helium has a smaller atomic mass than the molecular mass of nitrogen. Lighter particles move more rapidly than heavier particles at the same temperature.

CHANGES IN STATE (SECTION 6.11)

☑6.64	a.	Condensation	(g) → (l); remove heat	exothermic
	b.	Liquefaction	(g) → (l); remove heat	exothermic
	c.	Boiling	(l) → (g); add heat	endothermic

6.66 A change in state is the process by which a material gains enough energy or loses enough energy to change between two of the three phases of matter (solid, liquid, or gas).

EVAPORATION AND VAPOR PRESSURE (SECTION 6.12)

☑6.68 Methylene chloride is a volatile liquid. When it was sprayed in the mouth, the methylene chloride absorbed heat from the tissue and evaporated. The tissue became cold and was anesthetized.

BOILING AND THE BOILING POINT (SECTION 6.13)

☑6.70 Water and ethylene glycol differ in their boiling points. Water has a boiling point of 100°C, while ethylene glycol has a boiling point that is higher than 100°C. To determine the identity of the two boiling liquids, use a thermometer to measure the boiling point.

6.72 I would throw the potato into the campfire. The temperature of the burning wood will be higher than the temperature of the boiling water. Water boils at a lower temperature at higher elevations because the barometric pressure is lower at higher elevations. The temperature of combusting wood is not affected by elevation.

SUBLIMATION AND MELTING (SECTION 6.14)

☑6.74 To obtain pure solid iodine from a mixture of solid iodine and sand, heat the mixture until the iodine sublimes and provide a cold surface above the mixture so the iodine can be deposited as a solid. The sand will not sublime and the solid on the cold surface will be pure iodine.

ENERGY AND THE STATES OF MATTER (SECTION 6.15)

☑6.76 a. 50 g of aluminum from 25°C to 55°C

$$\text{heat} = (50\text{ g})\left(0.24\tfrac{\text{cal}}{\text{g °C}}\right)(55°\text{C} - 25°\text{C}) = 3.6\times10^2\text{ cal}$$

b. 2500 g of ethylene glycol from 80°C to 85°C

$$\text{heat} = (2500\text{ g})\left(0.57\tfrac{\text{cal}}{\text{g °C}}\right)(85°\text{C} - 80°\text{C}) = 7\times10^3\text{ cal}$$

c. 500 g of steam from 110°C to 120°C

$$\text{heat} = (500\ \text{g})\left(0.48\ \tfrac{\text{cal}}{\text{g}\ °\text{C}}\right)(120°\text{C} - 110°\text{C}) = 2.4\times10^3\ \text{cal}$$

☑6.78 a. $CaCl_2 \cdot 6H_2O$: melting point = 30.2°C, heat of fusion = 40.7 cal/g

$$\text{heat} = (1000\ \text{kg})\left(\frac{1000\ \text{g}}{1\ \text{kg}}\right)\left(\frac{40.7\ \text{cal}}{\text{g}}\right) = 4.07\times10^7\ \text{cal}$$

b. $LiNO_3 \cdot 3H_2O$: melting point = 29.9°C, heat of fusion = 70.7 cal/g

$$\text{heat} = (1000\ \text{kg})\left(\frac{1000\ \text{g}}{1\ \text{kg}}\right)\left(\frac{70.7\ \text{cal}}{\text{g}}\right) = 7.07\times10^7\ \text{cal}$$

c. $Na_2SO_4 \cdot 10H_2O$: melting point = 32.4°C, heat of fusion = 57.1 cal/g

$$\text{heat} = (1000\ \text{kg})\left(\frac{1000\ \text{g}}{1\ \text{kg}}\right)\left(\frac{57.1\ \text{cal}}{\text{g}}\right) = 5.71\times10^7\ \text{cal}$$

6.80

$$\text{heat} = (2.00\ \text{kg})\left(\frac{1000\ \text{g}}{1\ \text{kg}}\right)\left(\frac{38.6\ \text{cal}}{\text{g}}\right) = 7.72\times10^4\ \text{cal}$$

ADDITIONAL EXERCISES

6.82

$$\text{volume of 1 mole of gas at STP}: 22.4\ \text{L}\left(\frac{1000\ \text{mL}}{1\ \text{L}}\right) = 2.24\times10^4\ \text{mL}$$

mass of 1 mole of argon : 39.9 g

$$\text{density} = \frac{\text{mass}}{\text{volume}} = \frac{39.9\ \text{g}}{2.24\times10^4\ \text{mL}} = 0.00178\ \tfrac{\text{g}}{\text{mL}} \text{ or } 1.78\times10^{-3}\ \tfrac{\text{g}}{\text{mL}}$$

6.84 $2\ H_2\ (g) + O_2\ (g) \rightarrow 2\ H_2O\ (l)$

$$2.31\ \text{L}\ H_2\left(\frac{1\ \text{L}\ O_2}{2\ \text{L}\ H_2}\right) = 1.16\ \text{L}\ O_2$$

This reaction is classified as both a combination reaction and a redox reaction.

6.86 On the top of a mountain the barometric pressure is lower than at the base of the mountain. The decrease in air pressure results in a decrease in the concentration of oxygen at the top of a mountain; therefore, there is less oxygen in each lungful of air at the top of the mountain.

ALLIED HEALTH EXAM CONNECTION

6.88 Gases do not (a) have definite volume and shape.

6.90 Molecules are most likely to move freely in the (c) gas phase of matter.

6.92 In the kinetic molecular theory of gases, (d) when the temperature increases, fewer molecules will move at the new average speed.

6.94 The differentiating factor between potential and kinetic energy is (d) activity—in motion or in storage.

6.96 Evaporation can best be described as (a) the process in which molecules may have enough energy to leave the liquid phase and escape into the gaseous phase.

6.98 When solid iodine becomes gaseous iodine with no apparent liquid phase in between, the process is called (c) sublimation.

6.100 When one liquid evaporates much faster than another liquid, the first liquid is said to be more (a) volatile.

6.102 (c) Gay-Lussac's law predicts that if the temperature (in Kelvin) doubles, the pressure will also double.

6.104 The answer is (a) 3.2 x 10^4 J, as shown in the calculation below:

$$\left(4.2\tfrac{J}{g\cdot K}\right)(100\ g)(100°C-25°C)=3.2\times10^4\ J$$

6.106 If the temperature and pressure are held constant for a sample of gas and the number of moles is doubled, then the volume will (a) double.

6.108 The volume of 3.0 moles of krypton gas at STP is (d) 67.2 L.

$$3.0\ \text{moles}\left(\frac{22.4\ L}{1\ \text{mole}}\right)=67.2\ L$$

6.110 One liter of (a) C_2H_2 (26.04 g/mole) weighs 1.16 g under standard conditions.

$$\left(1.16\frac{g}{L}\right)\left(\frac{22.4\ L}{1\ \text{mole}}\right)=25.984\frac{g}{\text{mole}}$$

6.112 The pressure will be (d) 669 mm Hg.

$$(730\ \text{mmHg})(875\ \text{mL})=P_2(955\ \text{mL})\Rightarrow P_2=669\ \text{mmHg}$$

6.114 At sea level, water boils at (d) 373 K.

6.116 When a liquid is at its boiling point, the vapor pressure of the liquid (b) is equal to the external pressure on the liquid.

CHEMISTRY FOR THOUGHT

6.118 Gas density decreases as the temperature increases because the volume of the gas increases. A hot air balloon works because the air in the balloon is heated and less dense than the surrounding air; therefore, the balloon rises because less dense materials rise above more dense materials.

6.120 $$\frac{V_1}{T_1}=\frac{V_2}{T_2}\Rightarrow\frac{4.0\ L}{25°C+273}=\frac{V_2}{77\ K}\Rightarrow V_2=1.0\ L$$

The pressure in the hot-air balloon remains constant because the balloon is not air tight. When the air is heated, the extra air molecules can escape. When the air is cooled, extra air molecules can be pulled into the balloon.

6.122 Helium diffuses more rapidly through rubber than plastic. The distances between the molecules in rubber must be larger than the distances between the molecules in plastic because the helium atoms are able to diffuse through the rubber faster than through the plastic as seen in Figure 6.9.

6.124 The heat from the burner does not increase the temperature of the water-containing cup to the ignition temperature because the cup is the same temperature as the water. The energy from the burner must first heat the water to the boiling point and completely boil the water to dryness before the cup reaches a temperature above the boiling point of water.

If a sample of water were heated to the boiling point in a glass beaker using a single burner and then a second burner were added, the temperature of the boiling water would stay the same. The boiling point of water is not dependent on the amount of energy added to the water. The time required to reach the boiling point of water may decrease when using two burners rather than just one.

ADDITIONAL ACTIVITIES

Section 6.1 Review:

(1) Can the density of 3 g of gold be greater than the density of 150 g of water? Explain.
(2) Compare and contrast the shape and compressibility of gold and water at room temperature.
(3) Compare and contrast the thermal expansion of gold and water.
(4) Describe the density, shape, compressibility, and thermal expansion of helium gas relative to water and gold.

Section 6.2 Review:

(1) Are the ions that occupy lattice sites in a sodium chloride crystal lattice in motion? If so, how does the movement compare with the movement of water molecules in a glass of water and the movement of helium atoms inside a balloon?
(2) Relate the movement within the three states of matter to forms of dancing.
(3) Which phase of matter has the strongest cohesive forces?
(4) Which phase of matter has the strongest disruptive forces?

Section 6.3 Review:

(1) Refer to Figure 6.3 or draw your own representation of the solid state. Make sure that the picture reflects the molecular nature of matter.
(2) Using the picture from (1), determine why solids have a high density, definite shape, small compressibility, and very small thermal expansion.

Section 6.4 Review:

(1) Refer to Figure 6.3 or draw your own representation of the liquid state. Make sure that the picture reflects the molecular nature of matter.
(2) Using the picture from (1), determine why liquids have a high density, indefinite shape, small compressibility, and small thermal expansion.

Section 6.5 Review:

(1) Refer to Figure 6.3 or draw your own representation of the gaseous state. Make sure that the picture reflects the molecular nature of matter.
(2) Using the picture from (1), determine why gases have a low density, indefinite shape, large compressibility, and moderate thermal expansion.

Section 6.6 Review:

(1) If a barometer were made with water instead of mercury, would the glass tube need to be taller or shorter than for a mercury barometer? Mercury is more dense than water.
(2) Identify which unit of the following pairs is larger: atm or torr, torr or mmHg, mmHg or atm.
(3) If the temperature in degrees Celsius doubles from 20°C to 40°C, does the Kelvin temperature also double?
(4) If the pressure in atmospheres doubles from 1 atm to 2 atm, does the pressure in mmHg also double?
(5) If the volume in liters doubles from 0.500 L to 1.00 L, does the volume in milliliters also double?
(6) What are the differences between the conversions in (3), (4), and (5)?

Section 6.7 Review:

(1) If the temperature is held constant and the pressure on a helium balloon is increased, what will happen to the volume of the balloon?
(2) If the temperature is held constant as a valve is opened on tank of nitrogen, what will happen to the pressure of the gas as it leaves the tank and enters the room?
(3) What will happen to the volume of a helium balloon that is filled inside an air conditioned building and then placed in a hot delivery car? Assume the pressure in the building and the car is the same.
(4) Write an equation like 6.8 in the textbook for Boyle's law and for Charles's law.
(5) Do any conversions need to be made as long as the same units are used for both the "before" and "after" sides of the combined gas law? Explain.

Section 6.8 Review:

(1) Rearrange the ideal gas law to isolate each of the variables (P, V, n, T). You will need to write four separate equations, one for each variable.
(2) Could you calculate the molecular weight of a gas if you knew the mass, pressure, volume, and temperature of the sample? Explain using Equations 6.9 and 6.10.
(3) Why would interparticle forces cause gases to deviate from ideal behavior?

Section 6.9 Review:

(1) Rearrange the ideal gas law to isolate the ratio of pressure to number of moles.
(2) What happens to the pressure as the number of moles increases? Assume the volume and temperature of the container remain constant.
(3) If the all of the gases in a mixture are assumed to be ideal gases, what will have more effect on the pressure: the identities of the gases or the total number of moles of gas?

Section 6.10 Review:

(1) Label the following pictures as examples of effusion or diffusion.

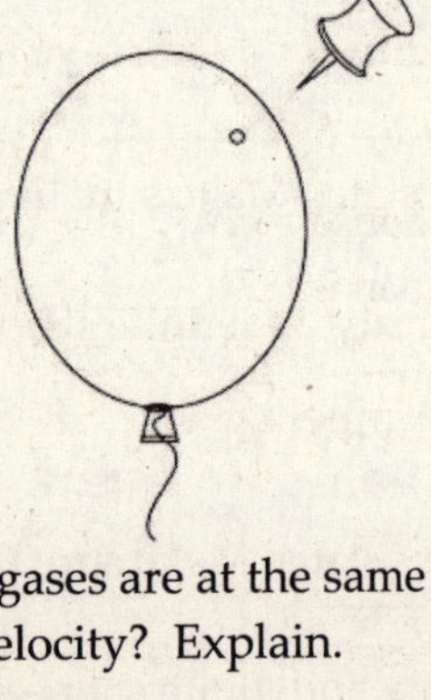

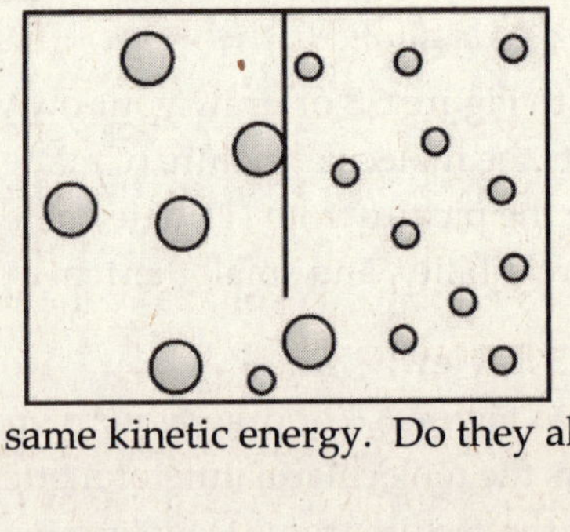

(2) When two gases are at the same temperature, they have the same kinetic energy. Do they also have the same velocity? Explain.

Section 6.11 Review:

(1) Redraw Figure 6.10 to show the relative kinetic energy levels associated with the different states of matter.
(2) Which direction do the arrows point in the endothermic picture from (1)?
(3) Which direction do the arrows point in the exothermic picture from (1)?

Section 6.12 Review:

(1) Draw a graph to indicate the rates of evaporation and condensation in a closed container over time.
(2) What is the relationship between interparticle forces and vapor pressure?

Section 6.13 Review:

(1) What is the relationship between interparticle forces and boiling point?
(2) Water boils at 91.4°C in La Paz, Bolivia, and at 100.0°C in San Francisco, CA. Approximately how much longer will it take for food to cook in La Paz, Bolivia?

Section 6.14 Review:

(1) What are the differences between melting, sublimation, and decomposition?
(2) How are vapor pressures of solids and interparticle forces related?
(3) Which would you expect to have a higher vapor pressure: an ionic solid or a molecular solid? Explain.

Section 6.15 Review:

(1) Figure 6.14 contains two plateaus and three angled sections. Briefly describe the process that occurs in each section of the curve.
(2) Why does the unit specific heat include a temperature term, but the units for the heat of fusion and heat of vaporization do not?

Tying It All Together with a Laboratory Application:

A deflated balloon is placed over the mouth of an Erlenmeyer flask that contains a 1.578 g ice cube at -10°C. The density of ice at -10°C is 0.99815 g/cm^3; therefore the volume of the ice is (1) _______. The shape of the ice (2) _______ (is, is not) affected by the Erlenmeyer flask. The specific heat of ice is 2.1 $\frac{J}{g \cdot °C}$, so as the ice cube is gently warmed to 0°C, the ice absorbs (3) _______ J. The density of ice at 0°C is 0.99987 g/cm^3. Solids have a (4) _______ thermal expansion.

The heat of (5) _______ for ice is 330 J/g. The ice will (6) _______ into water if (7) _______ J are added at 0°C. During this transition (called a (8) _______ (state of matter or phase change)), the temperature of the water (9) _______. The vapor pressure of the water (10) _______ the vapor pressure of the ice. The water is slowly heated to 100°C. Water has a specific heat of 4.184 $\frac{J}{g \cdot °C}$. The water absorbs (11) _______ J as the temperature increases. The heat of (12) _______ for water is 2.3 kJ/g. In order to create steam, (13) _______ kJ is added to the water. This transition is called (14) _______ or _______.

The steam has a(n) (15) _______ shape and a(n) (16) _______ volume. Consequently, the balloon starts to fill as the water becomes steam. The pressure inside the flask and balloon is (17) _______ the pressure in the laboratory. The pressure in the laboratory is 0.9845 atm or (18) _______ mmHg. Once all of the water is converted to steam, the volume of the steam will (19) _______ if the temperature is increased because the pressure will remain (20) _______.

If the experimental apparatus is cooled, the steam will (21) _______ or _______ into water. Further cooling would return the water to ice by (22) _______ or _______ it. These phase transitions are (23) _______ because heat is removed.

At temperatures below STP, H_2O is in the (24) _______ state of matter.

The state of matter with the highest kinetic energy is (25) _______, regardless of the substance. (26) If this experiment was repeated with the same number of moles of dry ice (solid carbon dioxide), would the same phase changes occur? (27) Will the values for density and other quantitative values given for water be the same for carbon dioxide? (Hint: Carbon dioxide is a gas at 1 atm and 195 K or (28) _______ °C.)

If the balloons filled with steam and carbon dioxide were both punctured with the same size pin hole, the (29) _______ would effuse more quickly because it has a (30) _______ molecular mass. (31) Is all of the pressure inside the balloons from the steam and carbon dioxide?

SOLUTIONS FOR THE ADDITIONAL ACTIVITIES

Section 6.1 Review:

(1) Yes, the density of gold is greater than the density of water because the densities of gold and water are constant regardless of their sample size. Density is an intensive physical property.

(2) Gold is a solid at room temperature. It will maintain its shape, regardless of the container it occupies. Water, on the other hand, is a liquid at room temperature and it will take the shape of the container it occupies. Neither gold nor water is very compressible. Water should be more compressible than gold because water is a liquid and liquids are generally more compressible than solids.

(3) The thermal expansion of solids and liquids is small. Neither gold nor water will expand much on heating.

(4) Helium gas has a much lower density than water or gold. Like water, helium gas will take the shape of its container. Helium gas can be highly compressed unlike gold or water. Helium gas will noticeably expand on heating, unlike gold or water.

Section 6.2 Review:

(1) Yes, the ions in a sodium chloride crystal lattice are in motion. Their motion is restricted, but they are still in motion. The movement of water molecules in a glass of water is less restricted than the movement of the ions in the crystal lattice. The movement of helium atoms inside a balloon is much less restricted than the movement of the water molecules or the ions in the crystal.

(2) The movement of the ions in the sodium chloride resembles any type of dancing that involves people being close together and moving very little. People at a concert who are swaying to the music, but keep their feet planted on the floor are an example of the type of restricted movement observed in a crystal. The movement of water molecules in a glass of water resembles any type of dancing that involves people moving relatively quickly and in a group of dancers. Waltzing is similar to the type of movement associated with the water molecules in a glass of water because the molecules revolve as they slip past other molecules. The movement of helium atoms in a balloon resembles any type of dancing that involves people moving extremely fast and dancing alone. Breakdancing is an example of this type of movement.

(3) Solids have the greatest cohesive forces. The attractive forces associated with potential energy are greater than the disruptive forces of kinetic energy.

(4) Gases have the strongest disruptive forces. Gases experience few attractive forces associated with potential energy because of their high kinetic energy and constant motion.

Section 6.3 Review:

(1) 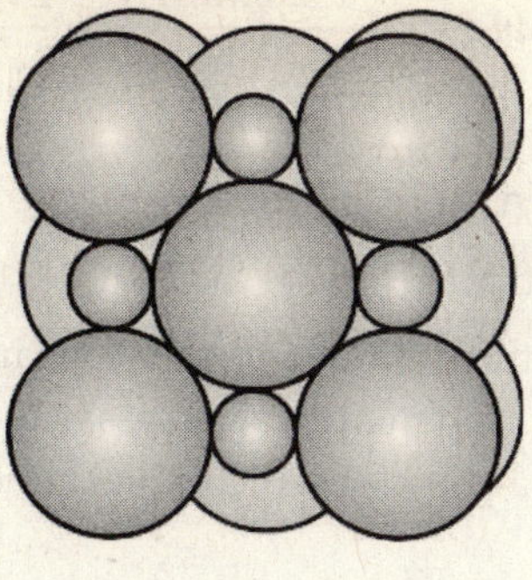

(2) Solids have a high density because the atoms, ions, or molecules they contain are close together. They have a definite shape because the particles they contain occupy specific lattice sites. The particles do not leave the lattice sites because of the interparticle forces discussed in Chapter 4. Solids have small compressibility because the particles are already extremely close together. Solids have very small thermal expansion because the particles cannot deviate very far from their original lattice sites without the solid breaking apart. The interparticles forces keep the particles close to their original lattice sites.

Section 6.4 Review:

(1) 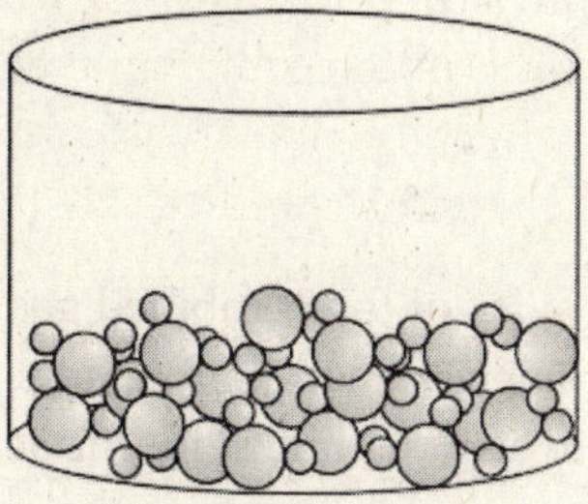

(2) Liquids have a high density because the atoms or molecules they contain are close together. They have an indefinite shape because the particles they contain are not connected in an extended structure. The particles do experience weak interparticle forces. Liquids have small compressibility because the particles are already close together. Liquids have small thermal expansion because the interparticle forces hold the particles together. If these cohesive forces are overcome, the particles would escape into the gas phase.

Section 6.5 Review:

(1) 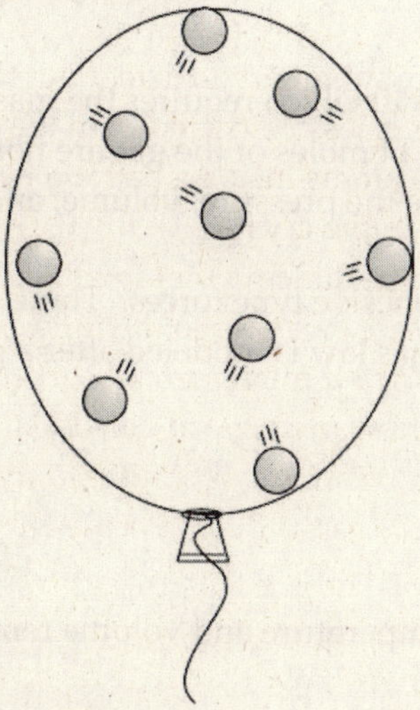

(2) Gases have a low density because the atoms or molecules they contain are far apart. They have an indefinite shape because the particles they contain are not connected in an extended structure and the particles experience only very weak interparticle forces. The disruptive forces dominate the gaseous state. Gases have large compressibility because the particles are far apart. Gases have moderate thermal expansion because particles are already far apart and are not held together by strong interparticle forces. Increasing the temperature of a gas provides additional kinetic energy which in turn increases the disruptive forces within the gas.

Section 6.6 Review:

(1) A water based barometer would need to be taller than a mercury based barometer. The weight of the water in the tube will be balanced by the weight of air pressing on the water pool. The volume of water needed to equal the mass of air pushing on the pool will be larger than the volume of mercury needed because water is less dense than mercury. Consequently, a taller tube is needed for the water than the mercury.

(2) atm or torr: atm is larger; torr or mmHg: they are equal; mmHg or atm: atm is larger;
1 atm = 760 torr = 760 mmHg

(3) No, the Kelvin temperature does not double. The Kelvin temperature only increases by 1.07 times.

$$\mathrm{K} = 20°\mathrm{C} + 273 = 293\ \mathrm{K} \qquad \mathrm{K} = 40°\mathrm{C} + 273 = 313\ \mathrm{K} \qquad \frac{313\ \mathrm{K}}{293\ \mathrm{K}} = \frac{1.07}{1}$$

(4) Yes, the pressure in mmHg also doubles.

$$1\text{ atm}\left(\frac{760\text{ mmHg}}{1\text{ atm}}\right)=760\text{ mmHg} \qquad 2\text{ atm}\left(\frac{760\text{ mmHg}}{1\text{ atm}}\right)=1520\text{ mmHg} \qquad \frac{1520\text{ mmHg}}{760\text{ mmHg}}=\frac{2}{1}$$

(5) Yes, the volume in milliliters also doubles.

$$0.500\text{ L}\left(\frac{1000\text{ mL}}{1\text{ L}}\right)=500\text{ mL} \qquad 1.00\text{ L}\left(\frac{1000\text{ mL}}{1\text{ L}}\right)=1000\text{ mL} \qquad \frac{1000\text{ mL}}{500\text{ mL}}=\frac{2}{1}$$

(6) The temperature conversion involves addition, but the pressure and the milliliter conversions involve multiplication. The ratio of two temperatures in degrees Celsius and Kelvin are not necessarily the same, but the ratio of two pressures or two volumes are the same regardless of which units are used.

Section 6.7 Review:

(1) The volume of the balloon will decrease as the pressure increases.

(2) The pressure will decrease as the volume increases.

(3) The volume of the balloon will increase as the temperature increases.

(4) Boyle's Law: $P_1V_1 = P_2V_2$, Charles's Law: $\frac{V_1}{T_1}=\frac{V_2}{T_2}$

(5) The temperature must be in Kelvins on both the "before" and "after" sides of the combined gas law. For pressure and volume, the units are not important as long as both sides have the same units. The conversion for temperature is additive, but the conversion for pressure and volume is multiplicative as shown in the Section 6.6 Review.

Section 6.8 Review:

(1) $P=\frac{nRT}{V}$, $V=\frac{nRT}{P}$, $n=\frac{PV}{RT}$, $T=\frac{PV}{nR}$

(2) Yes, the molecular weight of a gas can be calculated using Equation 6.10 which requires the mass of the gas as well as the number of moles of the gas. While, the number of moles of the gas are not known, Equation 6.9 can be used to calculate the moles of gas because the pressure, volume, and temperature of the sample are known.

(3) Interparticle forces between gas molecules will cause them to have cohesive type forces. These cohesive forces are not accounted for in the ideal gas law. The ideal gas law is modeled after a gas that is governed by disruptive forces.

Section 6.9 Review:

(1) $\frac{P}{n}=\frac{RT}{V}$

(2) The pressure will increase as the number of moles increases, if the temperature and volume remain constant.

(3) The total number of moles of gas will have a greater effect on the pressure because all of the gases are governed by the ideal gas law, if they are ideal gases. Their actual identity is not important.

Section 6.10 Review:

(1)  = effusion; 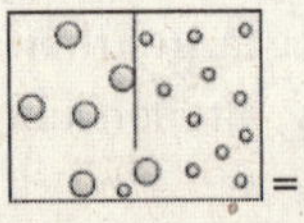 = diffusion

(2) Two gases that have the same kinetic energy will only have the same velocity if the mass of the gas molecules is the same. If the masses of the molecules are different, then the lighter gas will have a faster velocity than the heavier gas.

Section 6.11 Review:

(1) 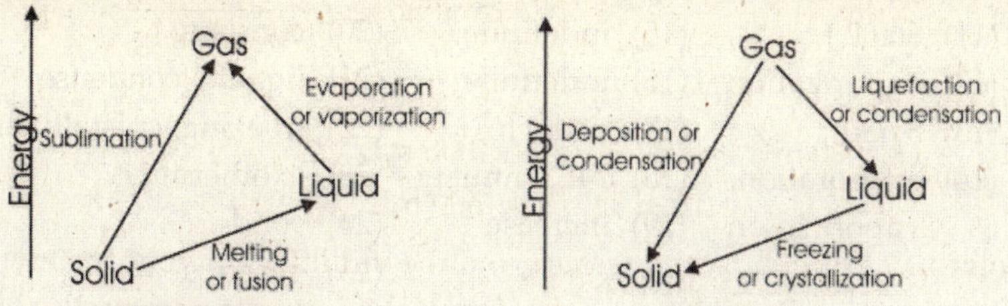

Endothermic **Exothermic**

(2) The arrows in the endothermic picture all point upwards because additional energy is required.
(3) The arrows in the exothermic picture all point downwards because energy is released.

Section 6.12 Review:

(1) 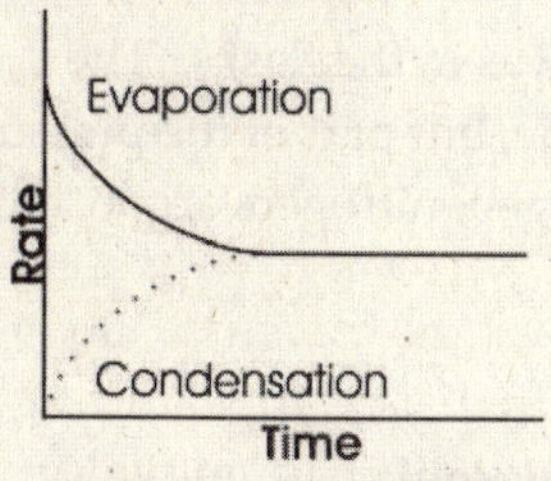

(2) Vapor pressure decreases as interparticle forces increase. Fewer molecules can escape from the liquid phase into the gas phase when they experience attraction for other liquid molecules.

Section 6.13 Review:

(1) As interparticle forces increase, the boiling point increases. Boiling occurs when the vapor pressure of the liquid is equal to atmospheric pressure. Vapor pressure decreases as interparticle forces increase; therefore, more energy is needed to counteract the cohesive nature of the interparticle forces and raise the vapor pressure to the atmospheric pressure.
(2) $(100.0°\text{C} - 91.4°\text{C})\left(\frac{2 \times \text{time}}{10°\text{C}}\right) = 1.72 \times \text{time}$; It takes approximately 1.72 times longer for food to cook in La Paz, Bolivia, than it does in San Francisco, California.

Section 6.14 Review:

(1) When heating a solid, it might turn into a liquid. This process is called melting and is a physical change.
When heating a solid, it might turn directly into a gas. This process is called sublimation and is a physical change.
When heating a solid, it might go through a chemical change to produce other compounds. This process is called decomposition.
(2) The vapor pressure of a solid is higher for solids with weak interparticle forces.
(3) A molecular solid will have a higher vapor pressure than an ionic solid because dispersion forces, dipolar forces, and hydrogen bonding are all weaker interparticle forces than ionic bonding.

Section 6.15 Review:

(1) Section AB = heating solid, Section BC = melting, Section CD = heating liquid, Section DE = boiling, Section EF = heating gas
(2) The specific heat is the amount of heat required to heat a gram of material by 1°C. The heat of fusion is the heat required per gram of material to melt a solid. The heat of vaporization is the heat required per gram of material to boil a liquid. Melting and boiling each occur at one temperature for a specific material. The material stays at the melting point or boiling point until the phase change has finished. Heating of a solid to its melting point, a liquid to its boiling point, or a gas occurs across a temperature range. Consequently, specific heat requires a temperature term in its unit.

Tying It All Together with a Laboratory Application:

(1) 1.581 cm^3
(2) is not
(3) 33 J
(4) small
(5) fusion
(6) melt
(7) 520 J
(8) phase change
(9) stays the same
(10) is larger than
(11) 660.2 J
(12) vaporization
(13) 3.6 kJ
(14) evaporation, vaporization
(15) indefinite
(16) indefinite
(17) equal to
(18) 748.2 mmHg
(19) increase
(20) constant
(21) liquefy, condense
(22) freezing, crystallizing
(23) exothermic
(24) solid
(25) gaseous

(26) No, carbon dioxide does not melt at atmospheric pressure, it sublimes.

(27) No, density, specific heat, heat of fusion, heat of vaporization, melting point, boiling point, and sublimation point are unique for each substance.

(28) -78°C
(29) steam
(30) smaller

(31) No, the flasks were not evacuated before the balloons were added, so air was in the flasks. The majority of the pressure in both flasks is from the water and carbon dioxide, but part of the pressure in both flasks is from air which is a mixture of several different gases (oxygen, nitrogen, argon, etc.).

SELF-TEST QUESTIONS

Multiple Choice

1. The density of ether is 0.736 g/mL. How much would 20.0 mL of ether weigh?
   a. 14.7 g
   b. 1.47 g
   c. 2.72 g
   d. 27.2 g

2. Calculate the density of a swimmer who weighs 40.0 kg and occupies a volume of 45.0 liters.
   a. 0.889 g/mL
   b. 0.0088 g/mL
   c. 1.12 g/mL
   d. 0.0112 g/mL

3. As a pure liquid is heated, its temperature increases and it becomes less dense. Therefore, which of the following is true?
   a. its potential energy increases
   b. its kinetic energy increases
   c. both its potential and kinetic energy increase
   d. its potential energy increases while kinetic energy decreases

4. Which of the following is an endothermic process?
   a. freezing of water
   b. melting of tin
   c. condensation of steam
   d. solidification of liquid sulfur

5. Which of the following is a property of the gaseous state?
   a. it has a lower density than the solid phase
   b. it has large compressibility
   c. it has moderate thermal expansion
   d. more than one response is correct

6. When the temperature is 0°C, a balloon has a volume of 5 liters. If the temperature is changed to 50°C and if we assume that the pressure inside the balloon equals atmospheric pressure at all times, which of the following will be true?
   a. the new volume will be larger
   b. the new volume will be smaller
   c. there will be no change in volume
   d. the new volume could be calculated from Boyle's law

7. A steel cylinder of CO_2 gas has a pressure of 20 atmospheres (atm) at 20°C. If the cylinder is heated to 70°C, what will the pressure be?
   a. it will be greater than 20 atm
   b. it will be exactly 20 atm
   c. it will be less than 20 atm
   d. it cannot be calculated

8. A balloon is filled with exactly 5.0 liters of helium gas in a room where the temperature is 20°C. What volume will the balloon have when it is taken outside and cooled to 10°C?
   a. 2.5 L   b. 4.8 L   c. 5.2 L   d. 10.0 L

9. If the specific heat of water is 1.00 calorie per gram °C $\left(\frac{\text{cal}}{\text{g·°C}}\right)$ and the heat of vaporization is 540 cal/g, how much heat would it take to raise the temperature of 10.0 g of water 10°C?
   a. 10 cal   b. 100 cal   c. 5400 cal   d. 54,000 cal

10. The heat of fusion of water is 80.0 cal/g. Therefore, what happens when 1.00 g of ice is melted?
   a. 80 calories of heat would be absorbed
   c. 80 calories of heat would be released
   b. no heat change would take place
   d. what would happen is not predictable

11. Calculate the amount of heat necessary to convert 5.00 g of water at 100°C into steam at 100°C. The heat of vaporization of water is 540 cal/g. The specific heat of steam is 0.480 calories per gram °C $\left(\frac{\text{cal}}{\text{g·°C}}\right)$.
   a. 24.0 cal   b. 545 cal   c. 2700 cal   d. 2724 cal

12. What volume would 1 mole of gaseous CH_4 occupy at standard temperature and 0.82 atmospheres of pressure? Remember that $R = 0.0821 \frac{\text{L·atm}}{\text{mol·K}}$.
   a. 0.100 L   b. 1.00 L   c. 2.70 L   d. 27.3 L

13. A sample of gas is found to occupy a volume of 6.80 L at a temperature of 30.0°C and a pressure of 640 torr. How many moles of gas are in the sample? Remember that $R = 0.0821 \frac{\text{L·atm}}{\text{mol·K}}$.
   a. 0.230 mol   b. 4.33 mol   c. 1769 mol   d. 2.33 mol

14. A 1.20 L sample of gas is under a pressure of 15.0 atm. What volume would the sample occupy if the pressure were lowered to 5.20 atm and the temperature was kept constant?
   a. 0.289 L   b. 0.416 L   c. 3.46 L   d. 34.6 L

True-False

15. According to Graham's law, a gas with molecules four times as heavy as a second gas will diffuse two time faster than the second gas.
16. Solids have a definite volume and a definite shape.
17. The density of a material depends on the amount of material.
18. The average particle speed decreases as temperature decreases.
19. A pressure of 1.50 atm can also be expressed as 1.14×10^3 torr.
20. If the volume of a gas is held constant, increasing the temperature would result in no increase in pressure.
21. The total pressure of a sample of oxygen saturated with water vapor is equal to the partial pressure of the oxygen plus the partial pressure of the water vapor.
22. The state of matter in which disruptive forces predominate is the gaseous state.
23. When the combined gas law is used in a calculation, the gas pressures must be expressed in atm.
24. When the vapor pressure of a liquid is equal to the atmospheric pressure, the liquid will freeze.
25. Boiling points of liquids decrease as atmospheric pressure decreases.
26. When a liquid is placed in a closed container, evaporation continues for a time and then stops.
27. The specific heat of a substance is the same for all three phases of matter.

Matching

Match the states of matter given as responses to the following descriptions.

28.	state has a high density	a.	solid
29.	state has a large compressibility	b.	liquid
30.	cohesive forces dominate slightly	c.	gas
31.	cohesive forces predominate over disruptive forces	d.	two or more states
32.	sample takes shape of its container		

Match the terms given as responses to the following descriptions.

33.	molten steel changes to a solid	a.	sublimation
34.	water in an open container disappears after a time	b.	evaporation
35.	ice on a car windshield disappears without melting	c.	freezing

Chapter 7: Solutions and Colloids

CHAPTER OUTLINE

7.1 Physical States of Solutions
7.2 Solubility
7.3 The Solution Process
7.4 Solution Concentrations
7.5 Solution Preparation
7.6 Solution Stoichiometry
7.7 Solution Properties
7.8 Colloids
7.9 Dialysis

LEARNING OBJECTIVES/ASSESSMENT

When you have completed your study of this chapter, you should be able to:

1. Classify mixtures as solutions or nonsolutions based on their appearance. (Section 7.1; Exercise 7.4)
2. Demonstrate your understanding of terms related to the solubility of solutes in solution. (Section 7.2; Exercises 7.6 and 7.12)
3. Predict in a general way the solubilities of solutes in solvents on the basis of molecular polarity. (Section 7.3; Exercise 7.16)
4. Calculate solution concentrations in units of molarity, weight/weight percent, weight/volume percent, and volume/volume percent. (Section 7.4; Exercises 7.22 b, 7.30 c, 7.34 a, and 7.38 c)
5. Describe how to prepare solutions of specific concentration using pure solutes and solvent, or solutions of greater concentration than the one desired. (Section 7.5; Exercises 7.46 and 7.48 b)
6. Do stoichiometric calculations based on solution concentrations. (Section 7.6; Exercise 7.56)
7. Do calculations based on the colligative solution properties of boiling point, freezing point, and osmotic pressure. (Section 7.7; Exercises 7.64 a & c and 7.74)
8. Describe the characteristics of colloids. (Section 7.8; Exercise 7.82)
9. Describe the process of dialysis, and compare it to the process of osmosis. (Section 7.9; Exercise 7.84)

SOLUTIONS FOR THE END OF CHAPTER EXERCISES

PHYSICAL STATES OF SOLUTIONS (SECTION 7.1)

7.2

		Solvent	Solutes
a.	Liquid laundry bleach: sodium hypochlorite 5.25%, inert ingredients 94.75%	water	sodium hypochlorite
b.	Rubbing alcohol: isopropyl alcohol 70%	isopropyl alcohol	water
c.	Hydrogen peroxide: 3% hydrogen peroxide	water	hydrogen peroxide
d.	After-shave: SD alcohol, water, glycerin, fragrance, menthol, benzophenone-1, coloring	SD alcohol	water, glycerin, fragrance, menthol, benzophenone-1, coloring

☑7.4

		Classification	Solvent	Solutes
a.	Foggy air	not a solution (cannot see through it)		
b.	Tears	solution	water	sodium chloride, other ions
c.	Freshly squeezed orange juice	not a solution (cannot see through it)		
d.	Strained tea	solution	water	extractions from tea leaves (color, caffeine, flavor)
e.	Creamy hand lotion	not a solution (cannot see through it)		

SOLUBILITY (SECTION 7.2)

☑7.6
a. 25 mL of cooking oil and 25 mL of vinegar – the resulting mixture is cloudy and gradually separates into two layers. — immiscible
b. 25 mL of water and 10 mL of rubbing alcohol – the resulting mixture is clear and colorless. — soluble
c. 25 mL of chloroform and 1 g of roofing tar – the resulting mixture is clear but dark brown in color. — soluble

7.8
a. A solution to which a small piece of solute is added, and it dissolves. — unsaturated
b. A solution to which a small piece of solute is added, and much more solute comes out of solution. — supersaturated
c. The final solution resulting from the process in part b. — saturated

7.10 This solution could become supersaturated by slowly lowering the temperature of the solution or by allowing some of the solvent to evaporate. The solution must be handled very gently as a supersaturated solution is unstable.

☑7.12
a. barium nitrate, $Ba(NO_3)_2$ (8.7 g at 20°C) — soluble
b. aluminum oxide, Al_2O_3 (9.8×10^{-5} g at 29°C) — insoluble
c. calcium sulfate, $CaSO_4$ (0.21 g at 30°C) — slightly soluble
d. manganese chloride, $MnCl_2$ (72.3 g at 25°C) — very soluble
e. lead bromide, $PbBr_2$ (0.46 g at 0°C) — slightly soluble

THE SOLUTION PROCESS (SECTION 7.3)

7.14 I would add water to the sample, stir the mixture, and filter out the solid from the heterogeneous mixture. The calcium carbonate is insoluble in water and would be the solid (residue) from the filtration of the heterogeneous mixture. Calcium chloride is very soluble and would dissolve in the water.

☑7.16
a. 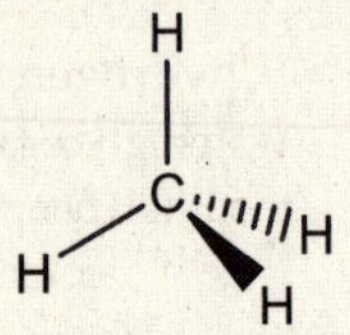

(tetrahedral) — soluble in benzene

b. Ne — soluble in benzene

c. 

(triangular-based pyramid) — soluble in water

d. 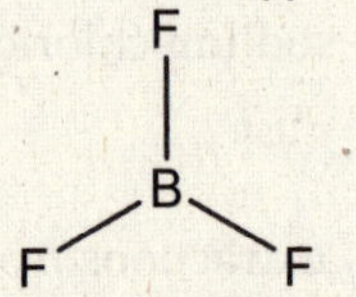

(flat triangle) — soluble in benzene

7.18 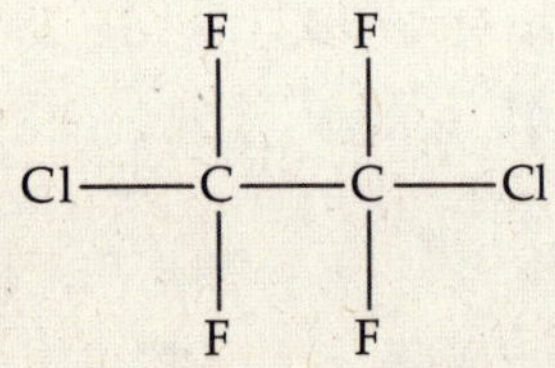

This molecule is mostly nonpolar and grease is nonpolar. Like dissolves like; therefore, Freon-114 will dissolve grease.

SOLUTION CONCENTRATIONS (SECTION 7.4)

7.20 a. 1.25 L of solution contains 0.455 mol of solute.

$$\frac{0.455\text{ mol}}{1.25\text{ L}} = 0.364\text{ M}$$

b. 250 mL of solution contains 0.215 mol of solute.

$$250\text{ mL}\left(\frac{1\text{ L}}{1000\text{ mL}}\right) = 0.250\text{ L}$$

$$M = \frac{\text{moles}}{\text{L}} \Rightarrow \frac{0.215\text{ mol}}{0.250\text{ L}} = 0.860\text{ M}$$

or

$$\frac{0.215\text{ mol}}{250\text{ mL}\left(\frac{1\text{ L}}{1000\text{ mL}}\right)} = 0.860\text{ M}$$

c. 0.175 mol of solute is put into a container and enough distilled water is added to give 100 mL of solution.

$$100\text{ mL}\left(\frac{1\text{ L}}{1000\text{ mL}}\right) = 0.100\text{ L}$$

$$M = \frac{\text{moles}}{\text{L}} \Rightarrow \frac{0.175\text{ mol}}{0.100\text{ L}} = 1.75\text{ M}$$

or

$$\frac{0.175\text{ mol}}{100\text{ mL}\left(\frac{1\text{ L}}{1000\text{ mL}}\right)} = 1.75\text{ M}$$

7.22 a. A sample of solid KBr weighing 11.9 g is put in enough distilled water to give 200 mL of solution.

$$11.9\text{ g}\left(\frac{1\text{ mole}}{119.00\text{ g}}\right) = 0.100\text{ mol}$$

$$200\text{ mL}\left(\frac{1\text{ L}}{1000\text{ mL}}\right) = 0.200\text{ L}$$

$$M = \frac{\text{mole}}{\text{L}} \Rightarrow \frac{0.100\text{ mol}}{0.200\text{ L}} = 0.500\text{ M}$$

or

$$\frac{11.9\text{ g}\left(\frac{1\text{ mole}}{119.00\text{ g}}\right)}{200\text{ mL}\left(\frac{1\text{ L}}{1000\text{ mL}}\right)} = 0.500\text{ M}$$

☑ b. A 14.2 g sample of solid Na_2SO_4 is dissolved in enough water to give 500 mL of solution.

$$14.2\text{ g}\left(\frac{1\text{ mole}}{142.05\text{ g}}\right) = 0.0100\text{ moles}$$

$$500\text{ mL}\left(\frac{1\text{ L}}{1000\text{ mL}}\right) = 0.500\text{ L}$$

$$M = \frac{\text{moles}}{\text{L}} \Rightarrow \frac{0.0100\text{ mol}}{0.500\text{ L}} = 0.0200\text{ M}$$

or

$$\frac{14.2\text{ g}\left(\frac{1\text{ mole}}{142.05\text{ g}}\right)}{500\text{ mL}\left(\frac{1\text{ L}}{1000\text{ mL}}\right)} = 0.200\text{ M}$$

c. A 10.0-mL sample of solution is evaporated to dryness and leaves 0.29 g of solid residue that is identified as Li_2SO_4.

$$0.29\text{ g}\left(\frac{1\text{ mole}}{109.95\text{ g}}\right) = 0.0026\text{ moles}$$

$$10.0\text{ mL}\left(\frac{1\text{ L}}{1000\text{ mL}}\right) = 0.0100\text{ L}$$

$$M = \frac{\text{moles}}{\text{L}} \Rightarrow \frac{0.0026\text{ mol}}{0.0100\text{ L}} = 0.26\text{ M}$$

or

$$\frac{0.29\text{ g}\left(\frac{1\text{ mole}}{109.95\text{ g}}\right)}{10.0\text{ mL}\left(\frac{1\text{ L}}{1000\text{ mL}}\right)} = 0.26\text{ M}$$

7.24 a. How many moles of solute are contained in 1.75 L of 0.215 M solution?

$$1.75\text{ L}\left(\frac{0.215\text{ moles}}{1\text{ L}}\right)=0.376\text{ moles}$$

b. How many moles of solute are contained in 250 mL of 0.300 M solution?

$$250\text{ mL}\left(\frac{1\text{ L}}{1000\text{ mL}}\right)\left(\frac{0.300\text{ moles}}{1\text{ L}}\right)=0.0750\text{ moles}$$

c. How many mL of 0.350 M solution contains 0.200 mol of solute?

$$0.200\text{ mol}\left(\frac{1\text{ L}}{0.350\text{ moles}}\right)\left(\frac{1000\text{ mL}}{1\text{ L}}\right)=571\text{ mL}$$

7.26 a. How many grams of solid would be left behind if 25.0 mL of 0.400 M NaCl solution was evaporated to dryness?

$$25.0\text{ mL}\left(\frac{1\text{ L}}{1000\text{ mL}}\right)\left(\frac{0.400\text{ moles}}{1\text{ L}}\right)\left(\frac{58.44\text{ g}}{1\text{ mole}}\right)=0.584\text{ g}$$

b. How many liters of 0.255 M HCl solution is needed to provide 0.0400 mol of HCl?

$$0.0400\text{ mol}\left(\frac{1\text{ L}}{0.255\text{ moles}}\right)=0.157\text{ L}$$

c. How many mL of 0.500 M $NaNO_3$ solution contains 50.0 g of solute?

$$50.0\text{ g}\left(\frac{1\text{ mole}}{85.00\text{ g}}\right)\left(\frac{1\text{ L}}{0.500\text{ moles}}\right)\left(\frac{1000\text{ mL}}{1\text{ L}}\right)=1.18\times10^{3}\text{ mL}$$

7.28 a. 5.3 g of sugar and 100 mL of water

$$100\text{ mL water}\left(\frac{1.00\text{ g}}{1\text{ mL}}\right)=100\text{ g water}$$

solution mass = 5.3 g sugar + 100 g water = 105.3 g (without SF)

$$\%(\text{w/w})=\frac{\text{solute mass}}{\text{solution mass}}\times100\Rightarrow\frac{5.3\text{ g}}{105.3\text{ g}}\times100=5.0\%(\text{w/w})$$

or

$$\frac{5.3\text{ g sugar}}{5.3\text{ g sugar}+100\text{ mL water}\left(\frac{1.00\text{ g}}{1\text{ mL}}\right)}\times100=5.0\%(\text{w/w})$$

b. 5.3 g of any solute and 100 mL of water

$$100\text{ mL water}\left(\frac{1.00\text{ g}}{1\text{ mL}}\right)=100\text{ g water}$$

solution mass = 5.3 g solute + 100 g water = 105.3 g (without SF)

$$\%(\text{w/w})=\frac{\text{solute mass}}{\text{solution mass}}\times100\Rightarrow\frac{5.3\text{ g}}{105.3\text{ g}}\times100=5.0\%(\text{w/w})$$

or

$$\frac{5.3\text{ g solute}}{5.3\text{ g solute}+100\text{ mL water}\left(\frac{1.00\text{ g}}{1\text{ mL}}\right)}\times100=5.0\%(\text{w/w})$$

c. 5.3 g of any solute and 100 g of any solvent

$$\frac{5.3\text{ g solute}}{5.3\text{ g solute}+100\text{ g solvent}}\times100=5.0\%(\text{w/w})$$

7.30 a. 15.0 g of salt is dissolved in 200 mL of water.

$$200\text{ mL water}\left(\frac{1.00\text{ g}}{1\text{ mL}}\right) = 200\text{ g water}$$

$$\text{solution mass} = 15.0\text{ g salt} + 200\text{ g water} = 215.0\text{ g (without SF)}$$

$$\%(\text{w/w}) = \frac{\text{solute mass}}{\text{solution mass}} \times 100 \Rightarrow \frac{15.0\text{ g}}{215.0\text{ g}} \times 100 = 6.98\%(\text{w/w})$$

or

$$\frac{15.0\text{ g salt}}{15.0\text{ g salt} + 200\text{ mL water}\left(\frac{1.00\text{ g}}{1\text{ mL}}\right)} \times 100 = 6.98\%(\text{w/w})$$

b. 0.150 mol of solid NH_4Cl is dissolved in 150 mL of water.

$$\text{mass solute} = 0.150\text{ mol}\left(\frac{53.49\text{ g}}{1\text{ mole}}\right) = 8.0235\text{ g (without SF)}$$

$$\text{mass of water} = 150\text{ mL}\left(\frac{1.00\text{ g}}{1\text{ mL}}\right) = 150\text{ g}$$

$$\text{mass of solution} = 8.0235\text{ g} + 150\text{ g} = 158.0235\text{ g}$$

$$\%(\text{w/w}) = \frac{\text{mass of solute}}{\text{mass of solution}} \times 100 \Rightarrow \frac{8.0235\text{ g}}{158.0235\text{ g}} \times 100 = 5.08\%(\text{w/w})$$

or

$$\frac{0.150\text{ mol}\left(\frac{53.49\text{ g}}{1\text{ mole}}\right)}{0.150\text{ mol}\left(\frac{53.49\text{ g}}{1\text{ mole}}\right) + 150\text{ mL water}\left(\frac{1.00\text{ g}}{1\text{ mL}}\right)} \times 100 = 5.08\%(\text{w/w})$$

☑ c. 75.0 g of solid is dissolved in 500 mL of water.

$$\text{mass of water} = 500\text{ mL water}\left(\frac{1.00\text{ g}}{1\text{ mL}}\right) = 500\text{ g}$$

$$\text{mass of solution} = 75.0\text{ g solute} + 500\text{ g water} = 575.0\text{ g (without SF)}$$

$$\%(\text{w/w}) = \frac{\text{mass solute}}{\text{mass solution}} \times 100 \Rightarrow \frac{75.0\text{ g}}{575.0\text{ g}} \times 100 = 13.0\%(\text{w/w})$$

or

$$\frac{75.0\text{ g solute}}{75.0\text{ g solute} + 500\text{ mL water}\left(\frac{1.00\text{ g}}{1\text{ mL}}\right)} \times 100 = 13.0\%(\text{w/w})$$

d. 20.0 mL of ethyl alcohol (density = 0.789 g/mL) is mixed with 15.0 mL of water.

$$\text{mass of ethyl alcohol} = 20.0\text{ mL ethyl alcohol}\left(\frac{0.789\text{ g}}{1\text{ mL}}\right) = 15.78\text{ g}$$

$$\text{mass of water} = 15.0\text{ mL water}\left(\frac{1.00\text{ g}}{1\text{ mL}}\right) = 15.0\text{ g}$$

$$\text{mass of solution} = 15.78\text{ g ethyl alcohol} + 15.0\text{ g water} = 30.78\text{ g}$$

$$\%(\text{w/w}) = \frac{\text{mass of solute}}{\text{mass of solution}} \times 100 \Rightarrow \frac{15.78\text{ g}}{30.78\text{ g}} \times 100 = 51.3\%(\text{w/w})$$

or

$$\frac{20.0\text{ mL ethyl alcohol}\left(\frac{0.789\text{ g}}{1\text{ mL}}\right)}{20.0\text{ mL ethyl alcohol}\left(\frac{0.789\text{ g}}{1\text{ mL}}\right) + 15.0\text{ mL water}\left(\frac{1.00\text{ g}}{1\text{ mL}}\right)} \times 100 = 51.3\%(\text{w/w})$$

7.32 a. 15.0 g of solute is dissolved in enough water to give 100 mL of solution. The density of the resulting solution is 1.20 g/mL.

$$\text{mass of solution} = 100 \text{ mL solution}\left(\tfrac{1.20\text{ g}}{1\text{ mL}}\right) = 120 \text{ g}$$

$$\%(\text{w/w}) = \frac{\text{mass of solute}}{\text{mass of solution}} \times 100 \Rightarrow \frac{15.0\text{ g}}{120\text{ g}} \times 100 = 12.5\%(\text{w/w})$$

or

$$\frac{15.0\text{ g solute}}{100\text{ mL solution}\left(\tfrac{1.20\text{ g}}{1\text{ mL}}\right)} \times 100 = 12.5\%(\text{w/w})$$

b. A 10.0-mL solution sample with a density of 1.15 g/mL leaves 1.54 g of solid residue when evaporated.

$$\text{mass of solution} = 10.0 \text{ mL solution}\left(\tfrac{1.15\text{ g}}{1\text{ mL}}\right) = 11.5 \text{ g}$$

$$\%(\text{w/w}) = \frac{\text{mass of solute}}{\text{mass of solution}} \times 100 \Rightarrow \frac{1.54\text{ g}}{11.5\text{ g}} \times 100 = 13.4\%(\text{w/w})$$

or

$$\frac{1.54\text{ g solute}}{10.0\text{ mL solution}\left(\tfrac{1.15\text{ g}}{1\text{ mL}}\right)} \times 100 = 13.4\%(\text{w/w})$$

c. A 50.0-g sample of solution on evaporation leaves a 3.12 g residue of $MgCl_2$.

$$\frac{3.12\text{ g MgCl}_2}{50.0\text{ g solution}} \times 100 = 6.24\%(\text{w/w})$$

7.34 ☑a. 150 mL of solution contains 12.0 mL of alcohol.

$$\frac{12.0\text{ mL alcohol}}{150\text{ mL solution}} \times 100 = 8.00\%(\text{v/v})$$

b. 150 mL of solution contains 12.0 mL of any soluble liquid solute.

$$\frac{12.0\text{ mL solute}}{150\text{ mL solution}} \times 100 = 8.00\%(\text{v/v})$$

c. 6.0 fluid ounces of oil is added to 2.0 gallons (256 fluid ounces) of gasoline.

$$\frac{6.0\text{ fluid ounces oil}}{6.0\text{ fluid ounces oil} + 256\text{ fluid ounces gasoline}} \times 100 = 2.3\%(\text{v/v})$$

d. A solution of alcohol and water is separated by distillation. A 150-mL solution sample gives 51.3 mL of alcohol.

$$\frac{51.3\text{ mL alcohol}}{150\text{ mL solution}} \times 100 = 34.2\%(\text{v/v})$$

7.36

$$(5.0\text{ L})(0.50\%(\text{v/v})) = (5.0\text{ L})(0.0050) = 0.025\text{ L}\left(\frac{1000\text{ mL}}{1\text{ L}}\right) = 25\text{ mL}$$

Note: This number may seem small when compared with the volume of a shot of alcohol. (Typically, 1 shot is 1.5 fl. oz. or 44 mL.) This question assumed that the drink was pure ethanol and that it was completely absorbed into the bloodstream. These assumptions do not take into account that most alcoholic beverages are not pure ethanol (many spirits are approximately 40% alcohol by volume) and that the alcohol is distributed throughout the body (not just the bloodstream).

7.38 a. 150 mL of solution contains 7.50 g of dissolved solid Na_2SO_4.

$$\frac{7.50\text{ g Na}_2\text{SO}_4}{150\text{ mL solution}} \times 100 = 5.00\%(\text{w/v})$$

b. 150 mL of solution contains 7.50 g of any dissolved solid solute.
$$\frac{7.50\text{ g solute}}{150\text{ mL solution}}\times 100 = 5.00\%(\text{w/v})$$

☑ c. 350 mL of solution contains 30.7 g of dissolved solid solute.
$$\frac{30.7\text{ g solute}}{350\text{ mL solution}}\times 100 = 8.77\%(\text{w/v})$$

7.40
$$\frac{65\text{ g KBr}}{65\text{ g KBr}+100\text{ g H}_2\text{O}}\times 100 = 39\%(\text{w/w})$$

SOLUTION PREPARATION (SECTION 7.5)

7.42 a.
$$200\text{ mL}\left(\frac{1\text{ L}}{1000\text{ mL}}\right)\left(\frac{0.150\text{ moles Na}_2\text{SO}_4}{1\text{ L solution}}\right)\left(\frac{142.05\text{ g Na}_2\text{SO}_4}{1\text{ mole Na}_2\text{SO}_4}\right) = 4.26\text{ g Na}_2\text{SO}_4$$

I would mass 4.26 g Na_2SO_4 and add it to a 200 mL volumetric flask. I would add enough water to dissolve the Na_2SO_4, then add water up to the mark on the volumetric flask, cap, and shake to ensure the solution is homogeneous.

b.
$$250\text{ mL}\left(\frac{1\text{ L}}{1000\text{ mL}}\right)\left(\frac{0.250\text{ moles Zn(NO}_3)_2}{1\text{ L solution}}\right)\left(\frac{189.41\text{ g Zn(NO}_3)_2}{1\text{ mole Zn(NO}_3)_2}\right) = 11.8\text{ g Zn(NO}_3)_2$$

I would mass 11.8 g $Zn(NO_3)_2$ and add it to a 250 mL volumetric flask. I would add enough water to dissolve the $Zn(NO_3)_2$, then add water up to the mark on the volumetric flask, cap, and shake to ensure the solution is homogeneous.

c. $150\text{ g}\times 2.25\%(\text{w/w}) = 150\text{ g}\times 0.0225 = 3.38\text{ g NaCl}$

$150\text{ g} - 3.38\text{ g} = 146.62\text{ g H}_2\text{O}\left(\frac{1\text{ mL}}{1.00\text{ g H}_2\text{O}}\right) = 146.62\text{ mL H}_2\text{O} \approx 147\text{ mL H}_2\text{O}$

I would mass 3.38 g NaCl and add it to a 250 mL Erlenmeyer flask. I would measure 147 mL of water and add enough of it to the Erlenmeyer flask to dissolve the salt, then I would add the rest of the water and swirl the flask to ensure that the solution is homogeneous.

d. $125\text{ mL}\times 0.75\%(\text{w/v}) = 0.938\text{ g KCl} \approx 0.94\text{ g KCl}$

I would add 0.94 g KCl to a 125 mL volumetric flask. I would add enough water to dissolve the KCl, then dilute the mixture to the mark on the volumetric flask. I would then cap the flask and shake it to ensure the solution is homogeneous.

7.44 mass of solution = 15.0 g ethyl alcohol + 45.0 g water = 60.0 g

$$\%(\text{w/w}) = \frac{\text{mass of solute}}{\text{mass of solution}}\times 100 \Rightarrow \frac{15.0\text{ g}}{60.0\text{ g}}\times 100 = 25.0\%(\text{w/w})$$

or

$$\frac{15.0\text{ g ethyl alcohol}}{15.0\text{ g ethyl alcohol}+45.0\text{ g water}}\times 100 = 25.0\%(\text{w/w})$$

$$\text{volume of solution} = \frac{\text{mass of solution}}{\text{density of solution}} \Rightarrow \frac{60.0\text{g}}{0.952\,\frac{\text{g}}{\text{mL}}} = 63.025 \text{ mL (without SF)}$$

$$\%(\text{w/v}) = \frac{\text{mass of solute}}{\text{volume of solution}} \times 100 \Rightarrow \frac{15.0\text{ g}}{63.025\text{ mL}} \times 100 = 23.8\%(\text{w/v})$$

or

$$\frac{15.0\text{ g ethyl alcohol}}{(15.0\text{ g ethyl alcohol} + 45.0\text{ g water})\left(\frac{1\text{ mL}}{0.952\text{ g}}\right)} \times 100 = 23.8\%(\text{w/v})$$

☑7.46 a. $$250\text{ mL}\left(\frac{1\text{ L}}{1000\text{ mL}}\right)\left(\frac{1.75\text{ moles Li}_2\text{CO}_3}{1\text{ L solution}}\right)\left(\frac{73.89\text{ g Li}_2\text{CO}_3}{1\text{ mole Li}_2\text{CO}_3}\right) = 32.3\text{ g Li}_2\text{CO}_3$$

b. $$200\text{ mL}\left(\frac{1\text{ L}}{1000\text{ mL}}\right)\left(\frac{3.50\text{ moles NH}_3}{1\text{ L solution}}\right) = 0.700\text{ moles NH}_3$$

c. $250\text{ mL} \times 12.5\%(\text{v/v}) = 250\text{ mL} \times 0.125 = 31.3\text{ mL}$

d. $50.0\text{ mL} \times 4.20\%(\text{w/v}) = 50.0\text{ mL} \times 0.0420 = 2.10\text{ g CaCl}_2$

7.48 a. $(6.00\text{ M})(5.00\text{ L}) = (18.0\text{ M})(V_c)$

$V_c = 1.67\text{ L}$

I would add 3.00 L of water to a 5.00 L volumetric flask, then add 1.67 L of 18.0 M H_2SO_4 to the flask, let it cool, and dilute to the mark with water. (Always add acid to water, not the reverse!)

☑b. $(0.500\text{ M})(250\text{ mL}) = (3.00\text{ M})(V_c)$

$V_c = 41.7\text{ mL}$

I would add 41.7 mL of 3.00 M $CaCl_2$ to a 250 mL volumetric flask, then dilute to the mark with water. I would be sure to shake well.

c. $(1.50\%(\text{w/v}))(200\text{ mL}) = (10.0\%(\text{w/v}))(V_c)$

$V_c = 30.0\text{ mL}$

I would add 30.0 mL of 10.0%(w/v) KBr to a 200 mL volumetric flask, then dilute to the mark with water. I would be sure to shake well.

d. $(10.0\%(\text{v/v}))(500\text{ mL}) = (50.0\%(\text{v/v}))(V_c)$

$V_c = 100\text{ mL}$

I would add 100 mL of 50.0%(v/v) alcohol to a 500 mL volumetric flask, then dilute to the mark with water. I would be sure to shake well.

7.50 a. 2.00 L $(100\text{ mL})(0.225\text{ M}) = (2.00\text{ L})\left(\frac{1000\text{ mL}}{1\text{ L}}\right)(C_d)$

$C_d = 0.0113\text{ M}$

b. 225 mL $(100\text{ mL})(0.225\text{ M}) = (225\text{ mL})(C_d)$

$C_d = 0.100\text{ M}$

c. 600 mL $$(100\text{ mL})(0.225\text{ M}) = (600\text{ mL})(C_d)$$
$$C_d = 0.0375\text{ M}$$

d. 850 mL $$(100\text{ mL})(0.225\text{ M}) = (850\text{ mL})(C_d)$$
$$C_d = 0.0265\text{ M}$$

SOLUTION STOICHIOMETRY (SECTION 7.6)

7.52 $Na_2CO_3\ (s) + 2\ HCl\ (aq) \rightarrow 2\ NaCl\ (aq) + CO_2\ (g) + H_2O\ (l)$

$$250\text{ mL HCl}\left(\frac{1\text{ L}}{1000\text{ mL}}\right)\left(\frac{1.25\text{ moles HCl}}{1\text{ L HCl}}\right)\left(\frac{1\text{ mole Na}_2\text{CO}_3}{2\text{ moles HCl}}\right)\left(\frac{106\text{ g Na}_2\text{CO}_3}{1\text{ moles Na}_2\text{CO}_3}\right)$$
$$= 16.6\text{ g Na}_2\text{CO}_3$$

7.54 $NaCl\ (aq) + AgNO_3\ (aq) \rightarrow NaNO_3\ (aq) + AgCl\ (s)$

$$25.0\text{ mL NaCl}\left(\frac{1\text{ L}}{1000\text{ mL}}\right)\left(\frac{0.200\text{ moles NaCl}}{1\text{ L NaCl}}\right)\left(\frac{1\text{ mole AgNO}_3}{1\text{ mole NaCl}}\right)\left(\frac{1\text{ L AgNO}_3}{0.250\text{ moles}}\right)\left(\frac{1000\text{ mL}}{1\text{ L AgNO}_3}\right)$$
$$= 20.0\text{ mL AgNO}_3$$

☑7.56 $2\ NH_3\ (aq) + H_2SO_4\ (aq) \rightarrow (NH_4)_2SO_4\ (aq)$

$$30.0\text{ mL H}_2\text{SO}_4\left(\frac{1\text{ L}}{1000\text{ mL}}\right)\left(\frac{0.190\text{ moles H}_2\text{SO}_4}{1\text{ L H}_2\text{SO}_4}\right)\left(\frac{2\text{ moles NH}_3}{1\text{ mole H}_2\text{SO}_4}\right)\left(\frac{1\text{ L NH}_3}{0.225\text{ moles NH}_3}\right)\left(\frac{1000\text{ mL NH}_3}{1\text{ L NH}_3}\right)$$
$$= 50.7\text{ mL NH}_3$$

7.58 $NaOH\ (aq) + HCl\ (aq) \rightarrow NaCl\ (q) + H_2O\ (l)$

$$30.0\text{ mL HCl}\left(\frac{1\text{ L}}{1000\text{ mL}}\right)\left(\frac{0.125\text{ moles HCl}}{1\text{ L HCl}}\right)\left(\frac{1\text{ mole NaOH}}{1\text{ mole HCl}}\right)\left(\frac{1\text{ L NaOH}}{0.135\text{ moles NaOH}}\right)\left(\frac{1000\text{ mL NaOH}}{1\text{ L NaOH}}\right)$$
$$= 27.8\text{ mL NaOH}$$

7.60 $CaCO_3\ (s) + 2\ HCl\ (aq) \rightarrow CO_2\ (g) + CaCl_2\ (aq) + H_2O\ (l)$

$$250\text{ mL HCl}\left(\frac{1\text{ L}}{1000\text{ mL}}\right)\left(\frac{0.10\text{ moles HCl}}{1\text{ L HCl}}\right)\left(\frac{1\text{ mole CaCO}_3}{2\text{ moles HCl}}\right)\left(\frac{100\text{ g CaCO}_3}{1\text{ mole CaCO}_3}\right)$$
$$= 1.25\text{ g CaCO}_3 \approx 1.3\text{ g CaCO}_3$$

SOLUTION PROPERTIES (SECTION 7.7)

7.62 As solutes are added to a solvent, the freezing point of the solution is depressed when compared to the freezing point of the pure solvent. The ice cream mixture contains several solutes dissolved in water and the freezing point is less than the freezing point of water. By adding salt to the ice and water, the freezing point of this solution is also decreased. This allows the ice cream mixture to transfer heat to the salt and ice water mixture, and eventually, freeze. Ice by itself will not work because its temperature is above the freezing point of the ice cream.

7.64 ☑a. KCl, a strong electrolyte

$$BP_{solution} = BP_{solvent} + \Delta t_b \qquad \Delta t_b = nK_bM$$
$$= 100.00°C + (2)\left(0.52\tfrac{°C}{M}\right)(1.50\text{ M}) = 101.6°C$$
$$FP_{solution} = FP_{solvent} - \Delta t_f \qquad \Delta t_f = nK_fM$$
$$= 0.00°C - (2)\left(1.86\tfrac{°C}{M}\right)(1.50\text{ M}) = -5.58°C$$

b. glycerol, a nonelectrolyte

$BP_{solution} = BP_{solvent} + \Delta t_b \qquad \Delta t_b = nK_bM$
$= 100.00°C + (1)(0.52\,\tfrac{°C}{M})(1.50\ M) = 100.78°C$

$FP_{solution} = FP_{solvent} - \Delta t_f \qquad \Delta t_f = nK_fM$
$= 0.00°C - (1)(1.86\,\tfrac{°C}{M})(1.50\ M) = -2.79°C$

☑c. $(NH_4)_2SO_4$, a strong electrolyte

$BP_{solution} = BP_{solvent} + \Delta t_b \qquad \Delta t_b = nK_bM$
$= 100.00°C + (3)(0.52\,\tfrac{°C}{M})(1.50\ M) = 102.3°C$

$FP_{solution} = FP_{solvent} - \Delta t_f \qquad \Delta t_f = nK_fM$
$= 0.00°C - (3)(1.86\,\tfrac{°C}{M})(1.50\ M) = -8.37°C$

d. $Al(NO_3)_3$, a strong electrolyte

$BP_{solution} = BP_{solvent} + \Delta t_b \qquad \Delta t_b = nK_bM$
$= 100.00°C + (4)(0.52\,\tfrac{°C}{M})(1.50\ M) = 103.1°C$

$FP_{solution} = FP_{solvent} - \Delta t_f \qquad \Delta t_f = nK_fM$
$= 0.00°C - (4)(1.86\,\tfrac{°C}{M})(1.50\ M) = -11.2°C$

7.66 a. A 0.750 M solution of urea, a nonelectrolyte

$BP_{solution} = BP_{solvent} + \Delta t_b \qquad \Delta t_b = nK_bM$
$= 100.0°C + (1)(0.52\,\tfrac{°C}{M})(0.750\ M) = 100.39°C$

$FP_{solution} = FP_{solvent} - \Delta t_f \qquad \Delta t_f = nK_fM$
$= 0.0°C - (1)(1.86\,\tfrac{°C}{M})(0.750\ M) = -1.40°C$

b. A 0.150 M solution of $CaCl_2$, a strong electrolyte

$BP_{solution} = BP_{solvent} + \Delta t_b \qquad \Delta t_b = nK_bM$
$= 100.0°C + (3)(0.52\,\tfrac{°C}{M})(0.150\ M) = 100.23°C$

$FP_{solution} = FP_{solvent} - \Delta t_f \qquad \Delta t_f = nK_fM$
$= 0.0°C - (3)(1.86\,\tfrac{°C}{M})(0.150\ M) = -0.837°C$

c. A solution containing 120 g of ethylene glycol ($C_2H_6O_2$) per 400 mL of solution

$$\frac{120\ g\ C_2H_6O_2\left(\dfrac{1\ mole\ C_2H_6O_2}{62.08\ g\ C_2H_6O_2}\right)}{400\ mL\left(\dfrac{1L}{1000\ mL}\right)} = 4.83247\ M$$

$BP_{solution} = BP_{solvent} + \Delta t_b \qquad \Delta t_b = nK_bM$
$= 100.0°C + (1)(0.52\,\tfrac{°C}{M})(4.83247\ M) = 102.51°C$

$FP_{solution} = FP_{solvent} - \Delta t_f \qquad \Delta t_f = nK_fM$
$= 0.0°C - (1)(1.86\,\tfrac{°C}{M})(4.83247\ M) = -8.99°C$

7.68 a. A 0.20 M solution of glycerol, a nonelectrolyte

$\text{osmolarity} = nM = (1)(0.20\ M) = 0.20\ \tfrac{osmol}{L}$

b. A 0.20 M solution of $(NH_4)_2SO_4$, a strong electrolyte

$\text{osmolarity} = nM = (3)(0.20\ M) = 0.60\ \tfrac{osmol}{L}$

c. A solution containing 33.7 g of LiCl (a strong electrolyte) per 500 mL

$$\text{osmolarity} = \text{nM} = (2)\left(\frac{33.7\text{ g LiCl}\left(\frac{1\text{ mole LiCl}}{42.39\text{ g LiCl}}\right)}{500\text{ mL solution}\left(\frac{1\text{ L}}{1000\text{ mL}}\right)}\right) = 3.18\,\tfrac{\text{osmol}}{\text{L}}$$

7.70

$$\Pi = \text{nMRT} = (0.250\text{ M})(0.0821\,\tfrac{\text{L}\cdot\text{atm}}{\text{mol}\cdot\text{K}})((25.0+273)\text{K}) = 6.12\text{ atm}$$

$$= (0.250\,\tfrac{\text{moles}}{\text{L}})(0.0821\,\tfrac{\text{L}\cdot\text{atm}}{\text{mol}\cdot\text{K}})((25.0+273)\text{K})\left(\frac{760\text{ torr}}{1\text{ atm}}\right) = 4.65\times10^3\text{ torr}$$

$$= (0.250\,\tfrac{\text{moles}}{\text{L}})(0.0821\,\tfrac{\text{L}\cdot\text{atm}}{\text{mol}\cdot\text{K}})((25.0+273)\text{K})\left(\frac{760\text{ mmHg}}{1\text{ atm}}\right) = 4.65\times10^3\text{ mmHg}$$

7.72

$$\Pi = \text{nMRT} = 3(0.200\text{ M})(0.0821\,\tfrac{\text{L}\cdot\text{atm}}{\text{mol}\cdot\text{K}})((25.0+273)\text{K}) = 14.7\text{ atm}$$

$$= 3(0.200\,\tfrac{\text{moles}}{\text{L}})(0.0821\,\tfrac{\text{L}\cdot\text{atm}}{\text{mol}\cdot\text{K}})((25.0+273)\text{K})\left(\frac{760\text{ torr}}{1\text{ atm}}\right) = 1.12\times10^4\text{ torr}$$

$$= 3(0.200\,\tfrac{\text{moles}}{\text{L}})(0.0821\,\tfrac{\text{L}\cdot\text{atm}}{\text{mol}\cdot\text{K}})((25.0+273)\text{K})\left(\frac{760\text{ mmHg}}{1\text{ atm}}\right) = 1.12\times10^4\text{ mmHg}$$

☑7.74

$$\frac{95.0\text{g}\left(\frac{1\text{ mole}}{60.06\text{ g}}\right)}{500\text{ mL}\left(\frac{1\text{ L}}{1000\text{ mL}}\right)} = 3.1635\text{ M}$$

$$\Pi = \text{nMRT} = 1(3.1635\text{ M})(0.0821\,\tfrac{\text{L}\cdot\text{atm}}{\text{mol}\cdot\text{K}})((25.0+273)\text{K}) = 77.4\text{ atm}$$

$$= 1(3.1635\text{ M})(0.0821\,\tfrac{\text{L}\cdot\text{atm}}{\text{mol}\cdot\text{K}})((25.0+273)\text{K})\left(\frac{760\text{ torr}}{1\text{ atm}}\right) = 5.88\times10^4\text{ torr}$$

$$= 1(3.1635\text{ M})(0.0821\,\tfrac{\text{L}\cdot\text{atm}}{\text{mol}\cdot\text{K}})((25.0+273)\text{K})\left(\frac{760\text{ mmHg}}{1\text{ atm}}\right) = 5.88\times10^4\text{ mmHg}$$

7.76 The solvent is assumed to be water.

$$\text{FP}_{\text{solution}} = \text{FP}_{\text{solvent}} - \Delta t_f \qquad \Delta t_f = \text{n}K_f\text{M}$$

$$-0.35°\text{C} = 0.00°\text{C} - \text{nM}(1.86\,\tfrac{°\text{C}}{\text{M}})$$

$$\text{nM} = 0.188172\text{ M}$$

$$\Pi = \text{nMRT} = (0.188172\text{ M})(0.0821\,\tfrac{\text{L}\cdot\text{atm}}{\text{mol}\cdot\text{K}})((25.0+273)\text{K}) = 4.6\text{ atm}$$

$$= (0.188172\,\tfrac{\text{moles}}{\text{L}})(0.0821\,\tfrac{\text{L}\cdot\text{atm}}{\text{mol}\cdot\text{K}})((25.0+273)\text{K})\left(\frac{760\text{ torr}}{1\text{ atm}}\right) = 3.5\times10^3\text{ torr}$$

$$= (0.188172\,\tfrac{\text{moles}}{\text{L}})(0.0821\,\tfrac{\text{L}\cdot\text{atm}}{\text{mol}\cdot\text{K}})((25.0+273)\text{K})\left(\frac{760\text{ mmHg}}{1\text{ atm}}\right) = 3.5\times10^3\text{ mmHg}$$

7.78

$$\text{nM} = 2\left(\frac{5.30\text{ g NaCl}\left(\frac{1\text{ mole NaCl}}{58.5\text{ g NaCl}}\right)}{750\text{ mL}\left(\frac{1\text{ L}}{1000\text{ mL}}\right)}\right) + 2\left(\frac{8.20\text{ g KCl}\left(\frac{1\text{ mole KCl}}{74.6\text{ g KCl}}\right)}{750\text{ mL}\left(\frac{1\text{ L}}{1000\text{ mL}}\right)}\right) = 0.534714\,\tfrac{\text{osmol}}{\text{L}}$$

$$\Pi = nMRT = \left(0.534714\ \tfrac{\text{osmol}}{\text{L}}\right)\left(0.0821\ \tfrac{\text{L}\cdot\text{atm}}{\text{mol}\cdot\text{K}}\right)\left((25.0+273)\text{K}\right) = 13.1\ \text{atm}$$

$$= \left(0.534714\ \tfrac{\text{osmol}}{\text{L}}\right)\left(0.0821\ \tfrac{\text{L}\cdot\text{atm}}{\text{mol}\cdot\text{K}}\right)\left((25.0+273)\text{K}\right)\left(\frac{760\ \text{torr}}{1\ \text{atm}}\right) = 9.94\times10^{3}\ \text{torr}$$

$$= \left(0.534714\ \tfrac{\text{osmol}}{\text{L}}\right)\left(0.0821\ \tfrac{\text{L}\cdot\text{atm}}{\text{mol}\cdot\text{K}}\right)\left((25.0+273)\text{K}\right)\left(\frac{760\ \text{mmHg}}{1\ \text{atm}}\right) = 9.94\times10^{3}\ \text{mmHg}$$

7.80 The water will flow from the 5.00% sugar solution into the 10.0% sugar solution because the 5.00% sugar solution contains more solvent (water) than the 10.0% sugar solution does. The 10.0% sugar solution will become diluted as osmosis takes place. Allowed enough time, the two solutions will eventually have the same concentration.

COLLOIDS (SECTIONS 7.8)

☑7.82 Detergents or soaps are needed if water is to be used as a solvent for cleaning clothes and dishes because oil and grease are nonpolar substances that will repel water alone. The detergent or soap molecules have both polar and nonpolar regions; therefore, the oil and grease can be dissolved by the nonpolar portion of the detergent or soap molecules, which form a micelle around the oil or grease. The outer portion of the micelle consists of the outward-facing polar portions of the soap molecules, which are attracted to the polar water molecules; therefore, the entire micelle becomes suspended in the water and can be washed away.

DIALYSIS (SECTIONS 7.9)

☑7.84
- a. The hydrated sodium and chloride ions will pass through the dialyzing membrane, but the starch (colloid) will not.
- b. The urea will pass through the dialyzing membrane because it is a small organic molecule, but the starch (colloid) will not.
- c. The hydrated potassium and chloride ions as well as the glucose molecules will pass through the dialyzing membrane, but the albumin (colloid) will not.

ADDITIONAL EXERCISES

7.86 $$7.00\%(\text{w/v}) = \frac{7.00\ \text{g sodium chloride}}{100\ \text{mL solution}}\times100$$

volume solution = volume of water

$$\text{Density} = \frac{\text{mass solution}}{\text{volume of solution}} = \frac{7.00\ \text{g NaCl} + 100\ \text{mL H}_2\text{O}\left(\frac{1.00\ \text{g}}{1\ \text{mL}}\right)}{100\ \text{mL}} = 1.07\ \tfrac{\text{g}}{\text{mL}}$$

7.88 $Zn\ (s) + 2\ HCl\ (aq) \rightarrow ZnCl_2\ (aq) + H_2\ (g)$

$$0.500\ \text{g Zn}\left(\frac{1\ \text{mole Zn}}{65.4\ \text{g Zn}}\right)\left(\frac{2\ \text{moles HCl}}{1\ \text{mole Zn}}\right)\left(\frac{1\ \text{L HCl}}{1.50\ \text{moles HCl}}\right)\left(\frac{1000\ \text{mL HCl}}{1\ \text{L HCl}}\right) = 10.2\ \text{mL HCl}$$

ALLIED HEALTH EXAM CONNECTION

7.90 If salt is added to water, (a) the boiling point will increase and the freezing point will decrease.

7.92 If a semipermeable membrane is placed between two different concentrations of a NaCl solution, then (c) the solvent will move toward the higher concentration of solute.

7.94 (b) NH_3 is a nonelectrolyte when dissolved in water.

7.96 A solution that contains all the solute it can normally dissolve at a given temperature must be (c) saturated.

7.98 Cells that contain more dissolved salts and sugars than the surrounding solution are called (b) hypertonic.

7.100 Ice can be melted most effectively by (b) calcium chloride if 1 mole is used.

7.102 (b) 25 g of sugar are needed to make 500 mL of a 5% (weight/volume) solution of sugar.

$$500\text{ mL}\left(\frac{5\text{ g sugar}}{100\text{ mL solution}}\right) = 25\text{ g sugar}$$

7.104 In a dilute solution of sodium chloride in water, the sodium chloride is the (b) solute.

7.106 If 58.5 g of NaCl (1 mole of NaCl) are dissolved in enough water to make 0.500 L of solution, the molarity of this solution is (a) 2.0 M.

$$\frac{1\text{ mole}}{0.500\text{ L}} = 2.0\text{ M}$$

7.108 If a 2.0 M solution is diluted to 0.5 M, and the final volume is 100 mL, the original volume was (d) 25 mL.

$$\begin{aligned} C_cV_c &= C_dV_d \\ (2.0\text{ M})V_c &= (0.5\text{ M})(100\text{ mL}) \\ V_c &= 25\text{ mL} \end{aligned}$$

7.110 The molarity of a solution made by dissolving 29.2 g of NaCl in water is (a) 0.5 M.

$$\frac{29.2\text{ g NaCl}\left(\frac{1\text{ mole NaCl}}{58.4\text{ g NaCl}}\right)}{1\text{ L solution}} = 0.5\text{ M NaCl}$$

7.112 When placed in distilled water, a human red blood cell (d) swells to a larger size.

7.114 The primary intermolecular interactions between a K cation and H_2O molecules are (c) ion-dipole interactions.

CHEMISTRY FOR THOUGHT

7.116 Dialysis tubing is not submerged in pure water because more water would flow into the blood, while more of the solutes would flow out of the dialysis tubing than desired. The dialyzing membranes allow both solvent (water) molecules and small solute (Na^+, Cl^-, K^+, HCO_3^-, and glucose) ions and molecules to flow through the membrane. Passing the tubing through a bath that contains a small amount of sodium, chloride, potassium, and bicarbonate ions as well as glucose, allows the blood to be cleaned without stripping it of too many of the needed ions and small molecules in the blood.

7.118 Fish sometimes die when the temperature of the water in which they live increases because the solubility of salts increases and the solubility of gases (like oxygen) decrease with an increase in temperature. If the concentration of the salts in the water becomes too high or the concentration of gases becomes too low, the fish are unable to survive.

7.120 Sugar dissolves faster in hot tea than in iced tea. The solubility of sugar is higher in hot tea than in iced tea (see Figure 7.3).

7.122 Strips of fresh meat can be dried by coating them with table salt and then exposing them to air because the salt will draw out the water from the meat (the concentration of salt on the surface of the meat is greater than the concentration of salt inside the meat; therefore, the water will flow by osmosis out of the meat into the salt layer) and then the water will evaporate from the salt layer (assuming the meat is being dried in an environment that is not too humid or a closed container in which equilibrium between the liquid and vapor phases will be established), leaving the meat dry.

ADDITIONAL ACTIVITIES

Section 7.1 Review:

(1) Draw a before and after picture of the process of making a salt water solution.
(2) What is the solute in (1)? What is the solvent in (1)?
(3) What process occurs in (1)?
(4) What ways can you think of to speed up the process in (1)?

Section 7.2 Review:

(1) What do the following prefixes and suffix mean: *in-*, *im-*, *super-*, *un-*, *-ility*?
(2) What is the difference between soluble and miscible?
(3) What is the relationship between solubility and temperature?
(4) What is the relationship between solubility and pressure?

Section 7.3 Review:

(1) Both solutes and solvents have interparticle forces. What has to occur in order for a solution to form?
(2) What does "like" mean in "like dissolves like"?
(3) Why does "like dissolves like"?
(4) What patterns are evident in Table 7.4?
(5) Describe the fastest way to make a glass of iced tea sweetened with solid sugar.
(6) Have the definitions of *endothermic* and *exothermic* changed since Chapter 5?

Section 7.4 Review:

(1) What is the difference between the molarity of a solution and the density of a solution?
(2) What do all of the formulas for molarity, %(w/w), %(w/v), and %(v/v) have in common?
(3) How many grams of solvent are in 200 g of a 12.0% (w/w) sugar solution?

Section 7.5 Review:

(1) What is the advantage of selling orange juice concentrate?
(2) What process must occur in order for the orange juice concentrate to become a beverage ready to drink?
(3) If the volume of orange juice concentrate is 16 fluid ounces and the volume of ready to drink orange juice is 64 fluid ounces, what the ratio of the concentration of the concentrate to the ready to drink beverage?

Section 7.6 Review:

(1) Describe the process of converting liters of solution to moles of solute using the molarity.
(2) Describe the process of converting moles of solute to liters of solution using the molarity.
(3) A solution has a concentration of 5.00 M, write two factors that could be used in dimensional analysis for this solution.
(4) Identify which factor from (3) could be used in the factor-unit method to convert from liters of solution to moles of solute.
(5) Identify which factor from (3) could be used in the factor-unit method to convert from moles of solute to liters of solution.

Section 7.7 Review:

(1) What is the difference between *dissolving* and *dissociating*?
(2) Do both *dissolving* and *dissociating* always occur at the same time?
(3) Describe the level of dissolving and dissociating for strong electrolytes, weak electrolytes, and nonelectrolytes.
(4) Compare the vapor pressure, boiling points, and freezing points of solutions and pure solvents.
(5) During osmosis, what flows through the semipermeable membrane?
(6) During osmosis, what is the direction of flow through the semipermeable membrane?
(7) I once left a bunch of grapes to soak in a container of water. I accidentally left the grapes overnight. When I returned, the grapes had split open. Explain what happened using your knowledge of colligative properties.

Section 7.8 Review:

(1) When making a gelatin dessert, a packet of gelatin flavored with sugar is added to 2 cups of boiling water. What is the dispersing medium in a gelatin dessert? What is the dispersed phase in a gelatin dessert?
(2) The gelatin dessert from (1) is placed in the refrigerator and allowed to cool. It becomes semisolid. What are two other names from this section that can be used to describe this dessert?
(3) What type of process (endothermic or exothermic) is described in (2)?
(4) What should happen if a flashlight is shown through the gelatin dessert after it cools? What is the name of this phenomenon?
(5) Mix half a cup of water and half a cup of vegetable oil in a clear container. Stir rapidly for 30 seconds. Stop stirring and observe the oil and water mixture. What happens?
(6) Add the yolk from an egg to the mixture from (1). Stir rapidly for 30 seconds. Stop stirring and observe the oil, water, and egg-yolk mixture. What happens?
(7) What is the difference between the mixtures in (1) and (2)?

Section 7.9 Review:

(1) Why is it important to use a dialyzing membrane rather than a semipermeable membrane when treating people suffering from a kidney malfunction?
(2) How could you test to see if grape skin acts as a semipermeable membrane or a dialyzing membrane?

Tying It All Together with a Laboratory Application:

A solution is made with iodine and benzene. Iodine is a (1) _______ molecule that dissolves well in benzene because benzene is (2) _______. Iodine is less (3) _______ in water than benzene because water is a (4) _______ molecule.

The actual (5) _______ of iodine in benzene is 16 g iodine per 100 g benzene which is a concentration of (6) _______ %(w/w). (7) _______ is the solute and (8) _______ is the solvent in this solution. If 0.81 g of iodine are added to 4.0 g of benzene, a(n) (9) _______ solution will form. (10) _______ of the iodine will dissolve. The (11) _______ can be increased by crushing or grinding the iodine and stirring the solution. The solubility will (12) _______ if the solution is heated. The solution cools and at room temperature, the solution is (13) _______. (14) What will happen if another crystal of iodine is added to the solution after it cools?

Iodine is a (15) _______ in benzene because the solution does not conduct electricity. This means that iodine (16) _______ (dissociates or does not dissociate) in benzene. The boiling point of pure benzene is 80.1°C and the K_b for benzene is 2.53°C/M. The boiling point of a 0.10 M iodine in benzene solution is (17) _______°C. The freezing point of the solution is (18) _______ (greater than, less than, or equal to) the freezing point of pure benzene. The vapor pressure of the solution is (19) _______ (greater than, less than, or equal to) the vapor pressure of pure benzene. At 22°C, the osmotic pressure of the 0.10 M iodine in benzene solution is (20) _______ torr.

In order to make 250 mL of a 0.250 M iodine in benzene solution, (21) _______ g of iodine should be added to a 250 mL volumetric flask and dissolve in a small amount of benzene before benzene is added to the mark on the flask. This solution could be used to prepare a 500 mL of a 0.10 M, if (22) _______ mL of the 0.25 M solution are (23) _______ to 500 mL.

Iodine reacts with aluminum to form aluminum iodide as shown below:

(24) _______ Al (s) + _______ I_2 (benzene) → _______ AlI_3 (s)

If a coil of 0.35 g of thin aluminum wire is added to 25 mL of the 0.10 M iodine in benzene solution, (25) _______ g of AlI_3 can be formed because (26) _______ is the limiting reactant.

SOLUTIONS FOR THE ADDITIONAL ACTIVITIES

Section 7.1 Review:

(1) 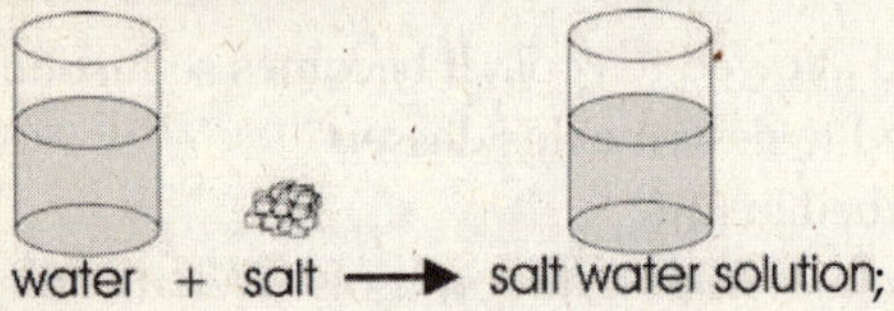

(2) Salt is the solute. Water is the solvent. (Note: The solute is the smaller part of a solution and has six letters. The solvent is the larger part of a solution and has 7 letters.)

(3) Dissolving is the process that occurs when making a solution. Because salt is an ionic compound, it will also dissociate when it dissolves.

(4) Heating the water, increasing the surface area of the salt, and stirring the components are a few methods for speeding up the solution process.

Section 7.2 Review:

(1) *in-* = no, not, without, non-; *im-* = same as *in-*, but used before *b*, *m*, or *p*; *super-* = to a degree greater than normal; *un-* = not, lack of, the opposite of; *-ility* = the quality of being (as specified)

(2) Soluble means that two substances can form a solution. Miscible means that two liquids can form a solution.

(3) For solid and liquid solutes, solubility usually increases as temperature increases. For gaseous solutes, solubility usually decreases as temperature increases.

(4) For solid and liquid solutes, solubility is not influenced by pressure. For gaseous solutes, solubility usually increases as pressure increases.

Section 7.3 Review:

(1) In order for a solution to form, the interparticle forces between solute particles must be overcome by the solvent particles and the interparticle forces between the solvent particles must be overcome by the solute particles. Overcoming the existing interparticle forces is easiest when the solute and solvent particles will establish strong interparticle forces as they interact.

(2) Similar interparticle forces in the solute and the solvent are the "like" in "like dissolves like."

(3) "Like dissolves like" because the new interparticle forces between the solute and the solvent will be strong enough to overcome the interparticle forces within the solute and within the solvent. Polar solvents are attracted to polar or ionic solutes because both the solvent and the solute have either partial or full charges. Nonpolar solvents are attracted to nonpolar solutes because neither the solvent nor the solute has any partial or full charges.

(4) The two ions that primarily form water insoluble compounds are the carbonates and the phosphates. Both of these ions have large charges. Compounds that contain these ions would have strong ionic bonds because of the differences in the charges. The ions that are associated with water soluble compounds have small charges. The chlorides and the sulfates are mainly water soluble, but when they combine with heavy ions, they can form insoluble compounds. Water is unable to break the strong bonds these compounds form.

(5) While the tea is still hot, stir in the desired amount of solid sugar. For the fastest results, use extremely fine sugar. Once the sugar has dissolved, add ice. The sugar will dissolve much more rapidly in a warm solution than a cold solution.

(6) The definitions of *endothermic* and *exothermic* have not changed since Chapter 5. Endothermic solution processes require energy just like endothermic chemical reactions. Exothermic solution processes release energy just like exothermic chemical reactions.

Section 7.4 Review:

(1) The molarity of a solution is the ratio of moles of solute per liter of solution. The density of a solution is the ratio of mass of solution per milliliter of solution. Density is calculated based on the mass of the entire solution. Molarity is calculated based on the moles of solute.

(2) All of the formulas for molarity, %(w/w), %(w/v), and %(v/v) contain a fraction with solute in the numerator and solution in the denominator.

$$\frac{\text{solute}}{\text{solution}};\ M=\frac{\text{moles }\textbf{solute}}{\text{L }\textbf{solution}};\ \%(w/w)=\frac{\text{mass }\textbf{solute}}{\text{mass }\textbf{solution}}\times 100;$$

$$\%(w/v)=\frac{\text{mass }\textbf{solute}\text{ in grams}}{\text{volume }\textbf{solution}\text{ in mL}}\times 100;\ \%(v/v)=\frac{\text{volume }\textbf{solute}}{\text{volume }\textbf{solution}}\times 100$$

(3) $200\text{ g solution}(12.0\%\text{ sugar solution})=24.0\text{ g sugar};\ \ 200\text{ g solution}-24.0\text{ g sugar}=176\text{ g solvent}$

Section 7.5 Review:

(1) Orange juice concentrate does not take up as much space as ready to drink orange juice. Refrigeration space is costly for a grocer, so it is advantageous to sell orange juice concentrate because it takes less refrigeration space.

(2) The concentrate must be diluted before it is ready to drink.

(3) $(C_c)(V_c)=(C_d)(V_d)\ \Rightarrow\ \frac{(C_c)}{(C_d)}=\frac{(V_d)}{(V_c)}\ \Rightarrow\ \frac{(C_c)}{(C_d)}=\frac{64\text{ fl. oz.}}{16\text{ fl. oz.}}=\frac{4}{1}$

The orange juice concentrate is four times more concentrated than the ready to drink beverage because it has ¼ the volume of ready to drink orange juice.

Section 7.6 Review:

(1) To convert liters of solution to moles of solute using the molarity, multiply the volume in liters by the molarity.

(2) To convert moles of solute to liters of solution using the molarity, divide the moles of solute by the molarity (or multiply by the reciprocal of the molarity).

(3) $\frac{5.00\text{ moles solute}}{1\text{ L solution}}$ and $\frac{1\text{ L solution}}{5.00\text{ moles solute}}$

(4) $\text{liters}\left(\frac{5.00 \text{ moles solute}}{1 \text{ L solution}}\right) = \text{moles}$

(5) $\text{moles}\left(\frac{1 \text{ L solution}}{5.00 \text{ moles solute}}\right) = \text{liters}$

Section 7.7 Review:

(1) Dissolving is the process by which a solute and solvent form a solution. Dissociating is the process by which a solute breaks apart into ions in solution.

(2) Dissolving does not necessarily involve a dissociation process; however, when dissociation occurs, it accompanies the dissolving process. Ionic compounds and highly polar covalent compounds dissociate in polar solvents like water. Nonpolar covalent compounds or weakly polar covalent compounds do not dissociate in polar solvents like water; however, some of these compounds dissolve in water.

(3) Strong electrolytes both dissolve and dissociate completely. Weak electrolytes dissolve completely, but they do not dissociate completely. Nonelectrolytes dissolve completely, but they do not dissociate.

(4) Solutions have lower vapor pressures, higher boiling points, and lower freezing points than pure solvents.

(5) Solvent flows through the semipermeable membrane during osmosis.

(6) Solvent flows from the solution of lower solute concentration to the solution of higher solute concentration during osmosis.

(7) The "skin" on the grapes functions as a semipermeable membrane. The concentration of solutes was higher inside the grapes than in the container of water. The solvent (water) flowed through the semipermeable membranes into the grapes. Osmosis continued until the skin of the grapes could not hold any more water and the skin split.

Section 7.8 Review:

(1) The dispersing medium is water. The dispersed phase is the gelatin flavored with sugar. (Notice: Medium is the larger component in a colloid and it has 6 letters. Phase is the smaller component in a colloid and it has 5 letters.)

(2) The dessert is a sol and a gel. It is a sol because the dispersing medium is a liquid and the dispersed phase is a solid. It is a gel because it is a sol that becomes semisolid.

(3) The process of forming this gel is exothermic. The heat in the dessert is released as the gel forms.

(4) The path of the beam of light should be visible. This is called the Tyndall effect.

(5) The oil and water begin to separate. An unstable emulsion exists briefly as a middle layer between the oil on top and the water on the bottom. Eventually, only two layers (oil and water) remain.

(6) The mixture becomes cloudy when the egg yolk is added. The mixture does not separate into layers. The consistency of the mixture is relatively homogenous.

(7) The egg yolk acts as an emulsifying agent (also known as a stabilizing agent) in the second test. The first mixture formed an unstable emulsion that coalesced into the original materials. Because the second mixture had the egg yolk to serve as an emulsifying agent it was more stable and the emulsion lasted longer.

Section 7.9 Review:

(1) Properly functioning kidneys filter out waste products from the blood which include small molecules and hydrated ions. If a semipermeable membrane were used to treat people suffering from a kidney malfunction, water could flow into the membrane and dilute the blood, but the waste products could not leave the blood. A dialyzing membrane allows for small molecules and hydrated ions to flow

through the membrane as well as for water to flow into and out of the membrane. The dialyzing membrane functions more like a kidney than a semipermeable membrane does.

(2) After cleaning a grape, it could be placed into a container of distilled water. A "control" sample of distilled water should also be kept. The grape should be left in the container of distilled water for a length of time, but not long enough to split the grape skin. Once the grape is removed from the container, the water should be tested for small molecules. If the water that interacted with the grape contains small molecules or ions that are not present in the control distilled water, then the grape cell membranes are functioning as a dialyzing membrane. If the water that interacted with the grape is identical to the control water, then the grape skin is functioning as a semipermeable membrane.

Tying It All Together with a Laboratory Application:

(1) nonpolar
(2) nonpolar
(3) soluble
(4) polar
(5) solubility
(6) 14%(w/w)
(7) Iodine
(8) benzene
(9) saturated
(10) Most (0.64 g, 79%)
(11) rate of dissolving
(12) increase
(13) supersaturated

(14) The additional crystal of iodine will act as a seed crystal. All of the extra dissolved iodine will precipitate out of solution, leaving a saturated solution.

(15) nonelectrolyte
(16) does not dissociate
(17) 80.4°C
(18) less than
(19) less than
(20) 1.8×10^3 torr
(21) 15.9 g
(22) 200 mL
(23) diluted
(24) 2, 3, 2
(25) 0.68 g
(26) I_2 in benzene

SELF-TEST QUESTIONS

Multiple Choice

1. A solution is prepared by dissolving a small amount of sugar in a large amount of water. In this case sugar would be the:
   a. filtrate.
   b. solute.
   c. precipitate.
   d. solvent.

2. The freezing point of a solution:
   a. is lower than that of pure solvent.
   b. is higher than that of pure solvent.
   c. is the same as that of pure solvent.
   d. cannot be measured.

3. Which of the following would show the Tyndall effect?
   a. a solution of salt in water
   b. a solution of sugar in water
   c. a solution of CO_2 gas in water
   d. a colloidal suspension

4. A crystal of solid magnesium sulfate is placed into a solution of magnesium sulfate in water. It is observed that the crystal dissolves slightly. The original solution was:
   a. saturated.
   b. unsaturated.
   c. supersaturated.
   d. cannot determine from the data given.

5. The weight/weight percent of sugar in a solution containing 25 grams of sugar and 75 grams of water would be:
   a. 25%
   b. 33%
   c. 75%
   d. 80%

6. How many moles of $Mg(NO_3)_2$ are contained in 500 mL of 0.400 M solution?
   a. 0.400
   b. 0.200
   c. 0.800
   d. 2.00

7. How many grams of $Mg(NO_3)_2$ must be dissolved in water to give 500 mL of 0.400 M solution?
   a. 29.7
   b. 17.3
   c. 59.3
   d. 172.6

8. Which of the following aqueous solutions would be expected to have the highest boiling point at 1 atm pressure?
   a. 1 M NaCl
   b. 1 M $C_{12}H_{22}O_{11}$ (sucrose)
   c. pure water
   d. 1 M AlF_3

9. Calculate the freezing point of a water solution that contains 1.60 grams of methyl alcohol (a nonelectrolyte) CH_3OH, in each 100.0 mL. The K_f for water is 1.86°C/M.
   a. 0.93°C
   b. -0.93°C
   c. -1.86°C
   d. -3.72°C

10. How many moles of solute would be needed to prepare 250 mL of 0.150 M solution?
    a. 37.5
    b. 0.150
    c. 0.0375
    d. 0.600

11. What volume of 0.200 M silver nitrate solution ($AgNO_3$) would have to be diluted to form 500 mL of 0.050 M solution?
    a. 200 mL
    b. 2.0×10^{-5}
    c. 80.0 mL
    d. 125 mL

12. Calculate the osmolarity of a 0.400 M NaCl solution.
    a. 0.400
    b. 0.800
    c. 0.200
    d. 0.100

13. What is the osmotic pressure (in torr) of a 0.0015 M NaCl solution at 25°C? $R = 62.4 \frac{\text{L·torr}}{\text{mol·K}}$.
    a. 4.68
    b. 2.34
    c. 27.9
    d. 55.8

True-False

14. Paint is a colloid.
15. Gases dissolve better in liquids when the liquids are warm.
16. A solution can contain only one solute.
17. A solution is a homogeneous mixture.
18. A colligative property is used to prevent winter freeze-up of cars.
19. Water is a good solvent for both ionic and polar covalent materials.
20. A nonpolar solute would be relatively insoluble in water.
21. On the basis of their solubility in each other, water and gasoline molecules have similar polarities.
22. Potassium nitrate is soluble in water.
23. Soluble solutes always dissolve rapidly.
24. The larger a solute particle, the faster it will dissolve in a solvent.
25. An exothermic solution process releases heat.
26. A solution has a higher boiling point than the pure solvent.
27. The number of moles of sodium chloride in 1.00 L of 0.300 M NaCl (aq) is greater than the number of moles of sodium chloride in 50.0 mL of 6.00 M NaCl (aq).
28. The concentration of a solution increases when it is diluted.
29. The colligative properties depend only on the concentration of solute particles in a solution, not the identity of the solute particles.
30. Semipermeable membranes allow solute molecules to pass from a solution of lower concentration to a solution of higher concentration.

Matching

31.	a liquid dispersed in a gas	a.	foam
32.	a liquid dispersed in a liquid	b.	emulsion
33.	mayonnaise is an example	c.	aerosol
34.	a gas dispersed in a liquid	d.	sol
35.	gelatin dessert is an example		

Chapter 8: Reaction Rates and Equilibrium

CHAPTER OUTLINE

8.1 Spontaneous and Nonspontaneous Processes
8.2 Reaction Rates
8.3 Molecular Collisions
8.4 Energy Diagrams
8.5 Factors that Influence Reaction Rates
8.6 Chemical Equilibrium
8.7 The Position of Equilibrium
8.8 Factors That Influence Equilibrium Position

LEARNING OBJECTIVES/ASSESSMENT

When you have completed your study of this chapter, you should be able to:

1. Use the concepts of energy and entropy to predict the spontaneity of processes and reactions. (Section 8.1; Exercise 8.6)
2. Calculate reaction rates from experimental data. (Section 8.2; Exercise 8.14)
3. Use the concept of molecular collisions to explain reaction characteristics. (Section 8.3; Exercise 8.20)
4. Represent and interpret the energy relationships for reactions by using energy diagrams. (Section 8.4; Exercise 8.26)
5. Explain how factors such as reactant concentrations, temperature, and catalysts influence reaction rates. (Section 8.5; Exercise 8.30)
6. Relate experimental observations to the establishment of equilibrium. (Section 8.6; Exercise 8.38)
7. Write equilibrium expressions based on reaction equations, and do calculations based on equilibrium expressions. (Section 8.7; Exercises 8.40 and 8.46)
8. Use Le Châtelier's principle to predict the influence of changes in concentration and reaction temperature on the position of equilibrium for a reaction. (Section 8.8; Exercise 8.52)

SOLUTIONS FOR THE END OF CHAPTER EXERCISES

SPONTANEOUS AND NONSPONTANEOUS PROCESSES (SECTION 8.1)

8.2	a.	The space shuttle leaves its pad and goes into orbit.	nonspontaneous	Rocket engines must continually operate to push the shuttle into an orbit.
	b.	The fuel in a booster rocket of the space shuttle burns.	spontaneous	Once the fuel is ignited, it will continue to burn. No additional energy has to be provided.
	c.	Water boils at 100°C and 1 atm pressure.	nonspontaneous	Heat must be continually supplied to maintain boiling.
	d.	Water temperature increases to 100°C at 1 atm pressure.	nonspontaneous	Increasing the temperature of water requires a continual supply of energy.
	e.	Your bedroom becomes orderly.	nonspontaneous	A room will not become orderly on its own. Cleaning requires energy.

8.4	a.	An automobile being pushed up a slight hill (from point of view of the one pushing)	exergonic	The person pushing the car gives energy to the car.
	b.	Ice melting (from point of view of the ice)	endergonic	Melting ice requires energy.
	c.	Ice melting (from point of view of surroundings of the ice)	exergonic	The surroundings release heat into the ice.
	d.	Steam condensing to liquid water (from point of view of the steam)	exergonic	Heat must be released from the steam.
	e.	Steam condensing to liquid water (from point of view of surroundings of the steam)	endergonic	Heat must be absorbed by the surroundings.

☑8.6	a.	On a cold day, water freezes.	energy decreases entropy decreases	spontaneous (if the temperature is less than 0°C)
	b.	A container of water at 40°C cools to room temperature.	energy decreases entropy decreases slightly	spontaneous
	c.	The odor from an open bottle of perfume spreads throughout a room.	energy increases slightly entropy increases	spontaneous

8.8
a. The highest entropy example from this set is the two opposing football teams when the whistle is blown, ending the play. This is the highest entropy because the football players from the two teams are most interspersed.
b. From the gold's perspective, the highest entropy example is a 10% copper/gold alloy because the gold atoms are more separated in this material.

From the copper's perspective, the highest entropy example is a 2% copper/gold alloy because the copper atoms are more separated in this material.
c. The purse on the ground with the contents scattered is the highest entropy example from this set. This is the highest entropy example because the disorder is highest in this example.
d. The coins in a piggy bank are the example from this set with the highest entropy. The random arrangement of coins in the piggy bank represents greater disorder than the other two examples.
e. The dozen loose pearls in a box are the example from this set with the highest entropy. The random order of the pearl represents greater disorder than the other two examples.

REACTION RATES (SECTION 8.2)

8.10	a.	The cooking of a pizza in a hot oven	slow*
	b.	The change in leaf color from green to red or orange in the autumn	very slow*
	c.	The drying of a wet swim suit hung out in the sun	slow*
	d.	The burning of a match after it has been ignited	fast*
	e.	The growing of grass during a warm summer	very slow*

*The definitions of very slow, slow, and fast are subjective. Your answers to this question may differ.

8.12	a.	The melting of a block of ice	The changing height of the block, the changing mass of the block, or the increasing volume of liquid formed could be measured.
	b.	The setting (hardening) of concrete	The ability of an object to penetrate or mark the surface of the concrete could be measured.
	c.	The burning of a candle	The changing height of the candle or the changing mass of the candle could be measured.

☑8.14
a. Pure A and B are mixed, and after 12.0 minutes the measured concentration of C is 0.396 mol/L.

$$\frac{0.396\text{ M} - 0.000\text{ M}}{12.0\text{ min}} = 0.0330\frac{\text{M}}{\text{min}}$$

b. Pure A, B, and C are mixed together at equal concentrations of 0.300 M. After 8.00 minutes, the concentration of C is found to be 0.455 M.

$$\frac{0.455\text{ M} - 0.300\text{ M}}{8.00\text{ min}} = 0.0194\frac{\text{M}}{\text{min}}$$

8.16

$$\frac{\left(\dfrac{2.97\times10^{-2}\text{ mol}}{250\text{ mL}\left(\frac{1\text{ L}}{1000\text{ mL}}\right)}\right)-0.000\text{ M}}{30.0\text{ min}}=3.96\times10^{-3}\ \frac{\text{M}}{\text{min}}$$

8.18

$$\left(2.77\times10^{-2}\text{ atm}\right)\left(500\text{ mL}\right)\left(\tfrac{1\text{ L}}{1000\text{ mL}}\right)=n\left(0.0821\tfrac{\text{L}\cdot\text{atm}}{\text{mol}\cdot\text{K}}\right)\left((25.0+273)\text{K}\right)$$

$$n=5.66\times10^{-4}\text{ moles}$$

$$\frac{\dfrac{5.66\times10^{-4}\text{ moles}}{500\text{ mL}\left(\frac{1\text{ L}}{1000\text{ mL}}\right)}-0\text{ M}}{750\text{ sec}}=1.51\times10^{-6}\ \frac{\text{M}}{\text{sec}}$$

MOLECULAR COLLISIONS (SECTION 8.3)

☑8.20
a. In order for a reaction to occur, the reactant molecules must collide. Fewer collisions will occur if the concentration is decreased.
b. In order for a reaction to occur, the reactant molecules must collide with the right amount of energy. If the temperature is lowered, the reactants will lose energy.
c. In order for a reaction to occur, the reactant molecules must collide. More collisions will occur if the concentration of either of the reactants is increased.

8.22 As temperature increases, the molecules move faster. The faster the molecules move, the more likely they are to collide.

As the temperature increases, the molecules gain energy. The more energy the molecules have, the more they are likely to react once they collide.

ENERGY DIAGRAMS (SECTION 8.4)

8.24
a. Endothermic (endergonic) reaction with activation energy.

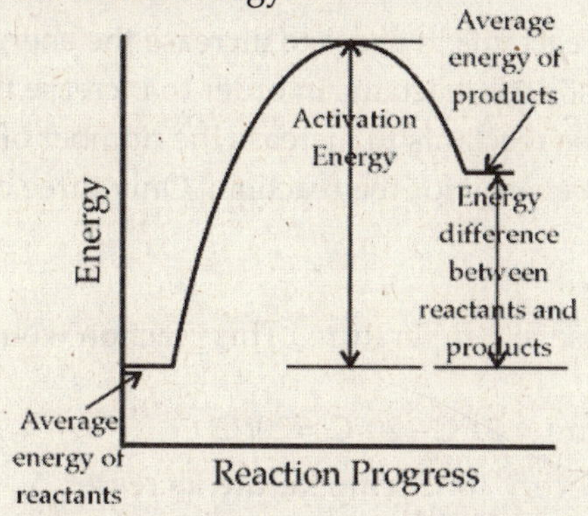

b. Endothermic (endergonic) reaction without activation energy.

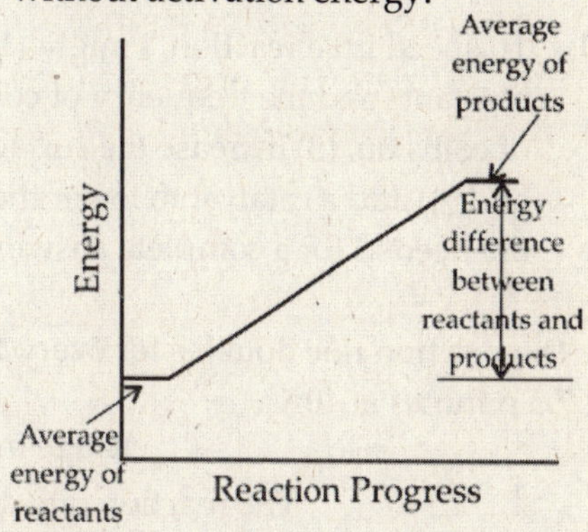

Both of these energy diagrams have the same average energy of the reactants, average energy of products, and energy difference between reactants and products. The main difference between these two energy diagrams is that the first diagram has an activation energy and the second diagram does not.

☑8.26

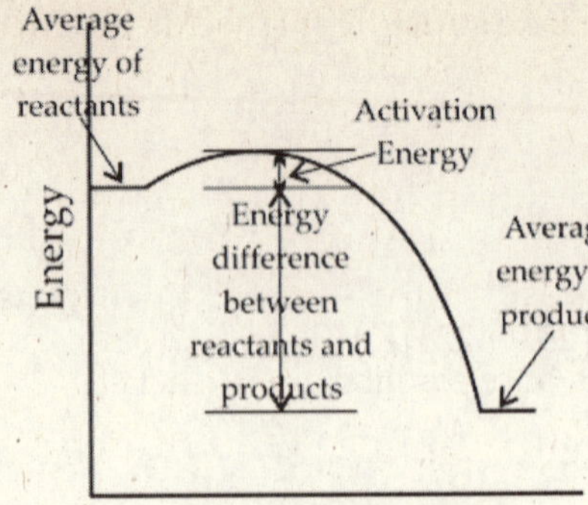

Reaction Progress
room temperature

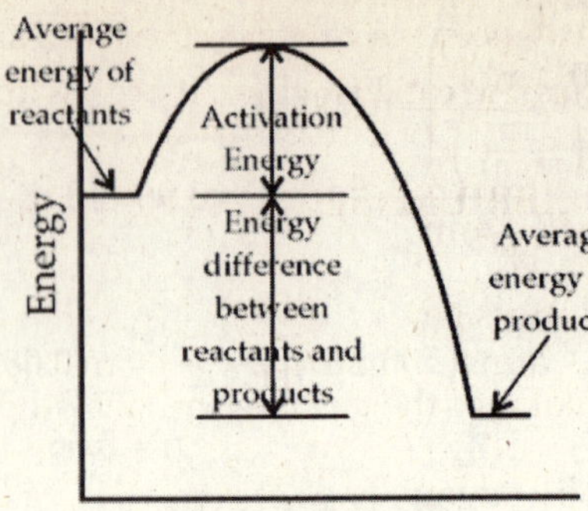

Reaction Progress
150°C

Both of these energy diagrams have the same average energy of the reactants, average energy of products, and energy difference between reactants and products (500 kJ/mol reactant). The reaction that occurs at room temperature has a small activation energy (or possibly does not require activation energy), but the reactants have sufficient energy to overcome the activation energy barrier. The reaction that occurs at 150°C has a significant activation energy.

FACTORS THAT INFLUENCE REACTION RATES (SECTION 8.5)

8.28 a. $2\ I^-\ (aq) + Pb^{2+}(aq) \rightarrow PbI_2\ (s)$
Rapid; The reactants are ions and the product is insoluble in water.

b. $Br^-\ (aq) + I^-\ (aq) \rightarrow BrI^{2-}\ (aq)$
Won't react; These ions are stable and do not have affinity for each other.

c. $NH_3\ (g) + HCl\ (g) \rightarrow NH_4Cl\ (s)$
Rapid; The reactant molecules are in the gas phase and will collide frequently.

d. $CaO\ (s) + CO_2\ (g) \rightarrow CaCO_3\ (s)$
The reaction rate depends on the surface area of the CaO. The larger the surface area of the solid, the faster the reaction will occur.

☑8.30 To speed up a reaction, I might (1) heat the reactants in order to increase the energy of the reactants and the frequency of collision, (2) stir the reactants in order to increase the frequency of collision, (3) increase the surface area of the reactants to increase the number of collisions, and (4) add a catalyst to lower the activation energy for the reaction. Only three of these steps are needed for a complete answer.

8.32 The reaction rate doubles for every 10°C increase in temperature. This reaction would only take 7.5 minutes at 40°C.

$$\text{change in temperature} = 40°\text{C} - 20°\text{C} = 20°\text{C}$$

The reaction rate doubles for every 10°C temperature increase.

$$20°\text{C change in temperature}\left(\frac{1\text{ double of the reaction rate}}{10°\text{C}}\right) = 2\text{ doubles of the reaction rate}$$

or 1 quadruple of the reaction rate

The reaction rate will be 4 times faster; therefore, the time required will be one – quarter as long.

$$30\text{ min}\left(\frac{1}{4}\right) = 7.5\text{ min}$$

8.34 Catalysts speed up reactions by lowering the activation energy required for a reaction. The activation energy may be lowered because the catalyst provides a surface on which the reaction can occur or may react to form an intermediate structure that yields products and regenerates the catalyst when it breaks apart.

CHEMICAL EQUILIBRIUM (SECTION 8.6)

8.36 a.

H_2	+	I_2	$\rightleftarrows$	2 HI
colorless gas		violet gas		colorless gas

The color of the gas mixture will stop changing once equilibrium is reached.

b.

solid sugar	+	water	$\rightleftarrows$	sugar solution

The amount of solid sugar will become constant once the mixture has reached equilibrium.

c.

N_2	+	$2\ O_2$	$\rightleftarrows$	$2\ NO_2$
colorless gas		colorless gas		red-brown gas

Both the color and the pressure of the gas mixture will stop changing once equilibrium is reached.

☑8.38 $H_2\ (g) + Br_2\ (g) \rightleftarrows 2\ HBr\ (g)$

If the sealed container initially contained only H_2 and red-brown Br_2, the concentrations of H_2 and Br_2 would decrease and the intensity of the color would decrease as equilibrium was established. Consequently, the concentration of HBr would increase as equilibrium was established. The color of the mixture would be constant once equilibrium was established.

THE POSITION OF EQUILIBRIUM (SECTION 8.7)

☑8.40 a. $2\ CO + O_2 \rightleftarrows 2\ CO_2$ $\quad K_{eq} = \frac{[CO_2]^2}{[CO]^2[O_2]}$

b. $N_2O_4 \rightleftarrows 2\ NO_2$ $\quad K_{eq} = \frac{[NO_2]^2}{[N_2O_4]}$

c. $2\ C_2H_6 + 7\ O_2 \rightleftarrows 4\ CO_2 + 6\ H_2O$ $\quad K_{eq} = \frac{[CO_2]^4[H_2O]^6}{[C_2H_6]^2[O_2]^7}$

d. $2\ NOCl \rightleftarrows 2\ NO + Cl_2$ $\quad K_{eq} = \frac{[NO]^2[Cl_2]}{[NOCl]^2}$

e. $2\ Cl_2O_5 \rightleftarrows O_2 + 4\ ClO_2$ $\quad K_{eq} = \frac{[O_2][ClO_2]^4}{[Cl_2O_5]^2}$

8.42 a. $Fe^{3+} + 6\ CN^- \rightleftarrows Fe(CN)_6^{3-}$ $\quad K_{eq} = \frac{[Fe(CN)_6^{3-}]}{[Fe^{3+}][CN^-]^6}$

b. $Ag^+ + 2\ NH_3 \rightleftarrows Ag(NH_3)_2^+$ $\quad K_{eq} = \frac{[Ag(NH_3)_2^+]}{[Ag^+][NH_3]^2}$

c. $Au^{3+} + 4\ Cl^- \rightleftarrows AuCl_4^-$ $\quad K_{eq} = \frac{[AuCl_4^-]}{[Au^{3+}][Cl^-]^4}$

8.44 a. $K = \frac{[CO_2][H_2O]^2}{[CH_4][O_2]^2}$ $CH_4 + 2\,O_2 \rightleftarrows CO_2 + 2\,H_2O$

b. $K = \frac{[CH_4][H_2O]}{[H_2]^3[CO]}$ $3\,H_2 + CO \rightleftarrows CH_4 + H_2O$

c. $K = \frac{[O_2]^3}{[O_3]^2}$ $2\,O_3 \rightleftarrows 3\,O_2$

d. $K = \frac{[NH_3]^4[O_2]^7}{[NO_2]^4[H_2O]^6}$ $4\,NO_2 + 6\,H_2O \rightleftarrows 4\,NH_3 + 7\,O_2$

☑8.46 $K_{eq} = \frac{[Br_2][Cl_2]}{[BrCl]^2} = \frac{[0.26\ M][0.26\ M]}{[0.38\ M]^2} = 0.47$

8.48 $K_{eq} = \frac{[NO]^2[Cl_2]}{[NOCl]^2} = \frac{[0.92\ M]^2[0.20\ M]}{[1.31\ M]^2} = 0.099$

8.50 a. $K = 5.9$ [reactants] smaller than [products]

b. $K = 3.3 \times 10^6$ [reactants] smaller than [products]

c. $K = 2.7 \times 10^{-4}$ [reactants] larger than [products]

d. $K = 0.0000558$ [reactants] larger than [products]

FACTORS THAT INFLUENCE EQUILIBRIUM POSITION (SECTION 8.8)

☑8.52 a. $Ag^+\ (aq) + Cl^-\ (aq) \rightleftarrows AgCl\ (s)$; some Ag^+ is removed. shift to the left

b. $2\,HI\ (g) + heat \rightleftarrows H_2\ (g) + I_2\ (g)$; the system is heated. shift to the right

c. $6\,Cu\ (s) + N_2\ (g) + heat \rightleftarrows 2\,Cu_3N\ (s)$; the system is cooled and some N_2 is removed. shift to the left

8.54 a. Cu^{2+} (aq) blue + $4\,NH_3$ (aq) colorless $\rightleftarrows$ $Cu(NH_3)_4^{2+}$ (aq); dark purple

some NH_3 is added to the equilibrium mixture.

The equilibrium will shift to the right and the mixture will become less blue and more purple.

b. Pb^{2+} (aq) colorless + $2\,Cl^-$ (aq) colorless $\rightleftarrows$ $PbCl_2$ (s) white solid + heat;

the equilibrium mixture is cooled.

The equilibrium will shift to the right and more precipitate will form. Heat will also be generated and the temperature of the container will increase.

c. C_2H_4 colorless gas + I_2 violet gas $\rightleftarrows$ $C_2H_4I_2$ colorless gas + heat;

some $C_2H_4I_2$ is removed from the equilibrium mixture.

The equilibrium will shift to the right and the mixture will become less violet and more heat will be produced.

d. C_2H_4 (colorless gas) + I_2 (violet gas) $\rightleftarrows$ $C_2H_4I_2$ (colorless gas) + heat; the equilibrium mixture is cooled.

The equilibrium will shift to the right and the mixture will become less violet and more heat will be produced.

e. heat + $4\ NO_2$ (brown gas) + $6\ H_2O$ (colorless gas) $\rightleftarrows$ $7\ O_2$ (colorless gas) + $4\ NH_3$ (colorless gas); a catalyst is added, and NH_3 is added to the equilibrium mixture.

The catalyst does not have any effect on the equilibrium; however, adding NH_3 shifts the equilibrium to the left and this produces heat, lowers the pressure because there will be fewer moles of gas present, and increases the brown color of the equilibrium mixture.

8.56 a. $2\ N_2O\ (g) + 3\ O_2\ (g) \rightleftarrows 4\ NO_2\ (g) + \text{heat}$; O_2 is added

The equilibrium will shift to the right. During the shift to the right, the concentrations of N_2O and O_2 will decrease and the concentration of NO_2 will increase. (Note: The O_2 equilibrium concentration will be higher than the initial O_2 concentration because the stress to the system was adding more O_2 and not all of the "new" O_2 will be used in the shift to the right.)

b. $2\ SO_3\ (g) + \text{heat} \rightleftarrows 2\ SO_2\ (g) + O_2\ (g)$; the system is heated

The equilibrium will shift to the right. The concentration of SO_3 will decrease and the concentrations of SO_2 and O_2 will increase.

c. $2\ CO\ (g) + O_2\ (g) \rightleftarrows 2\ CO_2\ (g) + \text{heat}$; the system is cooled

The equilibrium will shift to the right. The concentrations of CO and O_2 will decrease and the concentration of CO_2 will increase.

8.58 $N_2\ (g) + 3\ H_2\ (g) \rightleftarrows 2\ NH_3\ (g) + \text{heat}$

a. Some N_2 is added. — to the right
b. The temperature is increased. — to the left
c. Some NH_3 is removed. — to the right
d. Some H_2 is removed. — to the left
e. A catalyst is added. — no shift
f. The temperature is increased, and some H_2 is removed. — to the left

ADDITIONAL EXERCISES

8.60 A (g) + B (g) $\rightarrow$ C (s)

If gases A and B are used to fill a balloon, the concentration of A and B could be increased by decreasing the volume of the balloon or increasing the pressure on the balloon. Increasing the concentration of A and B will speed up the reaction.

8.62 a. Evaporation of a liquid — The energy is increasing and the entropy is increasing. Since this is spontaneous, the entropy increase must be enough to compensate for the energy increase.

b. Condensation of a gas to a liquid	The entropy is decreasing and the energy is decreasing. Since this is spontaneous, the energy decrease must be enough to compensate for the entropy decrease.
c. Sublimation of a solid to a gas	The energy is increasing and the entropy is increasing. Since this is spontaneous, the entropy increase must be enough to compensate for the energy increase.
d. Liquefaction of a gas to a liquid	The entropy is decreasing and the energy is decreasing. Since this is spontaneous, the energy decrease must be enough to compensate for the entropy decrease.
e. Crystallization of a liquid to a solid	The entropy is decreasing and the energy is decreasing. Since this is spontaneous, the energy decrease must be enough to compensate for the entropy decrease.

ALLIED HEALTH EXAM CONNECTION

8.64 If the reaction $A + B \rightarrow C + D$ is designated as first order, the rate depends on (a) the concentration of only one reactant.

8.66 A book is held six feet above the floor and then dropped. (a) The potential energy of the book is converted to kinetic energy.

8.68 Stored energy is referred to as (c) potential energy.

8.70 An example of an exothermic change is (b) condensation.

8.72 The best example of potential energy changing to kinetic energy is (a) pushing a rock off a cliff.

8.74 (a) Ice melting is endothermic.

8.76 A catalyst operates by (a) decreasing the activation energy barrier for a reaction.

8.78 When there is an increase in pressure to the system of $2\ CO\ (g) + O_2\ (g) \rightleftarrows 2\ CO_2\ (g)$, one would expect (a) an increase in the amount of carbon dioxide.

8.80 For $N_2\ (g) + 3\ H_2\ (g) \rightleftarrows 2\ NH_3\ (g) + \text{heat}$, it is incorrect that (a) an increase in temperature will shift the equilibrium to the right.

8.82 The effect of the addition of a catalyst to a reaction in equilibrium is (c) there is no change in composition of the reaction.

8.84 The equilibrium constant for $2\ SO_2\ (g) + O_2\ (g) \rightleftarrows 2\ SO_3\ (g)$ is (d) $K_c = \dfrac{[SO_3]^2}{[SO_2]^2[O_2]}$.

CHEMISTRY FOR THOUGHT

8.86 $2\ NOCl\ (g) \rightleftarrows 2\ NO\ (g) + Cl_2\ (g)$

$$K_c = \frac{[NO]^2[Cl_2]}{[NOCl]^2} = \frac{\left[\frac{0.70\ mole}{1.50\ L}\right]^2\left[\frac{0.35\ mole}{1.50\ L}\right]}{\left[\frac{1.80\ mole}{1.50\ L}\right]^2} = 0.035$$

8.88 $H_2 + I_2 \rightleftarrows 2\ HI \qquad K_{eq} = 50.5$

$$K_{eq} = \frac{[HI]^2}{[H_2][I_2]}$$

$$50.5 = \frac{[0.500\ M]^2}{[0.050][I_2]}$$

$$[I_2] = 0.099\ M$$

8.90 In the equation $N_2O_4 \rightleftarrows 2\ NO_2$, heat is a reactant because the system favors the reactants when the system is cooled and favors the products when the system is heated. LeChatelier's principle states that the equilibrium will shift to counteract the stress added to the system; therefore, if the system is cooled and shifts toward the reactants, that means heat is a reactant that could counteract the stress of cooling. If the system is heated and shifts toward the products, that means heat is a reactant that needs to be used in order to counteract the stress of heating. Heat is a reactant; therefore, the system is endothermic. The presence of a catalyst in the tube would not influence the equilibrium concentrations of the two gases, it would merely allow equilibrium to be reached sooner. The catalyst lowers the activation energy for both the forward and reverse reactions equally, and therefore, only increases reaction rates.

8.92 Smoking is dangerous in the presence of oxygen gas. The abundance of oxygen (an oxidizing agent) would increase the reaction rate for a redox reaction occurring between any reducing agent and the oxygen gas. A lit cigarette or even a small amount of ash containing an ember could provide the activation energy needed for an explosive redox reaction.

8.94 An unscrambled egg has less entropy than a scrambled egg.

ADDITIONAL ACTIVITIES

Section 8.1 Review:

(1) Place the following pictures in the order of increasing entropy.

(2) The three pictures below can be put into 6 different sequences. Identify the sequence that depicts the most nonspontaneous transitions from first picture to second picture and from second picture to third picture. Explain your order using the terms endergonic or exergonic and entropy.

(3) Does the reverse order of (2) show a spontaneous process? Explain.

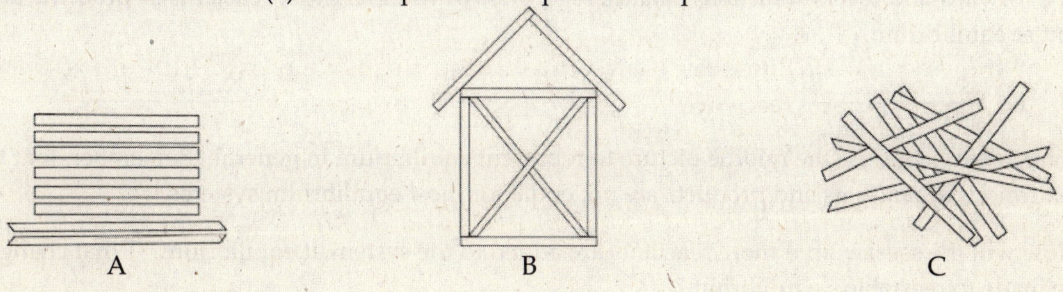

Section 8.2 Review:

(1) What is meant by the "rate of travel" in a car? What units are used?
(2) What is meant by an "interest rate"? What units are used?
(3) What do (1) and (2) have in common with a reaction rate?
(4) What is different about (1), (2), and reaction rates?

Section 8.3 Review:

(1) What are the three assumptions that serve as a basis for most reaction mechanisms?
(2) Why does orientation not matter for spherical reactants with uniform charges?
(3) What is the difference between *internal energy* and *activation energy*?

Section 8.4 Review:

Assume activation energy is required for both processes and draw/label energy diagrams for the following:

(1) $CH_4 (g) + 2 O_2 (g) \rightarrow CO_2 (g) + 2 H_2O (l) + 890 kJ$
(2) $NH_4NO_3 (s) + 26.5 kJ \rightarrow NH_4NO_3 (aq)$

Section 8.5 Review:

(1) What is meant by the "nature of the reactants"?
(2) What is the relationship between concentration, collisions, and reaction rates?
(3) What is the relationship between temperature and reaction rates?
(4) How does a catalyst affect reaction rates?
(5) How does an inhibitor affect reaction rates?

Section 8.6 Review:

(1) What is equal at equilibrium?
(2) What is not equal at equilibrium?
(3) Why is a double arrow used for an equilibrium reaction?

Section 8.7 Review:

(1) Write an equilibrium expression using a double arrow and the words *reactants* and *products*.
(2) Where are the reactants found in (1) (relative to the arrow)?
(3) Where are the products found in (1) (relative to the arrow)?
(4) Write an equilibrium expression for (1).
(5) What are the exponents in (4)?
(6) If K were 5×10^7 for (4), would there be more reactants or products at equilibrium?
(7) Would the equilibrium in (6) lie to the left or to the right?
(8) If K were 7×10^{-5} for (4), would there be more reactants or products at equilibrium?
(9) Would the equilibrium in (8) lie to the left or to the right?

Section 8.8 Review:

Equilibrium can be thought of as a seesaw. The pivot point (fulcrum) of the seesaw can be adjusted so that the forward and reverse reactions balance regardless of whether more reactants or products are present at equilibrium.

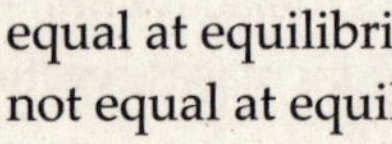

For convenience, let's use the middle picture to represent equilibrium in general. Remember, that the concentrations of reactants and products are not equal for most equilibrium systems.

(1) How will the seesaw tip if more reactants are added to the system at equilibrium? What change must be made to reestablish equilibrium?

(2) How will the seesaw tip if more products are added to the system at equilibrium? What change must be made to reestablish equilibrium?
(3) How will the seesaw tip if some reactants are removed from the system at equilibrium? What change must be made to reestablish equilibrium?
(4) How will the seesaw tip if some products are removed from the system at equilibrium? What change must be made to reestablish equilibrium?
(5) If the forward reaction is endothermic and heat is added, which way will the seesaw tip? What change must be made to reestablish equilibrium?
(6) If the forward reaction is exothermic and heat is added, which way will the seesaw tip? What change must be made to reestablish equilibrium?
(7) If the reverse reaction is endothermic and heat is removed, which way will the seesaw tip? What change must be made to reestablish equilibrium?
(8) If the reverse reaction is exothermic and heat is removed, which way will the seesaw tip? What change must be made to reestablish equilibrium?
(9) If a catalyst is added, which way will the seesaw tip? What change must be made to reestablish equilibrium?
(10) If an inhibitor is added, which way will the seesaw tip? What change must be made to reestablish equilibrium?
(11) Which of the scenarios above will change the equilibrium constant, K?

Tying It All Together with a Laboratory Application:
A beaker is placed in a small puddle of water on a piece of wood. A solution is prepared in the beaker by mixing 3.8 g of solid NH_4Cl with 50 mL of water. When the laboratory student attempts to lift the beaker from the piece of wood, she discovers that the beaker is stuck to the wood because the puddle of water has frozen. (1) Sketch an energy diagram for the solution formation. The formation of this solution is a(n) (2) _______ (endothermic or exothermic) process. It is also a spontaneous process. The level of entropy (3) _______ as the solution is prepared.

The ammonium ion establishes an equilibrium in the solution:

$$NH_4^+(aq) \rightleftharpoons H^+(aq) + NH_3\,(aq)$$

The equilibrium constant expression for this reaction is (4) _______. The equilibrium constant has a value of 5.6×10^{-10}. Equilibrium lies to the (5) _______ because the concentration of reactants is (6) _______ (greater than, less than, or equal to) the concentration of products at equilibrium. The reaction rate of the forward reaction is (7) _______ (greater than, less than, or equal to) the reaction rate of the reverse reaction at equilibrium.

A solution of silver nitrate is added to the reaction mixture, solid silver chloride forms, and precipitates to the bottom of the beaker.

$$AgNO_3\,(aq) + Cl^-(aq) \rightarrow AgCl\,(s) + NO_3^-(aq)$$

This reaction occurs (8) _______ than if solid silver nitrate were mixed with solid ammonium chloride because the reactants are both aqueous ions. The ions are spheres of uniform charge, thus the orientation of the ions (9) _______ important for an effective collision. The addition of silver nitrate (10) _______ (does, does not) affect the equilibrium established by the ammonium ion.

A solution of hydrochloric acid (HCl) is added to the reaction mixture. Hydrochloric acid dissociates in water to produce (11) _______ and _______ ions. This affects the equilibrium established by the ammonium ion because (12) _______ have been added. In order to reestablish equilibrium, the equilibrium will shift to the (13) _______. The concentration of NH_3 will (14) _______.

SOLUTIONS FOR THE ADDITIONAL ACTIVITIES

Section 8.1 Review:

(1) increasing entropy: B < A < C

(2) nonspontaneous process: C → A → B; Picture C is like a disordered stack of wood. Picture C has the highest entropy. To stack the wood as in picture A requires energy and is an endergonic process. The entropy is decreased. Turning a stack of wood (A) into a specific design (B) requires energy (is endergonic) and decreases entropy.

(3) The progression B → A → C is not spontaneous because the transition between picture B and picture A requires a significant activation energy. Both picture A and picture B are highly ordered. Picture B does have higher potential energy than picture A. The processes A → C and B → C are spontaneous processes because the entropy increases and the energy decreases.

Section 8.2 Review:

(1) The speed (or velocity) of the car is the "rate of travel." Speed is usually measured in miles per hour or kilometers per hour.

(2) The cost of borrowed money expressed as a percentage for a given period of time, usually one year, is an "interest rate." The units for interest rate are typically % APR (annual percentage rate). The % APR is divided by 12 to determine how much interest will be added per month.

(3) Speeds, interest rates, and reaction rates all represent the change in a unit divided by time.

(4) For car speeds, the unit that changes is miles or distance per unit time. For interest rates, the unit that changes is percentage per unit time. For reaction rates, the unit that changes is the concentration per unit time.

Section 8.3 Review:

(1) Molecules must collide in order to react. Molecules must collide with enough energy to react. Some molecules must collide at the correct orientation to react.

(2) No matter how the two spherical reactants collide, the orientation is the same.

(3) Internal energy is the energy of molecular vibrations. Chemical bonds can stretch, bend, and wag. Activation energy is the minimum energy required for molecules to react.

Section 8.4 Review:

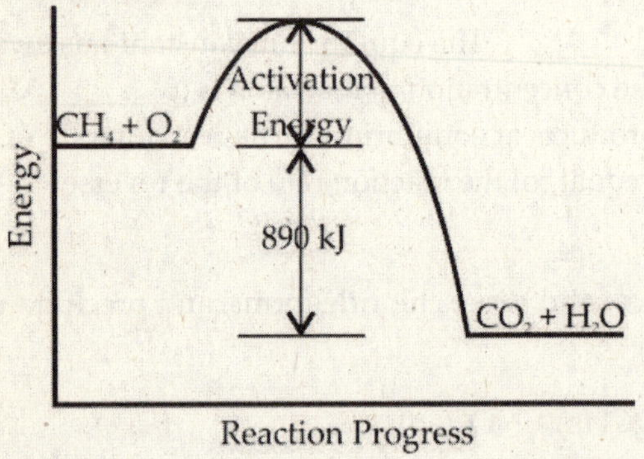

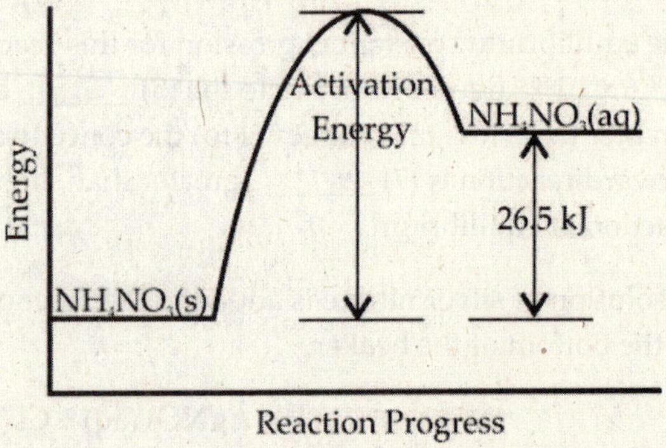

Section 8.5 Review:

(1) The "nature of the reactants" refers to the state of matter as well as the type of bonding in the reactants. Ions in solution and gases usually react very quickly. Solids and liquids react more slowly, unless the surface area of the solid is very large and the liquids are well mixed.

(2) As concentration increases, the number of collisions increases. The more collisions that occur, the higher the reaction rate is. As concentration increases, reaction rate increases.

(3) As temperature increases, the number of collisions increases. Also, the number of collisions with enough energy to be effective collisions increases. Consequently, the reaction rate increases.

(4) A catalyst increases reaction rates by lowering the activation energy.

(5) An inhibitor increases reaction rates by increasing the activation energy or by decreasing the number of effective collisions.

Section 8.6 Review:

(1) The rates of the forward and the reverse reactions are equal at equilibrium.
(2) The concentrations of reactants and products may not be equal at equilibrium.
(3) In equilibrium, the reaction proceeds from left to right and from right to left. The double arrow indicates both the forward and reverse reactions occur simultaneously.

Section 8.7 Review:

(1) reactants $\rightleftharpoons$ products
(2) Reactants are to the left of the arrow.
(3) Products are to the right of the arrow.
(4) $K = \frac{[\text{products}]}{[\text{reactants}]}$
(5) The exponents are both equal to 1 because neither the reactants nor the products in (1) had a coefficient. The exponents in the equilibrium constant expression are always equal to the coefficients from the chemical equation.
(6) More products would be present at equilibrium, if K were 5×10^7 for (4).
(7) The equilibrium would lie to the right.
(8) More reactants would be present at equilibrium, if K were 7×10^{-5} for (4).
(9) The equilibrium would lie to the left.

Section 8.8 Review:

(1) 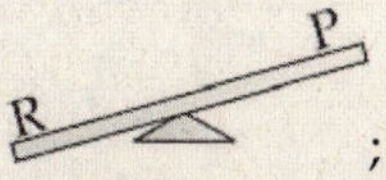

; In order to rebalance the seesaw, some of the reactants will need to convert to products. The equilibrium must shift to the right.

(2) 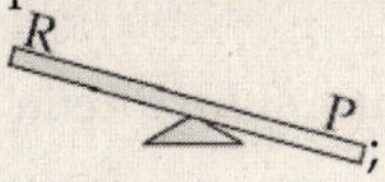

; In order to rebalance the seesaw, some of the products will need to convert to reactants. The equilibrium must shift to the left.

(3) 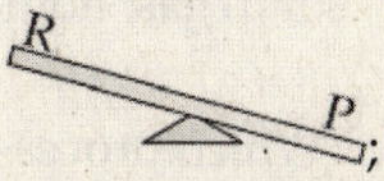

; In order to rebalance the seesaw, some of the products will need to convert to reactants. The equilibrium must shift to the left.

(4) 

; In order to rebalance the seesaw, some of the reactants will need to convert to products. The equilibrium must shift to the right.

(5) R + heat $\rightleftharpoons$ P; 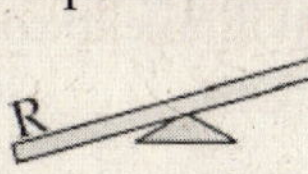

; In order to rebalance the seesaw, the heat must be used and some of the reactants will need to convert to products. The equilibrium must shift to the right.

(6) R $\rightleftharpoons$ P+ heat; 

; In order to rebalance the seesaw, the heat must be used and some of the products will need to convert to reactants. The equilibrium must shift to the left.

(7) R $\rightleftharpoons$ P+ heat ; 

; In order to rebalance the seesaw, heat must be produced and some of the reactants will need to convert to products. The equilibrium must shift to the right.

(8) R + heat $\rightleftharpoons$ P; 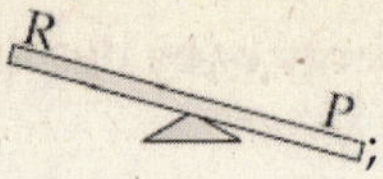

; In order to rebalance the seesaw, heat must be produced and some of the products will need to convert to reactants. The equilibrium must shift to the left.

(9) 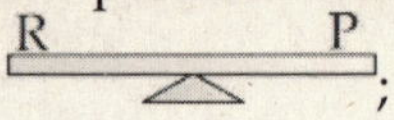

; A catalyst will not tip the seesaw. A catalyst will speed up the forward and reverse reaction rates, but it will not affect the concentration of reactants or products.

(10) 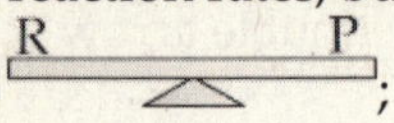

; An inhibitor will not tip the seesaw. An inhibitor will slow down the forward and reverse reaction rates, but it will not affect the concentration of reactants or products.

(11) All of the scenarios involving heat (5, 6, 7, and 8) will change the value of K.

Tying It All Together with a Laboratory Application:

(1) 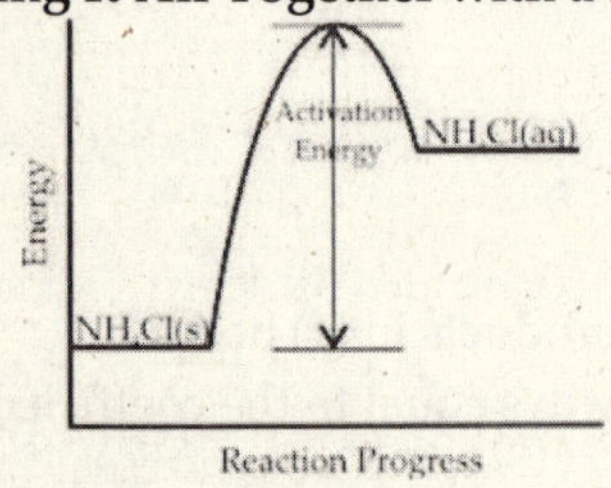

(2) endothermic

(3) increases

(4) $K = \dfrac{[H^+][NH_3]}{[NH_4^+]}$

(5) left

(6) greater than

(7) equal to

(8) faster

(9) is not

(10) does not

(11) H^+, Cl^-

(12) products

(13) left

(14) decrease

SELF-TEST QUESTIONS

Multiple Choice

1. A catalyst:
   a. is not used up in a reaction.
   b. changes the rate of reaction.
   c. affects the forward reaction the same as it affects the reverse reaction.
   d. all of the above.

2. Which of the following responses correctly arranges the states of matter for a pure substance in the order of decreasing entropy?
   a. gas, liquid, solid b. liquid, solid, gas c. solid, liquid, gas d. solid, gas, liquid

3. Four processes occur as the following changes take place in energy and entropy. Which process is definitely nonspontaneous?
   a. energy decrease and entropy increase
   b. energy decrease and entropy decrease
   c. energy increase and entropy increase
   d. energy increase and entropy decrease

4. A carrot cooks in 15 minutes in boiling water (100°C). How long will it take to cook a carrot inside a pressure cooker where the temperature is 10°C greater (110°C)?
   a. 3.7 minutes b. 5.0 minutes c. 7.5 minutes d. 30 minutes

5. The gases A and B react as follows: $A + B \rightarrow C$
   What is the average rate of reaction if pure A and B are mixed and after 30.0 seconds the concentration of C is found to be 4.75×10^{-3} M?
   a. 1.58×10^{-4} M/sec b. 2.38×10^{-3} M/sec c. 9.50×10^{-3} M/sec d. 1.43×10^{-1} M/sec

Questions 6 and 7 refer to the following reaction, which is assumed to be at equilibrium:

$$\text{heat} + 2\,NO + O_2 \rightleftarrows 2\,NO_2$$

In each case choose the response which best indicates the effects resulting from the described change in conditions.

6. The reaction mixture is heated.
   a. equilibrium shifts to the left
   b. equilibrium shifts to the right
   c. equilibrium does not shift
   d. the effect cannot be predicted

7. A catalyst is added to the reaction mixture.
   a. equilibrium shifts to the left
   b. equilibrium shifts to the right
   c. equilibrium does not shift
   d. the effect cannot be predicted

8. In the equilibrium constant expression for the reaction: $2\,N_2O_5 \rightleftarrows 4\,NO_2 + O_2$, the exponent on the concentration of N_2O_5 is:
   a. 1.
   b. 2.
   c. 0.
   d. can't be determined.

9. A sample of ICl is placed in a container and equilibrium is established according to the reaction:

   $$2\,ICl \rightleftarrows I_2 + Cl_2$$

   where all the materials are gases. Analysis of the equilibrium mixture gas the following molar concentrations: [ICl] = 0.26, $[I_2] = [Cl_2]$=0.09. What is the value of K, the equilibrium constant for the reaction?
   a. 0.031
   b. 0.12
   c. 0.35
   d. 14.8

10. The equilibrium shifts to the left when an equilibrium mixture is cooled. The forward reaction must be:
    a. exothermic.
    b. endothermic.
    c. high entropy.
    d. low entropy.

True-False

11. The entropy of a cluttered room is higher than that of an orderly room.
12. A reaction rate can be thought of as the speed of a reaction.
13. Effective molecular collisions are those that allow molecules to collide but not react.
14. Increasing the concentration of reactants usually increases the reaction rate.
15. Activation energy is the energy difference between reactants and products.
16. If an endothermic reaction is spontaneous, then entropy must have decreased.
17. Catalysts act by lowering the activation energy.
18. Substances that slow reactions are called inhibitors.
19. In a reaction at equilibrium, the forward and reverse reactions have both stopped.
20. In a reaction at equilibrium, the concentrations of reactants and products are equal.

Matching

For each of the following processes choose the appropriate response from those on the right.

21. a match burns	a. both entropy and energy increase
22. perspiration evaporates	b. both entropy and energy decrease
23. melted lead becomes a solid	c. entropy increases; energy decreases
24. an explosive detonates	d. entropy decreases; energy increases

Three liquid fuels are to be tested. A 1.0 gram sample of each fuel is weighed out and heated to its ignition temperature. When the fuel burns, the total heat liberated is measured. The results of this experiment are given in the table below. For Questions 25-30 choose the answer from the column on the right that best fits each statement on the left.

Fuel	Ignition Temperature	Heat Liberated
X	210°C	1680 cal
Y	110°C	1410 cal
Z	285°C	1206 cal

25. it has the highest activation energy
26. it has the second highest activation energy
27. it has the lowest activation energy
28. it has the smallest energy difference between reactants and products
29. it has the largest energy difference between reactants and products
30. the described reaction is exothermic (exergonic)

a. fuel X
b. fuel Y
c. fuel Z
d. two or more reactant fuels fit this category

Use the following equilibrium expression and match the effects on the equilibrium from the right with the changes made to the equilibrium system listed on the left.

$$\text{heat} + CO + H_2 \rightleftarrows CH_3OH$$

31. add CO
32. remove some H_2
33. add CH_3OH
34. heat the system
35. add a catalyst

a. shift left
b. shift right
c. no effect on equilibrium
d. cannot determine from the information given

Chapter 9: Acids, Bases, and Salts

CHAPTER OUTLINE

9.1 The Arrhenius Theory
9.2 The Brønsted Theory
9.3 Naming Acids
9.4 The Self-Ionization of Water
9.5 The pH Concept
9.6 Properties of Acids
9.7 Properties of Bases
9.8 Salts
9.9 The Strengths of Acids and Bases
9.10 Analyzing Acids and Bases
9.11 Titration Calculations
9.12 Hydrolysis Reactions of Salts
9.13 Buffers

LEARNING OBJECTIVES/ASSESSMENT

When you have completed your study of this chapter, you should be able to:

1. Write reaction equations that illustrate Arrhenius acid-base behavior. (Section 9.1; Exercise 9.2)
2. Write reaction equations that illustrate Brønsted acid-base behavior, and identify Brønsted acids and bases from written reaction equations. (Section 9.2; Exercises 9.6 and 9.10)
3. Name common acids. (Section 9.3; Exercise 9.22)
4. Do calculations using the concept of the self-ionization of water. (Section 9.4; Exercises 9.28 a & b, and 9.30 a & b)
5. Do calculations using the pH concept. (Section 9.5; Exercises 9.36 and 9.40)
6. Write reaction equations that illustrate the characteristic reactions of acids. (Section 9.6; Exercise 9.50)
7. Write reaction equations that represent neutralization reactions between acids and bases. (Section 9.7; Exercise 9.60)
8. Write reaction equations that illustrate various ways to prepare salts, and do calculations using the concept of an equivalent of salt. (Section 9.8; Exercises 9.68 and 9.74)
9. Demonstrate an understanding of the words *weak* and *strong* as applied to acids and bases. (Section 9.9; Exercise 9.86)
10. Demonstrate an understanding of the titration technique used to analyze acids and bases. (Section 9.10; Exercise 9.92)
11. Do calculations related to the analysis of acids and bases by titration. (Section 9.11; Exercises 9.98 and 9.100 a)
12. Explain the concept of salt hydrolysis, and write equations to illustrate the concept. (Section 9.12; Exercise 9.108)
13. Explain how buffers work, and write equations to illustrate their action. (Section 9.13; Exercise 9.116)

SOLUTIONS FOR THE END OF CHAPTER EXERCISES

THE ARRHENIUS THEORY (SECTION 9.1)

☑9.2
a. $HBrO_2$ — $HBrO_2\,(aq) \rightarrow H^+(aq) + BrO_2^-(aq)$
b. HS^- — $HS^-(aq) \rightarrow H^+(aq) + S^{2-}(aq)$
c. HBr — $HBr\,(aq) \rightarrow H^+(aq) + Br^-(aq)$
d. $HC_2H_3O_2$ — $HC_2H_3O_2\,(aq) \rightarrow H^+(aq) + C_2H_3O_2^-(aq)$

9.4
a. $LiOH$ — Arrhenius base — $LiOH\,(aq) \rightarrow Li^+(aq) + OH^-(aq)$
b. $C_2H_5NH_2$ — not an Arrhenius base
c. $Sr(OH)_2$ — Arrhenius base — $Sr(OH)_2\,(aq) \rightarrow Sr^{2+}(aq) + 2\,OH^-(aq)$
d. $NaNH_2$ — not an Arrhenius base

THE BRØNSTED THEORY (SECTION 9.2)

			Brønsted Acids	Brønsted Bases
☑9.6	a.	$HC_2O_4^- (aq) + H_2O (l) \rightleftharpoons H_3O^+ (aq) + C_2O_4^{2-} (aq)$	$HC_2O_4^-$, H_3O^+	H_2O, $C_2O_4^{2-}$
	b.	$HNO_2 (aq) + H_2O (l) \rightleftharpoons H_3O^+(aq) + NO_2^- (aq)$	HNO_2, H_3O^+	H_2O, NO_2^-
	c.	$PO_4^{3-} (aq) + H_2O (l) \rightleftharpoons HPO_4^{2-} (aq) + OH^- (aq)$	H_2O, HPO_4^{2-}	PO_4^{3-}, OH^-
	d.	$H_2SO_3 (aq) + H_2O (l) \rightleftharpoons HSO_3^- (aq) + H_3O^+ (aq)$	H_2SO_3, H_3O^+	H_2O, HSO_3^-
	e.	$F^-(aq) + H_2O (l) \rightleftharpoons HF (aq) + OH^-(aq)$	H_2O, HF	F^-, OH^-

Conjugate Acid-Base Pairs

9.8
- a. $HC_2O_4^-$ and $C_2O_4^{2-}$, H_3O^+ and H_2O
- b. HNO_2 and NO_2^-, H_3O^+ and H_2O
- c. HPO_4^{2-} and PO_4^{3-}, H_2O and OH^-
- d. H_2SO_3 and HSO_3^-, H_3O^+ and H_2O
- e. H_2O and OH^-, HF and F^-

☑9.10
- a. $HF (aq) + H_2O (l) \rightleftharpoons F^-(aq) + H_3O^+(aq)$
- b. $HClO_3 (aq) + H_2O (l) \rightleftharpoons ClO_3^-(aq) + H_3O^+(aq)$
- c. $HClO (aq) + H_2O (l) \rightleftharpoons ClO^-(aq) + H_3O^+(aq)$
- d. $HS^-(aq) + H_2O (l) \rightleftharpoons S^{2-}(aq) + H_3O^+(aq)$

			Conjugate Base
9.12	a.	$HSO_4^-(aq) + H_2O (l) \rightleftharpoons SO_4^{2-}(aq) + H_3O^+(aq)$	SO_4^{2-}
	b.	$CH_3NH_3^+(aq) + H_2O (l) \rightleftharpoons CH_3NH_2 (aq) + H_3O^+(aq)$	CH_3NH_2
	c.	$HClO_4 (aq) + H_2O (l) \rightleftharpoons ClO_4^-(aq) + H_3O^+(aq)$	ClO_4^-
	d.	$NH_4^+(aq) + H_2O (l) \rightleftharpoons NH_3 (aq) + H_3O^+(aq)$	NH_3
	e.	$HCl (aq) + H_2O (l) \rightleftharpoons Cl^-(aq) + H_3O^+(aq)$	Cl^-

			Conjugate Acid
9.14	a.	$HCO_3^-(aq) + H_2O (l) \rightleftharpoons H_2CO_3 (aq) + OH^-(aq)$	H_2CO_3
	b.	$S^{2-}(aq) + H_2O (l) \rightleftharpoons HS^-(aq) + OH^-(aq)$	HS^-
	c.	$HS^-(aq) + H_2O (l) \rightleftharpoons H_2S (aq) + OH^-(aq)$	H_2S
	d.	$HC_2O_4^- (aq) + H_2O (l) \rightleftharpoons H_2C_2O_4 (aq) + OH^-(aq)$	$H_2C_2O_4$
	e.	$HN_2O_2^-(aq) + H_2O (l) \rightleftharpoons H_2N_2O_2 (aq) + OH^-(aq)$	$H_2N_2O_2$

			Missing Formula
9.16	a.	$H_2AsO_4^-(aq) + ? \rightarrow NH_4^+ (aq) + HAsO_4^{2-}(aq)$	NH_3 (aq)
	b.	$C_6H_5NH_2 (aq) + ? \rightarrow C_6H_5NH_3^+(aq) + OH^-(aq)$	H_2O (l)
	c.	$S^{2-}(aq) + H_2O (l) \rightarrow ? + OH^-(aq)$	HS^- (aq)
	d.	$(CH_3)_2NH (aq) + ? \rightarrow (CH_3)_2NH_2^+ (aq) + Br^-(aq)$	HBr (aq)
	e.	$H_2PO_4^- (aq) + CH_3NH_2 (aq) \rightarrow HPO_4^{2-} (aq) + ?$	$CH_3NH_3^+$ (aq)

		Acid	Base	Equation
9.18	a.	HS^-	NH_3	$HS^-(aq) + NH_3(aq) \rightarrow S^{2-}(aq) + NH_4^+(aq)$
	b.	H_2O	ClO_3^-	$H_2O(l) + ClO_3^-(aq) \rightarrow HClO_3(aq) + OH^-(aq)$
	c.	H_2O	NH_2^-	$H_2O(l) + NH_2^-(aq) \rightarrow NH_3(aq) + OH^-(aq)$
	d.	HBO_3^{2-}	H_2O	$HBO_3^{2-}(aq) + H_2O(l) \rightarrow BO_3^{3-}(aq) + H_3O^+(aq)$
	e.	HNO_2	NH_3	$HNO_2(aq) + NH_3(aq) \rightarrow NH_4^+(aq) + NO_2^-(aq)$

NAMING ACIDS (SECTION 9.3)

9.20 HCN (aq) = hydrocyanic acid

☑9.22
a. H_2Te (aq) hydrotelluric acid
b. HClO hypochlorous acid
c. H_2SO_3 sulfurous acid
d. HNO_2 nitrous acid

9.24 $H_2C_4H_4O_4$ = succinic acid

9.26 carbonic acid = H_2CO_3

THE SELF-IONIZATION OF WATER (SECTION 9.4)

9.28	$[H_3O^+]$	Calculation	$[OH^-]$
☑a.	0.044	$[OH^-]=\frac{1.0\times10^{-14}}{0.044}=2.\overline{27}\times10^{-13}$ M	2.3 x 10^{-13} M
☑b.	1.3 x 10^{-4}	$[OH^-]=\frac{1.0\times10^{-14}}{1.3\times10^{-4}}=7.6923\times10^{-11}$ M	7.7 x 10^{-11} M
c.	0.0087	$[OH^-]=\frac{1.0\times10^{-14}}{0.0087}=1.1494\times10^{-12}$ M	1.1 x 10^{-12} M
d.	7.9 x 10^{-10}	$[OH^-]=\frac{1.0\times10^{-14}}{7.9\times10^{-10}}=1.2658\times10^{-5}$ M	1.3 x 10^{-5} M
e.	3.3 x 10^{-2}	$[OH^-]=\frac{1.0\times10^{-14}}{3.3\times10^{-2}}=3.\overline{03}\times10^{-13}$ M	3.0 x 10^{-13} M

9.30	$[OH^-]$	Calculation	$[H_3O^+]$
☑a.	6.9 x 10^{-5}	$[H_3O^+]=\frac{1.0\times10^{-14}}{6.9\times10^{-5}}=1.449275\times10^{-10}$ M	1.4 x 10^{-10} M
☑b.	0.074	$[H_3O^+]=\frac{1.0\times10^{-14}}{0.074}=1.35\overline{135}\times10^{-13}$ M	1.4 x 10^{-13} M
c.	4.9	$[H_3O^+]=\frac{1.0\times10^{-14}}{4.9}=2.0408\times10^{-15}$ M	2.0 x 10^{-15} M
d.	1.7 x 10^{-3}	$[H_3O^+]=\frac{1.0\times10^{-14}}{1.7\times10^{-3}}=5.88235\times10^{-12}$ M	5.9 x 10^{-12} M
e.	9.2 x 10^{-9}	$[H_3O^+]=\frac{1.0\times10^{-14}}{9.2\times10^{-9}}=1.0869565\times10^{-6}$ M	1.1 x 10^{-6} M

9.32	9.28		9.30	
a.	$[OH^-]$ = 2.3 x 10^{-13} M	acidic	$[H_3O^+]$ = 1.4 x 10^{-10} M	basic
b.	$[OH^-]$ = 7.7 x 10^{-11} M	acidic	$[H_3O^+]$ = 1.4 x 10^{-13} M	basic

c.	$[OH^-] = 1.1 \times 10^{-12}$ M	acidic	$[H_3O^+] = 2.0 \times 10^{-15}$ M	basic
d.	$[OH^-] = 1.3 \times 10^{-5}$ M	basic	$[H_3O^+] = 5.9 \times 10^{-12}$ M	basic
e.	$[OH^-] = 3.0 \times 10^{-13}$ M	acidic	$[H_3O^+] = 1.1 \times 10^{-6}$ M	acidic

THE pH CONCEPT (SECTION 9.5)

9.34
a. pH = 4.7 acidic
b. pH = 8.8 basic
c. pH = 1.5 acidic
d. pH = 11 basic
e. pH = 5.9 acidic

☑9.36
a. $[H^+] = 4.1 \times 10^{-9}$ — $pH = -\log[4.1\times10^{-9}] = 8.39$ — basic
b. $[OH^-] = 9.4 \times 10^{-4}$ — $pH = 14 - (-\log[9.4\times10^{-4}]) = 10.97$ — basic
c. $[OH^-] = 10\,[H^+]$ — $pH = -\log[\sqrt{\frac{1.0\times10^{-14}}{10}}] = 7.50$ — basic
d. $[H^+] = 2.3 \times 10^{-2}$ — $pH = -\log[2.3\times10^{-2}] = 1.64$ — acidic
e. $[OH^-] = 5.1 \times 10^{-10}$ — $pH = 14 - (-\log[5.1\times10^{-10}]) = 4.71$ — acidic

9.38
a. $[H^+] = 2.2 \times 10^{-3}$ — $pH = -\log[2.2\times10^{-3}] = 2.66$ — acidic
b. $[H^+] = 3.9 \times 10^{-12}$ — $pH = -\log[3.9\times10^{-12}] = 11.41$ — basic
c. $[H^+] = 7.5 \times 10^{-6}$ — $pH = -\log[7.5\times10^{-6}] = 5.12$ — acidic
d. $[OH^-] = 2.5 \times 10^{-4}$ — $pH = 14 - (-\log[2.5\times10^{-4}]) = 10.40$ — basic
e. $[OH^-] = 8.6 \times 10^{-10}$ — $pH = 14 - (-\log[8.6\times10^{-10}]) = 4.93$ — acidic

☑9.40
a. pH = 9.27 — $[H^+] = 10^{-9.27} = 5.4\times10^{-10}$ M
b. pH = 2.55 — $[H^+] = 10^{-2.55} = 2.8\times10^{-3}$ M
c. pH = 5.42 — $[H^+] = 10^{-5.42} = 3.8\times10^{-6}$ M

9.42
a. pH = 3.95 — $[H^+] = 10^{-3.95} = 1.1\times10^{-4}$ M — $[OH^-] = \frac{10^{-14}}{10^{-3.95}} = 8.9\times10^{-11}$ M
b. pH = 4.00 — $[H^+] = 10^{-4.00} = 1.0\times10^{-4}$ M — $[OH^-] = \frac{10^{-14}}{10^{-4.00}} = 1.0\times10^{-10}$ M
c. pH = 11.86 — $[H^+] = 10^{-11.86} = 1.4\times10^{-12}$ M — $[OH^-] = \frac{10^{-14}}{10^{-11.86}} = 7.2\times10^{-3}$ M

9.44
a. Bile, pH = 8.05 — $[H^+] = 10^{-8.05} = 8.9\times10^{-9}$ M — basic
b. Vaginal fluid, pH = 3.93 — $[H^+] = 10^{-3.93} = 1.2\times10^{-4}$ M — acidic
c. Semen, pH = 7.38 — $[H^+] = 10^{-7.38} = 4.2\times10^{-8}$ M — basic
d. Cerebrospinal fluid, pH = 7.40 — $[H^+] = 10^{-7.40} = 4.0\times10^{-8}$ M — basic
e. Perspiration, pH = 6.23 — $[H^+] = 10^{-6.23} = 5.9\times10^{-7}$ M — acidic

9.46
a. Soft drink, pH = 2.91 — $[H^+] = 10^{-2.91} = 1.2\times10^{-3}$ M — acidic
b. Tomato juice, pH= 4.11 — $[H^+] = 10^{-4.11} = 7.8\times10^{-5}$ M — acidic

c. Lemon juice, pH = 2.32 $[H^+] = 10^{-2.32} = 4.8\times10^{-3}$ M acidic

d. Grapefruit juice, pH = 3.07 $[H^+] = 10^{-3.07} = 8.5\times10^{-4}$ M acidic

PROPERTIES OF ACIDS (SECTION 9.6)

9.48 a. $(2\ L)(3.0\ M) = (V_c)(6\ M) \Rightarrow V_c = 1\ L$

I would add approximately 1 L of water to a 2 L graduated cylinder, then dilute up to the 2 L mark with 6 M nitric acid. I would make sure to stir the solution.

b. $(500\ mL)(1.5\ M) = (V_c)(15\ M) \Rightarrow V_c = 50\ mL$

I would add approximately 450 mL of water to a 500 mL graduated cylinder, then dilute up to the 500 mL mark with 15 M aqueous ammonia. I would make sure to stir the solution.

c. $(5\ L)(0.2\ M) = (V_c)(12\ M) \Rightarrow V_c = 83\ mL$

I would add approximately 4.9 L of water to a 5 L graduated cylinder, then dilute up to the 5 L mark with 12 M hydrochloric acid. I would make sure to stir the solution.

☑9.50 a. H_2SO_4 (aq) + 2 H_2O (l) → 2 H_3O^+(aq) + SO_4^{2-}(aq)
b. H_2SO_4 (aq) + CaO (s) → H_2O (l) + $CaSO_4$ (aq)
c. H_2SO_4 (aq) + $Mg(OH)_2$ (s) → 2 H_2O (l) + $MgSO_4$ (aq)
d. H_2SO_4 (aq) + $CuCO_3$ (s) → H_2O (l) + $CuSO_4$ (aq) + CO_2 (g)
e. H_2SO_4 (aq) + 2 $KHCO_3$ (s) → 2 H_2O (l) + K_2SO_4 (aq) + 2 CO_2 (g)
f. H_2SO_4 (aq) + Mg (s) → H_2 (g) + $MgSO_4$ (aq)

*M = molecular equation, **TIE = total ionic equation, ***NIE = net ionic equation

9.52 a. TIE**: 2 H^+ (aq) + SO_4^{2-} (aq) + 2 H_2O (l) → 2 H_3O^+ (aq) + SO_4^{2-}(aq)
NIE***: H^+ (aq) + H_2O (l) → H_3O^+ (aq)

b. TIE**: 2 H^+ (aq) + SO_4^{2-} (aq) + CaO (s) → H_2O (l) + Ca^{2+}(aq) + SO_4^{2-}(aq)
NIE***: 2 H^+ (aq) + CaO (s) → H_2O (l) + Ca^{2+}(aq)

c. TIE**: 2 H^+ (aq) + SO_4^{2-} (aq) + $Mg(OH)_2$ (s) → 2 H_2O (l) + Mg^{2+}(aq) + SO_4^{2-} (aq)
NIE***: 2 H^+ (aq) + $Mg(OH)_2$ (s) → 2 H_2O (l) + Mg^{2+}(aq)

d. TIE**: 2 H^+ (aq) + SO_4^{2-} (aq) + $CuCO_3$ (s) → H_2O (l) + Cu^{2+}(aq) + SO_4^{2-} (aq) + CO_2 (g)
NIE***: 2 H^+ (aq) + $CuCO_3$ (s) → H_2O (l) + Cu^{2+}(aq) + CO_2 (g)

e. TIE**: 2 H^+ (aq) + SO_4^{2-} (aq) + 2 $KHCO_3$ (s) → 2 H_2O (l) + 2 K^+(aq) + SO_4^{2-} (aq) + 2 CO_2 (g)
NIE***: H^+ (aq) + $KHCO_3$ (s) → H_2O (l) + K^+(aq) + CO_2 (g)

f. TIE**: 2 H^+ (aq) + SO_4^{2-} (aq) + Mg (s) → H_2 (g) + Mg^{2+}(aq) + SO_4^{2-} (aq)
NIE***: 2 H^+ (aq) + Mg (s) → H_2 (g) + Mg^{2+}(aq)

9.54 a. 2 HCl (aq) + MgO (s) → H_2O (l) + $MgCl_2$ (aq)
b. 2 HCl (aq) + $Mg(OH)_2$ (s) → 2 H_2O (l) + $MgCl_2$ (aq)
c. 2 HCl (aq) + $MgCO_3$ (s) → H_2O (l) + $MgCl_2$ (aq) + CO_2 (g)

d. $2\ HCl\ (aq) + Mg(HCO_3)_2\ (s) \rightarrow 2\ H_2O\ (l) + MgCl_2\ (aq) + 2\ CO_2\ (g)$

e. $2\ HCl\ (aq) + Mg\ (s) \rightarrow H_2\ (g) + MgCl_2\ (aq)$

9.56 a. M*: $Sn\ (s) + H_2SO_3\ (aq) \rightarrow SnSO_3\ (aq) + H_2\ (g)$

TIE**: $Sn\ (s) + 2\ H^+(aq) + SO_3^{2-}(aq) \rightarrow Sn^{2+}(aq) + SO_3^{2-}(aq) + H_2\ (g)$

NIE***: $Sn\ (s) + 2\ H^+(aq) \rightarrow Sn^{2+}(aq) + H_2\ (g)$

b. M*: $3\ Mg\ (s) + 2\ H_3PO_4\ (aq) \rightarrow Mg_3(PO_4)_2\ (s) + 3\ H_2\ (g)$

TIE**: $3\ Mg\ (s) + 6\ H^+\ (aq) + 2\ PO_4^{3-}\ (aq) \rightarrow Mg_3(PO_4)_2\ (s) + 3\ H_2\ (g)$

NIE***: $3\ Mg\ (s) + 6\ H^+\ (aq) + 2\ PO_4^{3-}\ (aq) \rightarrow Mg_3(PO_4)_2\ (s) + 3\ H_2\ (g)$

c. M*: $Ca\ (s) + 2\ HBr\ (aq) \rightarrow CaBr_2\ (aq) + H_2\ (g)$

TIE**: $Ca\ (s) + 2\ H^+(aq) + 2\ Br^-(aq) \rightarrow Ca^{2+}(aq) + 2\ Br^-(aq) + H_2\ (g)$

NIE***: $Ca\ (s) + 2\ H^+(aq) \rightarrow Ca^{2+}(aq) + H_2\ (g)$

PROPERTIES OF BASES (SECTION 9.7)

9.58 a. M*: $RbOH\ (aq) + HBr\ (aq) \rightarrow RbBr\ (aq) + H_2O\ (l)$

TIE**: $Rb^+(aq) + OH^-\ (aq) + H^+(aq) + Br^-(aq) \rightarrow Rb^+(aq) + Br^-(aq) + H_2O\ (l)$

NIE***: $OH^-\ (aq) + H^+(aq) \rightarrow\ H_2O\ (l)$

b. M*: $2\ RbOH\ (aq) + H_2SO_3\ (aq) \rightarrow Rb_2SO_3\ (aq) + 2\ H_2O\ (l)$

TIE**: $2\ Rb^+(aq) + 2\ OH^-\ (aq) + 2\ H^+(aq) + SO_3^{2-}(aq) \rightarrow 2\ Rb^+(aq) + SO_3^{2-}(aq) + 2\ H_2O\ (l)$

NIE***: $OH^-\ (aq) + H^+(aq) \rightarrow\ H_2O\ (l)$

c. M*: $2\ RbOH\ (aq) + H_3PO_3\ (aq) \rightarrow Rb_2HPO_3\ (aq) + 2\ H_2O\ (l)$

TIE**: $2\ Rb^+(aq) + 2\ OH^-(aq) + 2\ H^+(aq) + HPO_3^{2-}\ (aq) \rightarrow 2\ Rb^+(aq) + HPO_3^{2-}(aq) + 2\ H_2O\ (l)$

NIE***: $OH^-\ (aq) + H^+(aq) \rightarrow\ H_2O\ (l)$

☑9.60 a. M*: $2\ KOH\ (aq) + H_3PO_4\ (aq) \rightarrow K_2HPO_4\ (aq) + 2\ H_2O\ (l)$

TIE**: $2\ K^+(aq) + 2\ OH^-(aq) + 2\ H^+(aq) + HPO_4^{2-}\ (aq) \rightarrow 2\ K^+(aq) + HPO_4^{2-}(aq) + 2\ H_2O\ (l)$

NIE***: $OH^-\ (aq) + H^+(aq) \rightarrow\ H_2O\ (l)$

b. M*: $3\ KOH\ (aq) + H_3PO_4\ (aq) \rightarrow K_3PO_4\ (aq) + 3\ H_2O\ (l)$

TIE**: $3\ K^+(aq) + 3\ OH^-(aq) + 3\ H^+(aq) + PO_4^{3-}\ (aq) \rightarrow 3\ K^+(aq) + PO_4^{3-}(aq) + 3\ H_2O\ (l)$

NIE***: $OH^-(aq) + H^+(aq) \rightarrow H_2O\ (l)$

c. M*: $KOH\ (aq) + H_2C_2O_4\ (aq) \rightarrow KHC_2O_4\ (aq) + H_2O\ (l)$

TIE**: $K^+(aq) + OH^-(aq) + H^+(aq) + HC_2O_4^-\ (aq) \rightarrow K^+(aq) + HC_2O_4^-(aq) + H_2O\ (l)$

NIE***: $OH^-\ (aq) + H^+(aq) \rightarrow\ H_2O\ (l)$

SALTS (SECTION 9.8)			Cations	Anions
9.62	a.	$CuCl_2$	Cu^{2+}	$2\ Cl^-$
	b.	$(NH_4)_2SO_4$	$2\ NH_4^+$	SO_4^{2-}
	c.	Li_3PO_4	$3\ Li^+$	PO_4^{3-}

	Cations	Anions
d. $MgCO_3$	Mg^{2+}	CO_3^{2-}
e. $Ca(C_2H_3O_2)_2$	Ca^{2+}	$2\ C_2H_3O_2^-$
f. KNO_3	K^+	NO_3^-

		Cations	Base	Anions	Acids
9.64	a. $CuCl_2$	Cu^{2+}	$Cu(OH)_2$	$2\ Cl^-$	HCl
	b. $(NH_4)_2SO_4$	$2\ NH_4^+$	NH_3 or NH_4OH	SO_4^{2-}	H_2SO_4
	c. Li_3PO_4	$3\ Li^+$	LiOH	PO_4^{3-}	H_3PO_4
	d. $MgCO_3$	Mg^{2+}	$Mg(OH)_2$	CO_3^{2-}	H_2CO_3
	e. $Ca(C_2H_3O_2)_2$	Ca^{2+}	$Ca(OH)_2$	$2\ C_2H_3O_2^-$	$HC_2H_3O_2$
	f. KNO_3	K^+	KOH	NO_3^-	HNO_3

9.66 a. Epsom salt = $MgSO_4 \cdot 7H_2O$

$$MgSO_4 \cdot 7\ H_2O\ (s) \xrightarrow{\Delta} MgSO_4\ (s) + 7\ H_2O\ (g)$$

$$1.0 \text{ mol Epsom salt}\left(\frac{7 \text{ moles } H_2O}{1 \text{ mole epsom salt}}\right)\left(\frac{18.0 \text{ g } H_2O}{1 \text{ mole } H_2O}\right) = 1.3\times10^2 \text{ g } H_2O$$

b. Borax = $Na_2B_4O_7 \cdot 10H_2O$

$$Na_2B_4O_7 \cdot 10\ H_2O\ (s) \xrightarrow{\Delta} Na_2B_4O_7\ (s) + 10\ H_2O\ (g)$$

$$1.0 \text{ mol Borax}\left(\frac{10 \text{ moles } H_2O}{1 \text{ mole borax}}\right)\left(\frac{18.0 \text{ g } H_2O}{1 \text{ mole } H_2O}\right) = 1.8\times10^2 \text{ g } H_2O$$

Both of these reactions will produce anhydrous salts as well as water. For every mole of Epsom salt, 7 moles of water are released. For every mole of borax, 10 moles of water are released.

		Acid	Solid
☑9.68	a. $Mg(NO_3)_2$	HNO_3	$MgCO_3$
	b. $CaCl_2$	HCl	CaO
	c. Rb_2SO_4	H_2SO_4	$RbHCO_3$

9.70
a. $2HNO_3\ (aq) + MgCO_3\ (s) \rightarrow H_2O\ (l) + CO_2\ (g) + Mg(NO_3)_2\ (aq)$
b. $2\ HCl\ (aq) + CaO\ (s) \rightarrow H_2O\ (l) + CaCl_2\ (aq)$
c. $H_2SO_4\ (aq) + 2\ RbHCO_3\ (s) \rightarrow 2\ H_2O\ (l) + 2\ CO_2\ (g) + Rb_2SO_4\ (aq)$

9.72		
a.	$CaCl_2$	1 eq $CaCl_2$ = 0.5 moles $CaCl_2$
b.	$Cu(NO_3)_2$	1 eq $Cu(NO_3)_2$ = 0.5 moles $Cu(NO_3)_2$
c.	$CrCl_3$	1 eq $CrCl_3$ = 0.33 moles $CrCl_3$

☑9.74 a. 0.22 mol $ZnCl_2$

$$0.22 \text{ mol } ZnCl_2\left(\frac{2 \text{ eq. } ZnCl_2}{1 \text{ mole } ZnCl_2}\right) = 0.44 \text{ eq } ZnCl_2$$

$$0.22 \text{ mol } ZnCl_2\left(\frac{2 \text{ eq. } ZnCl_2}{1 \text{ mole } ZnCl_2}\right)\left(\frac{1000 \text{ meq}}{1 \text{ eq.}}\right) = 4.4\times10^2 \text{ meq } ZnCl_2$$

b. 0.45 mol CsCl

$$0.45 \text{ mol CsCl}\left(\frac{1 \text{ eq. CsCl}}{1 \text{ mole CsCl}}\right) = 0.45 \text{ eq CsCl}$$

$$0.45 \text{ mol CsCl}\left(\frac{1 \text{ eq. CsCl}}{1 \text{ mole CsCl}}\right)\left(\frac{1000 \text{ meq}}{1 \text{ eq.}}\right) = 4.5\times10^2 \text{ meq CsCl}$$

c. 3.12×10^{-2} mol $Fe(NO_3)_2$

$$3.12\times10^{-2} \text{ mol } Fe(NO_3)_2\left(\frac{2 \text{ eq. } Fe(NO_3)_2}{1 \text{ mole } Fe(NO_3)_2}\right) = 6.24\times10^{-2} \text{ eq } Fe(NO_3)_2$$

$$3.12\times10^{-2} \text{ mol } Fe(NO_3)_2\left(\frac{2 \text{ eq. } Fe(NO_3)_2}{1 \text{ mole } Fe(NO_3)_2}\right)\left(\frac{1000 \text{ meq}}{1 \text{ eq.}}\right) = 62.4 \text{ meq } Fe(NO_3)_2$$

9.76 a. $5.00\text{ g Na}_2\text{CO}_3\cdot 10\text{ H}_2\text{O}\left(\frac{1\text{ mole Na}_2\text{CO}_3\cdot 10\text{ H}_2\text{O}}{286\text{ g Na}_2\text{CO}_3\cdot 10\text{ H}_2\text{O}}\right)\left(\frac{2\text{ eq Na}_2\text{CO}_3\cdot 10\text{ H}_2\text{O}}{1\text{ mole Na}_2\text{CO}_3\cdot 10\text{ H}_2\text{O}}\right) = 3.50\times 10^{-2}\text{ eq Na}_2\text{CO}_3\cdot 10\text{ H}_2\text{O}$

$5.00\text{ g Na}_2\text{CO}_3\cdot 10\text{ H}_2\text{O}\left(\frac{1\text{ mole Na}_2\text{CO}_3\cdot 10\text{ H}_2\text{O}}{286\text{ g Na}_2\text{CO}_3\cdot 10\text{ H}_2\text{O}}\right)\left(\frac{2\text{ eq Na}_2\text{CO}_3\cdot 10\text{ H}_2\text{O}}{1\text{ mole Na}_2\text{CO}_3\cdot 10\text{ H}_2\text{O}}\right)\left(\frac{1000\text{ meq}}{1\text{ eq.}}\right) = 35.0\text{ meq Na}_2\text{CO}_3\cdot 10\text{ H}_2\text{O}$

b. $5.00\text{ g CuSO}_4\cdot 5\text{ H}_2\text{O}\left(\frac{1\text{ mole CuSO}_4\cdot 5\text{ H}_2\text{O}}{250\text{ g CuSO}_4\cdot 5\text{ H}_2\text{O}}\right)\left(\frac{2\text{ eq CuSO}_4\cdot 5\text{ H}_2\text{O}}{1\text{ mole CuSO}_4\cdot 5\text{ H}_2\text{O}}\right) = 4.00\times 10^{-2}\text{ eq CuSO}_4\cdot 5\text{ H}_2\text{O}$

$5.00\text{ g CuSO}_4\cdot 5\text{ H}_2\text{O}\left(\frac{1\text{ mole CuSO}_4\cdot 5\text{ H}_2\text{O}}{250\text{ g CuSO}_4\cdot 5\text{ H}_2\text{O}}\right)\left(\frac{2\text{ eq CuSO}_4\cdot 5\text{ H}_2\text{O}}{1\text{ mole CuSO}_4\cdot 5\text{ H}_2\text{O}}\right)\left(\frac{1000\text{ meq}}{1\text{ eq.}}\right) = 40.0\text{ meq CuSO}_4\cdot 5\text{ H}_2\text{O}$

c. $5.00\text{ g Li}_2\text{CO}_3\left(\frac{1\text{ mole Li}_2\text{CO}_3}{73.9\text{ g Li}_2\text{CO}_3}\right)\left(\frac{2\text{ eq Li}_2\text{CO}_3}{1\text{ mole Li}_2\text{CO}_3}\right) = 1.35\times 10^{-1}\text{ eq Li}_2\text{CO}_3$

$5.00\text{ g Li}_2\text{CO}_3\left(\frac{1\text{ mole Li}_2\text{CO}_3}{73.9\text{ g Li}_2\text{CO}_3}\right)\left(\frac{2\text{ eq Li}_2\text{CO}_3}{1\text{ mole Li}_2\text{CO}_3}\right)\left(\frac{1000\text{ meq}}{1\text{ eq.}}\right) = 135\text{ meq Li}_2\text{CO}_3$

d. $5.00\text{ g NaH}_2\text{PO}_4\left(\frac{1\text{ mole NaH}_2\text{PO}_4}{120\text{ g NaH}_2\text{PO}_4}\right)\left(\frac{1\text{ eq NaH}_2\text{PO}_4}{1\text{ mole NaH}_2\text{PO}_4}\right) = 4.17\times 10^{-2}\text{ eq NaH}_2\text{PO}_4$

$5.00\text{ g NaH}_2\text{PO}_4\left(\frac{1\text{ mole NaH}_2\text{PO}_4}{120\text{ g NaH}_2\text{PO}_4}\right)\left(\frac{1\text{ eq NaH}_2\text{PO}_4}{1\text{ mole NaH}_2\text{PO}_4}\right)\left(\frac{1000\text{ meq}}{1\text{ eq.}}\right) = 41.7\text{ meq NaH}_2\text{PO}_4$

9.78 $150\text{ mL i.c. fluid}\left(\frac{1\text{ L}}{1000\text{ mL}}\right)\left(\frac{133\text{ meq}}{1\text{ L}}\right)\left(\frac{1\text{ eq}}{1000\text{ meq}}\right)\left(\frac{1\text{ mole K}_2\text{SO}_4}{2\text{ eq K}_2\text{SO}_4}\right) = 9.98\times 10^{-3}\text{ moles K}_2\text{SO}_4$

$150\text{ mL i.c. fluid}\left(\frac{1\text{ L}}{1000\text{ mL}}\right)\left(\frac{133\text{ meq}}{1\text{ L}}\right)\left(\frac{1\text{ eq}}{1000\text{ meq}}\right)\left(\frac{1\text{ mole K}_2\text{SO}_4}{2\text{ eq K}_2\text{SO}_4}\right)\left(\frac{174.27\text{ g K}_2\text{SO}_4}{1\text{ mole K}_2\text{SO}_4}\right) = 1.74\text{ g K}_2\text{SO}_4$

THE STRENGTHS OF ACIDS AND BASES (SECTION 9.9)

9.80 (weakest) acid B < acid A < acid C < acid D (strongest)
The smaller the K_a, the weaker the acid. The larger the K_a, the stronger the acid.

9.82 a. (weakest) acid B < acid A < acid C < acid D (strongest)
The smaller the K_a, the weaker the acid. The larger the K_a, the stronger the acid.

b. (weakest) base D < base C < base A< base B (strongest)
The smaller the K_a, the weaker the acid and the stronger the conjugate base. The larger the K_a, the stronger the acid and the stronger the conjugate base.

9.84 a. HNO_2 $\quad HNO_2(aq) \rightleftharpoons H^+(aq) + NO_2^-(aq)$ $\quad K_a = \frac{[H^+][NO_2^-]}{[HNO_2]}$

b. HCO_3^- $\quad HCO_3^-(aq) \rightleftharpoons H^+(aq) + CO_3^{2-}(aq)$ $\quad K_a = \frac{[H^+][CO_3^{2-}]}{[HCO_3^-]}$

c. $H_2PO_4^-$ $\quad H_2PO_4^-(aq) \rightleftharpoons H^+(aq) + HPO_4^{2-}(aq)$ $\quad K_a = \frac{[H^+][HPO_4^{2-}]}{[H_2PO_4^-]}$

d. HS^- $\quad HS^-(aq) \rightleftharpoons H^+(aq) + S^{2-}(aq)$ $\quad K_a = \frac{[H^+][S^{2-}]}{[HS^-]}$

e. HBrO $\quad HBrO(aq) \rightleftharpoons H^+(aq) + BrO^-(aq)$ $\quad K_a = \frac{[H^+][BrO^-]}{[HBrO]}$

☑9.86 The 20% acetic acid solution is the weak acid solution. It is a weak acid because acetic acid does not completely dissociate.

The 0.05 M HCl is a dilute strong acid solution. If someone wanted this solution instead, they should use the term "dilute."

ANALYZING ACIDS AND BASES (SECTION 9.10)

9.88 A titration is performed to determine the concentration of a solution by reacting a measured volume of the solution with another solution of known concentration.

9.90 a. The endpoint of a titration occurs when an indicator changes color. The equivalence point of a titration is the point at which the unknown solution has completely reacted with the known solution. The resulting mixture from the titration will have a specific pH at the equivalence point. If the indicator selected changes color at the same pH as the pH of the resulting mixture, the endpoint and the equivalence point will be the same.

b. If the indicator selected changes color at a different pH than the pH of the mixture that results from the titration, then the endpoint and the equivalence point will not be the same.

☑9.92 a. $HBr + NaOH \rightarrow H_2O + NaBr$

$$250\text{ mL}\left(\frac{1\text{L}}{1000\text{ mL}}\right)\left(\frac{0.400\text{ moles HBr}}{1\text{ L}}\right)\left(\frac{1\text{ mole NaOH}}{1\text{ mole HBr}}\right) = 0.100\text{ moles NaBr}$$

b. $HClO_4 + NaOH \rightarrow H_2O + NaClO_4$

$$750\text{ mL}\left(\frac{1\text{L}}{1000\text{ mL}}\right)\left(\frac{0.300\text{ moles HClO}_4}{1\text{ L}}\right)\left(\frac{1\text{ mole NaOH}}{1\text{ mole HClO}_4}\right) = 0.225\text{ moles HClO}_4$$

TITRATION CALCULATIONS (SECTION 9.11)

9.94 a. $2\ NaOH\ (aq) + H_2CrO_4\ (aq) \rightarrow Na_2CrO_4\ (aq) + 2\ H_2O\ (l)$

b. $NaOH\ (aq) + HClO_3\ (aq) \rightarrow NaClO_3\ (aq) + H_2O\ (l)$

c. $3\ NaOH\ (aq) + H_3AsO_3\ (aq) \rightarrow Na_3AsO_3\ (aq) + 3\ H_2O\ (l)$

9.96 a. $2\ HCl\ (aq) + Sr(OH)_2\ (aq) \rightarrow SrCl_2\ (aq) + 2\ H_2O\ (l)$

b. $3\ HCl\ (aq) + Ni(OH)_3\ (s) \rightarrow NiCl_3\ (aq) + 3\ H_2O\ (l)$

c. $2\ HCl\ (aq) + Zn(OH)_2\ (s) \rightarrow ZnCl_2\ (aq) + 2\ H_2O\ (l)$

☑9.98 $H_2C_2O_4 + 2\ NaOH \rightarrow 2\ H_2O + Na_2C_2O_4$

$$\frac{43.88\text{ mL}\left(\frac{1\text{L}}{1000\text{ mL}}\right)\left(\frac{0.1891\text{ moles NaOH}}{1\text{ L}}\right)\left(\frac{1\text{ mole H}_2\text{C}_2\text{O}_4}{2\text{ mole NaOH}}\right)}{25.00\text{ mL}\left(\frac{1\text{ L}}{1000\text{ mL}}\right)} = 0.1660\text{ M H}_2\text{C}_2\text{O}_4$$

9.100 ☑a. $NaOH + HClO_4 \rightarrow H_2O + NaClO_4$

$$20.00\text{ mL}\left(\frac{1\text{L}}{1000\text{ mL}}\right)\left(\frac{0.200\text{ moles HClO}_4}{1\text{ L}}\right)\left(\frac{1\text{ mole NaOH}}{1\text{ mole HClO}_4}\right)\left(\frac{1\text{ L NaOH}}{0.120\text{ moles NaOH}}\right)\left(\frac{1000\text{ mL}}{1\text{ L}}\right) = 33.3\text{ mL NaOH}$$

b. $2\ NaOH + H_2SO_4 \rightarrow 2\ H_2O + Na_2SO_4$

$$20.00\text{ mL}\left(\frac{1\text{L}}{1000\text{ mL}}\right)\left(\frac{0.125\text{ moles H}_2\text{SO}_4}{1\text{ L}}\right)\left(\frac{2\text{ moles NaOH}}{1\text{ mole H}_2\text{SO}_4}\right)\left(\frac{1\text{ L NaOH}}{0.120\text{ moles NaOH}}\right)\left(\frac{1000\text{ mL}}{1\text{ L}}\right) = 41.7\text{ mL NaOH}$$

c. $4\ NaOH + H_4P_2O_6 \rightarrow 4\ H_2O + Na_4P_2O_6$

$$20.00\text{ mL}\left(\frac{1\text{L}}{1000\text{ mL}}\right)\left(\frac{0.150\text{ moles H}_4\text{P}_2\text{O}_6}{1\text{ L}}\right)\left(\frac{4\text{ moles NaOH}}{1\text{ mole H}_4\text{P}_2\text{O}_6}\right)\left(\frac{1\text{ L NaOH}}{0.120\text{ moles NaOH}}\right)\left(\frac{1000\text{ mL}}{1\text{ L}}\right) = 100.\text{ mL NaOH}$$

d. $3\ NaOH + H_3PO_4 \rightarrow 3\ H_2O + Na_3PO_4$

$$20.00\text{ mL}\left(\frac{0.120\text{ moles H}_3\text{PO}_4}{500\text{ mL}}\right)\left(\frac{3\text{ moles NaOH}}{1\text{ mole H}_3\text{PO}_4}\right)\left(\frac{1\text{ L NaOH}}{0.120\text{ moles NaOH}}\right)\left(\frac{1000\text{ mL}}{1\text{ L}}\right) = 120.\text{ mL NaOH}$$

e. $2\ NaOH + H_2SO_4 \rightarrow 2\ H_2O + Na_2SO_4$

$$20.00\ mL\left(\frac{6.25\ g\ H_2SO_4}{250\ mL}\right)\left(\frac{1\ mole\ H_2SO_4}{98.09\ g\ H_2SO_4}\right)\left(\frac{2\ moles\ NaOH}{1\ mole\ H_2SO_4}\right)\left(\frac{1\ L\ NaOH}{0.120\ moles\ NaOH}\right)\left(\frac{1000\ mL}{1\ L}\right) = 85.0\ mL\ NaOH$$

f. $NaOH + HClO_3 \rightarrow H_2O + NaClO_3$

$$20.00\ mL\left(\frac{1L}{1000\ mL}\right)\left(\frac{0.500\ moles\ HClO_3}{1\ L}\right)\left(\frac{1\ mole\ NaOH}{1\ mole\ HClO_3}\right)\left(\frac{1\ L\ NaOH}{0.120\ moles\ NaOH}\right)\left(\frac{1000\ mL}{1\ L}\right) = 83.3\ mL\ NaOH$$

9.102 a. $H_2SO_4 + 2\ NaOH \rightarrow 2\ H_2O + Na_2SO_4$

$$\frac{29.88\ mL\left(\frac{1L}{1000\ mL}\right)\left(\frac{1.17\ moles\ NaOH}{1L}\right)\left(\frac{1\ mole\ H_2SO_4}{2\ moles\ NaOH}\right)}{5.00\ mL\left(\frac{1\ L}{1000\ mL}\right)} = 3.50\ M\ H_2SO_4$$

b. $HC_2H_3O_2 + KOH \rightarrow H_2O + KC_2H_3O_2$

$$\frac{35.62\ mL\left(\frac{1L}{1000\ mL}\right)\left(\frac{0.250\ moles\ KOH}{1L}\right)\left(\frac{1\ mole\ HC_2H_3O_2}{1\ mole\ KOH}\right)}{10.00\ mL\left(\frac{1\ L}{1000\ mL}\right)} = 0.891\ M\ HC_2H_3O_2$$

c. $HCl + NaOH \rightarrow H_2O + NaCl$

$$\frac{20.63\ mL\left(\frac{1L}{1000\ mL}\right)\left(\frac{6.00\ moles\ NaOH}{1\ L}\right)\left(\frac{1\ mole\ HCl}{1\ mole\ NaOH}\right)}{10.00\ mL\left(\frac{1\ L}{1000\ mL}\right)} = 12.4\ M\ HCl$$

9.104 benzoic acid + $NaOH \rightarrow H_2O$ + sodium benzoate

$$46.75\ mL\left(\frac{1\ L}{1000\ mL}\right)\left(\frac{0.1021\ moles\ NaOH}{1\ L}\right)\left(\frac{1\ mole\ benzoic\ acid}{1\ mole\ NaOH}\right) = 0.004773175\ moles\ benzoic\ acid$$

$$\frac{0.5823\ g\ benzoic\ acid}{0.004773175\ moles\ benzoic\ acid} = 122.0\frac{g}{mole}$$

or

$$\frac{0.5823\ g\ benzoic\ acid}{46.75\ mL\left(\frac{1\ L}{1000\ mL}\right)\left(\frac{0.1021\ moles\ NaOH}{1\ L}\right)\left(\frac{1\ mole\ benzoic\ acid}{1\ mole\ NaOH}\right)} = 122.0\frac{g}{mole}$$

HYDROLYSIS REACTIONS OF SALTS (SECTION 9.12)

9.106 Na_2CO_3 is a salt that could be produced by the reaction of a strong base (NaOH) and a weak acid (H_2CO_3). The hydrolysis of Na_2CO_3 produces an alkaline solution.

☑9.108 a. NaOCl — NaOCl could be formed from NaOH and HOCl. The pH is greater than 7 because NaOH is a strong base and HOCl is a weak acid.

$$OCl^-(aq) + H_2O\ (l) \rightleftharpoons HOCl\ (aq) + OH^-(aq)$$

b. $NaCHO_2$ — $NaCHO_2$ could be formed from NaOH and $HCHO_2$. The pH is greater than 7 because NaOH is a strong base and $HCHO_2$ is a weak acid.

$$CHO_2^-(aq) + H_2O\ (l) \rightleftharpoons HCHO_2\ (aq) + OH^-(aq)$$

c. KNO_3 — KNO_3 could be formed from KOH and HNO_3. The pH equals 7 because KOH is a strong base and HNO_3 is a strong acid.

d. Na_3PO_4 — Na_3PO_4 could be formed from NaOH and H_3PO_4. The pH is greater than 7 because NaOH is a strong base and H_3PO_4 is a weak acid.

$$PO_4^{3-}(aq) + H_2O\ (l) \rightleftharpoons HPO_4^{2-}\ (aq) + OH^-(aq)$$

9.110 More than one indicator should be available in the laboratory because the pH at the equivalence point of a titration will vary depending on the salt produced by the reaction. Not all indicators change color at the same pH. Each indicator has its own unique pH range over which its color changes.

BUFFERS (SECTION 9.13)

9.112 $HPO_4^{2-}(aq) + H_3O^+(aq) \rightleftharpoons H_2PO_4^-(aq) + H_2O\ (l)$

$H_2PO_4^-(aq) + OH^-(aq) \rightleftharpoons HPO_4^{2-}(aq) + H_2O\ (l)$

9.114 $HCO_3^-(aq) + H^+(aq) \rightleftharpoons H_2CO_3\ (aq)$

The bicarbonate ion reacts with the excess hydrogen ions to form carbonic acid, which combats acidosis.

☑9.116 a. $pH = 3.85 + \log\frac{(0.1)}{(0.1)} = 3.85$

b. $pH = 3.85 + \log\frac{(1)}{(1)} = 3.85$

c. The solution in part b has greater buffer capacity than the solution in part a because the higher concentration of the buffer components will allow it to react with larger added amounts of acid or base.

9.118 a. $pH = 4.74 + \log\frac{(0.25)}{(0.40)} = 4.54$

b. $pH = 7.21 + \log\frac{(0.40)}{(0.10)} = 7.81$

c. $pH = 7.00 + \log\frac{(0.20)}{(1.50)} = 6.12$

9.120 $7.65 = 7.21 + \log\frac{[HPO_4^{2-}]}{[H_2PO_4^-]} \Rightarrow 0.44 = \log\frac{[HPO_4^{2-}]}{[H_2PO_4^-]} \Rightarrow \frac{[HPO_4^{2-}]}{[H_2PO_4^-]} = 10^{0.44} = 2.8$

The concentration of Na_2HPO_4 has to be 2.8 times the concentration of NaH_2PO_4.

ADDITIONAL EXERCISES

9.122 $Cl^-\ (aq) + H_3O^+\ (aq) \rightarrow HCl\ (aq) + H_2O\ (l)$

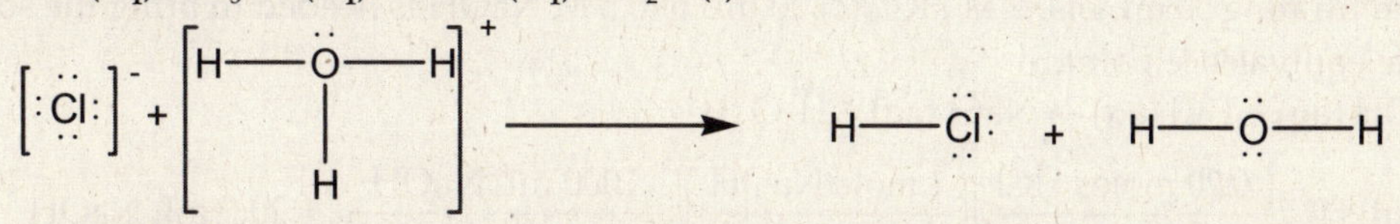

The chloride ion is a Brønsted base because it is a proton acceptor. The hydronium ion is a Brønsted acid because it is a proton donor. An alternative way to look at this is reaction is that the chloride ion has a pair of electrons that it can donate to one of the hydrogen atoms in the hydronium ion in order to form a covalent bond and HCl. The electron pair donor (chloride ion) is acting as a base. The electron pair acceptor (hydronium ion) is acting as an acid.

9.124 $K_w = [H_3O^+][OH^-]$ $K_w = 5.5 \times 10^{-14}$ at 50°C and for water, $[H_3O^+] = [OH^-]$

$$5.5 \times 10^{-14} = [H_3O^+]^2$$

$$[H_3O^+] = \sqrt{5.5 \times 10^{-14}}$$

$$pH = -\log[H_3O^+] = -\log\left(\sqrt{5.5 \times 10^{-14}}\right) = 6.63$$

ALLIED HEALTH EXAM CONNECTION

9.126 An acid is a substance that dissociates in water into one or more (a) hydrogen ions and one or more anions.

9.128 Alkaline solutions contain (c) more OH^- ion than H^+ ion.

9.130 The formula for the hydronium ion is (c) H_3O^+.

9.132 Dissolving H_2SO_4 in water creases an acid solution by increasing the (c) hydrogen ions.

9.134 A common detergent has a pH of 11.0, so the detergent is (c) alkaline.

9.136 The pH of a blood sample is 7.40 at room temperature. The pOH is therefore (a) 6.60 (pOH = 14-7.40).

9.138 Blood with a pH of 7.4 indicates that the blood sample is (d) weakly basic.

9.140 (d) $HC_2H_3O_2$ is likely to taste sour because it is an acid.

9.142 A substance that functions to prevent rapid, drastic changes in the pH of a body fluid by changing strong acids and bases into weak acids and bases is called a (b) buffer.

9.144 (c) $2\ HNO_3 + Mg(OH)_2 \rightarrow 2\ H_2O + Mg(NO_3)_2$ is a neutralization reaction.

9.146 When titrating 50 mL of 0.2 M HCl, (c) 20 mL of 0.5 M NaOH is needed to bring the solution to the equivalence point.

$$NaOH\ (aq) + HCl\ (aq) \rightarrow NaCl\ (aq) + H_2O\ (l)$$

$$50\text{ mL HCl}\left(\frac{0.20\text{ moles HCl}}{1000\text{ mL HCl}}\right)\left(\frac{1\text{ mole NaOH}}{1\text{ mole HCl}}\right)\left(\frac{1000\text{ mL NaOH}}{0.50\text{ moles NaOH}}\right) = 20.0\text{ mL NaOH}$$

CHEMISTRY FOR THOUGHT

9.148 A saturated solution of solid $Ca(OH)_2$ in water has a $[OH^-]$ of only 2.50×10^{-2} M, and yet $Ca(OH)_2$ is a strong base. $Ca(OH)_2$ has limited solubility in water; however, it is a hydroxide of a group IIA(2) metal which is defined to be a strong base because it dissociates completely when dissolved to form a solution. In other words, the small amount which is dissolved dissociates completely.

9.150 a. $2.61 = pK_a + \log\frac{(2.48\times10^{-3})}{(0.0975)}$

$pK_a = 4.20$

$K_a = 10^{-4.20} = 6.2\times10^{-5}$

b. $4.16 = pK_a + \log\frac{(6.93\times10^{-5})}{(0.200)}$

$pK_a = 7.62$

$K_a = 10^{-7.62} = 2.4\times10^{-8}$

c. $3.55 = pK_a + \log\frac{(0.000284)}{(0.150)}$

$pK_a = 6.27$

$K_a = 10^{-6.27} = 5.4\times10^{-7}$

9.152 When marble reacts with HCl, carbon dioxide gas is produced. Marble structures exposed to acid rain will eventually deteriorate as the marble reacts with acid to produce a soluble salt ($CaCl_2$), water, and carbon dioxide.

9.154 Phenolphthalein would not be a useful indicator to differentiate between two solutions with pH values of 5 and 7 because it is clear at both of those pH values. The color change in phenolphthalein occurs between pH 8 and 9.
Methyl red changes from red to orange around pH 6 and would be a useful indicator to differentiate between solutions with pH values of 5 and 7.

Bromothymol blue changes from yellow to blue around pH 7 and would be a useful indicator to differentiate between solutions with pH vales of 5 and 7.

9.156 The Na^+ ion is the cation from a strong base; therefore, it does not hydrolyze. Two other ions that would have behaved like the Na^+ ion are Li^+ and Rb^+ (any group IA or IIA metal ion could have been selected).

ADDITIONAL ACTIVITIES

Section 9.1 Review:

(1) How many protons, neutrons, and electrons are in this hydrogen ion: ${}_1^1H^+$?

(2) Write a general formula like Equation 9.1 for an Arrhenius acid (HA) dissolved in water.

(3) Write a general formula like Equation 9.2 for an Arrhenius base (MOH) dissolved in water.

Section 9.2 Review:

(1) Write a general formula like Equation 9.5 for a Brønsted acid (HA) dissolved in water.

(2) Identify the conjugate acid-base pairs in (1).

Section 9.3 Review:

(1) Can H_2O be named as an acid? Explain.

(2) Assume the following are dissolved in water. Name them as acids.

H_2Se	H_2SO_3	HF	H_3P	HClO	$HClO_3$	$HC_2H_3O_2$	$H_2Cr_2O_7$

Section 9.4 Review:

(1) Add the corresponding $[OH^-]$ concentrations to the diagram below.

(2) Label the acidic, basic, and neutral regions of the diagram.

(3) Mark the position of $[H_3O^+] = 0.125$ M. Is this acidic, basic, or neutral?

(4) Mark the position of $[OH^-] = 6.7 \times 10^{-7}$ M. Is this acidic, basic, or neutral?
(5) Mark the position of $[H_3O^+] = 2.5 \times 10^{-9}$ M. Is this acidic, basic, or neutral?
(6) Mark the position of $[OH^-] = 3.4 \times 10^{-3}$ M. Is this acidic, basic, or neutral?

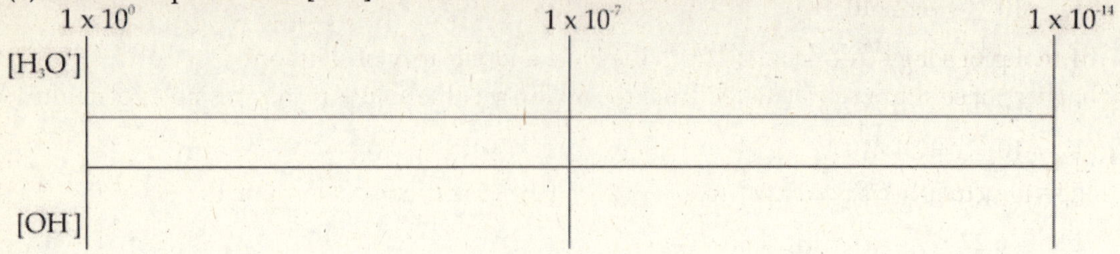

Section 9.5 Review:

(1) Calculate the pH for each of the solutions described in (3)-(6) of the Section 9.4 Review.
(2) Classify each of these solutions as acidic, basic, or neutral.
(3) Does the classification from (2) match the classification from the Section 9.4 Review?
(4) What is the $[H_3O^+]$ for a solution with a pH of 4.76? Is this acidic, basic, or neutral?
(5) What is the $[OH^-]$ for a solution with a pH of 8.27? Is this acidic, basic, or neutral?

Section 9.6 Review:

(1) The labels on three containers in the laboratory have fallen off. The containers hold iron (II) oxide, iron (II) carbonate, and iron (II) hydroxide. Can reactions with an acid be performed to tell these compounds apart? If so, describe how. If not, explain why not.
(2) Write the total ionic equations and net ionic equations for any reactions described in (1). Identify spectator ions.
(3) Will iron react with 6.00 M hydrochloric acid? If so, write a molecular equation, a total ionic equation, and a net ionic equation. Also, identify any spectator ions.

Section 9.7 Review:

(1) For water soluble Arrhenius bases and acids, what is the net ionic equation for neutralization?
(2) Why do neutralization reactions occur?

Section 9.8 Review:

(1) Referring back to the net ionic equation in the Section 9.7 Review, identify the reactant ions as cations or anions.
(2) Did the OH^- ion in (1) come from the acid or the base? What was charge on the counterion in this compound?
(3) Did the H^+ ion in (1) come from the acid or the base? What was charge on the counterion in this compound?
(4) Why is the salt excluded from the net ionic equation in the Section 9.7 Review?

Section 9.9 Review:

(1) Does a dilute strong acid have a small value of K_a? Explain.
(2) Does a concentrated weak acid have a small value of K_a? Explain.
(3) If a base is weak, what is the strength of its conjugate acid?
(4) If a base is strong, what is the strength of its conjugate acid?

Section 9.10 Review:

(1) Write an equilibrium expression for the dissociation of hydrofluoric acid. The K_a is 6.46×10^{-4}.
(2) Referring back to the seesaw analogy from the Section 8.8 Review, what happens to the seesaw as a weak acid like hydrofluoric acid reacts with a strong base?
(3) How is equilibrium reestablished after the hydrofluoric acid reacts with a strong base?

(4) What is the moment called when all of the acid has reacted with the base?
(5) What is the moment called when the reaction mixture changes color?

Section 9.11 Review:
(1) How can the moles of solute be calculated from the liters and molarity of solution?
(2) How can the liters of solution be calculated from the molarity of the solution and the moles of solute?

Section 9.12 Review:
Consider the following "tug-of-war" competitors. For each pair of reactants, determine the pH of the resulting salts.

(1)

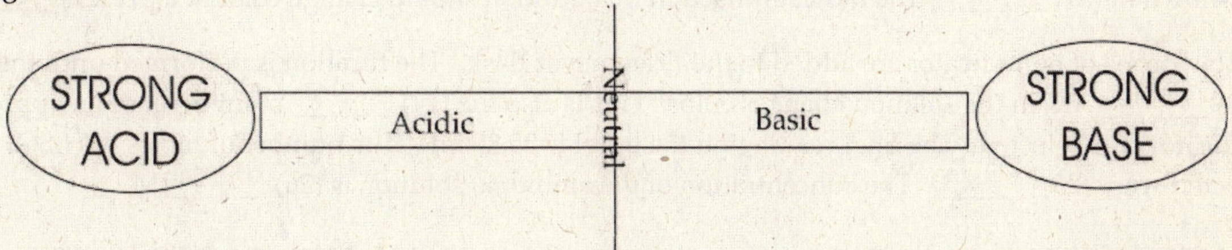

(2)

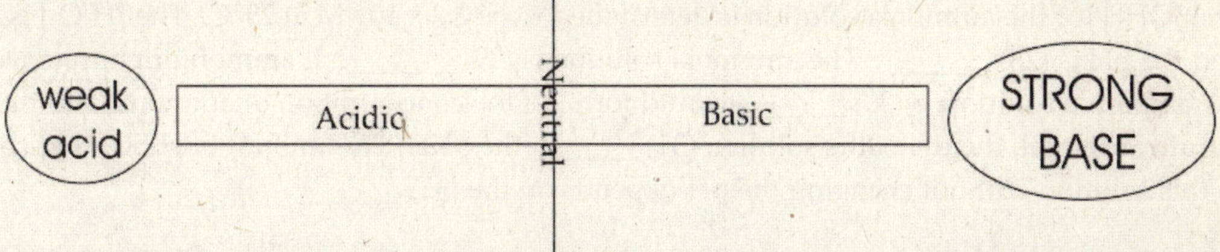

(3)

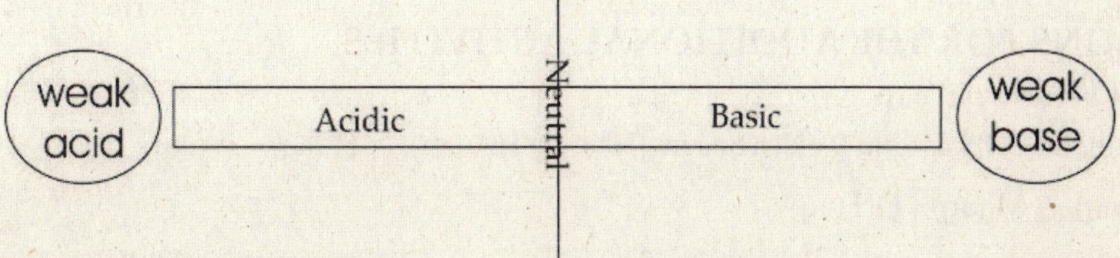

(4)

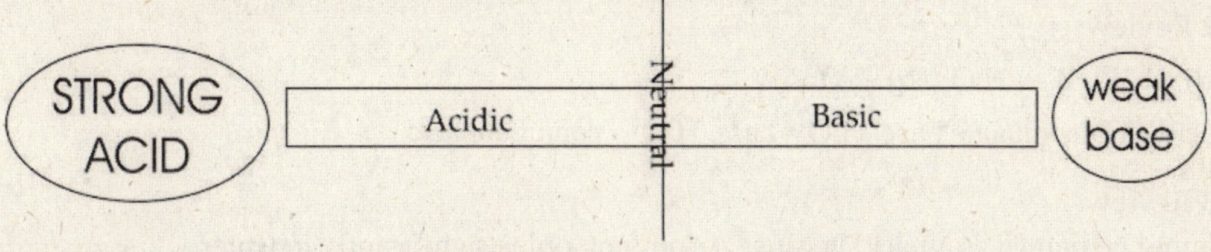

Section 9.13 Review: Refer back to the dissociation equation written for HF in the Section 9.10 Review.
(1) What ion must be present in a salt added to HF in order to form a buffer?
(2) Should the other ion in the salt come from a strong base or a weak base?
(3) Will the salt completely dissociate or establish an equilibrium of its own?
(4) How will adding the salt affect the seesaw for the HF equilibrium? What will happen to reestablish equilibrium?
(5) For a buffer with pH = pKa, place HF, H^+, and F^- in order of increasing concentration.
(6) Which component of this buffer will react with acids? How will this affect the equilibrium?
(7) Which component of this buffer will react with bases? How will this affect the equilibrium?
(8) What concentration levels ensure a large buffer capacity?

Tying It All Together with a Laboratory Application:
A (1) _______ is performed to determine the concentration of an ammonia (NH_3) solution. Ammonia (2) _______ (choose one: is, is not) a base according the Arrhenius theory because (3) _______; however, ammonia will react with water to produce (4) _______ and _______ ions. Ammonia is a(n) (5) _______ according to the Brønsted theory because (6) _______.

An aliquot of 25.00 mL is pipetted from the ammonia solution of unknown concentration into a 125 mL Erlenmeyer flask. A 0.125 M H_2SO_4 solution is added to the (7) _______. The initial volume reading is 5.87 mL. H_2SO_4 is a(n) (8) _______ according to the Arrhenius theory because (9) _______. H_2SO_4 is a(n) (10) _______ according to the Brønsted theory because (11) _______. The name of H_2SO_4 is (12) _______.

The molecular equation for the reaction is (13) _______. The total ionic equation for the reaction is (14) _______. The spectator ion(s) is(are) (15) _______. The net ionic equation for the reaction is (16) _______. This is the reaction between a (17) _______ acid and a (18) _______ base. The solution at the equivalence point will have a pH (19) _______ 7 because of the (20) _______ reaction of the salt with water as shown in (21) _______. The indicator used in this titration should change color at a pH (22) _______ 7.

A few drops of an indicator are added to the Erlenmeyer flask. The titration is performed until the (23) ________ when the solution changes color. This is also the (24) _______ point because an appropriate indicator was selected. The final reading on the buret is 35.27 mL. The number of moles of H_2SO_4 that reacted were (25) _______. The concentration of the ammonia solution is (26) _______ M.

The $[OH^-]$ for the ammonia solution is determined to be 2.3×10^{-3} M at 25°C. The $[H_3O^+]$ is (27) _______ and the pH is (28) _______. The ammonia solution is (29) _______. If ammonium nitrate were added to the ammonia solution, a (30) _______ would form. If the concentration of ammonia and the ammonium nitrate are equal, the pH of the solution (31) _______ the pKa. The amount of H_2SO_4 that could be added to this solution without changing the pH depends on the (32) _______.

SOLUTIONS FOR THE ADDITIONAL ACTIVITIES

Section 9.1 Review:

(1) 1 proton, 0 neutrons, and 0 electrons in ${}^{1}_{1}H^+$; (2) HA (aq) → H^+(aq) + A^-(aq);

(3) MOH (aq) → M^+(aq) + OH^-(aq)

Section 9.2 Review:

(1) HA (aq) + H_2O (l) → H_3O^+(aq) + A^-(aq)

(2) HA = acid, A^- = conjugate base; H_2O = base, H_3O^+ = conjugate acid

Section 9.3 Review:

(1) H_2O cannot be named as an acid because it does not ionize significantly in water.

(2) H_2Se = hydroselenic acid; H_2SO_3 = sulfurous acid; HF = hydrofluoric acid; H_3P = hydrophosphoric acid; HClO = hypochlorous acid; $HClO_3$ = chloric acid; $HC_2H_3O_2$ = acetic acid; $H_2Cr_2O_7$ = dichromic acid

Section 9.4 Review:

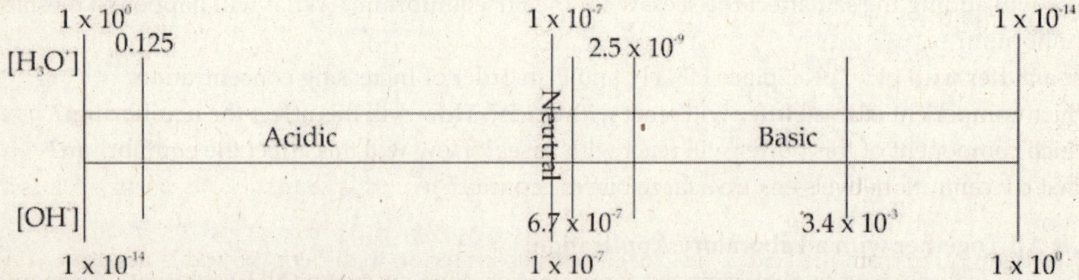

(3) acidic; (4) basic; (5) basic; (6) basic

Section 9.5 Review:

(1) $[H_3O^+] = 0.125$ M → pH = 0.903; $[OH^-] = 6.7 \times 10^{-7}$ M → pH = 7.83; $[H_3O^+] = 2.5 \times 10^{-9}$ M → pH = 8.60; $[OH^-] = 3.4 \times 10^{-3}$ M → pH = 11.53; (2) acidic; basic; basic; basic

(3) Yes, the classification from (2) matches the classification from the Section 9.4 Review.

(4) pH = 4.76 → $[H_3O^+] = 1.7 \times 10^{-5}$ M; acidic; (5) pH = 8.27 → $[OH^-] = 1.9 \times 10^{-6}$ M; basic

Section 9.6 Review:

(1) When iron (II) carbonate reacts with an acid, it will produce carbon dioxide gas in addition to salt and water. When iron (II) oxide and iron (II) hydroxide react with acid, they produce a salt and water. The iron (II) carbonate could be identified by a reaction with an acid because bubbles are produced; however, the other two materials cannot be identified unless stoichiometric calculations are performed. Both the iron (II) oxide and the iron (II) hydroxide react in a 1:2 mole ratio with a monoprotic acid. The molecular weights of these compounds are different, though. Consequently, 1 g FeO should react completely with 27.8 mL of a 1 M monoprotic acid and 1 g $Fe(OH)_2$ should react completely with 22.3 mL of a 1 M monoprotic acid. All three solids can be identified.

(2) Molecular = 2 HA (aq) + $FeCO_3$ (s) → CO_2 (g) + FeA_2 (aq) + H_2O (l)
TIE = 2 H^+(aq) + 2 A^- (aq) + $FeCO_3$ (s) → CO_2 (g) + Fe^{2+} (aq) + 2 A^- (aq) + H_2O (l)
NIE = 2 H^+(aq) + $FeCO_3$ (s) → CO_2 (g) + Fe^{2+} (aq) + H_2O (l)
Spectator ions = 2 A^- (aq)

Molecular = 2 HA (aq) + FeO (s) → FeA_2 (aq) + H_2O (l)
TIE = 2 H^+(aq) + 2 A^- (aq) + FeO (s) → Fe^{2+} (aq) + 2 A^- (aq) + H_2O (l)
NIE = 2 H^+(aq) + FeO (s) → Fe^{2+} (aq) + H_2O (l)
Spectator ions = 2 A^- (aq)
Molecular = 2 HA (aq) + $Fe(OH)_2$ (s) → FeA_2 (aq) + 2 H_2O (l)
TIE = 2 H^+(aq) + 2 A^- (aq) + $Fe(OH)_2$ (s) → Fe^{2+} (aq) + 2 A^- (aq) + 2 H_2O (l)
NIE = 2 H^+(aq) + $Fe(OH)_2$ (s) → Fe^{2+} (aq) + 2 H_2O (l)
Spectator ions = 2 A^- (aq)

(3) Yes, iron will react with 6.00 M HCl.
Molecular = 2 HCl (aq) + Fe (s) → $FeCl_2$ (aq) + H_2 (g)
TIE = 2 H^+(aq) + 2 Cl^- (aq) + Fe (s) → Fe^{2+} (aq) + 2 Cl^- (aq) + H_2 (g)
NIE = 2 H^+(aq) + Fe (s) → Fe^{2+} (aq) + H_2 (g)
Spectator ions = 2 Cl^- (aq)

Section 9.7 Review:

(1) OH^-(aq) + H^+ (aq) → H_2O (l)

(2) Neutralization reactions occur because hydroxide and hydrogen ions (really, hydronium ions) react to form liquid water.

Section 9.8 Review:

(1) OH^- = anion; H^+ = cation (Note: An anion is ***a n****egative* ***ion****.* A cation is "paws"itive and cats have paws!)

(2) The OH^- came from the base. The charge on the counterion in this compound is positive. It is a cation.

(3) The H^+ ion came from the acid. The charge on the counterion in this compound is negative. It is an anion.

(4) The positive cation (from the base) and the negative anion (from the acid) formed a water soluble salt. Consequently, these spectator ions were removed from the net ionic equation.

Section 9.9 Review:

(1) No, a dilute strong acid does not have a small value of K_a. A strong acid will completely dissociate in water regardless of whether it is concentrated or diluted.

(2) Yes, a concentrated weak acid has a small value of K_a. A weak acid will not completely dissociate in water regardless of whether it is concentrated or diluted.

(3) If a base is weak, it has a strong conjugate acid.

(4) If a base is strong, it has a weak conjugate acid.

Section 9.10 Review:

(1) $HF\ (aq) \rightleftharpoons H^{+}(aq) + F^{-}\ (aq)$

(2) During a reaction with a base, the free H^+ ions will react. The products side of equilibrium between undissociated HF and the hydrogen and fluoride ions will become lighter. More HF will dissociate in order to reestablish equilibrium. Consequently, the concentration of undissociated HF will decrease.

(3) More HF will dissociate into hydrogen and fluoride ions.

(4) equivalence point

(5) endpoint

Section 9.11 Review:

(1) The moles of solute can be calculated by multiplying the liters of solution by the molarity of the solution.

(2) The liters of solution can be calculated by dividing the moles of solute by the molarity of the solution.

Section 9.12 Review:

(1) strong acid + strong base → neutral salt (pH ~ 7); (2) weak acid + strong base → basic salt (pH > 7)

(3) weak acid + weak base → neutral salt (pH ~ 7); (4) strong acid + weak base → acidic salt (pH < 7)

Section 9.13 Review:

(1) The salt must contain the fluoride ion.

(2) The other ion (cation) should come from a strong base. This ensures that the cation will not undergo hydrolysis and establish an additional equilibrium system in the buffer.

(3) The salt will completely dissociate in water.

(4) The addition of the salt to water will increase the concentration of the F^- ion. Because the F^- ion is a product, the seesaw will tip toward the products. In order to reestablish equilibrium, some of the additional F^- will react with the H^+ to produce more HF. Equilibrium will shift to the left.

(5) $H^+ \ll F^- = HF$; In a buffer with pH = pKa, the concentrations of the undissociated acid and its conjugate base are equal. Both of these chemical species are in much higher concentration than the hydrogen (hydronium) ion.

(6) The F^- ion will react with acids. This will remove some of the F^- from solution and create more HF.

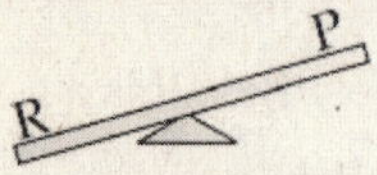

To reestablish equilibrium, some of the HF will dissociate and equilibrium will shift to the right.

(7) The HF will react with bases. This will remove some of the HF from solution.

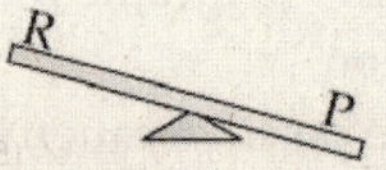

To reestablish equilibrium, some of the H^+ and F^- will combine to form HF and equilibrium will shift to the left.

(8) High concentrations of both the HF and salt containing F^- will ensure a large buffer capacity.

Tying It All Together with a Laboratory Application:

(1) titration (2) is not (3) it does not dissociate to produce OH^- ions in water
(4) OH^-, NH_4^+ (5) base (6) it is a proton acceptor
(7) buret (8) acid (9) it dissociates to produces H^+ ions in water
(10) acid (11) it is a proton donor (12) sulfuric acid

(13) $H_2SO_4 (aq) + 2\ NH_3 (aq) \rightarrow (NH_4)_2SO_4 (aq)$

(14) $2\ H^+ (aq) + SO_4^{2-} (aq) + 2\ NH_3 (aq) \rightarrow 2\ NH_4^+(aq) + SO_4^{2-} (aq)$

(15) SO_4^{2-} (aq) (16) $H^+ (aq) + NH_3 (aq) \rightarrow NH_4^+(aq)$ (17) strong
(18) weak (19) less than (20) hydrolysis

(21) $NH_4^+(aq) + H_2O (l) \rightleftharpoons NH_3 (aq) + H_3O^+(aq)$

(22) less than (23) endpoint (24) equivalence (25) 3.68×10^{-3} moles
(26) 0.294 (27) 4.3×10^{-12} M (28) 11.36 (29) basic
(30) buffer (31) equals (32) buffer capacity

SELF-TEST QUESTIONS

Multiple Choice

1. The Arrhenius definition of a base focuses on:
   a. the acceptance of H^+.
   b. the production of OH^-.
   c. the formation of covalent bonds.
   d. more than one response is correct.

2. What is the pH of solution that is 0.100 M in OH^-?
   a. 0.794 b. 1.000 c. 1.259 d. 13.000

3. A beaker contains 100 mL of a liquid with a pH of 7.0. When 0.5 mL of 0.2 M acid is added, the pH changes to 6.88. When 0.5 mL of 0.1 M base is added to another 100 mL sample of the liquid, the pH changes to 7.20. The liquid in the beaker is:
   a. water. b. an acid solution. c. a base solution. d. a buffer solution.

Refer to the following reaction for Questions 4 through 7: $HCl + NaOH \rightarrow NaCl + H_2O$

4. The above reaction between hydrochloric acid and sodium hydroxide is correctly classified as:
   a. combustion.
   b. decomposition.
   c. neutralization.
   d. more than one response is correct.

5. The HCl solution is prepared to be 0.120 M. A 20.00 mL sample requires 18.50 mL of NaOH solution for complete reaction. What is the molarity of the NaOH solution?
   a. 0.0800 b. 0.110 c. 0.130 d. 0.200

6. How many moles of HCl would be contained in the 20.00 mL sample used in Question 5?
   a. 0.204 b. 2.40 c. 0.120 d. 0.0024

7. How many grams of HCl would be contained in the 20.00 mL sample used in Question 5?
   a. 0.0875 b. 4.38 c. 8.75 d. 36.5

8. A 25.00 mL sample of monoprotic acid is titrated with a standard 0.100 M base. Exactly 20.00 mL of base is required to titrate to the proper endpoint. What is the molarity of the acid?
   a. 0.0500 b. 0.100 c. 0.0800 d. 0.125

9. A certain solution has a pH of 1. This solution is best described as:
   a. very basic. b. neutral. c. slightly acidic. d. very acidic.

10. If the pH of an aqueous solution cannot be changed significantly by adding small amounts of strong acid or strong base, the solution contains:
    a. an indicator. c. a protective colloid.
    b. a buffer. d. a strong acid and a strong base.

True-False

11. The terms *weak acid* and *dilute acid* can be used interchangeably.
12. Some H_3O^+ ions are present in pure water.
13. One of the products formed in the titration of a strong acid by a strong base is water.
14. A solution with a pH of 3 is correctly classified as acidic.
15. A solution with a pH of 6.00 has a concentration of 6.00 M H^+.
16. Three different sodium salts of phosphoric acid, H_3PO_4, are possible.
17. The anion produced by the first step in the dissociation of sulfurous acid, H_2SO_3, is SO_3^{2-}.
18. The second step in the dissociation of H_3PO_4 produces $H_2PO_4^-$.
19. In pure water, $[H_3O^+] = [OH^-]$.
20. In an acid-base titration, the point at which the acid and base have exactly reacted is called the equivalence point.
21. The pH is 7 at the equivalence point of all acid-base titrations.

Matching

Choose the response that best completes each reaction below. The acid involved in each reaction is represented by HA.

22. $2\,HA + ? \rightarrow H_2 + MgA_2$ a. NaA
23. $HA + H_2O \rightarrow\ ? + A^-$ b. Na_2CO_3
24. $HA + NaOH \rightarrow H_2O + ?$ c. Mg
25. $2\,HA + ? \rightarrow CO_2 + H_2O + 2\,NaA$ d. H_3O^+

Match the classifications given on the right to the species listed on the left. The species are involved in the following reversible reaction. Responses can be used more than once.

$$Na^+\,(aq) + C_2H_3O_2^-\,(aq) + H_2O\,(l) \rightleftharpoons Na^+\,(aq) + HC_2H_3O_2\,(aq) + OH^-\,(aq)$$

26. Na^+ a. behaves as a Brønsted acid
27. $C_2H_3O_2^-$ b. behaves as a Brønsted base
28. H_2O c. behaves as neither a Brønsted acid nor base
29. $HC_2H_3O_2$ d. behaves as a both Brønsted acid and base

Choose a silver compound formula from the right to complete each of the reactions used to prepare silver salts. In each reaction, HA represents an acid.

30. $2\,HA + ? \rightarrow 2\,AgA + H_2O$
31. $2\,HA + ? \rightarrow H_2O + 2\,AgA + CO_2$
32. $HA + ? \rightarrow H_2O + AgA$
33. $HA + ? \rightarrow CO_2 + H_2O + AgA$

a. AgOH
b. Ag_2O
c. $AgHCO_3$
d. Ag_2CO_3

Classify the systems described on the left into one of the pH ranges given as responses.

34. the $[OH^-]=[H_3O^+]$ in pure H_2O
35. oven cleaners are strongly basic
36. the active ingredient in the stomach, digestive juice, is 0.01 M hydrochloric acid
37. a carbonated soft drink has a tart taste

a. the pH is much lower than 7
b. the pH is much higher than 7
c. the pH is near 7
d. the pH is exactly 7

Choose a description from the right that best characterizes the solution made by dissolving in water each of the salts indicated on the left.

38. Na_2SO_4
39. Na_3PO_4
40. NH_4Cl

a. it is acidic because hydrolysis occurs
b. it is basic because hydrolysis occurs
c. it is neutral because no hydrolysis occurs
d. more than one of the above is correct

Chapter 10: Radioactivity and Nuclear Processes

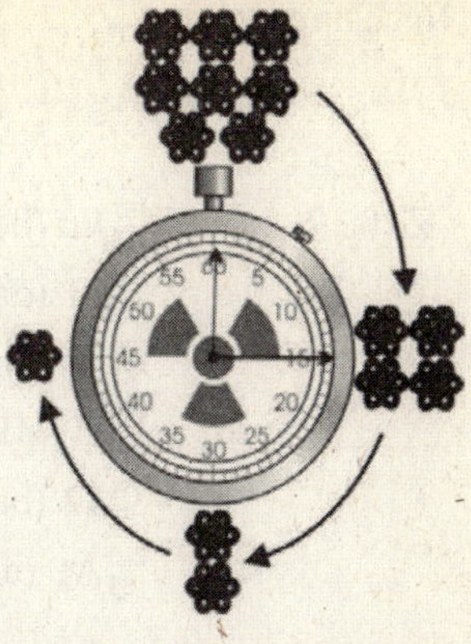

CHAPTER OUTLINE

10.1 Radioactive Nuclei
10.2 Equations for Nuclear Reactions
10.3 Isotope Half-Life
10.4 The Health Effects of Radiation
10.5 Measurement Units for Radiation
10.6 Medical Uses of Radioisotopes
10.7 Nonmedical Uses of Radioisotopes
10.8 Induced Nuclear Reactions
10.9 Nuclear Energy

LEARNING OBJECTIVES/ASSESSMENT

When you have completed your study of this chapter, you should be able to:

1. Describe and characterize the common forms of radiation emitted during radioactive decay and other nuclear processes. (Section 10.1; Exercise 10.2)
2. Write balanced equations for nuclear reactions. (Section 10.2; Exercise 10.12)
3. Solve problems using the half-life concept. (Section 10.3; Exercise 10.16)
4. Describe the effects of radiation on health. (Section 10.4; Exercise 10.22)
5. Describe and compare the units used to measure quantities of radiation. (Section 10.5; Exercise 10.24)
6. Describe, with examples, medical uses of radioisotopes. (Sections 10.6; Exercise 10.30)
7. Describe, with examples, nonmedical uses of radioisotopes. (Sections 10.7; Exercise 10.36)
8. Show that you understand the concept of induced nuclear reactions. (Section 10.8; Exercise 10.38)
9. Describe the differences between nuclear fission and nuclear fusion reactions. (Section 10.9; Exercise 10.48)

SOLUTIONS FOR THE END OF CHAPTER EXERCISES

RADIOACTIVE NUCLEI (SECTION 10.1)

☑10.2
a. mass number = 0: beta, gamma, positron
b. positive charge: alpha, positron
c. charge = 0: gamma, neutron

10.4
a. A beta particle = an electron
b. An alpha particle = 2 protons and 2 neutrons
c. A positron = positive electron

EQUATIONS FOR NUCLEAR REACTIONS (SECTION 10.2)

			Atomic number change	Mass number change
10.6	a.	An alpha particle is emitted.	decrease by 2	decrease by 4
	b.	A beta particle is emitted.	increase by 1	no change
	c.	An electron is captured.	decrease by 1	no change
	d.	A gamma ray is emitted.	no change	no change
	e.	A positron is emitted.	decrease by 1	no change

10.8	a.	An aluminum-27 nucleus	${}^{27}_{13}\text{Al}$
	b.	A nucleus of element number 38 with a mass number of 90	${}^{90}_{38}\text{Sr}$
	c.	A nucleus of the krypton isotope that contains 40 neutrons	${}^{76}_{36}\text{Kr}$

10.10
a. ${}^{234}_{90}\text{Th} \rightarrow ? + {}^{234}_{91}\text{Pa}$ $? = {}^{0}_{-1}\beta$ (beta)
b. ${}^{35}_{16}\text{S} \rightarrow ? + {}^{0}_{-1}\beta$ $? = {}^{35}_{17}\text{Cl}$
c. $? \rightarrow {}^{4}_{2}\alpha + {}^{244}_{98}\text{Cf}$ $? = {}^{248}_{100}\text{Fm}$
d. ${}^{226}_{88}\text{Ra} \rightarrow ? + {}^{4}_{2}\alpha$ $? = {}^{222}_{86}\text{Rn}$

e. $^{113}_{47}Ag \rightarrow ^{113}_{48}Cd + ?$? = $^{0}_{-1}\beta$ (beta)   f. $^{41}_{18}Ar + ^{0}_{-1}e \rightarrow ?$? = $^{41}_{17}Cl$

☑10.12
a. $^{157}_{63}Eu$ (beta emission) $^{157}_{63}Eu \rightarrow ^{0}_{-1}\beta + ^{157}_{64}Gd$
b. $^{190}_{78}Pt$ (daughter = osmium-186) $^{190}_{78}Pt \rightarrow ^{4}_{2}\alpha + ^{186}_{76}Os$
c. $^{138}_{62}Sm$ (electron capture) $^{138}_{62}Sm + ^{0}_{-1}e \rightarrow ^{138}_{61}Pm$
d. $^{188}_{80}Hg$ (daughter Au-188) $^{188}_{80}Hg \rightarrow ^{0}_{+1}\beta + ^{188}_{79}Au$
e. $^{234}_{90}Th$ (beta emission) $^{234}_{90}Th \rightarrow ^{0}_{-1}\beta + ^{234}_{91}Pa$
f. $^{218}_{85}At$ (alpha emission) $^{218}_{85}At \rightarrow ^{4}_{2}\alpha + ^{214}_{83}Bi$

ISOTOPE HALF-LIFE (SECTION 10.3)

10.14 Half-life is the amount of time required for half of a sample to undergo a specific process. For example, if the half-life of a cake is one day, then half of a cake will be eaten the first day, the next day half of the remaining cake (¼ of the original cake) would be eaten, the following day half of the remaining cake (⅛ of the original cake) would be eaten, etc.

☑10.16

$$\text{Amount remaining} = 9.0\text{ ng}\left(\frac{1}{2}\right)^{24\text{ hours}\left(\frac{1\text{ half-life}}{6\text{ hours}}\right)} = 0.56\text{ ng}$$

10.18

$$12.5\% = 100\%\left(\frac{1}{2}\right)^{\text{half-lives}} \Rightarrow \log(0.125) = \log\left(\left(\frac{1}{2}\right)^{\text{half-lives}}\right) \Rightarrow \log(0.125) = \text{half-lives}\left(\log\left(\frac{1}{2}\right)\right)$$

$$\text{half-lives} = 3.000$$

$$\text{Time elapsed} = 3\text{ half-lives}\left(\frac{5600\text{ years}}{1\text{ half-life}}\right) = 16800\text{ years} = 1.68\times10^4\text{ years}$$

10.20

$$\left(\frac{1}{2}\right)^{\text{half-lives}} = \frac{1}{16} \Rightarrow \log\left(\left(\frac{1}{2}\right)^{\text{half-lives}}\right) = \log\left(\frac{1}{16}\right) \Rightarrow \text{half-lives}\left(\log\left(\frac{1}{2}\right)\right) = \log\left(\frac{1}{16}\right)$$

$$\text{half-lives} = 4$$

$$\text{Time elapsed} = 4\text{ half-lives}\left(\frac{5600\text{ years}}{1\text{ half-life}}\right) = 22400\text{ years} = 2.24\times10^4\text{ years}$$

THE HEALTH EFFECTS OF RADIATION (SECTION 10.4)

☑10.22 Long-term, low-level exposure to radiation may lead to genetic mutations because ionizing radiation can produce free radicals in exposed tissues. Short-term exposure to intense radiation destroys tissue rapidly and causes radiation sickness. Both forms of exposure have negative effects on health.

MEASUREMENT UNITS FOR RADIATION (SECTION 10.5)

☑10.24 Physical units of radiation indicate the activity of a source of radiation; whereas, biological units of radiation indicate the damage caused by radiation in living tissue. Examples of physical units of radiation include the Curie and the Becquerel. Examples of biological units of radiation include the Roentgen, the Rad, the Gray, and the Rem.

10.26

$$3.1\text{ rads beta radiation}\left(\frac{1\text{ roentgen}}{0.96\text{ rad}}\right) = 3.2\text{ roentgen}$$

10.28 $$4.6\ \mu\text{Ci}\left(\frac{1\ \text{Ci}}{10^6\ \mu\text{Ci}}\right)\left(\frac{3.7\times10^{10}\ \frac{\text{disintegrations}}{\text{second}}}{1\ \text{Ci}}\right)=1.7\times10^5\ \frac{\text{disintegrations}}{\text{second}}$$

MEDICAL USES OF RADIOISOTOPES (SECTION 10.6)

☑10.30 Radioactive isotopes can be used for diagnostic work. When the radioactive isotope concentrates in a tissue under observation, the location is called a hot spot. When the radioactive isotope is excluded or rejected by a tissue under observation, the location is called a cold spot. Both hot spots and cold spots can be used for diagnostic work.

10.32 $^{51}_{24}\text{Cr} + ^{0}_{-1}\text{e} \rightarrow ^{51}_{23}\text{V}$; daughter nucleus = vanadium-51

NONMEDICAL USES OF RADIOISOTOPES (SECTION 10.7)

10.34 By using water that contains a radioactive isotope of oxygen, the oxygen gas produced could be analyzed to see if it contains the radioactive isotope of oxygen from the water or a nonradioactive isotope of oxygen from the hydrogen peroxide.

☑10.36 The radioactivity of a gallon of water that contains a radioisotope with a long half-life could be measured with the Geiger-Müller counter. The water could then be added to the pool and given time to circulate. A sample of pool water could then be taken and the level of radioactivity measured. The dilution formula could then be used to determine the volume of the pool based on the relative levels of radioactivity.

INDUCED NUCLEAR REACTIONS (SECTION 10.8)

☑10.38 $^{24}_{12}\text{Mg} + ^{4}_{2}\alpha \rightarrow ^{27}_{14}\text{Si} + ^{1}_{0}\text{n}$

10.40 $^{238}_{92}\text{U} + ^{1}_{0}\text{n} \rightarrow ^{239}_{94}\text{Pu} + 2\,^{0}_{-1}\beta$

10.42 A moderator is a material that can be placed in the path of a neutron in order to slow down the neutrons as they pass through the moderator. As the neutrons slow, their kinetic energy decreases and the nuclear forces allow the neutrons to be captured.

10.44 $^{66}_{30}\text{Zn} + ^{1}_{1}\text{p} \rightarrow ^{67}_{31}\text{Ga}$

NUCLEAR ENERGY (SECTION 10.9)

10.46 $^{1}_{1}\text{H} + ^{2}_{1}\text{H} \rightarrow ^{3}_{2}\text{He} + ^{0}_{0}\gamma$ or another reaction that indicates two nuclei come together to form a single larger nucleus.

☑10.48 Nuclear fission is the process by which a large nucleus divides into smaller nuclei. An example of induced fission is the following reaction of uranium-235:

$$^{235}_{92}\text{U} + ^{1}_{0}\text{n} \rightarrow ^{135}_{53}\text{I} + ^{97}_{39}\text{Y} + 4\,^{1}_{0}\text{n}$$

Nuclear fusion is the process by which small nuclei join to form a larger nucleus. An example is the reaction between hydrogen-1 and hydrogen-2 shown in the solution to exercise 10.46.

10.50 $^{238}_{94}\text{Pu} \rightarrow ^{4}_{2}\alpha + ^{234}_{92}\text{U}$

ADDITIONAL EXERCISES

10.52 The fusion reactions occurring in the sun only take place at extremely high temperatures because at low temperatures the nuclei do not have enough energy to overcome their repulsion for each other (they are both positive) in order to get close enough to each other to collide, let alone join together.

10.54 In 2.4 minutes, half of the atoms in the Zn-71 sample will decay and release one beta particle per atom that decays.

$$\frac{0.200\text{ g }^{71}\text{Zn}\left(\frac{1\text{ mole }^{71}\text{Zn}}{71\text{ g }^{71}\text{Zn}}\right)\left(\frac{6.02\times10^{23}\text{ atoms }^{71}\text{Zn}}{1\text{ mole }^{71}\text{Zn}}\right)\left(\frac{1\text{ atom }^{71}\text{Zn decays}}{2\text{ atoms }^{71}\text{Zn present}}\right)\left(\frac{1\text{ beta particle formed}}{1\text{ atom }^{71}\text{Zn decays}}\right)}{2.4\text{ min}} = 3.5\times10^{20}\ \frac{\text{beta particles}}{\text{min}}$$

$$\left(\frac{0.200\text{ g }^{71}\text{Zn}\left(\frac{1\text{ mole }^{71}\text{Zn}}{70.92773\text{ g }^{71}\text{Zn}}\right)\left(\frac{6.02\times10^{23}\text{ atoms }^{71}\text{Zn}}{1\text{ mole }^{71}\text{Zn}}\right)\left(\frac{1\text{ atom }^{71}\text{Zn decays}}{2\text{ atoms }^{71}\text{Zn present}}\right)}{2.4\text{ min}\left(\frac{60\text{ sec}}{1\text{ min}}\right)}\right)\left(\frac{1\text{ Ci}}{3.7\times10^{10}\ \frac{\text{decays}}{\text{second}}}\right) = 1.6\times10^{8}\text{ Ci}$$

ALLIED HEALTH EXAM CONNECTION

10.56 When a radioactive element decays, (a) the nucleus gives off particles.

10.58 (a) Radio waves have the least energy when compared with gamma rays, ultraviolet radiation, and visible light.

10.60 The least penetrating radiation given off by a radioactive substance consists of (a) alpha particles.

10.62 $^{234}_{92}\text{U} \rightarrow {}^{4}_{2}\alpha + {}^{230}_{90}\text{Th}$

When uranium-234 decays by alpha emission, the resulting isotope is (d) $^{230}_{90}\text{Th}$.

10.64 $^{232}_{90}\text{Th} \rightarrow {}^{228}_{88}\text{Ra} + \text{X}$

The particle X in the above equation is (d) an alpha particle.

10.66 $\text{B} \rightarrow ? + {}^{0}_{-1}\beta + {}^{0}_{0}\gamma$

$^{238}_{92}\text{B} \rightarrow {}^{238}_{93}\text{Np} + {}^{0}_{-1}\beta + {}^{0}_{0}\gamma$

The particle B in the above equation is (a) $^{238}_{93}\text{B}$.

10.68 $^{42}_{17}\text{X} \rightarrow {}^{42}_{18}\text{Y} + ?$

$? = {}^{0}_{-1}\beta$

The missing product in the above equation is (d) $^{0}_{-1}\beta$.

10.70

$$\text{Amount remaining} = 100\text{ g}\left(\frac{1}{2}\right)^{10\text{ days}\left(\frac{1\text{ half-life}}{5\text{ days}}\right)} = 25\text{ g}$$

or

The t1/2 is 5 days. After 10 days, 2 half-lives have passed since $\frac{10}{5} = 2$.

At this time the fraction remaining = ½ x ½ = $\left(\frac{1}{2}\right)^2 = \frac{1}{4}$.

The amount remaining is $100\text{ g}\left(\frac{1}{4}\right) = 25\text{ g}$.

After 10 days, (a) 25 g of the radioactive isotope will remain.

10.72

$$10\text{ g} = 40\text{ g}\left(\frac{1}{2}\right)^{16\text{ days}\left(\frac{1\text{ half-life}}{x\text{ days}}\right)} \Rightarrow \frac{10\text{ g}}{40\text{ g}} = \left(\frac{1}{2}\right)^{16\text{ days}\left(\frac{1\text{ half-life}}{x\text{ days}}\right)}$$

$$\log\frac{1}{4} = 16\text{ days}\left(\frac{1\text{ half-life}}{x\text{ days}}\right)\log\left(\frac{1}{2}\right)$$

$$\frac{\log\left(\frac{1}{4}\right)}{\log\left(\frac{1}{2}\right)} = 16\text{ days}\left(\frac{1\text{ half-life}}{x\text{ days}}\right)$$

$$2 = 16\text{ days}\left(\frac{1\text{ half-life}}{x\text{ days}}\right)$$

$$2\left(\frac{x\text{ days}}{1\text{ half-life}}\right) = 16\text{ days}$$

$$x = 8\text{ days}$$

or

If 40 g has decreased to 10 g, then the amount remaining is $\frac{10}{40} = \frac{1}{4}$, which will occur in 2 half-lives because ½ x ½ = $\left(\frac{1}{2}\right)^2 = \frac{1}{4}$. Since 16 days is 2 half-lives, the $t_{1/2}$ is 8 days because $\frac{16}{2} = 8$.

The half-life of the substance is (b) 8 days.

10.74

$$\text{Amount remaining} = \left(\frac{1}{2}\right)^{15\text{ years}\left(\frac{1\text{ half-life}}{5\text{ years}}\right)} = \left(\frac{1}{2}\right)^3 = \frac{1}{8}$$

or

The $t_{1/2}$ is 5 years. After 15 years, 3 half-lives have passed since $\frac{15}{5} = 3$.

At this time the fraction remaining = ½ x ½ x ½ = $\left(\frac{1}{2}\right)^3 = \frac{1}{8}$.

The fraction of the substance remaining is (b) $\frac{1}{8}$.

10.76

$$\text{Amount remaining} = 100\text{ mg}\left(\frac{1}{2}\right)^{8\text{ days}\left(\frac{1\text{ half-life}}{8\text{ days}}\right)} = 100\text{ mg}\left(\frac{1}{2}\right)^1 = 50\text{ mg}$$

or

The $t_{1/2}$ is 8 days. After 8 days, 1 half-life has passed since $\frac{8}{8} = 1$.

At this time the fraction remaining = ½ = $\left(\frac{1}{2}\right)^1 = \frac{1}{2}$.

The amount remaining is $100\text{ mg}\left(\frac{1}{2}\right) = 50.0\text{ mg}$.

The amount of the substance remaining is (a) 50 milligrams.

10.78

$$0.0625\text{ g} = 1\text{ g}\left(\frac{1}{2}\right)^{x\left(\frac{1\text{ half-life}}{5.73\times10^3\text{ years}}\right)}$$

$$\frac{0.0625\text{ g}}{1\text{ g}} = \left(\frac{1}{2}\right)^{x\left(\frac{1\text{ half-life}}{5.73\times10^3\text{ years}}\right)}$$

$$\log\left(\frac{0.0625\text{ g}}{1\text{ g}}\right) = x\left(\frac{1\text{ half}-\text{life}}{5.73\times10^3\text{ years}}\right)\log\left(\frac{1}{2}\right)$$

$$x = \left(\frac{\log\left(\frac{0.0625\text{ g}}{1\text{ g}}\right)}{\log\left(\frac{1}{2}\right)}\right)\left(\frac{5.73\times10^3\text{ years}}{1\text{ half}-\text{life}}\right) = 2.29\times10^4\text{ years}$$

or

The fraction remaining is $\frac{0.0625}{1} = \frac{1}{16}$, which is ½ x ½ x ½ x ½ = $\left(\frac{1}{2}\right)^4$; therefore, 4 half-lives have elapsed. The $t_{1/2}$ is 5.73 x 10^3 years, which means the time required is 2.29 x 10^4 years because $5.73\times10^3\times4 = 2.29\times10^4$.

The time required to decay a 1 g sample of carbon-14 from 1 g to 0.0625 g would be (c) 22.92 x 10^3 years.

CHEMISTRY FOR THOUGHT

10.80 While radioactive decay does occur naturally, it is unlikely that lead changes into gold naturally because lead has an atomic number of 82 and gold has an atomic number of 79, which makes a difference of 3. None of the common nuclear decay processes change the atomic number of the parent nucleus by 3.

10.82 In principle, a radioactive isotope never completely disappears by radioactive decay because only half of the sample decays per half life, so half of the initial sample remains. In reality, all of a sample will decay because eventually one atom will remain in a sample and when that one atom undergoes decay the entire atom will undergo decay, not half of the atom.

10.84 While at first sending nuclear waste into outer space might seem like an attractive possibility because it would remove the hazardous materials from the earth, it is unlikely to be the best solution for waste disposal. Unfortunately, successfully launching the materials into space would not be assured, and if the spacecraft used were to explode during or shortly after launch, radioactive material would be scattered and the results could be devastating. In addition, we have no idea what the eventual fate of the radioactive waste would be. The presence of pockets of radioactive materials in the cosmos could have a myriad of unintended effects on our universe.

10.86

$${}^{238}_{92}\text{U} \rightarrow {}^{4}_{2}\alpha + {}^{234}_{90}\text{Th}$$

$${}^{234}_{90}\text{Th} \rightarrow {}^{0}_{-1}\beta + {}^{234}_{91}\text{Pa}$$

$${}^{234}_{91}\text{Pa} \rightarrow {}^{0}_{-1}\beta + {}^{234}_{92}\text{U}$$

ADDITIONAL ACTIVITIES

Section 10.1 Review:

(1) Draw representations of the types of radiation in Table 10.1.
(2) Rank the types of radiation from Table 10.1 in order of increasing mass.
(3) Place the types of radiation on the "continuum of charge" shown below.

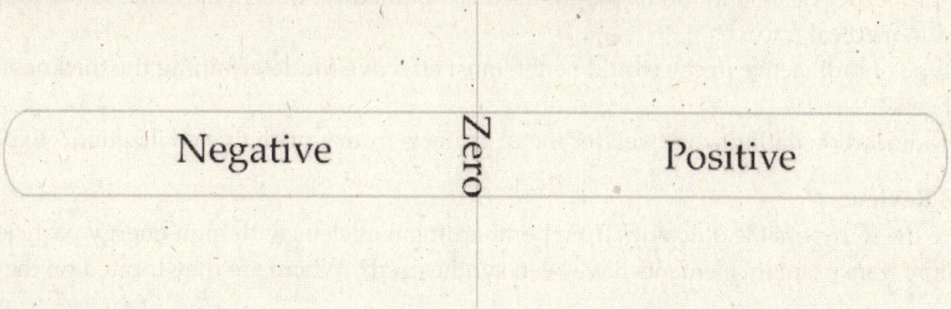

Section 10.2 Review:

Consider whether hydrogen-3 could theoretically undergo the following radioactive processes. Write an equation to represent each process or explain why the process is not possible.

(1) alpha decay
(2) beta decay*
(3) gamma emission
(4) positron emission
(5) electron capture

*Note: Beta decay is the actual decay process that hydrogen-3 undergoes.

Section 10.3 Review: All of the following questions refer to the picture of the 60-minute stopwatch at the beginning of this chapter of the study guide.

(1) What is the half-life for the element represented in the picture?
(2) How many half-lives have occurred at 7.5 seconds?
(3) Refer to Figure 10.3, predict what fraction of the sample will remain after 7.5 seconds.
(4) If each atom in the picture represents a mole of atoms, how many moles of atoms would remain after 7.5 seconds?
(5) Do answers (3) and (4) agree? Explain.

Section 10.4 Review:

(1) The inverse square law of radiation is mathematically similar to Graham's Law for gases. Rewrite Equation 10.2 in the same form as Equation 6.13.
(2) The inverse square law for radiation relates intensity of radiation to the distance from the radiation source. Graham's law relates the rate of effusion to the molecular mass of a gas. Based on (1), which variables in the two equations are similar?
(3) Why does an X-ray technician step behind the wall when taking dental X-rays?
(4) Why do dental patients wear a lead apron when having X-rays taken?
(5) A dental X-ray typically exposes patients to 2-3 mrem. Classify a dental X-ray as low-level or intense radiation.

Section 10.5 Review:

(1) Rank the physical units of radiation in order of increasing size.
(2) Divide the biological units into two categories: units for absorbed doses and units for effective doses.
(3) Why would one rad of alpha radiation have a more severe health effect than one rad of gamma radiation? (Note: 1 rad of alpha radiation has the same health effect as 10 rad of gamma radiation.)

Section 10.6 Review:

(1) What are the similarities in the properties desired in radioisotopes used as tracers and in therapy?
(2) What are the differences in the properties desired in radioisotopes used as tracers and in therapy?

Section 10.7 Review:

(1) Are the properties desired in radioisotopes used as nonmedical tracers the same as the properties required in medical tracers?
(2) Which type of radioactive decay would be the most effective for determining the thickness of thick metal sheets?
(3) Would radiocarbon dating work well for metal artifacts from a primitive civilization? Explain.

Section 10.8 Review:

(1) What are the four possible outcomes from bombarding a nucleus with high-energy particles?
(2) How many transuranium elements have been synthesized? Where are they located on the periodic table?

Section 10.9 Review:

(1) What is the major difference between fission and fusion?
(2) Why are fission reactors currently used to generate electricity, but fusion reactors are not?

Tying It All Together with a Laboratory Application:

Radioisotopes of chlorine are under investigation for use in nuclear medicine. Approximately 0.12% of the human body is made of the chloride ion. Stomach acid is hydrochloric acid. The following chart provides information for the radioisotopes of chlorine. (1) Complete the chart.

Isotope	Half-Life	Mode of Decay	Daughter Nuclei	Daughter Nuclei's Mode of Decay or Granddaughter identity
^{36}Cl	3.01×10^5 yrs	beta		stable
^{38}Cl	37.2 min		^{38}Ar	stable
^{39}Cl	55.6 min	beta		beta (268 yrs)
^{40}Cl	1.38 min		^{40}Ar	stable
^{41}Cl	34 sec	beta		beta (1.82 hr)
^{42}Cl	6.8 sec		^{42}Ar	^{42}K (33 yrs)
^{43}Cl	3.3 sec	beta		^{43}K (5.4 min)

On the basis of the daughter nuclei produced, the three best radioisotopes of chlorine for nuclear medicine are (2) _______ because they produce (3) _______. The (4) _______ of the daughter nuclei in the body should also be investigated. On the basis of a half-life that is long enough to prepare and administer a sample, the best radioisotopes of chlorine for nuclear medicine are (5) _______. On the basis of having a short enough half-life that the radioisotopes will decay during the diagnosis but emit little radiation after the procedure, the best radioisotopes of chlorine for nuclear medicine are (6) _______. Combining all of the qualifications for using a radioisotope of chlorine in nuclear medicine, (7) _______ is(are) the best choice(s). If the chlorine radioisotopes were used as tracers for stomach illnesses, the radioisotopes would form a (8) _______ spot. The isotopes of chlorine that emit beta rays would be better used in (9) _______ rather than as (10) _______.

The activity of a sample of radioactive chlorine could be measured using (11) _______, _______, or _______. The units to report the disintegrations per second are (12) _______ units, like the curie or (13) _______. The units to report the health effects of the radioisotopes of chlorine are (14) _______ units, like the rad. To compare the health effects of chlorine-41 to the health effects of an X-ray source, the best units to use are (15) _______.

The granddaughter isotope of chlorine-39 is (16) _______. The mode of decay for the daughter nuclei of chlorine-43 is (17) _______. If chlorine-36 was bombarded with an alpha particle and a new nucleus formed, the product would be (18) _______. If chlorine-36 were bombarded with a neutron, chlorine-(19) _______ could form which would be (20) _______ (stable or unstable) or nuclear (21) _______ or _______ could occur. Bombarding radioisotopes with energetic particles leads to (22) _______ nuclear reactions.

The percentage of chlorine-40 that will remain after 2.5 minutes is (23) _______. The amount of time required for only 2.5 mg to remain from a 40 mg sample of chlorine-40 is (24) _______. If the distance from a sample of chlorine-40 is tripled, the measurement of the intensity of radiation will (25) _______ (increase or decrease) by a factor of (26) _______.

Hospital personnel who are administering radioisotopes should be aware of symptoms like nausea, fatigue, vomiting, and general malaise. These are symptoms of (27) _______ which can occur from (28) _______ exposure to radiation.

SOLUTIONS FOR THE ADDITIONAL ACTIVITIES

Section 10.1 Review:

(1) alpha =  ; beta = ⊖; gamma = 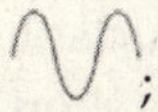 ; neutron = ⓪; positron = ⊕

(2) (least massive) gamma < beta = positron < neutron < alpha (most massive)

(3)

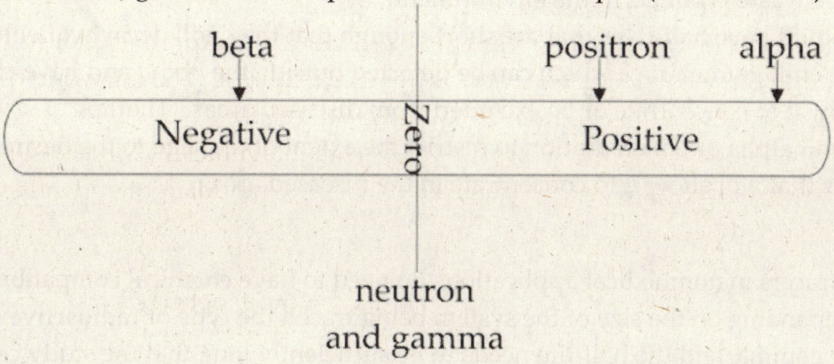

Section 10.2 Review:

(1) Hydrogen-3 could not undergo alpha decay because hydrogen-3 only contains 1 proton and two neutrons. An alpha particle contains two protons and two neutrons.

(2) ${}^{3}_{1}H \rightarrow {}^{0}_{-1}\beta + {}^{3}_{2}He$; (3) ${}^{3}_{1}H \rightarrow {}^{0}_{0}\gamma + {}^{3}_{1}H$; (4) ${}^{3}_{1}H \rightarrow {}^{0}_{+1}\beta + 3{}^{1}_{0}n$; (5) ${}^{3}_{1}H + {}^{0}_{-1}e \rightarrow 3{}^{1}_{0}n$

Section 10.3 Review:

(1) half-life = 15 seconds; (2) 0.50 half-lives at 7.5 seconds; (3) less than ¾ of the original sample will remain; (4) 5.7 moles of atoms would remain; (5) Yes, 5.7 moles out of 8 moles is 71% (from 4) which is less than 75% or ¾ (from (3)).

Section 10.4 Review:

(1) $\frac{I_x}{I_y} = \frac{d_y^2}{d_x^2} \Rightarrow \sqrt{\frac{I_x}{I_y}} = \frac{d_y}{d_x}$

(2) The distance from the radiation source in the inverse square law for radiation is equivalent to the rate of effusion in Graham's law. The intensity of radiation in the inverse square law for radiation is analogous to the molecular mass from Graham's law.

(3) The X-ray technician steps behind the wall when taking dental X-rays to increase his or her distance from the radiation source as well as to insert a barrier between himself or herself and the radiation source.

(4) Dental patients wear a lead apron when having X-rays taken to minimize the exposure of other tissues to the X-ray radiation.
(5) A dental X-ray is a low-level exposure to radiation.

Section 10.5 Review:

(1) becquerel < curie
(2) units for absorbed doses: roentgen, rad, gray; units for effective doses: rem
(3) Gamma rays have high energy; however, they do not have any mass. Alpha particles, on the other hand, have less energy than gamma rays; however, they have mass and are more easily stopped by tissues than gamma rays are. Consequently, alpha particles do not penetrate tissue as deeply as gamma rays. In order for alpha particles to transfer the energy associated with one rad (1×10^{-2} J of energy to 1 kg of tissue), more alpha particles are required to interact with the tissue than for gamma rays. Both alpha particles and gamma rays are forms of ionizing radiation. If more alpha particles strike the tissue, then more free radicals will likely be produced from the alpha particles than the number of free radicals that the gamma rays could produce.

Section 10.6 Review:

(1) Both tracer and therapy radioisotopes should have half-lives that should be long enough for the radioisotope to be prepared and used and the decay products should be nontoxic and give off little or no radiation. The half-life of the radioisotope should be as short as is consistent with use in order to minimize the radioactive waste products in the environment.
(2) Tracer radioisotopes should have half-lives that are short enough that they will decay while the diagnosis is performed, emit gamma rays which can be detected outside the body, and have chemical properties that will allow it to concentrate or be excluded from diseased areas. Therapy radioisotopes should emit alpha or beta radiation to restrict the extent of damage to the desired tissue and chemical properties that will allow it to concentrate in the diseased area.

Section 10.7 Review:

(1) Radioisotopes used as tracers in nonmedical applications do need to have chemical compatibility with the application, depending on the size of the system being traced the type of radioactive decay could be alpha, beta, or gamma, and the half-life needs to be sufficiently long that the study can be completed. The requirement of nontoxic, stable decay products is not as important for nonmedical applications; however, waste products that are generated will need to be disposable. Stable, nontoxic decay products are much safer (and cheaper) for waste disposal.
(2) For thick metal sheets, gamma rays would be the most effective at penetrating the material.
(3) Metal artifacts contain little very little carbon, so radiocarbon dating would not be effective. Looking for another radioisotope might prove to be more successful.

Section 10.8 Review:

(1) producing a stable nucleus, producing an unstable nucleus, nuclear fission occurring, nuclear fusion occurring
(2) At least twenty transuranium elements have been synthesized. They occupy sites in the d, f, and p blocks of the seventh period.

Section 10.9 Review:

(1) Fission is the process of heavier nuclei breaking apart into smaller nuclei. Fusion is the process of smaller nuclei coming together to form larger nuclei.
(2) Fission can occur at low temperatures; however, fusion only occurs at extremely high temperatures. Scientists are still working to determine how to contain the high temperatures needed to maintain a fusion reactor.

Tying It All Together with a Laboratory Application:

(1)

Isotope	Mode of Decay	Daughter Nuclei
^{36}Cl	beta	^{36}**Ar**
^{38}Cl	**beta**	^{38}Ar
^{39}Cl	beta	^{39}**Ar**
^{40}Cl	**beta**	^{40}Ar
^{41}Cl	beta	^{41}**Ar**
^{42}Cl	**beta**	^{42}Ar
^{43}Cl	beta	^{43}**Ar**

(2) ^{36}Cl, ^{38}Cl, ^{40}Cl
(3) stable daughter nuclei
(4) toxicity
(5) ^{36}Cl, ^{38}Cl, ^{39}Cl, ^{40}Cl
(6) ^{38}Cl, ^{39}Cl, ^{40}Cl, ^{41}Cl, ^{42}Cl, ^{43}Cl
(7) ^{38}Cl, ^{40}Cl
(8) hot
(9) therapy
(10) tracers
(11) Geiger-Müller tube, scintillation counter, film badge
(12) physical
(13) becquerel
(14) biological
(15) rem
(16) potassium-39
(17) beta
(18) potassium-40
(19) 37
(20) stable
(21) fission, fusion
(22) induced
(23) 28%
(24) 5.5 minutes
(25) decrease
(26) 9
(27) radiation sickness
(28) intense, short-term

SELF-TEST QUESTIONS

Multiple Choice

1. The three common types of radiation emitted by naturally radioactive elements are
   a. electrons, protons, and neutrons
   b. alpha rays, beta rays, and neutrons
   c. x-rays, gamma rays, and protons
   d. alpha particles, beta particles, and gamma rays

2. Which of the following types of radiation is composed of particles which carry a +2 charge?
   a. alpha b. beta c. gamma d. neutrons

3. Which of the following types of radiation is not composed of particles?
   a. alpha b. beta c. gamma d. neutrons

4. After four half-lives have elapsed, the amount of a radioactive sample which has not decayed is
   a. 40% of the original sample
   b. ¼ of the original sample
   c. ⅛ of the original sample
   d. $\frac{1}{16}$ of the original sample

5. If ⅛ of an isotope sample is present after 22 days, what is the half-life of the isotope?
   a. 10 days b. 5 days c. 2¾ days d. 7⅓ days

6. By doubling the distance between yourself and a source of radiation, the intensity of the radiation
   a. is ½ as great b. is ⅓ as great c. is ¼ as great d. is ⅛ as great

7. Which of the following would be the most convenient unit to use when determining the total dose of radiation received by an individual who was exposed to several different types of radiation?
   a. roentgen b. rad c. gray d. rem

8. Which type of radiation has the lowest penetration?
   a. α b. β c. γ d. X

9. The reaction, ${}^{24}_{11}Na \rightarrow {}^{24}_{12}Mg + {}^{0}_{-1}\beta$, is an example of
   a. alpha decay b. beta decay c. positron decay d. fission

10. The reaction, ${}^{235}_{92}U + {}^{1}_{0}n \rightarrow {}^{103}_{41}Nb + {}^{131}_{51}Sb + 2\ {}^{1}_{0}n$, is an example of
   a. alpha decay b. beta decay c. positron decay d. fission

True-False

11. Radioactive tracers are useful in both medical and nonmedical applications.
12. Radioactive isotopes are not taken into the body during medical uses of radioisotopes.
13. A Curie is a measurement technique used to determine the amount of radiation produced by a sample.
14. The rem is a biological radiation measurement unit.
15. Fission and fusion processes produce energy.
16. The mass number of potassium-40 is 19.
17. Radiation sickness is most commonly associated with long-term, low-level exposure to radiation.

Matching

18. ${}^{13}_{7}N \rightarrow {}^{13}_{6}C + ?$
19. ${}^{27}_{13}Al + {}^{2}_{1}H \rightarrow {}^{25}_{12}Mg + ?$
20. ${}^{9}_{4}Be + {}^{4}_{2}He \rightarrow {}^{12}_{6}C + ?$

a. beta, ${}^{0}_{-1}\beta$
b. neutron, ${}^{1}_{0}n$
c. positron, ${}^{0}_{1}\beta$
d. alpha, ${}^{4}_{2}\alpha$

Chapter 11: Organic Compounds: Alkanes

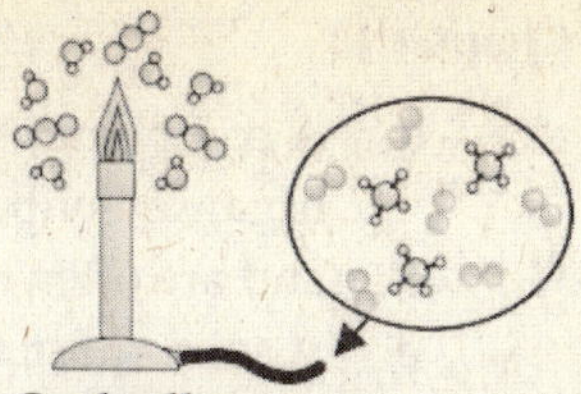

CHAPTER OUTLINE

- 11.1 Carbon: The Element of Organic Compounds
- 11.2 Organic and Inorganic Compounds Compared
- 11.3 Bonding Characteristics and Isomerism
- 11.4 Functional Groups: The Organization of Organic Chemistry
- 11.5 Alkane Structures
- 11.6 Conformations of Alkanes
- 11.7 Alkane Nomenclature
- 11.8 Cycloalkanes
- 11.9 The Shape of Cycloalkanes
- 11.10 Physical Properties of Alkanes
- 11.11 Alkane Reactions

LEARNING OBJECTIVES/ASSESSMENT

When you have completed your study of this chapter, you should be able to:

1. Show that you understand the general importance of organic chemical compounds. (Section 11.1; Exercise 11.2)
2. Recognize the molecular formulas of organic and inorganic compounds. (Section 11.1; Exercise 11.4)
3. Explain some general differences between inorganic and organic compounds. (Section 11.2; Exercise 11.8)
4. Use structural formulas to identify compounds that are isomers of each other. (Section 11.3; Exercise 11.20)
5. Write condensed or expanded structural formulas for compounds. (Section 11.4; Exercise 11.24)
6. Classify alkanes as normal or branched. (Section 11.5; Exercise 11.28)
7. Use structural formulas to determine whether compounds are structural isomers. (Section 11.6; Exercise 11.30)
8. Assign IUPAC names and draw structural formulas for alkanes. (Section 11.7; Exercise 11.34)
9. Assign IUPAC names and draw structural formulas for cycloalkanes. (Section 11.8; Exercise 11.44)
10. Name and draw structural formulas for geometric isomers of cycloalkanes. (Section 11.9; Exercise 11.54)
11. Describe the key physical properties of alkanes. (Section 11.10; Exercise 11.56)
12. Write alkane combustion reactions. (Section 11.11; Exercise 11.60)

SOLUTIONS FOR THE END OF CHAPTER EXERCISES

CARBON: THE ELEMENT OF ORGANIC COMPOUNDS (SECTION 11.1)

☑11.2 Fruits and vegetables, the family pet, plastics, sugar, cotton, and wood are a few of the many items composed of organic compounds.

11.4 All organic compounds contain carbon atoms.

ORGANIC AND INORGANIC COMPOUNDS COMPARED (SECTION 11.2)

11.6 Covalent bonding is the most prevalent type of bonding in organic compounds.

☑11.8
a. A liquid that readily burns is most likely an **organic compound** because organic compounds can exist in any of the states of matter and are often flammable.
b. A white solid with a melting point of 735°C is most likely an **inorganic compound** because inorganic compounds are usually high melting point solids.
c. A liquid that floats on the surface of water and does not dissolve is most likely an **organic compound** because organic compounds can exist in any of the states of matter and often have low water solubility.

d. A compound that exists as a gas at room temperature and ignites easily is most likely an **organic compound** because organic compounds can exist in any of the states of matter and are often flammable.

e. A solid substance that melts at 65°C is most likely an **organic compound** because organic compounds can exist as low melting point solids.

11.10 Organic compounds are nonconductors of electricity because they do not form ions readily on their own or in water. In fact, many organic compounds do not even dissolve in water, let alone, dissociate. Without ions to carry the charge, electricity cannot flow.

BONDING CHARACTERISTICS AND ISOMERISM (SECTION 11.3)

11.12 Numerous organic compounds exist because each carbon atom can form four covalent bonds, including bonds to other carbon atoms, and carbon containing molecules can exhibit isomerism. Carbon has the unique capacity to form extremely long chains which are stable. Since the same number of carbon atoms can link together to form both straight and branched chains, the number of possibilities rapidly increases as the number of carbon atoms increases.

11.14 The carbon-hydrogen bond in CH_4 results from the overlap of an sp^3 hybrid orbital on the carbon atom with the 1*s* orbital on the hydrogen atom.

11.16 An unhybridized p orbital has two lobes of equal size separated by a node; however, a hybridized sp^3 orbital has two lobes of unequal size separated by a node.

11.18 a.

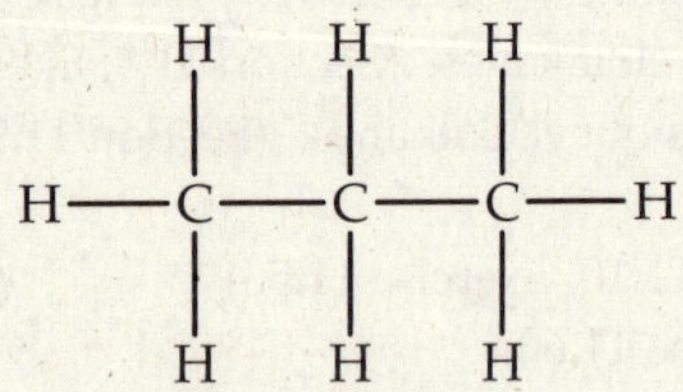

c.

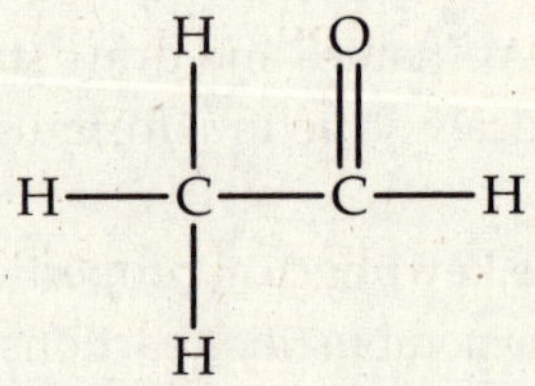

b.

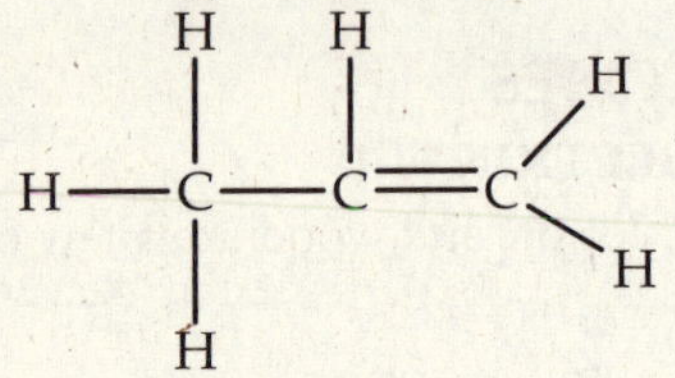

d.

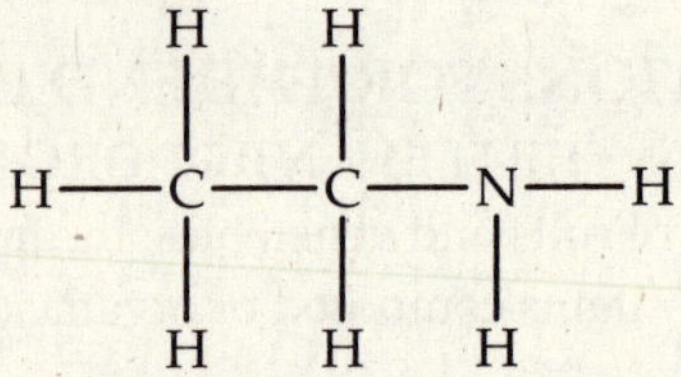

☑11.20
a. not isomers (different molecular formulas - C_4H_8 and C_4H_{10})
b. isomers (same molecular formula - C_5H_{12}, but different structures)
c. not isomers (different molecular formulas - $C_4H_{10}O$ and C_4H_8O)
d. isomers (same molecular formula - C_3H_6O, but different structures)
e. isomers (same molecular formula - C_3H_9N, but different structures)

11.22 a.

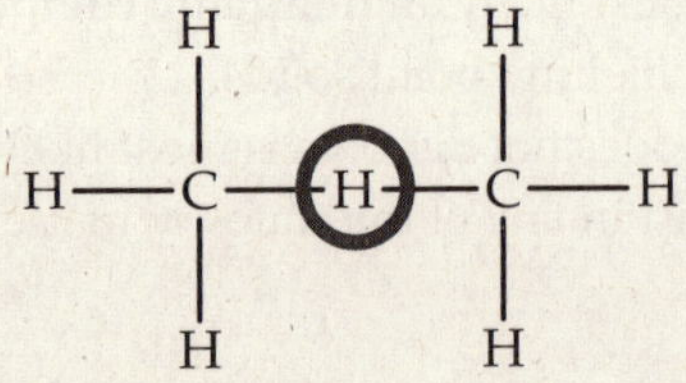

This structure is incorrect because the circled hydrogen atom is making 2 bonds. Hydrogen atoms are only able to make 1 bond.

b.

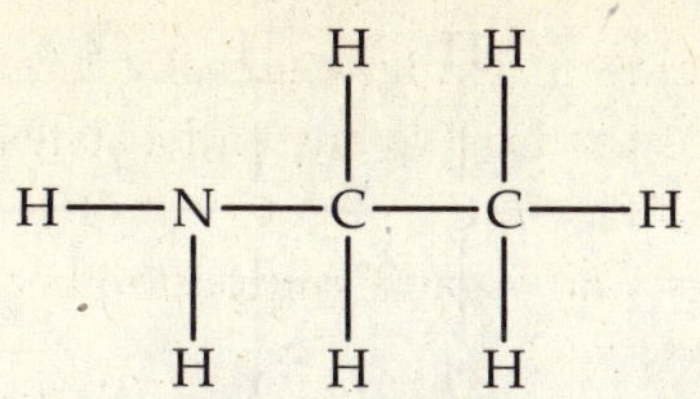

This structure is correct. Each hydrogen atom has 1 bond. Each carbon atom has 4 bonds. The nitrogen atom has 3 bonds.

c.

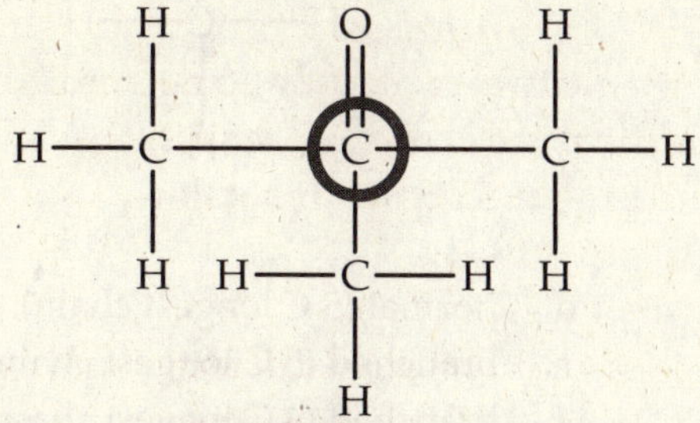

This structure is incorrect because the circled carbon atom is making 5 bonds. Carbon atoms are only able to make 4 bonds.

d.

This structure is incorrect because the circled carbon atoms are only making 3 bonds each. Each carbon atom must make 4 bonds.

e.

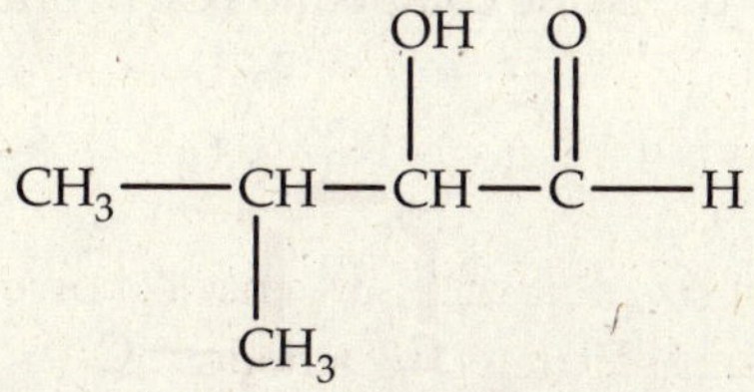

This structure is correct. Each carbon atom has 4 bonds. Each hydrogen atom has 1 bond. The oxygen atom has two bonds.

FUNCTIONAL GROUPS: THE ORGANIZATION OF ORGANIC CHEMISTRY (SECTION 11.4)

☑11.24 a.

Expanded

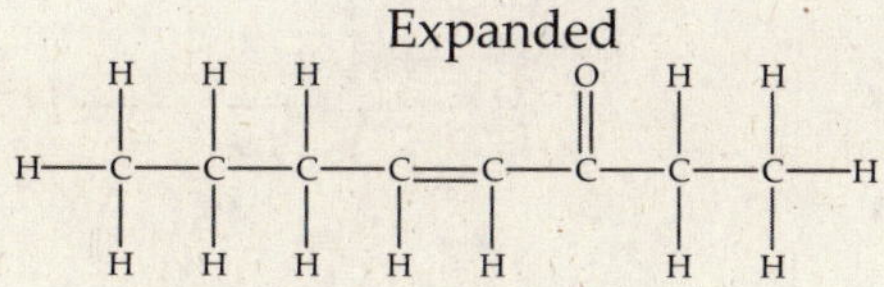

Condensed

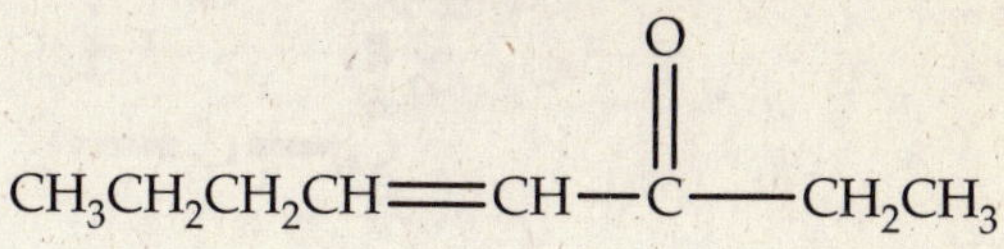

b.

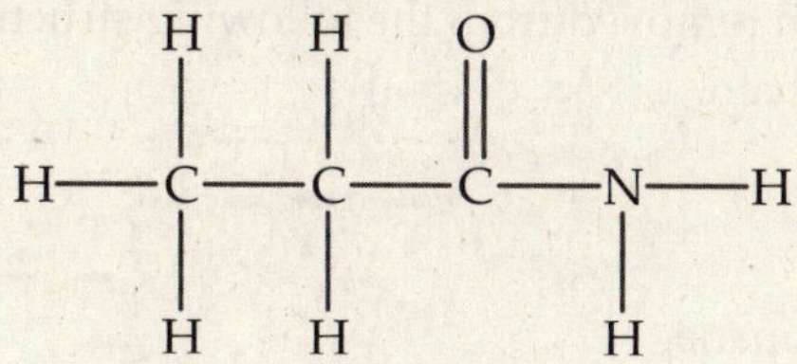

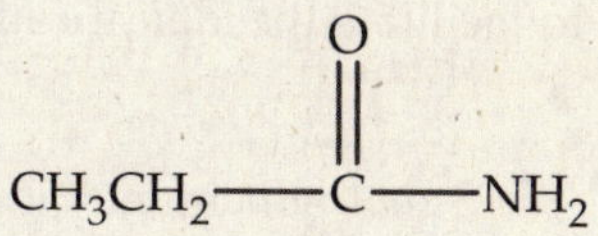

11.26 a.

Condensed

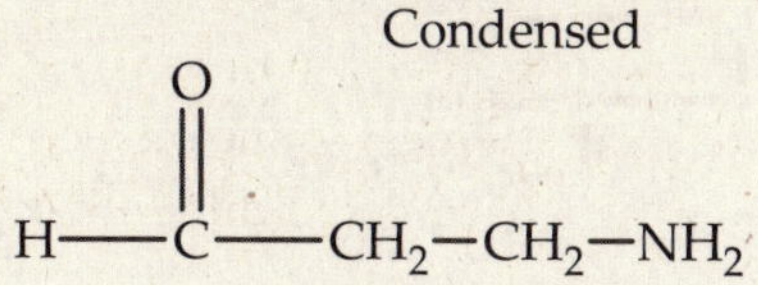

Expanded

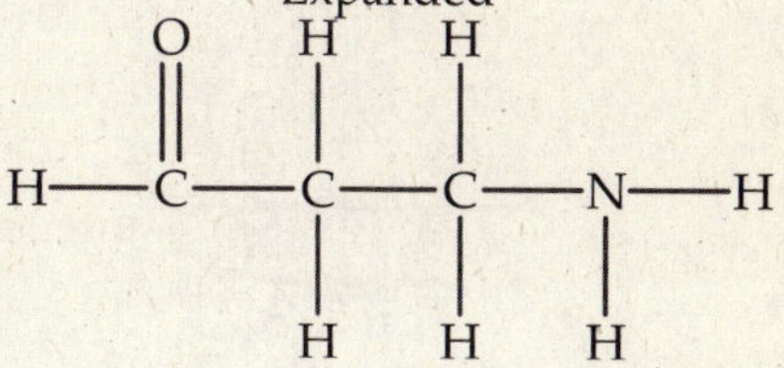

b. 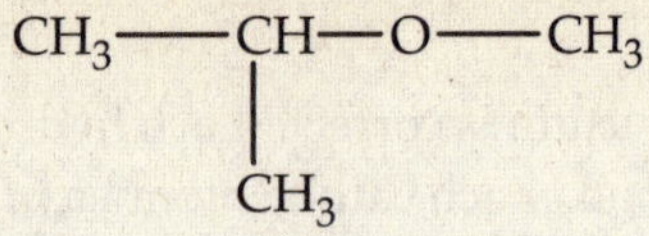 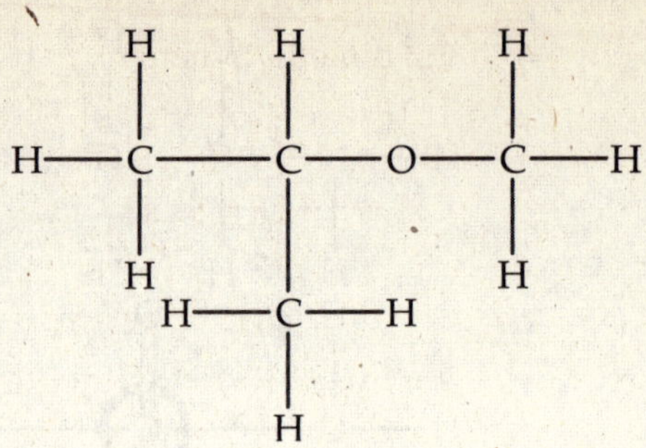

ALKANE STRUCTURES (SECTION 11.5)

☑11.28 a. branched (3 C longest chain)
b. normal (5 C longest chain)
c. normal (4 C longest chain)
d. normal (5 C longest chain)
e. branched (5 C longest chain)
f. branched (4 C longest chain)

CONFORMATIONS OF ALKANES (SECTION 11.6)

☑11.30 a. same compound (C_4H_{10} normal)
b. same compound (C_6H_{14} normal)
c. isomers (C_4H_{10} branched and normal)
d. same compound (C_5H_{12} branched)

ALKANE NOMENCLATURE (SECTION 11.7)

11.32 a.  6 C atoms

b. 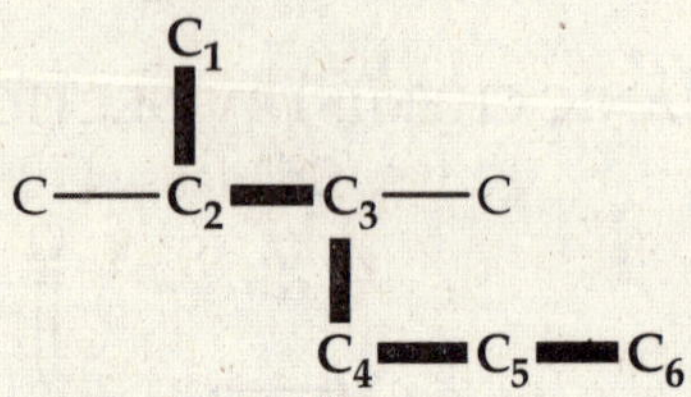 6 C atoms

c. 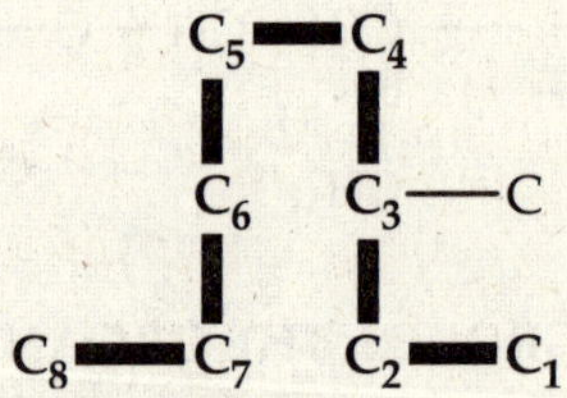 8 C atoms

☑11.34 Note: Most of the hydrogen atoms have been removed from the following structures in order to facilitate the identification of the parent chain.

a. 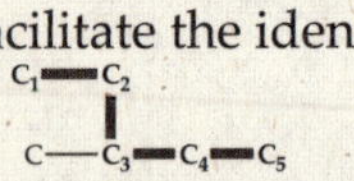
3-methylpentane

b. 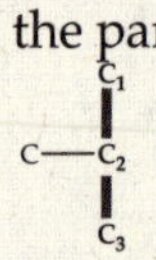
2-methylpropane

c. 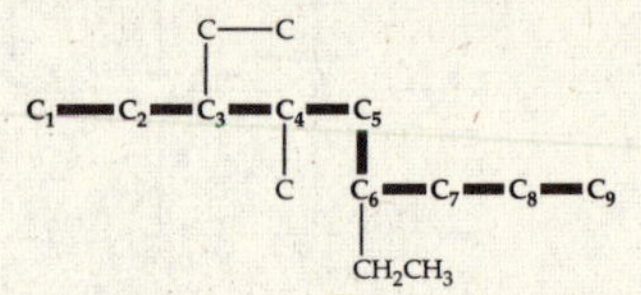
3,6-diethyl-4-methylnonane

d. 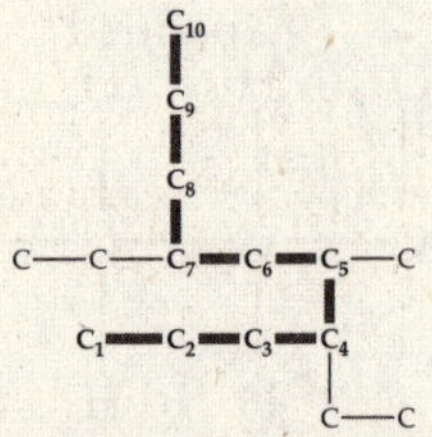
4,7-diethyl-5-methyldecane

e. 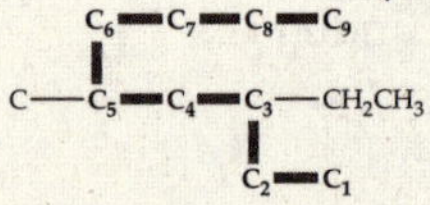
3-ethyl-5-methylnonane

11.36 a. 3-ethylpentane
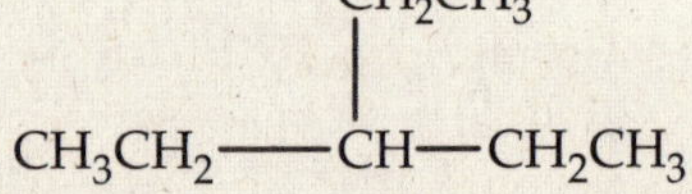

b. 2,2-dimethylbutane

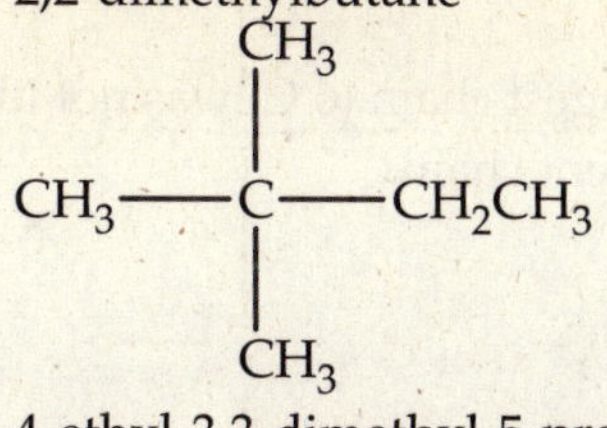

c. 4-ethyl-3,3-dimethyl-5-propyldecane

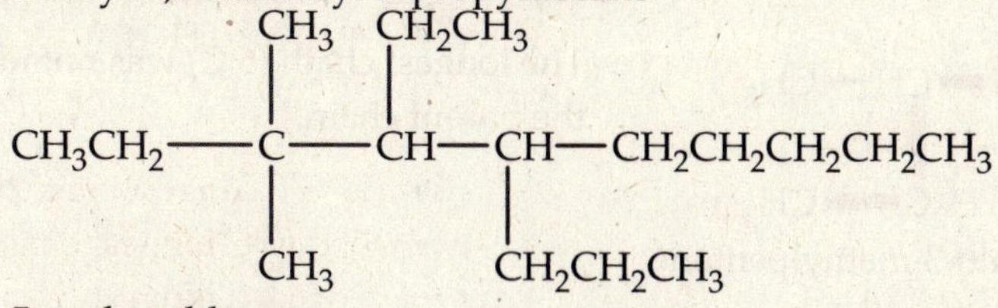

d. 5-*sec*-butyldecane

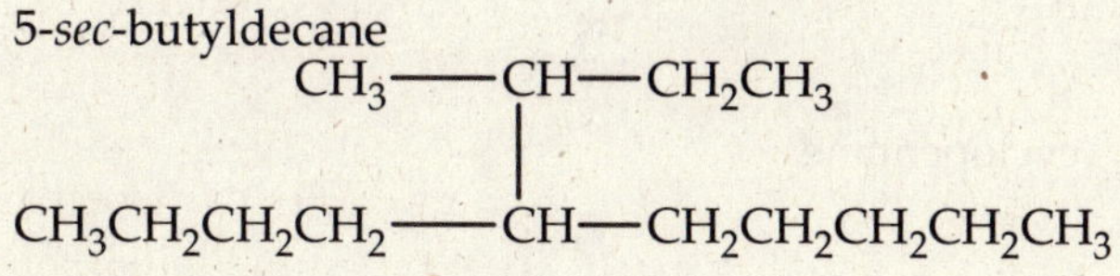

11.38

$CH_3CH_2CH_2CH_2CH_3$
pentane

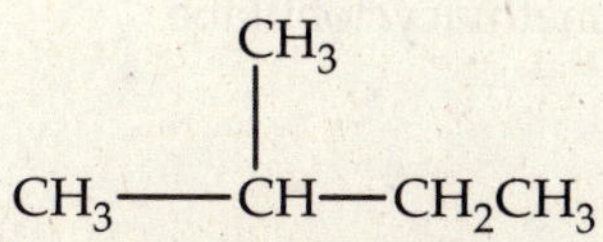

2-methylbutane

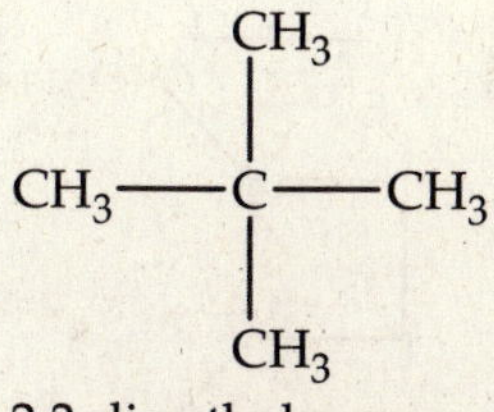

2,2-dimethylpropane

11.40

$CH_3CH_2CH_2CH_2CH_2CH_3$
hexane

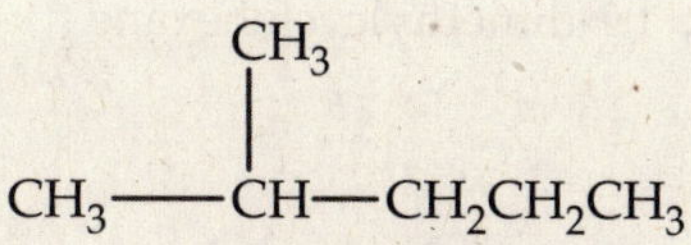

2-methylpentane

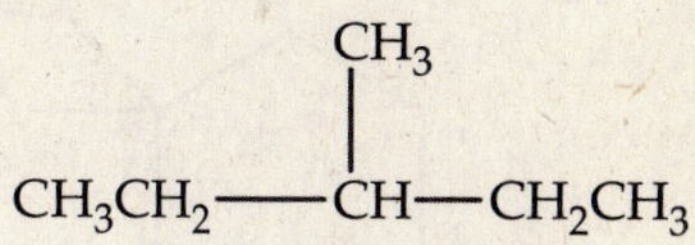

3-methylpentane

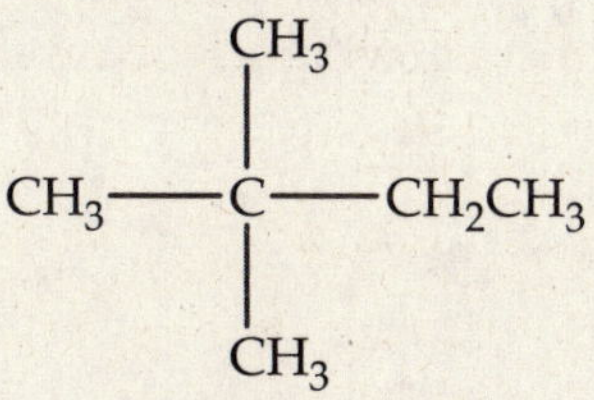

2,2-dimethylbutane

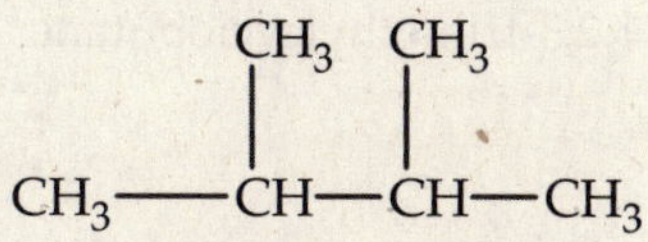

2,3-dimethylbutane

11.42 a. incorrect = 1,2-dimethylpropane

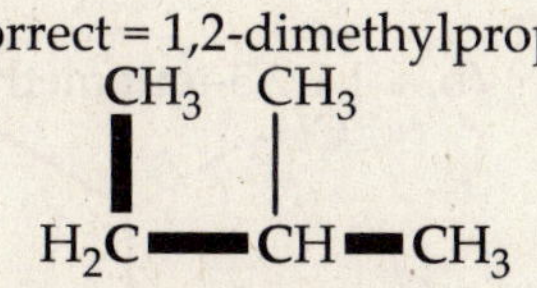

corrected = 2-methylbutane

The longest chain (4 C) was not identified as the parent chain.

b. incorrect = 3,4-dimethylpentane

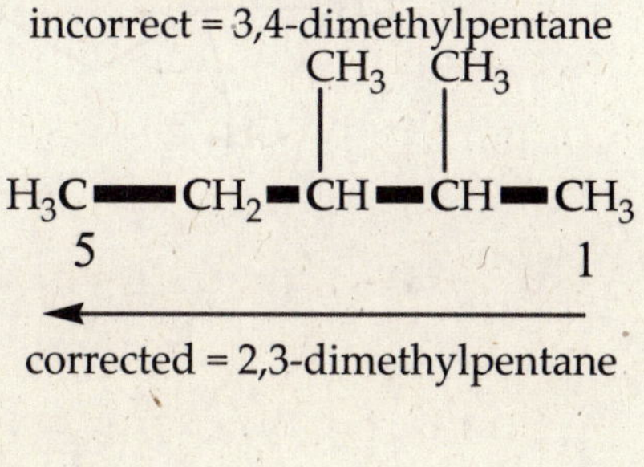

corrected = 2,3-dimethylpentane

The parent chain was not numbered so that the attached groups were on the lowest numbered carbon atoms.

c. incorrect = 2-ethyl-4-methylpentane

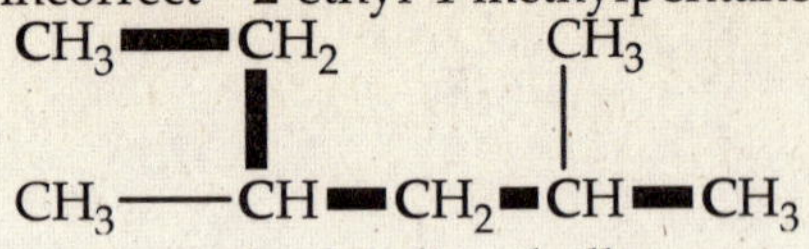

corrected = 2,4-dimethylhexane

The longest chain (6 C) was not identified as the parent chain.

d. incorrect = 2-bromo-3-ethylbutane

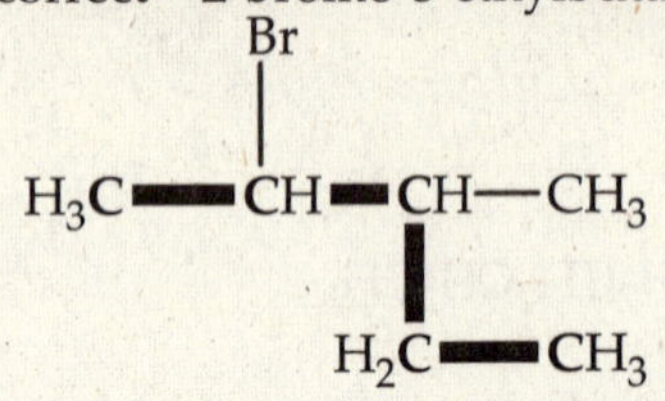

corrected = 2-bromo-3-methylpentane

The longest chain (5 C) was not identified as the parent chain.

CYCLOALKANES (SECTION 11.8)

☑11.44 a. cyclopentane

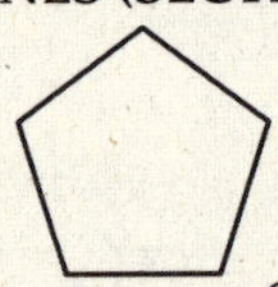

b. 1,2-dimethylcyclobutane

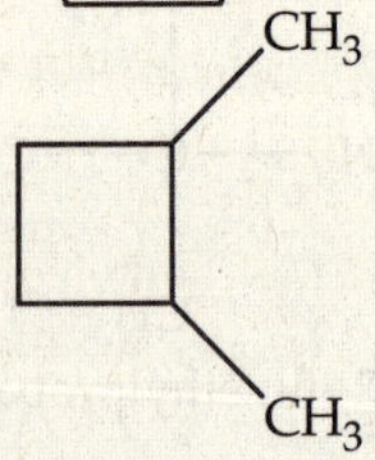

c. 1,1-dimethylcyclohexane

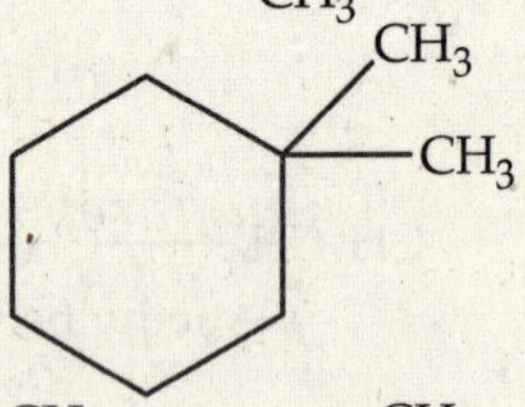

d. 1,2,3-trimethylcyclobutane

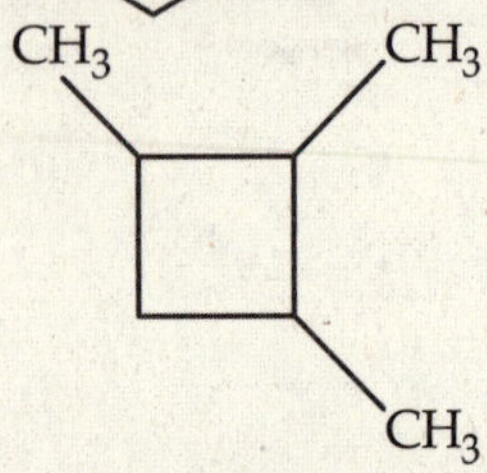

11.46 a. ethylcyclobutane

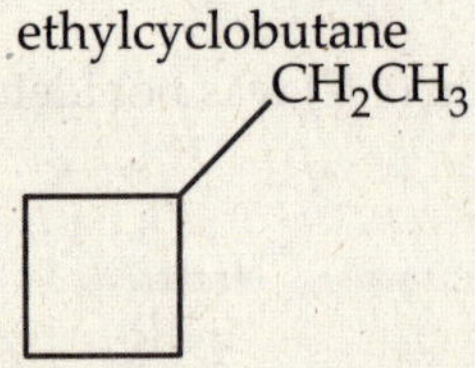

b. 1,1,2,5-tetramethylcyclohexane

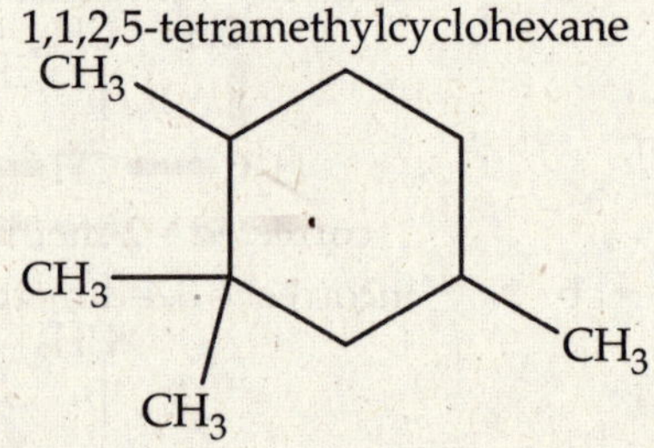

c. 1-butyl-3-isopropylcyclopentane

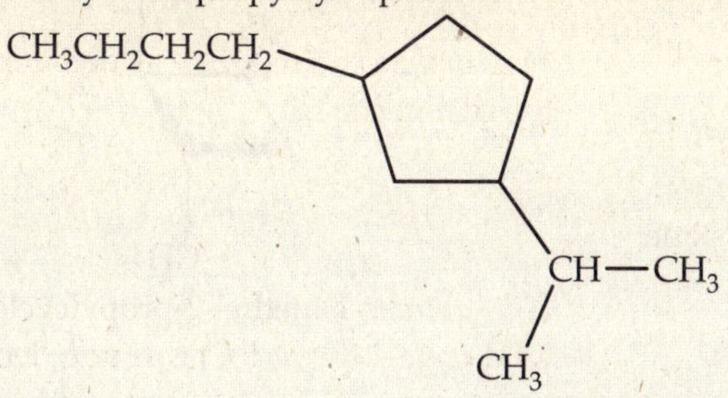

11.48 a. same compound (C_6H_{12}; 1,2-dimethylcyclobutane)
b. structural isomers (C_8H_{16}; 1-ethyl-2-methylcyclopentane & 1-ethyl-3-methylcyclopentane)
c. structural isomers (C_5H_{10}; ethylcyclopropane & 1,2-dimethylcyclopropane)
d. structural isomers (C_8H_{16}; 1,1-dimethylcyclohexane & 1,2-dimethylcyclohexane)

THE SHAPE OF CYCLOALKANES (SECTION 11.9)

11.50 Each carbon atom in cyclohexane has 4 single bonds and assumes a tetrahedral shape. The only way for six carbon atoms in a ring to each have a tetrahedral shape (with bond angles of 109.5°) is for the overall shape to be a chair rather than a planar hexagon (which would have bond angles of 120°).

11.52 a. Methylcyclohexane does not have any geometric isomers because it only has one group attached to the ring. In order for cycloalkanes to exhibit geometric isomerism, they must have 2 groups attached to the ring on different carbon atoms.
b. 1-ethyl-1-methylcyclopentane does not have any geometric isomers because it has two groups attached to the same carbon in the ring. In order for cycloalkanes to exhibit geometric isomerism, they must have 2 groups attached to the ring on different carbon atoms.
c.

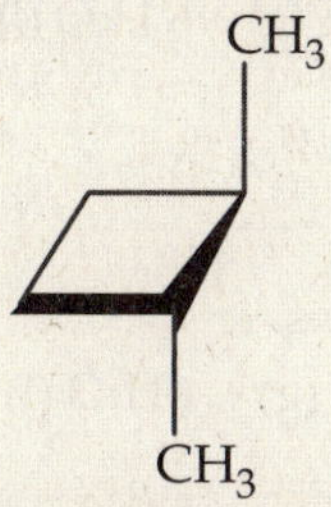

trans-1,2-dimethylcyclobutane

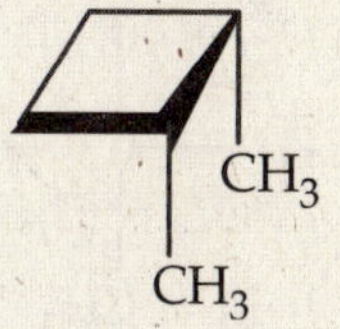

cis-1,2-dimethylcyclobutane

d.

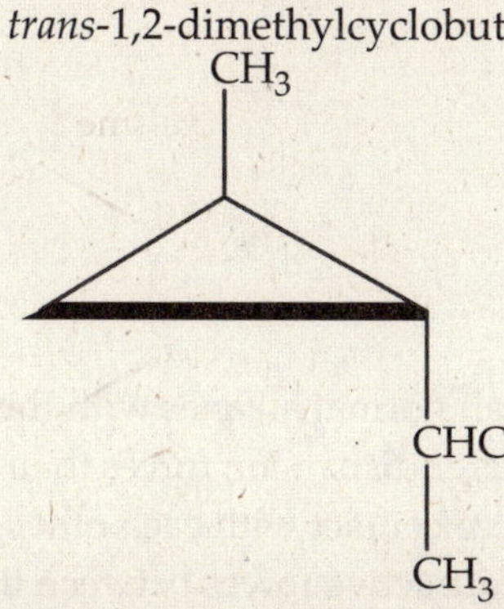

trans-1-isopropyl-2-methylcyclopropane

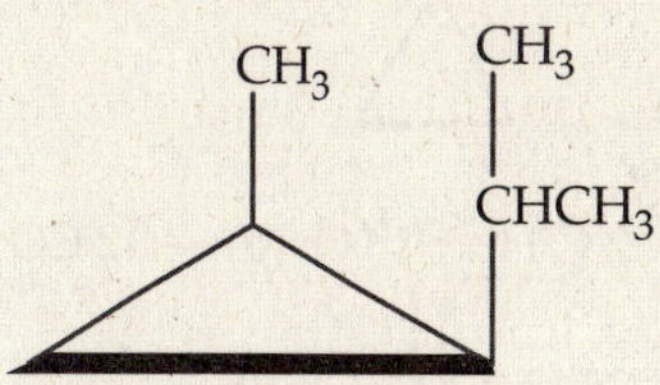

cis-1-isopropyl-2-methylcyclopropane

☑11.54 a.

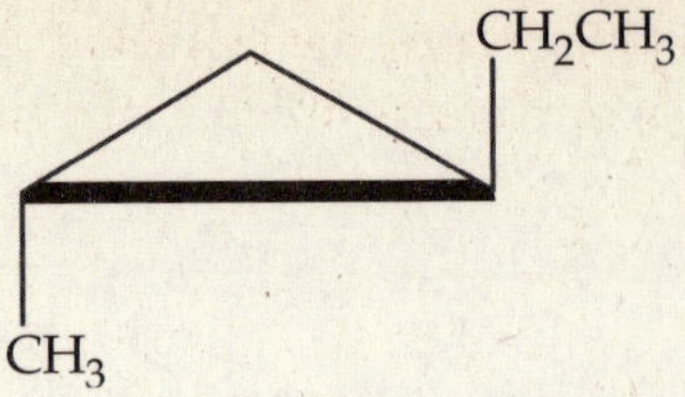

trans-1-ethyl-2-methylcyclopropane

c.

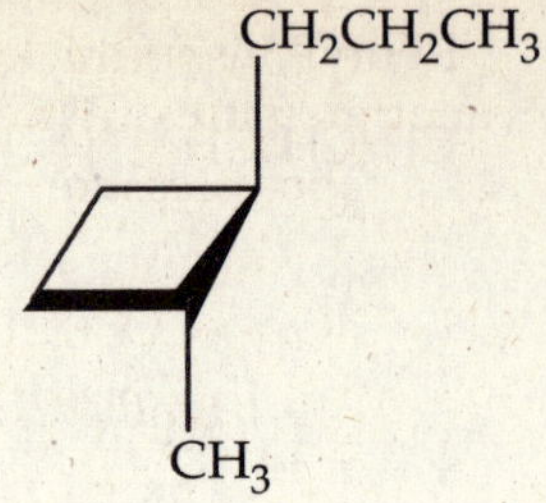

trans-1-methyl-2-propylcyclobutane

b.

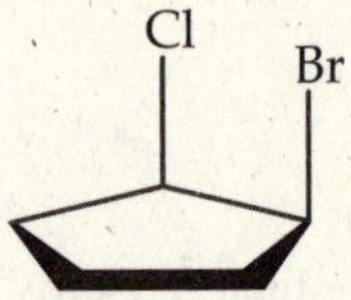

cis-1-bromo-2-chlorocyclopentane

d.

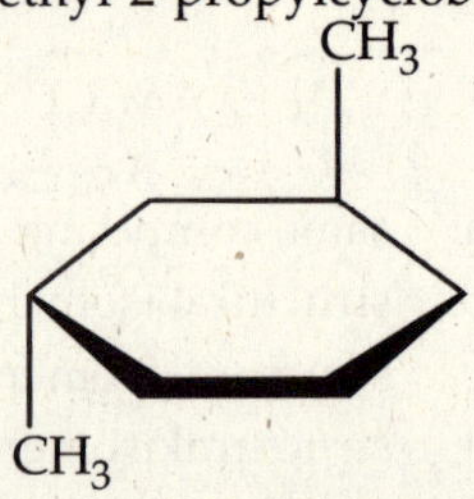

trans-1,3-dimethylcyclohexane

PHYSICAL PROPERTIES OF ALKANES (SECTION 11.10)

☑11.56 a. Decane is a liquid at room temperature. c. Decane is soluble in hexane.
b. Decane is not soluble in water. d. Decane is less dense than water.

11.58 The sample with the higher melting point and boiling point is 2-methylheptane. It has a higher molecular mass than 2-methylhexane. Both these compounds are nonpolar hydrocarbons which experience dispersion forces as their only intermolecular force. Dispersion forces increase with increasing molecular mass.

ALKANE REACTIONS (SECTION 11.11)

☑11.60 a. butane $2\ C_4H_{10} + 13\ O_2\ (g) \rightarrow 8\ CO_2\ (g) + 10\ H_2O\ (l)$

b.

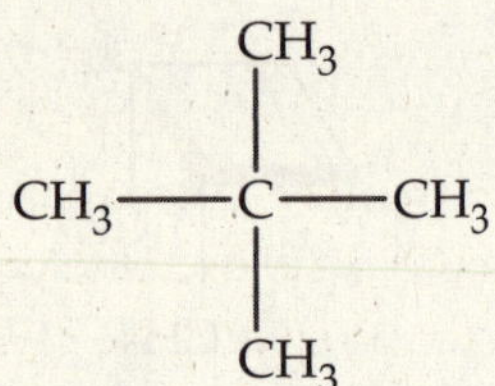

$C_5H_{12} + 8\ O_2\ (g) \rightarrow 5\ CO_2\ (g) + 6\ H_2O\ (l)$

c.

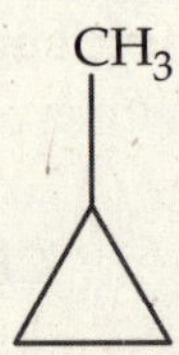

$C_4H_8 + 6\ O_2\ (g) \rightarrow 4\ CO_2\ (g) + 4\ H_2O\ (l)$

11.62 $2\ C_6H_{14} + 13\ O_2\ (g) \rightarrow 12\ CO\ (g) + 14\ H_2O\ (l)$

ADDITIONAL EXERCISES

11.64 Most cycloalkanes have higher boiling points than normal alkanes with the same number of carbon atoms because the cycloalkanes have greater dispersion forces than normal alkanes. As a result of their shape, cycloalkanes make more contact with each other than normal alkanes; this increased contact results in greater attractive forces between the molecules for cycloalkanes than for normal alkanes.

11.66 Vapor pressure decreases as the molecular weight increases; therefore, the pentane has a vapor pressure of 414.5 torr, the hexane has a vapor pressure of 113.9 torr, and the heptane has a vapor pressure of 37.2 torr.

11.68 $CH_4 + 2\ O_2 \rightarrow CO_2 + 2\ H_2O$

$$1.00 \text{ g } CH_4 \left(\frac{1 \text{ mole } CH_4}{16.0 \text{ g } CH_4}\right)\left(\frac{2 \text{ moles } O_2}{1 \text{ mole } CH_4}\right)\left(\frac{22.4 \text{ L } O_2}{1 \text{ mole } O_2 \text{ at STP}}\right)\left(\frac{100 \text{ L air}}{21 \text{ L } O_2}\right) = 13 \text{ L air}$$

ALLIED HEALTH EXAM CONNECTION

11.70
a. Propane, 3 carbon atoms — n = 3; therefore, C_3H_8 and $CH_3CH_2CH_3$

b. Octane, 8 carbon atoms — n = 8; therefore, C_8H_{18} and $CH_3(CH_2)_6CH_3$

c. Butane, 4 carbon atoms — n = 4; therefore, C_4H_{10} and $CH_3(CH_2)_2CH_3$

11.72 (a) C_2H_6 is an alkane because its formula follows the pattern of C_nH_{2n+2}.

11.74 The general products of a combustion reaction are (d) CO_2 and H_2O.

11.76 The deadly property of carbon monoxide, if inhaled, is due to its (c) high affinity for hemoglobin.

CHEMISTRY FOR THOUGHT

11.78 The study of organic compounds might be important to someone interested in the health or life sciences because living organisms are composed of mostly organic compounds and most medications are also organic compounds.

11.80 The low melting point of the ski wax indicates that the molecules in the ski wax have weak forces between the molecules. (Note: The forces are dispersion forces.)

11.82 If carbon did not form hybridized orbitals, the formula of the simplest compound of carbon and hydrogen would be C_2H_4 because each carbon atom would have two unpaired electrons that would each pair with the unpaired electron in a hydrogen atom and the carbon atoms could share their two lone pairs of electrons in a double bond. With this arrangement, both carbon atoms would have a complete octet and each hydrogen atom would have a complete valence shell with two electrons.

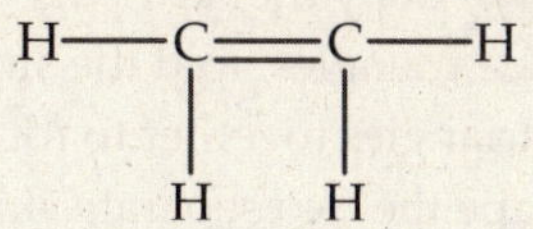

11.84 If a semi truck loaded with cyclohexane overturns during a rainstorm, spilling its contents over the road embankment and the rain continues, then the cyclohexane will float on top of the standing water and run off along the water drainage route. Some of the cyclohexane will probably also evaporate because it has weak intermolecular forces.

11.86 Alkanes do not mix with water and are less dense than water; therefore, when an oil spill occurs, the oil floats on top of the water. Alkanes are relatively unreactive; therefore, short of burning the oil, a chemical reaction cannot be performed to "neutralize" the oil spill. Burning the oil is not a good idea because of the wildlife that quickly becomes disabled by contact with the oil and because crude oil often contains compounds (other than the pure hydrocarbons) that may produce toxic products when burned. Birds that come into contact with the oil must be cleaned with soap before they are able to fly again. A bird will try to clean the oil off of its own feathers and ingest the oil, which does damage to its internal organs. Sea otters and killer whales as well as small organisms at the bottom of the food chain are also impacted by contact with oil.

ADDITIONAL ACTIVITIES

Section 11.1 Review: Link the ideas in the rectangles to the appropriate category.

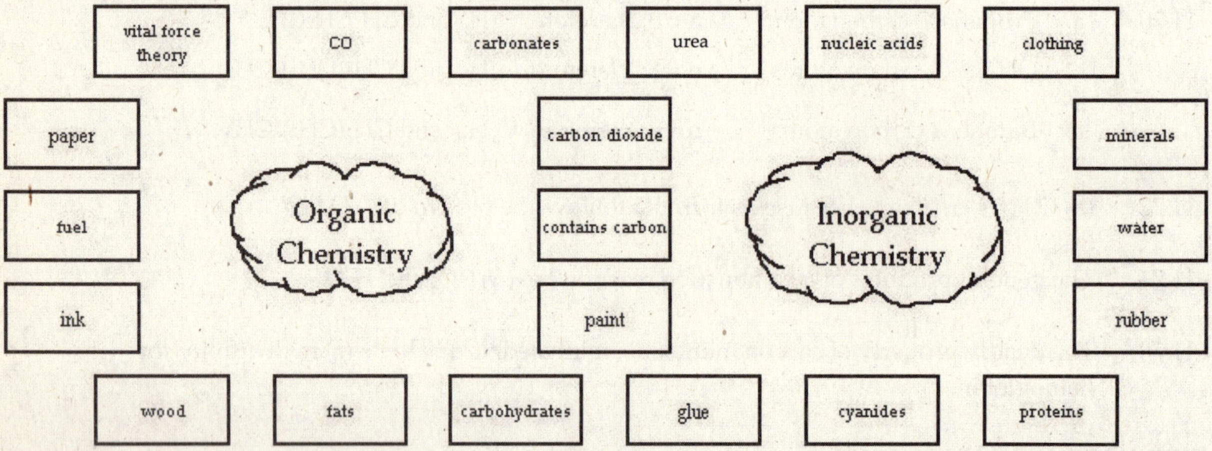

Section 11.2 Review: Identify the facts/properties from each of the following categories for "typical" organic compounds.

Category	Possibility A	Possibility B
Number of known compounds	250,000	> 6 million
Bonding within molecules	ionic	covalent
Interparticle forces	weak	strong
Flammability	flammable	nonflammable
Water solubility	high	low
Conductivity of water solutions	conductor	nonconductor
Rate of chemical reactions	slow	fast
Melting point	high	low

Section 11.3 Review: Copy the next page onto another sheet of paper or a transparency, then cut out the shapes. For the large triangles, fold the smaller triangles up from the small triangular base and tape the edges of the small triangles together to form a "closed" tetrahedron. For the bows, fold along the black-gray borders and tape the two separate skinny gray rectangles together to form 3 legs. These legs can be stabilized by adding pieces of wooden toothpicks behind them. To form an "open" tetrahedron, align two of the legs on two of the "bows" and tape. This structure will be stabilized by adding the other two "bows" in such a manner that all of the resulting endpoints have three connected bows. The open tetrahedron will fit inside the closed tetrahedron.

(1) What is the angle between unhybridized *p* orbitals?
(2) What type of hybridized orbital is represented by the legs on the bows?
(3) What is the angle between the legs on the bows?
(4) Why is this angle different from (1)?

(5) If carbon atoms can be represented as tetrahedrons, how many points of the tetrahedrons will touch for a carbon-carbon single bond?
(6) If carbon atoms can be represented as tetrahedrons, how many points of the tetrahedrons will touch for a carbon-carbon double bond?
(7) If carbon atoms can be represented as tetrahedrons, how many points of the tetrahedrons will touch for a carbon-carbon triple bond?
(8) Can carbon atoms make a quadruple bond? Explain using the image of carbon as a tetrahedron.

Section 11.4 Review:

Study Tip: Make flashcards of the class name and the corresponding key functional groups from Table 11.2. Practice with the flashcards until you can identify the key functional group by seeing the class name and the class name by seeing the key functional group.

Section 11.5 Review:

(1) List the similarities between methane, ethane, propane, and butane.
(2) What is similar in the structures of methane, ethane, and propane? How is butane different?
(3) Pentane is the fifth member of this series. What is its formula? Can it have a normal and/or a branched structure?

Section 11.6 Review:

Place a toothpick into the top point of one of your closed tetrahedrons from the Section 11.3 Review. Place the other end of the toothpick into the top point of the other closed tetrahedron from the Section 11.3 Review. Color one of the "free" points on each tetrahedron. These will represent CH_3– groups.

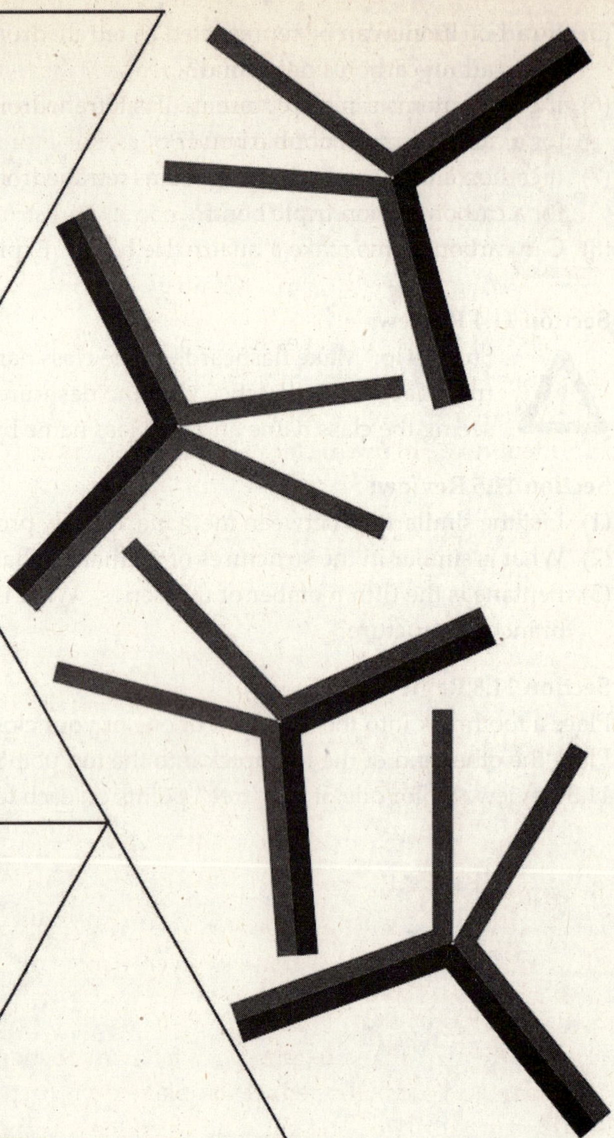

Section 11.6 Review (continued):

(1) Hold the tetrahedrons in such a way that both of the colored points are oriented toward the ceiling and you are looking at one end of the toothpick. Can you see the other tetrahedron?

(2) Tip the structure so that the "front" tetrahedron is lower than the "rear" tetrahedron. Are the CH_3- groups aligned?

(3) Rotate the "front" tetrahedron 90° clockwise. (The colored point should move from the "12 o'clock" position to the "3 o'clock" position). Are the CH_3-groups aligned? Are the points on the tetrahedrons aligned?

(4) Rotate the "front tetrahedron another 30° clockwise. (The colored point should move to the "4 o'clock" position.) Are the CH_3-groups aligned? Are the points on the tetrahedrons aligned?

(5) What are the different orientations seen in (2), (3), and (4) called?

(6) Has the bonding order changed in (2), (3), or (4)?

(7) Are (2), (3), and (4) structural isomers of each other?

Section 11.7 Review:

Study Tip: (1) Make flashcards for the number of carbon atoms in the longest chain and their corresponding root names. Practice with the flashcards until you can name the longest carbon chain based on the number of carbon atoms in the chain and determine how many carbon atoms are in the longest chain based on the name.

(2) Make flashcards of the names and structures of the alkyl groups as well as the common nonalkyl groups. Practice the flashcards until you can draw the structures from the names and name the structures within a compound without having to refer to your notes.

(1) Suggest a reason alkanes have an *–ane* ending.
(2) Suggest a reason that unbranched carbon chains are given the prefix *n-*.
(3) What is the common structural element for all *iso-* compounds?
(4) Identify the following alkyl groups by their names.

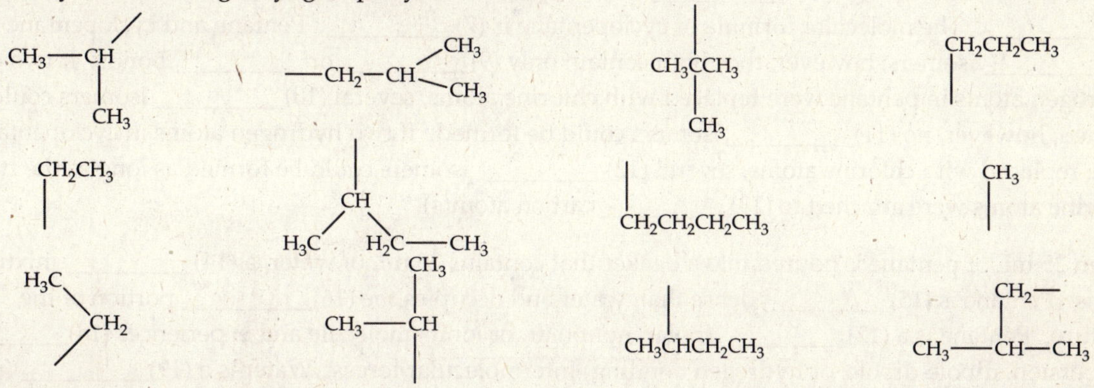

Section 11.8 Review:

(1) Is the prefix *cyclo-* appropriate to describe compounds that contain a ring? Explain.
(2) What is the general formula for cycloalkanes?

Section 11.9 Review:

(1) Can the two linked tetrahedrons from the Section 11.6 Review be placed on one of the small triangles from the closed tetrahedron template in the Section 11.3 Review in order to represent cyclopropane? If so, what is the bond angle between the carbon atoms? If not, can cyclopropane be represented if the tetrahedrons are removed from the toothpick? What is the associated bond angle?
(2) What are the angles in a planar square?
(3) What are the angles in a planar pentagon?
(4) What is the preferred bond angle for adjacent carbon-carbon bonds in a cycloalkane?
(5) What must the atoms in cyclobutane and cyclopentane do to obtain a more desirable bond angle?
(6) Can the atoms in cyclopropane make the same change as in (5)?
(7) What do the prefixes *cis-* and *trans-* mean?

Section 11.10 Review:

Box the following properties that apply to alkanes. Correct the properties that do not apply to alkanes as they are currently written.

unpleasant odor | high boiling point | low melting point | polar | dispersion forces

soluble in water | less dense than water | hydrophobic

Section 11.11 Review:

(1) What is the major type of reaction that alkanes undergo?
(2) Is the reaction endothermic or exothermic?
(3) What is an application for this type of alkane reaction?
(4) What is the limiting reactant when CO is a product of this type of reaction?
(5) What is the limiting reactant when only carbon dioxide and water are the products of this type of reaction?

Tying It All Together with a Laboratory Application:

The expanded structural formula of pentane is (1)_________. The condensed structural formula of pentane is (2)_________. Rotating the carbon-carbon bonds in the pentane will produce different (3)_________. The molecular formula of pentane is (4)_________. The relationship between pentane and 2-methylbutane is called (5)_________. The condensed structural formula of cyclopentane is (6)_________. The molecular formula of cyclopentane is (7)_________. Pentane and cyclopentane (8)_________ isomers; however, they both contain only (9)_________ or _________ bonds. If two of the hydrogen atoms in pentane were replaced with chlorine atoms, several (10)_________ isomers could be formed; however, no (11)_________ isomers could be formed. If two hydrogen atoms in cyclopentane were replaced with chlorine atoms, several (12)_________ isomers could be formed as long as the two chlorine atoms were attached to (13)_________ carbon atom(s).

When 25 mL of pentane is poured into a beaker that contains 25 mL of water, a (14)_________ mixture forms. Pentane is (15)_________ dense than water and occupies the (16)_________ portion of the mixture. Pentane is a (17)_________ (polar, nonpolar, or ionic) molecule and experiences (18)_________ (dispersion, dipole-dipole, or hydrogen bonding) intermolecular forces. Water is a (19)_________ (polar, nonpolar, or ionic) molecule and experiences (20)_________ (dispersion, dipole-dipole, or hydrogen bonding) intermolecular forces. These compounds do not fit the guideline for solubility which is (21)_________. Pentane is also said to be (22)_________ because it (23)_________ dissolve in water.

The beaker is placed in the fume hood and a smoldering splint is placed into the beaker above the liquid level. What happens? (24)_________

A. Nothing, because the pentane and water are not flammable.
B. The pentane begins to burn because it is flammable.
C. The pentane and water begin to burn because they are both flammable.

Flammability is a property of most (25)_________ compounds. The process of burning a compound in the presence of oxygen is called (26)_________. When oxygen is the limiting reactant, the reaction is classified as (27)_________. When oxygen is the excess reactant, the reaction is classified as (28)_________. Burning is an (29)_________ process because heat is released.

SOLUTIONS FOR THE ADDITIONAL ACTIVITIES

Section 11.1 Review:

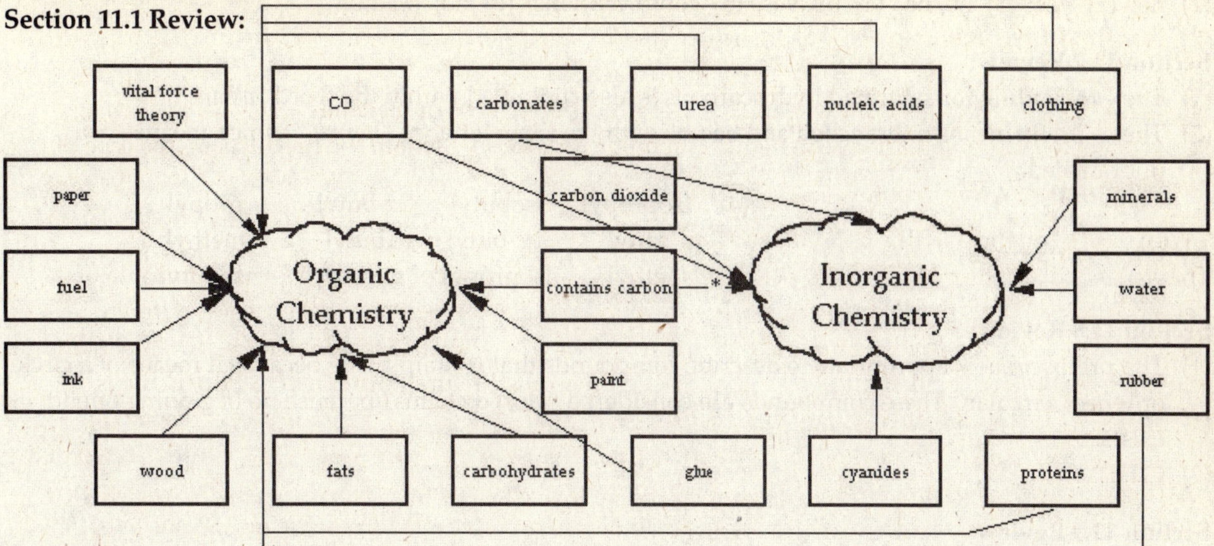

*sometimes

Section 11.2 Review:

Number of known compounds	250,000	> 6 million
Bonding within molecules	ionic	covalent
Interparticle forces	weak	strong
Flammability	flammable	nonflammable
Water solubility	high	low
Conductivity of water solutions	conductor	nonconductor
Rate of chemical reactions	slow	fast
Melting point	high	low

Section 11.3 Review:

(1) The angle between unhybridized *p* orbitals is 90°.

(2) The legs on the bows represent the larger lobe of the sp^3 hybridized orbitals.

(3) The angle between the legs on the bows is 109.5°.

(4) When a *s* orbital blends with 3 *p* orbitals to make 4 new sp^3 hybrid orbitals on a carbon atom, the four new orbitals share the same energy level. Each hybrid orbital contains one electron. These orbitals repel each other and establish a new angle of 109.5°. The reorganization of the orbitals results in a new angle between the orbitals.

(5) single bond – 1 point

(6) double bond – 2 points

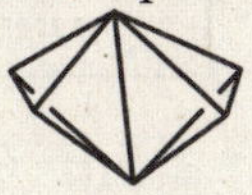

(7) triple bond – 3 points

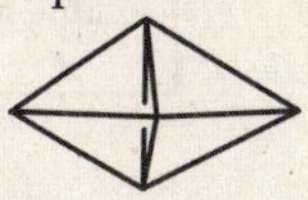

(8) Carbon atoms cannot make a quadruple bond because it is impossible to have all four corners of the tetrahedrons touch.

Section 11.5 Review:

(1) Methane, ethane, propane, and butane are saturated hydrocarbons with the general formula C_nH_{2n+2}. They have bond angles of 109.5°. They can be burned as fuels.

(2) Methane, ethane, and propane cannot not have branched structures. They are normal alkanes. Butane can have a branched or a normal structure.

(3) Pentane, C_5H_{12}, can have either a branched or a normal structure.

Section 11.6 Review:

(1) no; (2) yes; (3) no, no; (4) no, yes; (5) conformations; (6) no; (7) no

Section 11.7 Review:

(1) The *–ane* ending for saturated hydrocarbons is also contained within the word *"alkane."*

(2) The *n*- prefix means unbranched and begins with the same letter as *"normal"* which means unbranched.

(3) 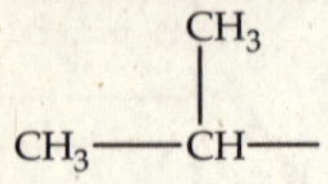

(4)

isopropyl	isobutyl	*t*-butyl	propyl
ethyl	*sec*-butyl	butyl	methyl
ethyl	isopropyl	*sec*-butyl	isobutyl

Section 11.8 Review:

(1) The prefix *cyclo-* is appropriate to describe compounds that contain a ring because it means of a circle or wheel, circular. These compounds are considered "closed chains" because no beginning or end exists.

(2) C_nH_{2n}

Section 11.9 Review:

(1) No, the two linked tetrahedrons from the Section 11.6 Review cannot be placed on one of the small triangles in order to represent cyclopropane. The distance between the tetrahedron points is too large. By removing the tetrahedrons from the toothpick and adjusting the adjacent carbon-carbon bond angles to 60°, a representation of cyclopropane can be made.

(2) 90°; (3) 108°; (4) 109.5°; (5) Cyclobutane and cyclopentane can both deviate from a planar arrangement of carbon atoms in order to obtain a more desirable bond angle.

(6) No, cyclopropane cannot deviate from a planar arrangement of carbon atoms because the three carbon atoms automatically form a plane.

(7) *cis-* = on this side of; *trans-* = on the other side of, to the other side of, over, across, through

Section 11.10 Review:

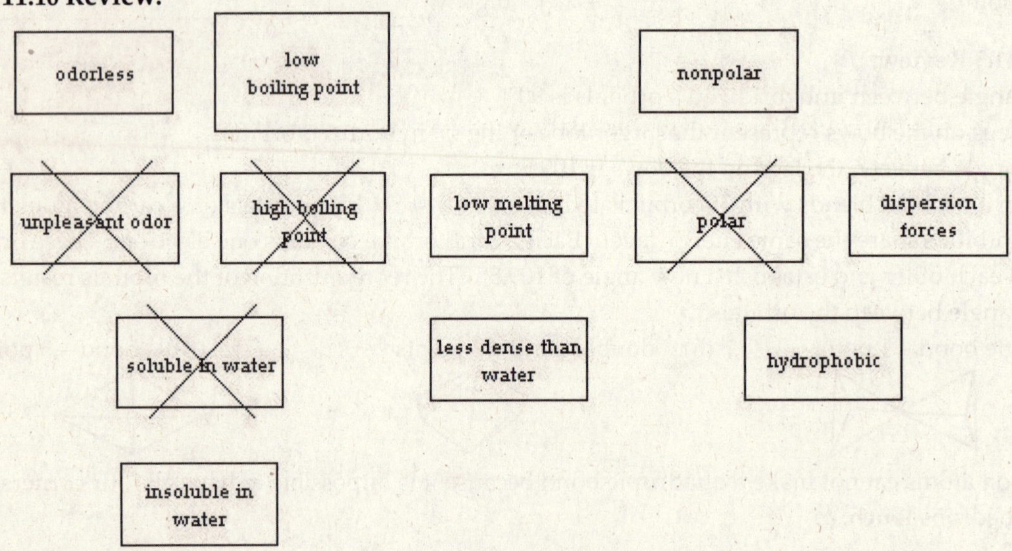

Section 11.11 Review:

(1) Alkanes undergo rapid oxidation (combustion) reactions.

(2) Combustion reactions are exothermic.

(3) Alkane combustion can be used as a source of heat, for cooking, or for running an engine.

(4) Oxygen is the limiting reactant when CO is produced. This is incomplete combustion.

(5) The alkane is the limiting reactant when only carbon dioxide and water are the products of combustion. This is complete combustion.

Tying It All Together with a Laboratory Application:

(1) 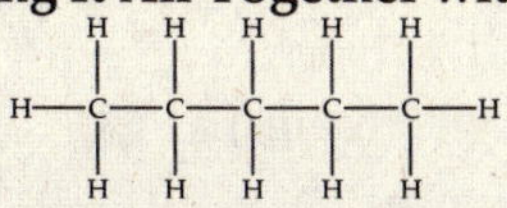
(2) $CH_3CH_2CH_2CH_2CH_3$
(3) conformers
(4) C_5H_{12}
(5) structural isomerism
(6) 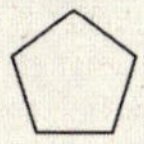
(7) C_5H_{10}
(8) are not
(9) single, sigma
(10) structural
(11) geometric
(12) geometric
(13) different
(14) heterogeneous
(15) less
(16) upper
(17) nonpolar
(18) dispersion
(19) polar
(20) hydrogen bonding
(21) "like dissolves like"
(22) hydrophobic
(23) does not
(24) B
(25) organic
(26) combustion
(27) incomplete combustion
(28) complete combustion
(29) exothermic

SELF-TEST QUESTIONS

Multiple Choice

1. What is the maximum number of covalent bonds which carbon can form?
   a. 1 b. 2 c. 3 d. 4

2. Which of the following is considered an organic compound?
   a. CH_4 b. NaOH c. Na_2CO_3 d. KCN

3. How many hydrogen atoms are needed to complete the following structure?
   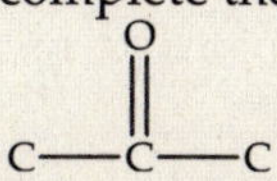
   a. 2 b. 4 c. 6 d. 8

4. A C—H bond in CH_4 is formed by the overlap of what orbitals?
   a. sp^3 and 1*s* b. 1*s* and 1*s* c. *p* and 1*s* d. *sp* and 1*s*

5. Which of the structures are two representations of the same compound?
   I. $CH_3-CH_2-CH_2-CH_3$
   IV. 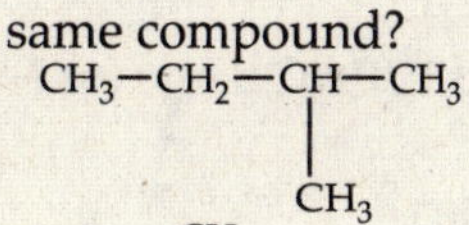
   II. 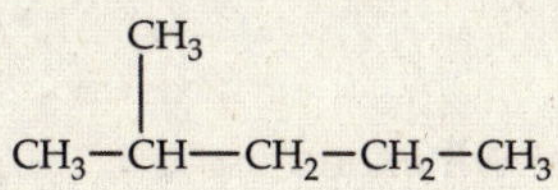
   V. 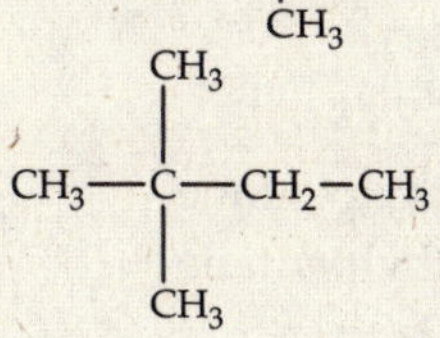
   III. 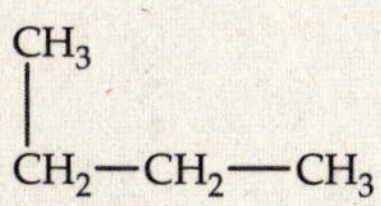
   a. I and II b. I and III c. II and III d. IV and V

6. What is the molecular formula for 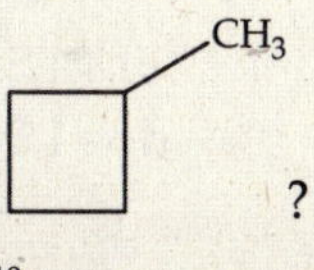 ?
   a. C_5H_{12} b. C_5H_{10} c. C_5H_9 d. C_4H_9

7. Which of the following compounds is a structural isomer of 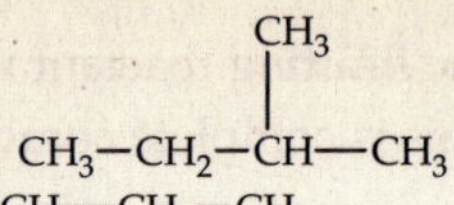 ?

a. $CH_3-CH_2-CH_2-CH_3$

b. 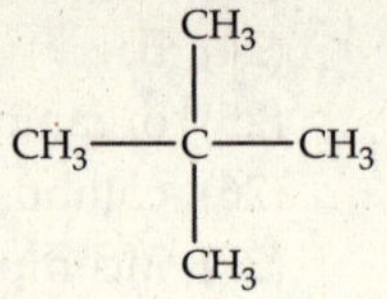

c. 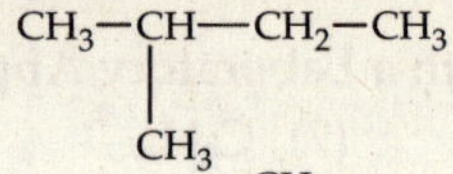

d. 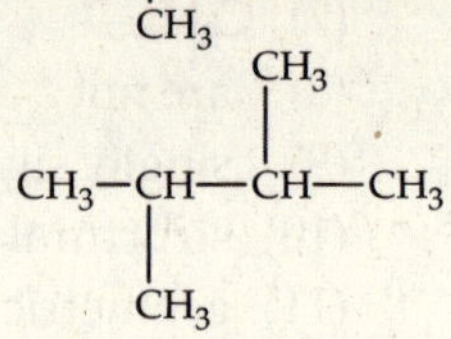

8. Which of the following compounds is a structural isomer of 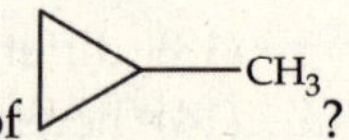 ?

a. 

b. 

c. $CH_3-CH_2-CH_2-CH_3$

d. 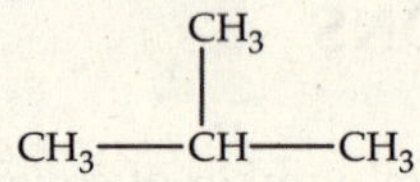

9. What is the unbranched alkane that contains eight carbon atoms called?

a. hexane b. heptane c. octane d. nonane

10. How many structural isomers have the formula C_4H_{10}?

a. 2 b. 3 c. 4 d. 5

11. How many carbon atoms are in the longest chain of 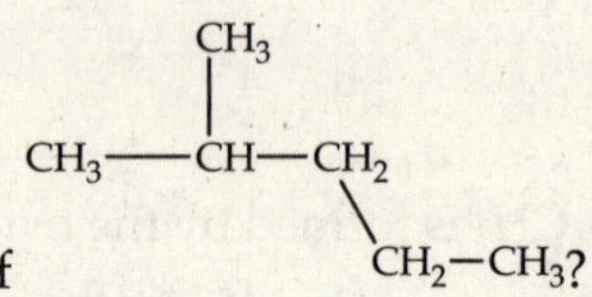

a. 3 b. 4 c. 5 d. 6

12. What is the correct IUPAC name for 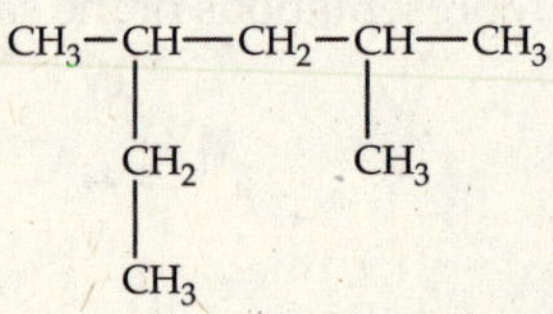 ?

a. 2-ethyl-4-methylpentane
b. 2-methyl-4-ethylpentane
c. 3,5-dimethylhexane
d. 2,4-dimethylhexane

13. What is the correct IUPAC name for 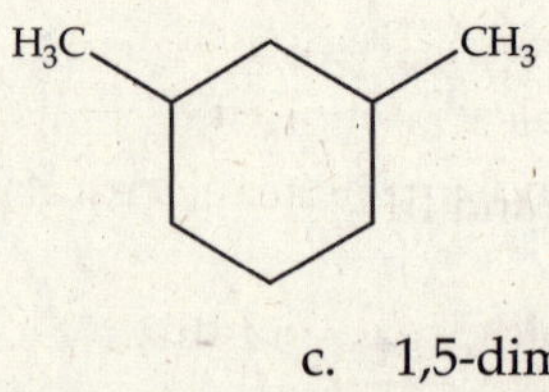 ?

a. 1,3-dimethylhexane
b. 1,3-methylcyclohexane
c. 1,5-dimethylcyclohexane
d. 1,3-dimethylcyclohexane

14. What is the position of the bromine atom in 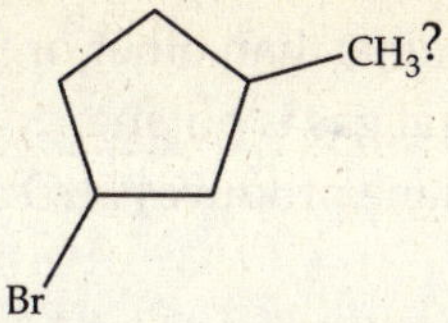

a. 1 b. 2 c. 3 d. 5

15. A 12-carbon alkane should be a _______ at room temperature.

a. solid b. liquid c. gas d. none of these

Matching

Match an alkyl group name to each structure on the left.

16. CH_3-CH_2-
17. $CH_3-CH_2-CH_2-$
18. $CH_3-\underset{\underset{CH_3}{|}}{CH}-CH_2-$

a. propyl
b. *sec*-butyl
c. isobutyl
d. ethyl

Match the structures on the left to the descriptions on the right.

19. 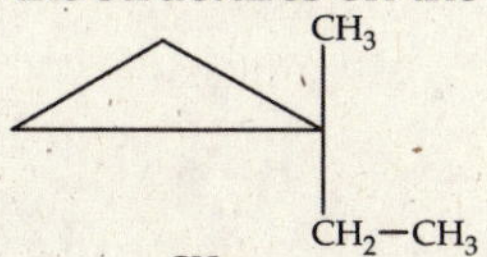

20. 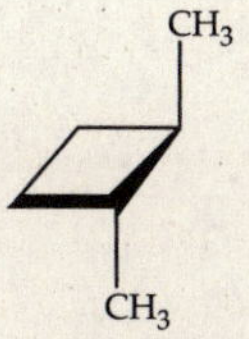

a. a *cis* compound
b. a *trans* compound
c. neither *cis* nor *trans*

21. 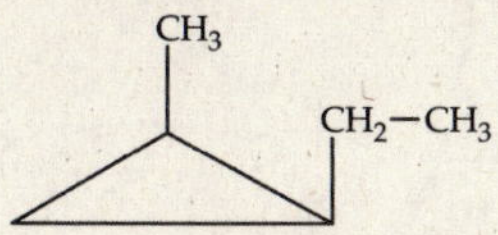

22. 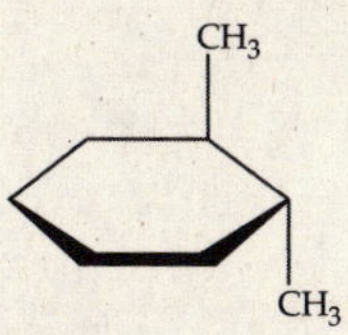

a. a *cis* compound
b. a *trans* compound
c. neither *cis* nor *trans*

True-False

23. Most organic compounds are very soluble in water.
24. Covalent bonds are more prevalent in inorganic compounds than in organic compounds.
25. Solutions of inorganic compounds are better electrical conductors than solutions of organic compounds.
26. There are more known inorganic compounds than organic compounds.
27. A few compounds of carbon, such as CO_2, are classified as inorganic.
28. Structural isomers always have the same molecular formula.
29. Structural isomers always have the same functional groups.
30. A condensed structural formula may show some bonds.
31. An expanded structural formula may not show all the bonds.
32. A molecule may have more than one functional group.
33. Pentane and cyclopentane are isomers of each other.

34. Alkanes have lower boiling points than other organic compounds.
35. The main component of natural gas is butane.
36. Complete combustion of pentane produces H_2O and CO_2.
37. Alkanes are polar molecules.
38. Cycloalkanes experience dispersion forces.
39. The general formula for cycloalkanes is C_nH_{2n+2}.
40. Alkanes are also known as paraffins.

Chapter 12: Unsaturated Hydrocarbons

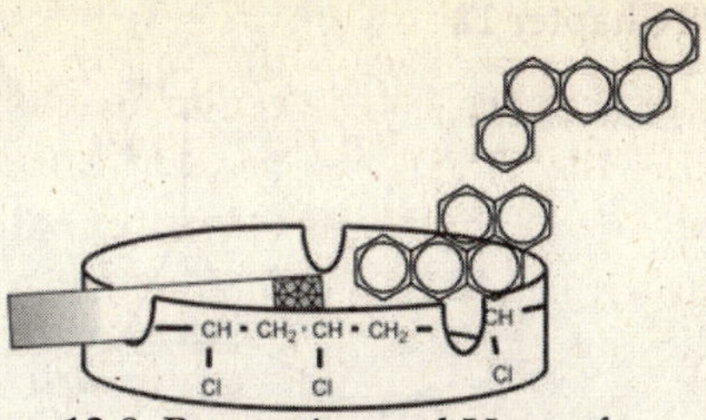

CHAPTER OUTLINE

12.1 The Nomenclature of Alkenes
12.2 The Geometry of Alkenes
12.3 Properties of Alkenes
12.4 Addition Polymers
12.5 Alkynes
12.6 Aromatic Compounds and the Benzene Structure
12.7 The Nomenclature of Benzene Derivatives
12.8 Properties and Uses of Aromatic Compounds

LEARNING OBJECTIVES/ASSESSMENT

When you have completed your study of this chapter, you should be able to:

1. Classify unsaturated hydrocarbons as alkenes, alkynes, or aromatics. (Section 12.1; Exercise 12.2)
2. Write the IUPAC names of alkenes from their molecular structures. (Section 12.1; Exercise 12.4)
3. Predict the existence of geometric (cis-trans) isomers from formulas of compounds. (Section 12.2; Exercise 12.18)
4. Write the names and structural formulas for geometric isomers. (Section 12.2; Exercise 12.20)
5. Write equations for addition reactions of alkenes, and use Markovnikov's rule to predict the major products of certain reactions. (Section 12.3; Exercise 12.26)
6. Write equations for addition polymerization, and list uses for addition polymers. (Section 12.4; Exercise 12.36)
7. Write the IUPAC names of alkynes from their molecular structures. (Section 12.5; Exercise 12.44)
8. Classify organic compounds as aliphatic or aromatic. (Section 12.6; Exercise 12.48)
9. Name and draw structural formulas for aromatic compounds. (Section 12.7; Exercises 12.52 and 12.54)
10. Recognize uses for specific aromatic compounds. (Section 12.8; Exercise 12.66)

SOLUTIONS FOR THE END OF CHAPTER EXERCISES

THE NOMENCLATURE OF ALKENES (SECTION 12.1) AND ALKYNES (SECTION 12.5)

☑12.2 An alkene is a hydrocarbon that contains at least one carbon-carbon double bond.
An alkyne is a hydrocarbon that contains at least one carbon-carbon triple bond.
An aromatic hydrocarbon is a compound that contains a benzene ring or other similar feature.

☑12.4

a. $CH_3CH{=}CHCH_3$
2-butene

b. $CH_3CH_2{-}C(CH_2CH_3){=}CHCH_3$
3-ethyl-2-pentene

c. $CH_3{-}C{\equiv}C{-}C(CH_3)_2{-}CH_2CH_3$
4,4-dimethyl-2-hexyne

d. 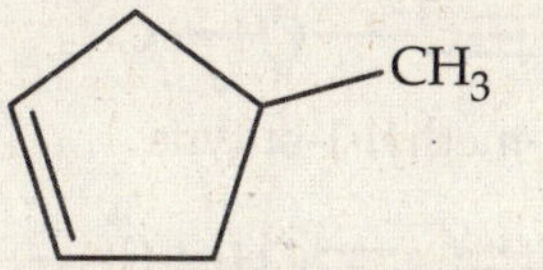
4-methylcyclopentene

e. $CH_3CH(Br)CH_2{-}C{\equiv}C{-}CH(CH_3){-}CH_3$
6-bromo-2-methyl-3-heptyne

f.

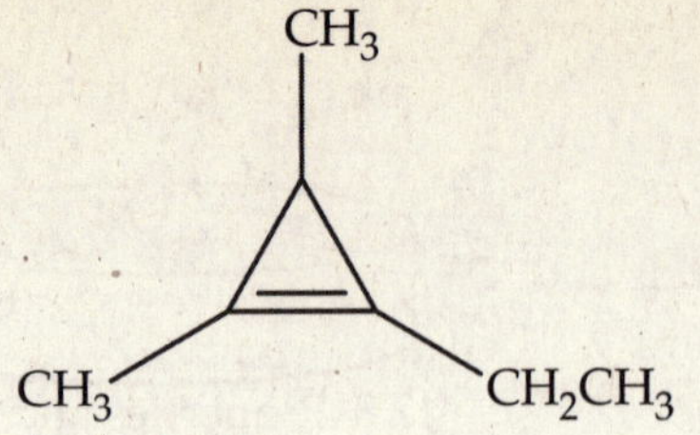

1-ethyl-2,3-dimethylcyclopropene

g.

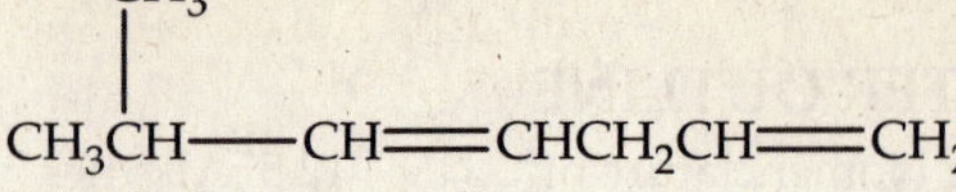

6-methyl-1,4-heptadiene

12.6 a. 3-ethyl-2-hexene

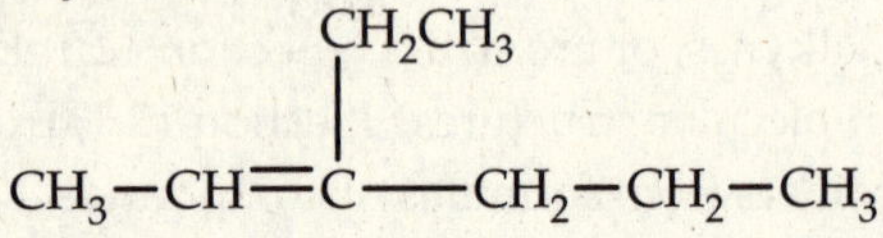

b. 3,4-dimethyl-1-pentene

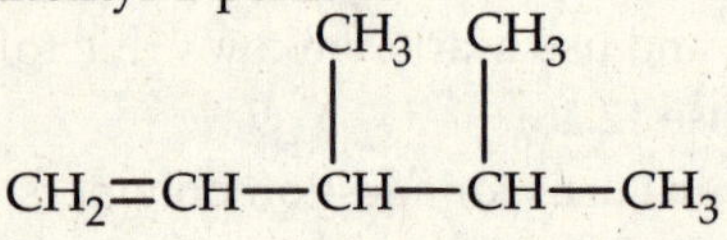

c. 3-methyl-1,3-pentadiene

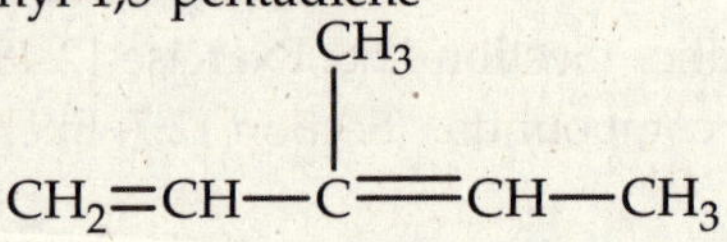

d. 2-isopropyl-4-methylcyclohexene

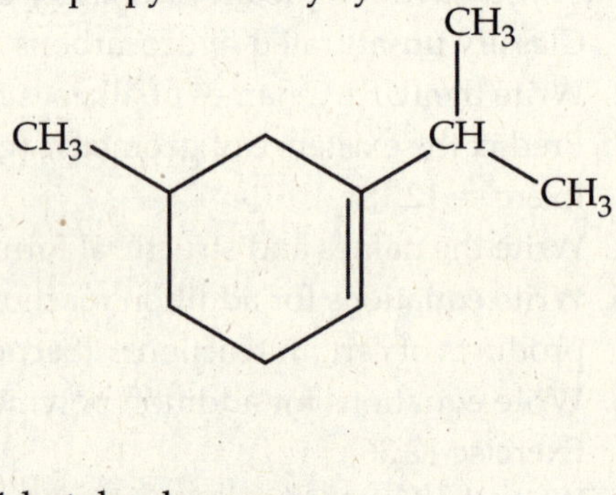

e. 1-butylcyclopropene

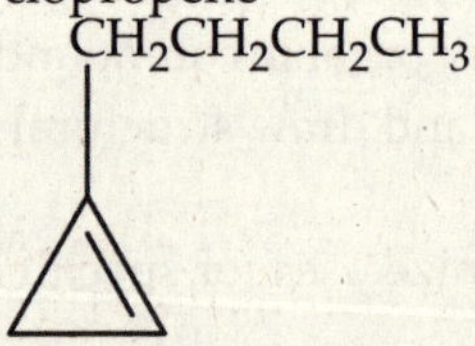

12.8 a. C_5H_8 alkyne

$CH{\equiv}C{-}CH_2{-}CH_2{-}CH_3$

1-pentyne

$CH_3{-}C{\equiv}C{-}CH_2{-}CH_3$

2-pentyne

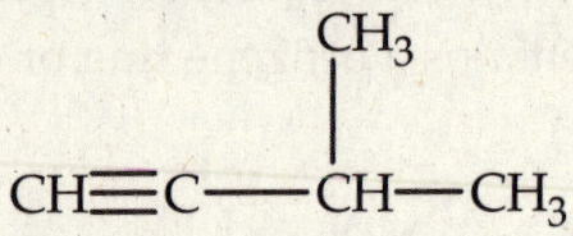

3-methyl-1-butyne

b. C_5H_8 diene

$CH_2{=}C{=}CH{-}CH_2{-}CH_3$

1,2-pentadiene

$CH_3{-}CH{=}C{=}CH{-}CH_3$

2,3-pentadiene

$CH_2{=}CH{-}CH{=}CH{-}CH_3$

1,3-pentadiene

$CH_2{=}CH{-}CH_2{-}CH{=}CH_2$

1,4-pentadiene

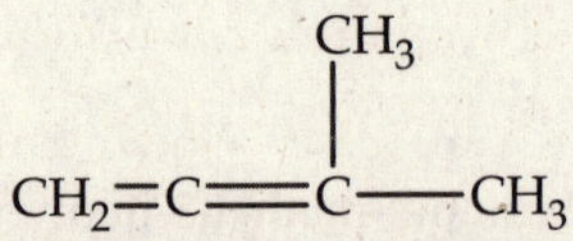

3-methyl-1,2-butadiene

c. C_5H_8 cyclic alkene

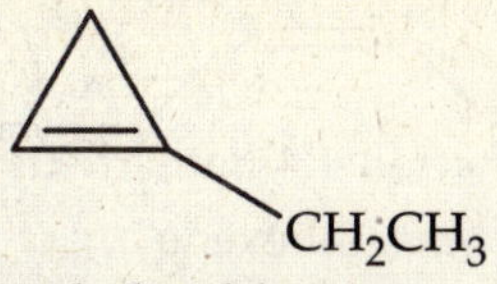
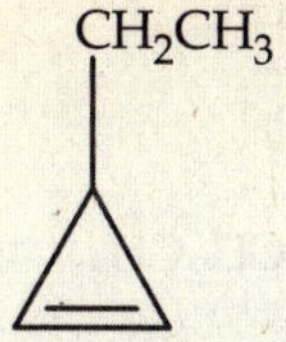
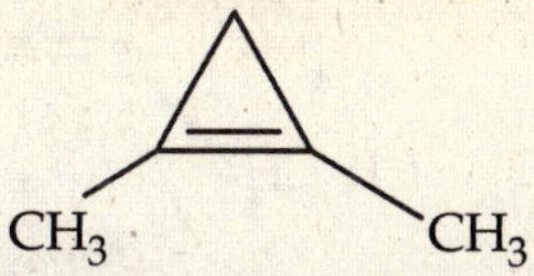

1-ethylcyclopropene 3-ethylcyclopropene 1,2-dimethylcyclopropene

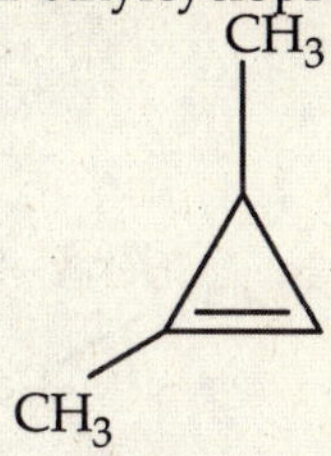
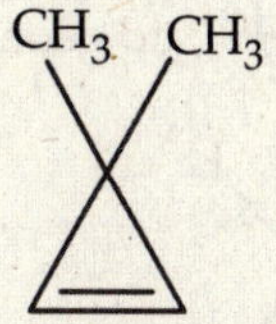
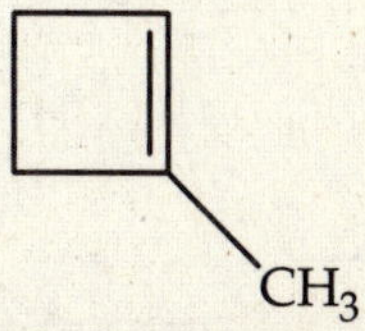

1,3-dimethylcyclopropene 3,3-dimethylcyclopropene 1-methylcyclobutene

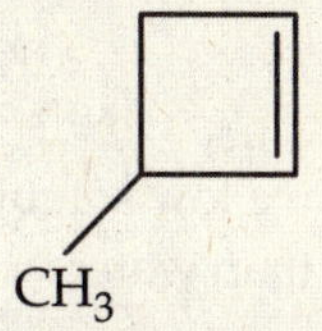
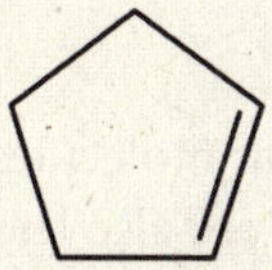

3-methylcyclobutene cyclopentene

12.10

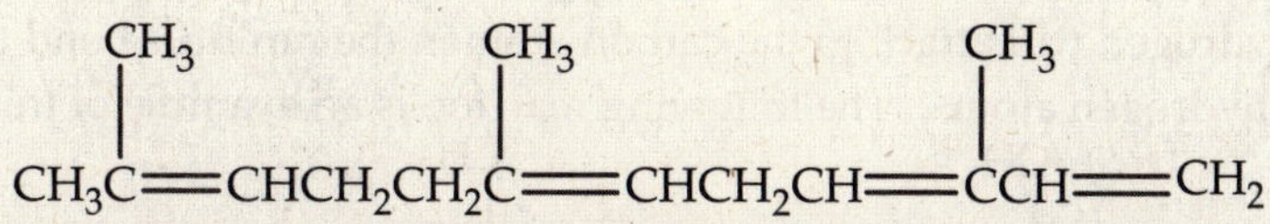

3,7,11-trimethyl-1,3,6,10-dodecatetraene

12.12 a. incorrect = 2-methyl-4-hexene
corrected = 5-methyl-2-hexene

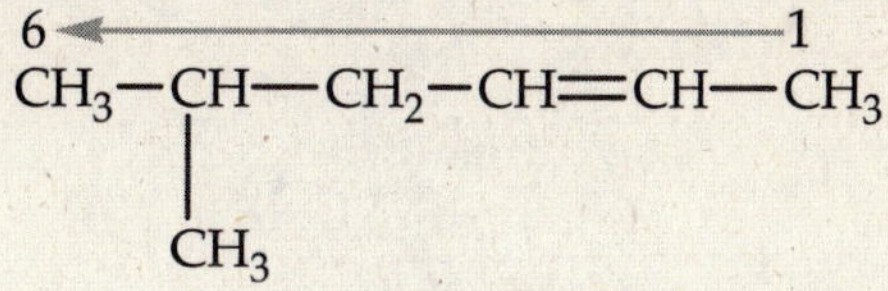

b. incorrect = 3,5-heptadiene
corrected = 2,4-heptadiene

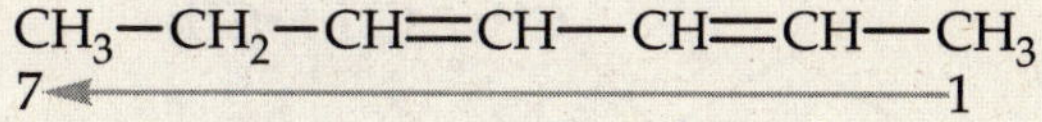

c. incorrect = 4-methylcyclobutene
corrected = 3-methylcyclobutene

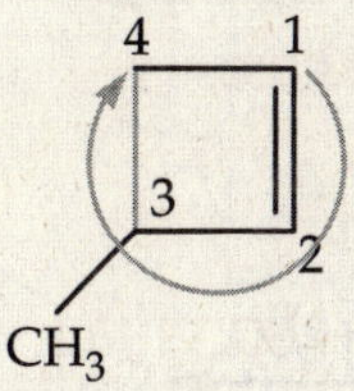

THE GEOMETRY OF ALKENES (SECTION 12.2)

12.14 Pi bonds are formed when unhybridized *p* orbitals overlap sideways. The pi bond is represented by π. The pi bond contains 2 electrons.

12.16 Structural isomers have a different order of linkage of atoms. Geometric isomers have the same order of linkage of atoms; however, the three dimensional structures are unique because of restricted rotation as a result of a ring or a double bond.

☑12.18 a. $CH_3CH_2CH_2CH_2CH{=}CH_2$; no geometric isomers

b.

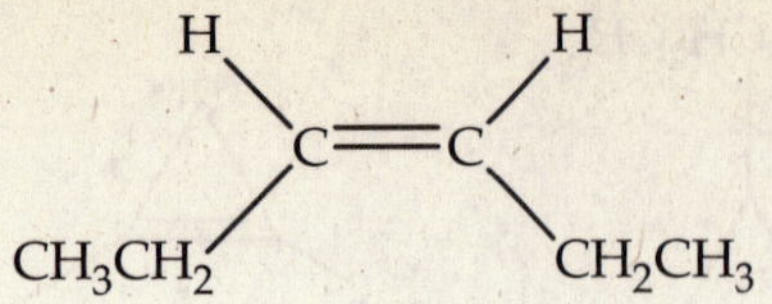

cis-3-hexene

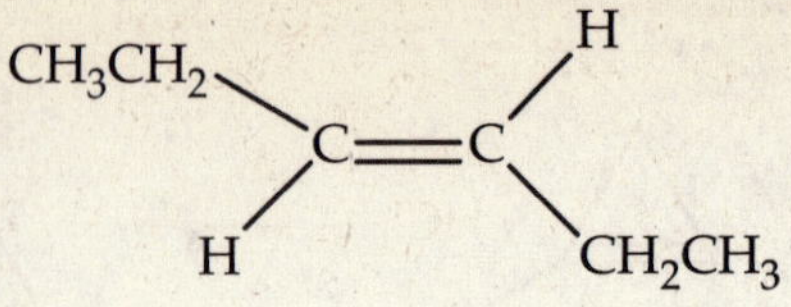

trans-3-hexene

c.

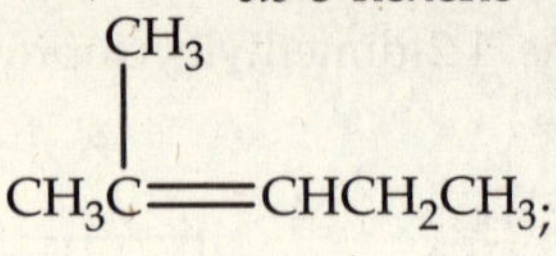

; no geometric isomers

☑12.20 a.

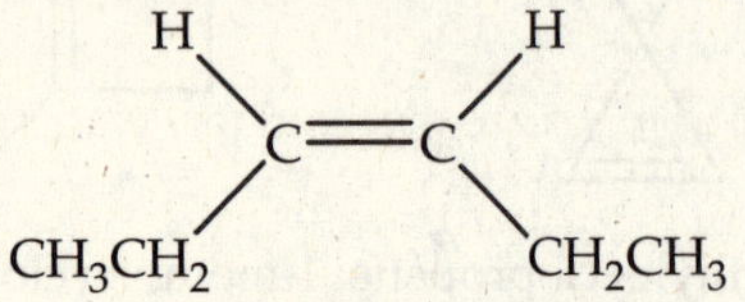

cis-3-hexene

b.

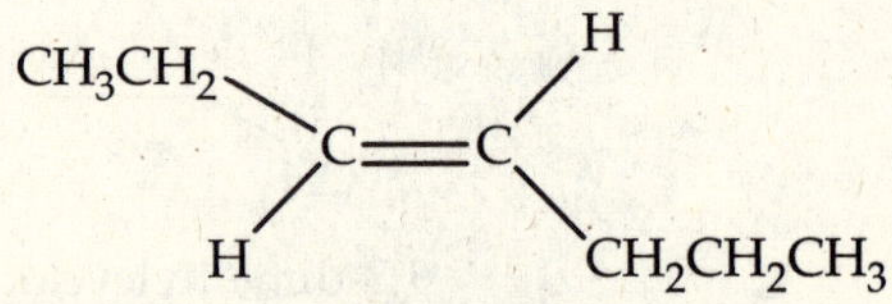

trans-3-heptene

PROPERTIES OF ALKENES (SECTION 12.3)

12.22 Alkenes and alkanes are both nonpolar molecules that have low solubility in water and high solubility in nonpolar solvents. They are also less dense than water.

12.24 Markovnikov's rule states that when a heteroatomic compound containing hydrogen is added to a multiple bond, the hydrogen will attach to the carbon atom in the multiple bond that is directly bonded to more hydrogen atoms. The following reaction is an example of this rule:

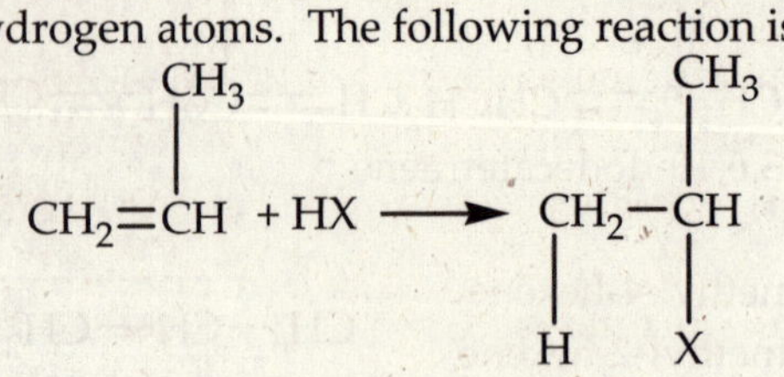

☑12.26 a.

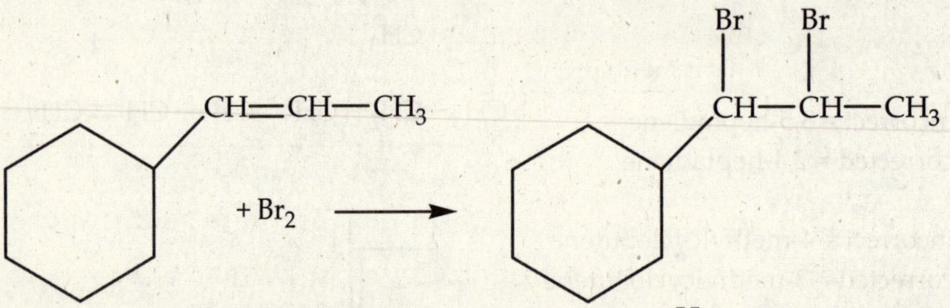

b.

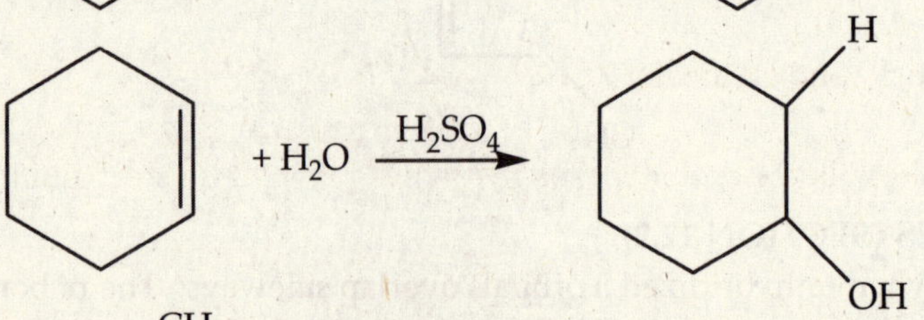

c.

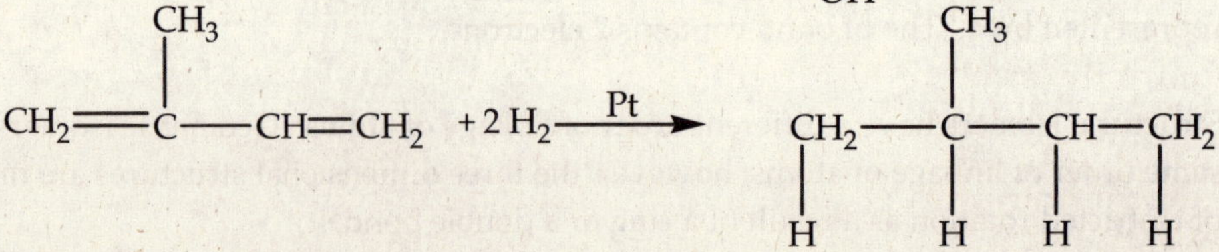

d. 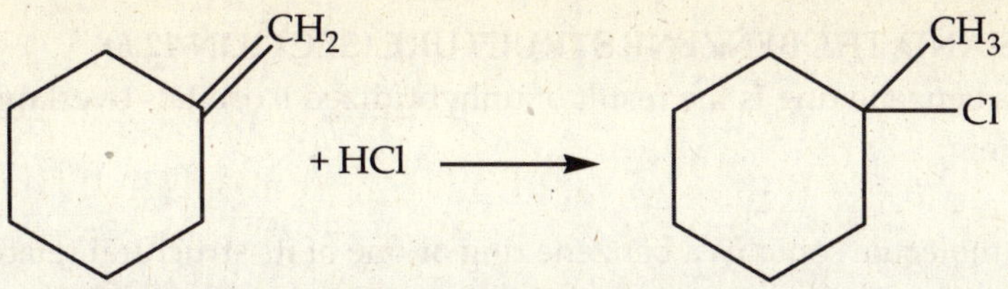

12.28 a. To prepare 3,4-dibromohexane from 3-hexene, I would use Br_2.
b. To prepare hexane from 3-hexene, I would use hydrogen gas and a platinum catalyst.
c. To prepare 3-chlorohexane from 3-hexene, I would use HCl.
d. To prepare 3-hydroxyhexane (3-hexanol) from 3-hexene, I would use water and a sulfuric acid catalyst.

12.30 Adding reddish-brown bromine to cyclohexane will produce a reddish-brown solution. Adding reddish-brown bromine to 2-hexene will produce a clear solution in a very short time, since the alkene undergoes an addition reaction. The differences in the chemical reactivity of these substances will allow them to be differentiated.

ADDITION POLYMERS (SECTION 12.4)

12.32 A monomer is the starting material for a polymer. It can be a small molecule.
A polymer is a large molecule made up of repeating units (often thousands of repeating units).
An addition polymer is a polymer formed by the reaction of monomers that contained multiple bonds to form the repeating units of a polymer.
A copolymer is a polymer formed by the reaction of at least two different types of monomers.

12.34 All of the monomers in Table 12.3 contain one carbon-carbon double bond.

☑12.36 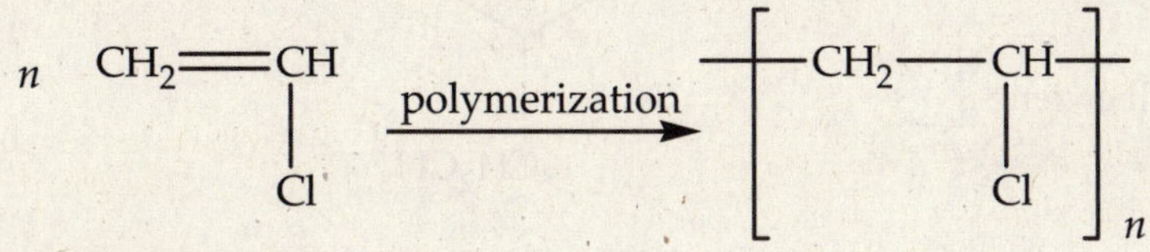

ALKYNES (SECTION 12.5)

12.38 Each carbon atom bonded with a triple bond has two *sp* hybrid orbitals.

12.40 The geometry of a triple bond is linear.

12.42 Acetylene is the simplest alkyne and it is used in torches for welding steel and in making plastics and synthetic fibers.

☑12.44 $CH{\equiv}C{-}CH_2{-}CH_2{-}CH_3$ 1-pentyne

$CH_3{-}C{\equiv}C{-}CH_2{-}CH_3$ 2-pentyne

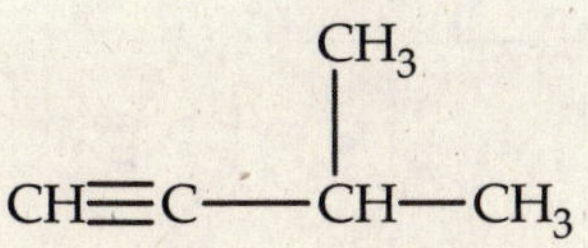

3-methyl-1-butyne

AROMATIC COMPOUNDS AND THE BENZENE STRUCTURE (SECTION 12.6)

12.46 The pi bonding in a benzene ring is the result of unhybridized *p* orbitals overlapping to form a delocalized π system.

☑12.48 Aromatic means a molecule contains a benzene ring or one of its structural relatives. Aliphatic means a molecule does not contain a benzene ring or one of its structural relatives.

12.50 a.

cis-1,2-dibromohexane

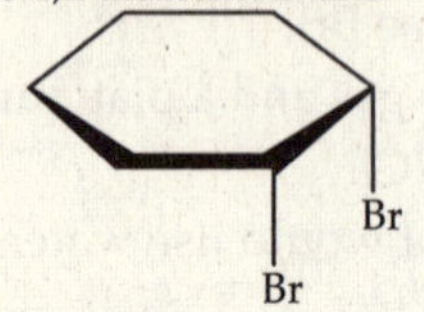

trans-1,2-dibromohexane

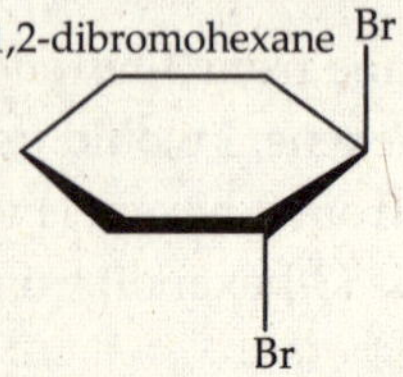

Dibromocyclohexane is not a planar molecule. Each carbon atom has two attached groups (H or another atom) in addition to two positions of attachment in the ring. In order to maintain a tetrahedral geometry around the carbon atoms, the attached groups can be "above" or "below" the ring. Consequently, the molecule can exhibit cis-trans isomerism when it has two attached groups (other than hydrogen) on two different carbon atoms in the ring.

b. 1,2-dibromobenzene is a planar molecule because each carbon atom in the ring is only bonded to 3 other atoms. Consequently, the attached bromine atoms cannot be "above" or "below" the plane of the molecule. This in turn means the molecule cannot exhibit cis-trans isomerism.

THE NOMENCLATURE OF BENZENE DERIVATIVES (SECTION 12.7)

☑12.52 a.

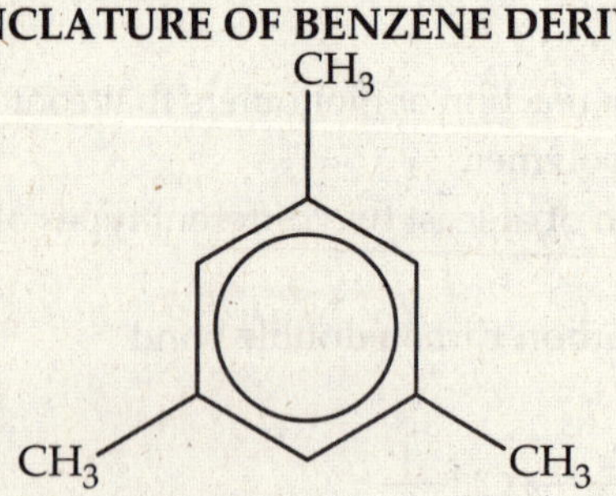

1,3,5-trimethylbenzene

b.

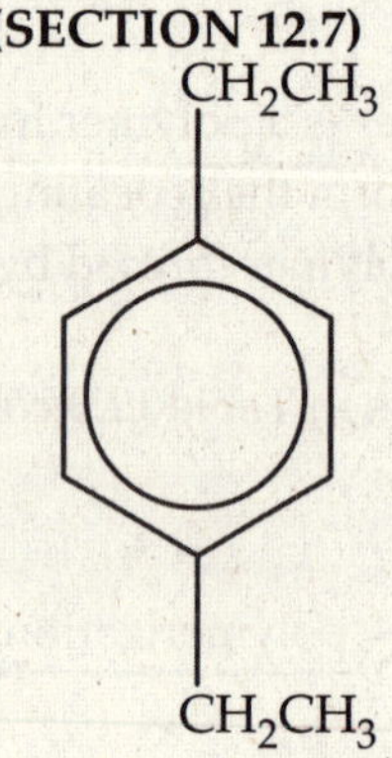

1,4-diethylbenzene
p-diethylbenzene

☑12.54 a.

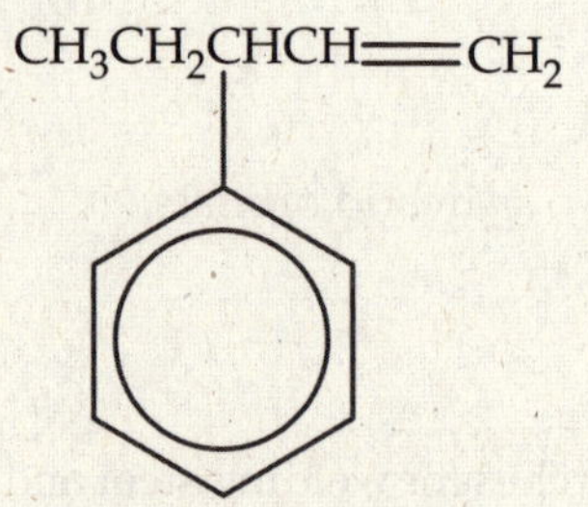

3-phenyl-1-pentene

b.

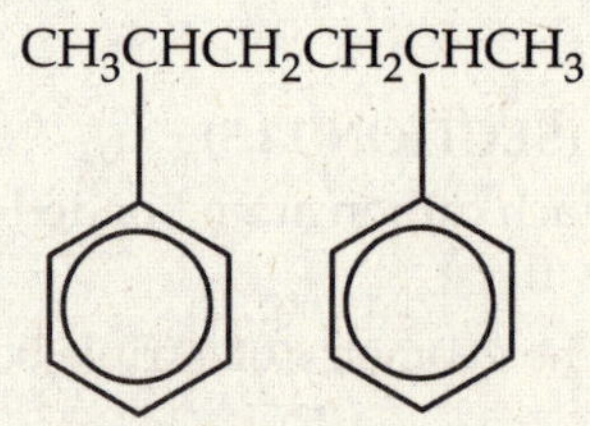

2,5-diphenylhexane

12.56 a.

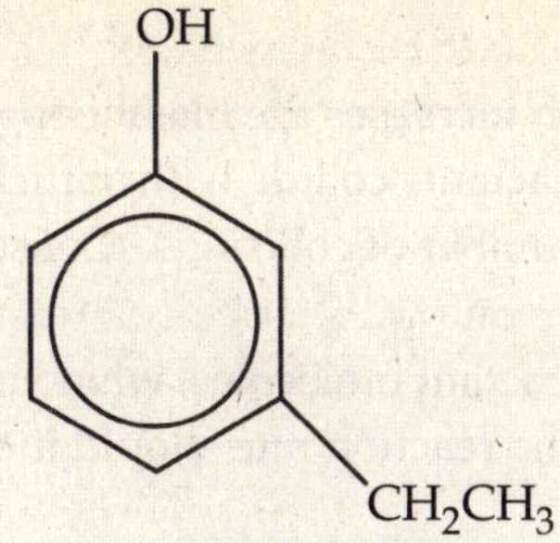

m-ethylphenol

b.

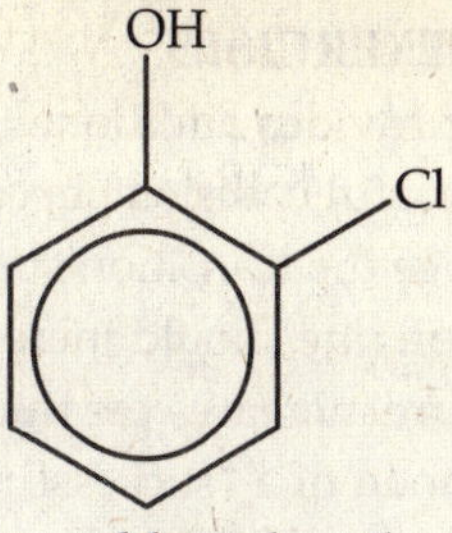

o-chlorophenol

12.58 a.

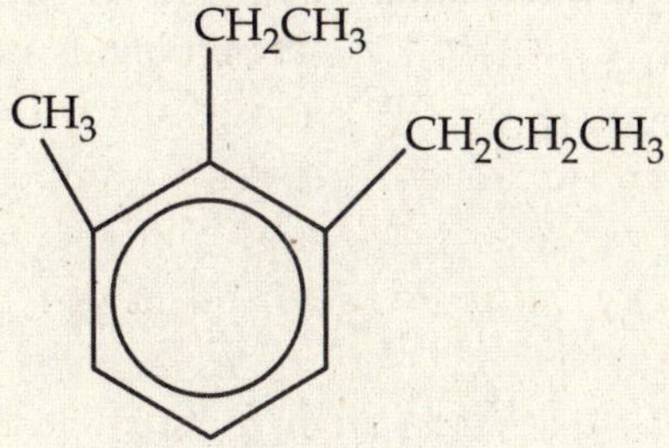

2-ethyl-3-propyltoluene
(1-ethyl-2-methyl-6-propylbenzene)

b.

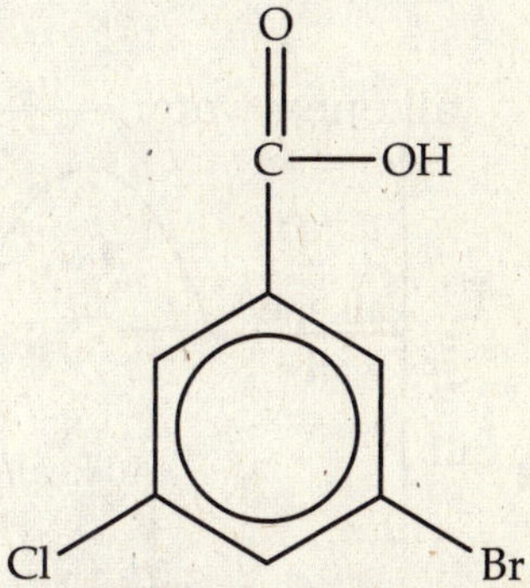

3-bromo-5-chlorobenzoic acid

12.60 a.

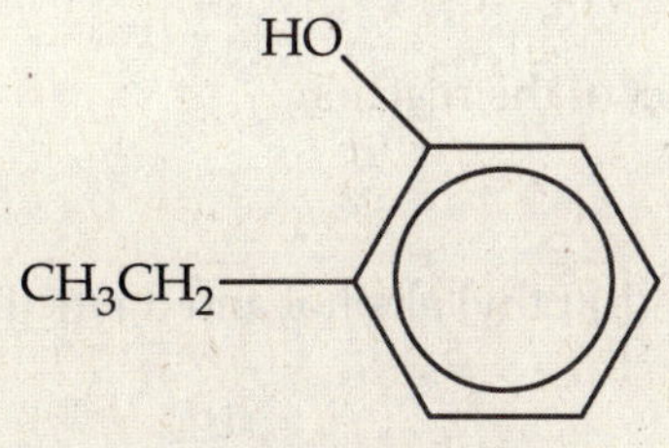

o-ethylphenol

b.

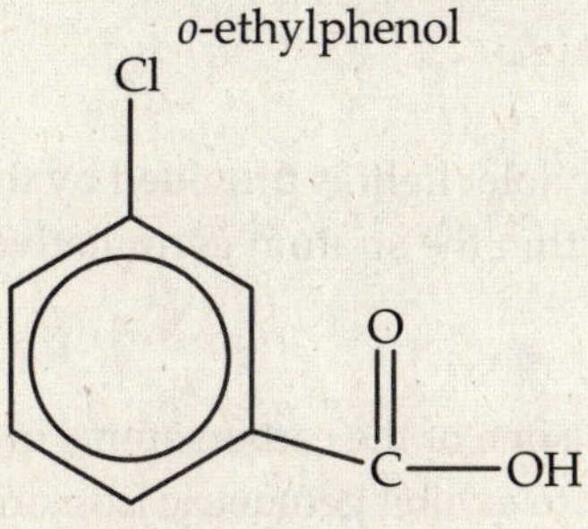

m-chlorobenzoic acid

c.

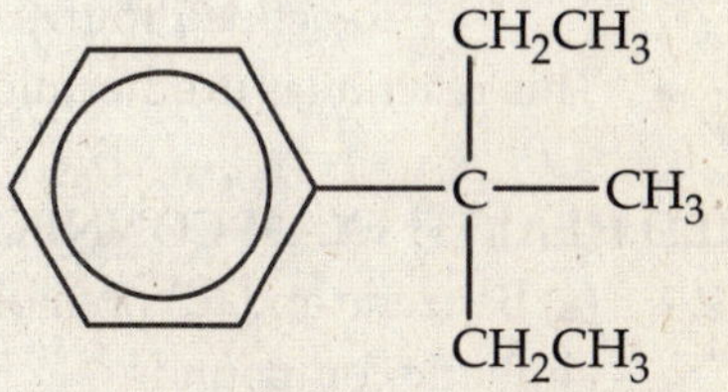

3-methyl-3-phenylpentane

PROPERTIES AND USES OF AROMATIC COMPOUNDS (SECTION 12.8)

12.62 Aromatic hydrocarbons are nonpolar molecules that are insoluble in water and soluble in nonpolar solvents. They are also less dense than water.

12.64 Cyclohexene readily undergoes addition reactions. Benzene resists addition reactions and favors substitution reactions. Both benzene and cyclohexene can undergo combustion.

☑12.66

a.	Used in the production of Formica®	phenol
b.	A starting material for polystyrene	styrene
c.	Used to manufacture drugs	aniline
d.	A starting material for Bakelite®	phenol

ADDITIONAL EXERCISES

12.68 Heat provides additional energy to the reactants which increases the likelihood of a "successful collision" because reactions occur when reactants collide with sufficient energy to produce the transition state. Heat also increases the number of collisions. Consequently, the reaction rate should increase as the reaction is heated.
Pressure increases the number of collisions between reactant molecules, which increases the likelihood of a "successful collision." Consequently, the reaction rate should increase as the pressure is increased.
Catalysts lower the activation energy required for a reaction to occur; therefore, the reaction rate should increase when a catalyst is added to the reaction.

12.70 $\text{alkene} + \text{water} \xrightarrow{H_2SO_4} \text{alcohol} + 10\ \text{kcal/mol}$

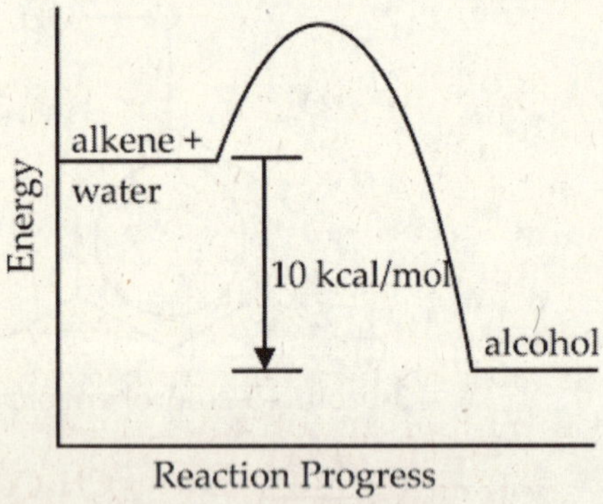

This reaction is exothermic because energy is a product of the reaction.

ALLIED HEALTH EXAM CONNECTION

12.72 (a) Benzene and (d) phenol are aromatic compounds. (b) Ethyl alcohol and (c) methane are aliphatic compounds.

12.74 $CH_2{=}CH{-}CH_2{-}CH_2{-}CH_3$ is an example of (c) an alkene.

12.76 Protection of the skin from the harmful effects of ultraviolet light is provided by the pigment (b) melanin, which is produced by specialized cells within the stratum germinativum.

CHEMISTRY FOR THOUGHT

12.78 Propene does not exhibit geometric isomerism because one of the carbon atoms in the double bond is also bonded to two hydrogen atoms. In order to exhibit geometric isomerism, both carbon atoms in the double bond must also be bonded to two unique groups.

12.80 $$5.0\times10^4\ \text{u}\left(\frac{1\ \text{ethylene monomer}}{28.0\ \text{u}}\right) = 1.8\times10^3\ \text{ethylene monomers}$$

12.82 Addition polymers form a large extended structure. Polyvinyl acetate prevents the formation of sharp fragments in automobile safety glass because it holds the glass together on a flexible surface after the glass is broken. The copolymer of styrene and 1,3-butadiene allows a large flexible bubble to be formed by the girl chewing gum.

12.84 The polymer that produces hydrogen cyanide, HCN, is polyacrylonitrile.
The polymer that produces hydrogen chloride, HCl, is poly (vinyl chloride).

12.86

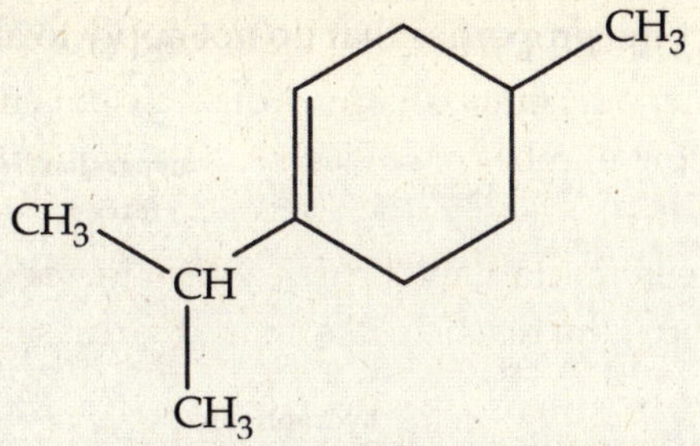

1-isopropyl-4-methylcyclohexene
(menthene)

ADDITIONAL ACTIVITIES

Section 12.1 Review:

(1) What is the relationship between the general formulas of alkenes and alkanes?
(2) What is the relationship between the general formulas alkenes and cycloalkanes?
(3) Is *–ene* a fitting ending for compounds that contain a carbon-carbon double bond? Explain.
(4) What is the significance of the ending *–diene*?
(5) In terms of numbering, what takes priority in the alkene longest chain? Explain.

Section 12.2 Review: Copy this page onto another sheet of paper or a transparency, then cut out the shapes. The circle is included for additional stability, but does not have any real physical meaning. Insert a pipe cleaner into the center of each circle. Center the structure on the pipe cleaner.

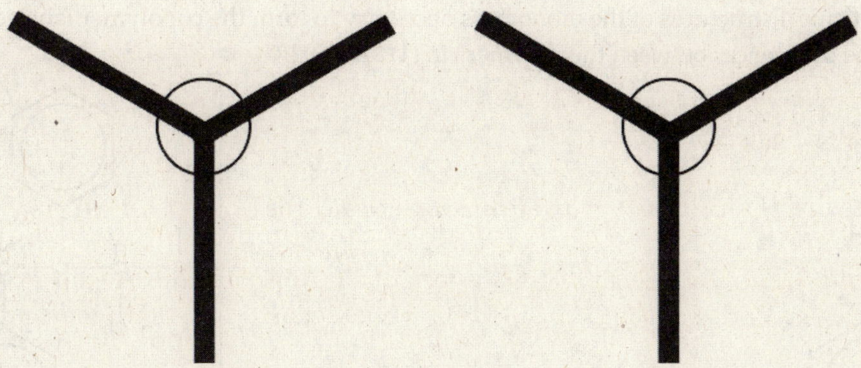

(1) What type of orbitals are represented by the straight black lines in the above structures?
(2) What type of orbital is represented by the pipe cleaner?
(3) What is the angle between the straight black lines in the above structures?
(4) What is the angle between the straight black lines and the pipe cleaner?
(5) Bring the two structures close enough for the sp^2 hybrid orbitals on the two carbon atoms to overlap. What type of bond does this form? How many bonds does this form?
(6) Bend the tops and bottoms of the pipe cleaner so they can also overlap. What type of bond does this form? How many bonds does this form?
(7) How does a pi bond differ from a sigma bond?

Section 12.3 Review:

Box the following properties that apply to alkenes. Correct the properties that do not apply to alkenes as they are currently written.

unpleasant odor

high boiling point

low melting point

polar

dispersion forces

soluble in water

less dense than water

hydrophobic

(1) List the four types of addition reactions alkenes can undergo.
(2) Divide these types of reactions into homoatomic and heteroatomic reactants based on the material added to the alkene.
(3) Which group from (2) requires knowledge and use of Markovnikov's rule?
(4) Divide the types of addition reactions from (1) into two groups: those that require a catalyst and those that do not.
(5) Are the groups the same in (2) and (4)?
(6) Divide the type of addition reactions from (1) into groups by product formed. For the types of reactions grouped together, can any distinction be made between their products?
(7) Which of the groupings (2), (4), or (6) will be most useful for remembering these reactions?

Section 12.4 Review:

(1) Write a polymerization reaction using the monomer labeled (1) below.
(2) Draw the chemical structures of the monomers necessary to form the copolymer labeled (2).
(3) What are the differences between the polymers in (1) and (2)?

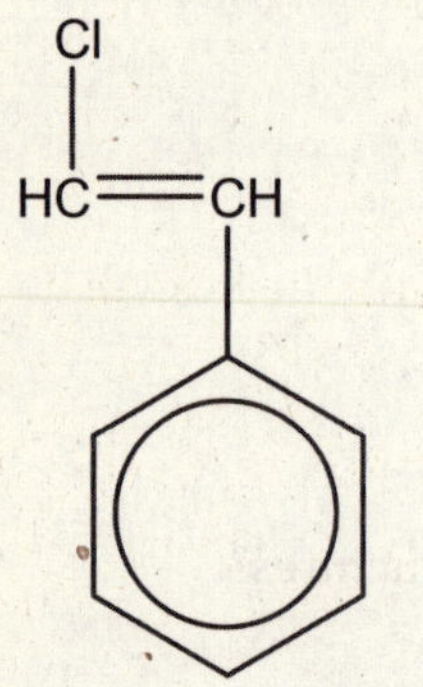

(1)

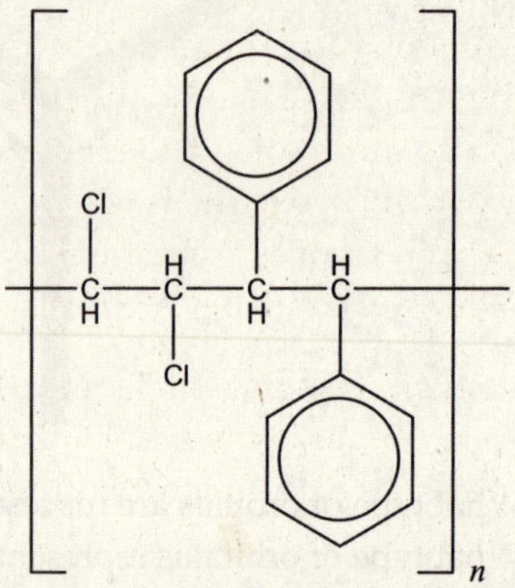

(2)

Section 12.5 Review:

(1) Based on the activities in the Section 11.3 Review and the Section 12.2 Review, describe how to model a carbon atom that has undergone *sp* hybridization.
(2) What is the angle between the *sp* hybrid orbitals?
(3) What is the angle between the *sp* hybrid orbitals and the unhybridized *p* orbitals?
(4) What is the angle between the unhybridized *p* orbitals?
(5) How many *sp* hybrid orbitals are formed?
(6) How many *p* orbitals are unhybridized?
(7) What type of bond is formed when the *sp* hybrid orbitals on two carbon atoms overlap?
(8) How many pi bonds will these carbon atoms form?
(9) Explain why molecules that contain a carbon-carbon triple bond have a –*yne* ending.

(10) Box the following properties that apply to alkynes. Correct the properties that do not apply to alkynes as they are currently written.

combustible	unpleasant odor	high boiling point	low melting point	polar	dispersion forces
do not undergo addition reactions	soluble in water		less dense than water		hydrophobic

Section 12.6 Review:

(1) Which class of hydrocarbons contains carbon atoms with the same type of hybridized orbitals as are found in benzene?
(2) What is the major difference in the bonding within the hydrocarbon class from (1) and benzene?
(3) What differences does (2) present in terms of structure?

Section 12.7 Review:

Study Tip: Make flashcards of the benzene derivatives known by common names that are also IUPAC accepted. Practice with the flashcards until you can correctly name the structures of the benzene derivatives as well as draw the structure of a given benzene derivative without referring to your notes.

(1) What are the numbers associated with the attached groups in a benzene structure with the prefix *o*-?
(2) What are the numbers associated with the attached groups in a benzene structure with the prefix *m*-?
(3) What are the numbers associated with the attached groups in a benzene structure with the prefix *p*-?

Section 12.8 Review:

(1) What physical properties do all hydrocarbons share?
(2) Which class of hydrocarbons has the most similar reactivity to the aromatic compounds?
(3) Could the difference identified in (2) of the Section 12.6 Review account for the difference in reactivity between alkenes and aromatic compounds? Explain.
(4) What is the major structural difference between benzopyrene and riboflavin?
(5) What is the difference between the interaction of benzopyrene and riboflavin with the human body?

Tying It All Together with a Laboratory Application:

A laboratory student has five unknown liquids to identify. The unknowns are labeled A, B, C, D, and E. The identities of the unknowns (not in order) are benzene, 1-hexene, 1-hexyne, cyclohexene, and cyclohexane.

The student also has the following information:

	benzene	1-hexene	1-hexyne	cyclohexene	cyclohexane
Color	colorless	colorless	colorless	colorless	colorless
Odor	paint thinner like	?	?	sweetish	mild sweet
Boiling Point (°C)	80.1	63	71	83	80.7
Freezing Point (°C)	5.5	-139.8	-131.9	-104	6.6
Density (g/mL)	0.8786	0.673	0.716	0.811	0.779

(1) Draw the structural formulas for each of the unknowns.
(2) Write the molecular formulas for each of the unknowns.
(3) Are any of the compounds structural isomers? If so, which ones?
(4) Are any of the compounds geometric isomers? If so, which ones?

The student wants to obtain 5.00 mL of each unknown and find the mass of each sample in order to determine the (5) _________ of the unknowns. The unknowns in order of increasing mass (for 5.00 mL samples) are:

(6) _________ < (7) _________ < (8) _________ < (9) _________ < (10) _________

(11) Comment on this order. The student takes the test tubes over to the balance and discovers that the balance is out of order. Another procedure is needed to distinguish the compounds.

All of the unknowns are (12) _________ because they contain only carbon and hydrogen atoms. They are also (13) _________ molecules that (14) _________ dissolve in water. All of the compounds are (15) _________ dense than water. The student decides that adding a small quantity of each unknown to separate beakers of water (16) _________ allow the unknowns to be identified by density.

(17) Write the reactions for the complete combustion of each of the unique molecular formulas. Aromatic compounds generally burn with more soot than (18) _________ compounds.
(19) Write the reactions for the addition of bromine to each of the compounds.

The student decides to add a bromine solution to 0.5 mL samples of each of the unknowns. The student determined the number of drops required to obtain a reddish color for each unknown:

unknown	A	B	C	D	E
drops	43	77	2	36	1

Based on this data, unknowns C and E (20) _________ react with bromine. The student considers distinguishing these compounds by smell, but unfortunately, this student has a cold. The student decides to combust a few drops of each of these compounds in watch glasses in the fume hood. The student observes that unknown C produces much more smoke than unknown E as it burns. A black residue is evident on both watch glasses; however, it is much thicker on the watch glass for unknown C. The identity of unknown C is (21) _________. The identity of unknown E is (22) _________.

Based on the remaining data obtained from the bromine reaction, the student identifies unknown B as (23) _________ because (24) _________. To double check this identification, the student decides to check the boiling points of unknowns A, B, and D. The student sets up a beaker of water on a hot plate in the fume hood and uses clamps and a ring stand to secure the three test tubes of unknowns in the water bath. The student also adds a secured thermometer to each test tube. The student does not use a Bunsen burner to check the boiling points of these compounds because they are (25) _________.

The boiling temperatures the student obtains are:

unknown	A	B	D
boiling point	~80°C	~70°C	~60°C

The identity of unknown A is (26) _________, unknown B is (27) _________, and unknown D is (28) _________.

SOLUTIONS FOR THE ADDITIONAL ACTIVITIES

Section 12.1 Review:

(1) alkene = C_nH_{2n}; alkane = C_nH_{2n+2}; If an alkene and an alkane contain the same number of carbon atoms, the alkene will contain 2 fewer hydrogen atoms than the alkane because it contains a double bond.

(2) alkene = C_nH_{2n}; cycloalkane = C_nH_{2n}; If an alkene and a cycloalkane contain the same number of carbon atoms, they will also contain the same number of hydrogen atoms. Forming a double bond or a ring requires two hydrogen atoms to be removed from the corresponding alkane structure.
(3) Compounds that contain a carbon-carbon double bond are alkenes. The ending of the group is also the ending for the IUPAC names of these compounds.
(4) The ending *–diene* means that a compound contains two double bonds. It is an alkene twice over.
(5) The double bond takes priority in the alkene longest chain. The carbon atoms in the longest chain will be numbered so that the double bond is associated with the carbon atom with the lowest possible number.

Section 12.2 Review:
(1) sp^2 hybrid orbitals; (2) unhybridized p orbital; (3) 120°; (4) 90°; (5) sigma bond, one bond;
(6) pi bond, one bond; (7) A sigma bond has one area of overlap along the internuclear axis. A pi bond has two areas of overlap, one above and one below the internuclear axis.

Section 12.3 Review:

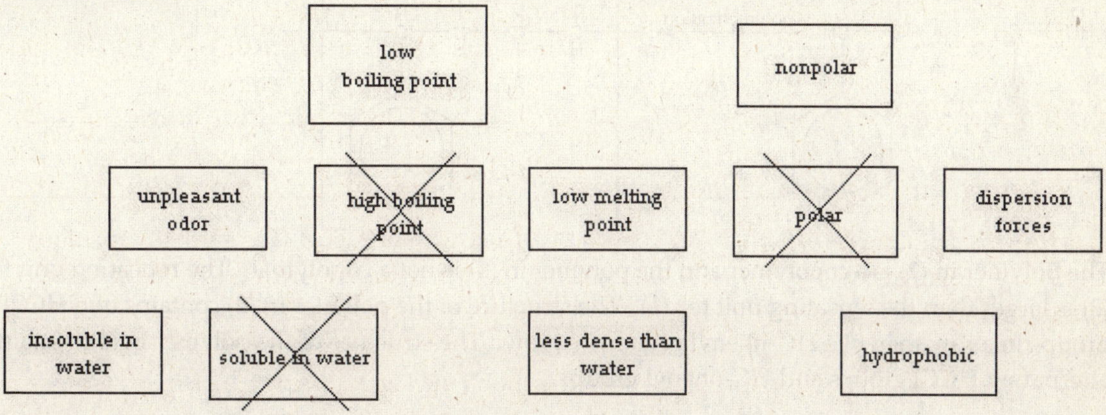

(1) halogenation, hydrogenation, addition of hydrogen halides, hydration
(2) homoatomic reactants: halogenation, hydrogenation
heteroatomic reactants: addition of hydrogen halides, hydration
(3) Heteroatomic reactants require knowledge and use of Markovnikov's rule.
(4) Require catalyst: hydrogenation, hydration
Do not require catalyst: halogenation, addition of hydrogen halides
(5) No, the groups are different in (2) and (4).
(6) haloalkane (alkyl halide): halogenation, addition of hydrogen halides
alkane: hydrogenation
alcohol: hydration
While both halogenation and the addition of hydrogen halides produce haloalkanes, halogenation produces disubstituted haloalkanes (contain 2 halogen atoms), while the addition of hydrogen halides produces monosubstituted haloalkanes (contain 1 halogen atom).
(7) The "best" grouping is a matter of personal study style. Markovnikov's rule is an important part of this chapter, so the grouping from (2) is useful. The need for a catalyst is important both in recording the details of a reaction and in performing reactions in the laboratory, so the grouping from (4) is useful. The ability to predict products for a reaction is also important, so the grouping from (6) is also useful.

Section 12.4 Review:

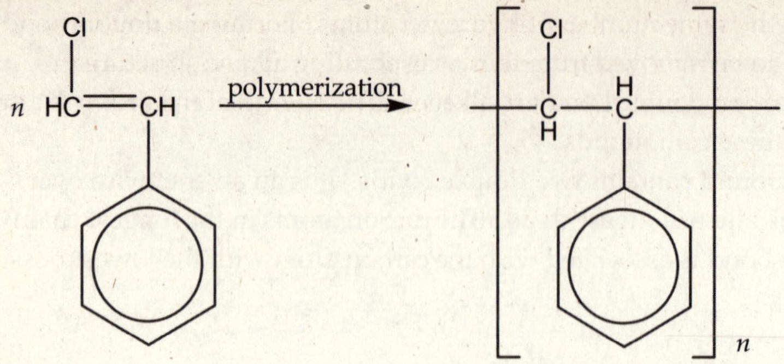

(1)

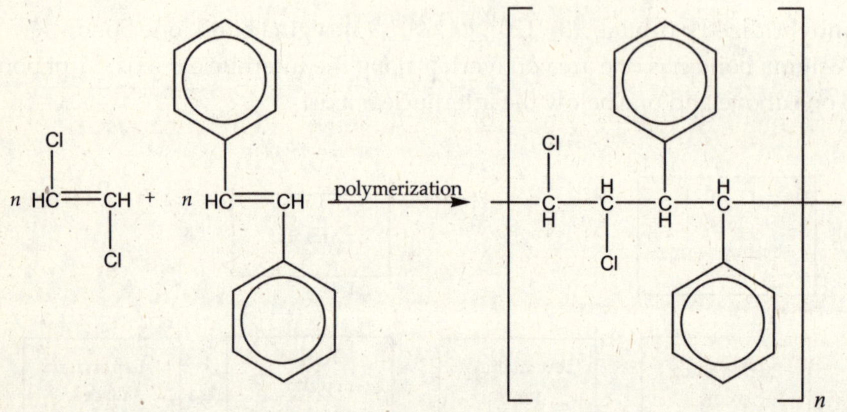

(2)

(3) The polymer in (2) is a copolymer and the polymer in (1) is not a copolymer. The repeating unit for (2) is larger than the repeating unit for (1). The structure of the polymer in (2) contains two HCCl groups in a row then two HC-phenyl groups in a row. The structure of the polymer in (1) contains alternating HCCl groups and HC-phenyl groups.

Section 12.5 Review:

(1) The *sp* hybrid orbitals could be represented as two different colored straws joined to make a straight line. The two unhybridized *p* orbitals could be represented using two different colored pipe cleaners at 90° angles to each other and the straws.

(2) 180°; (3) 90°; (4) 90°; (5) 2; (6) 2; (7) sigma; (8) 2

(9) Molecules that contain a carbon-carbon triple bond key functional group are alkynes. The molecules in this class are given *-yne* endings in the IUPAC nomenclature system to indicate they are alkynes.

(10)

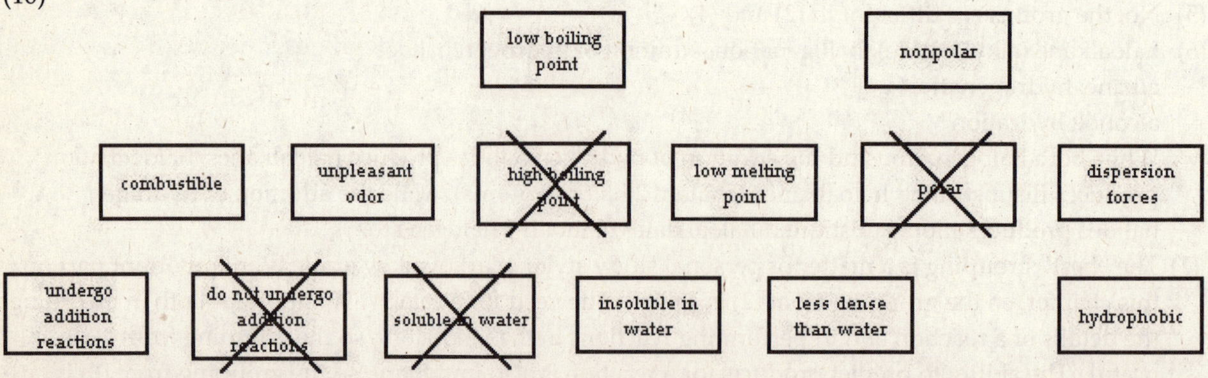

Section 12.6 Review:

(1) Alkenes and aromatic compounds contain sp^2 hybridized carbon atoms.

(2) The major difference in bonding between alkenes and benzene is that alkenes contain 1 localized pi bond, while benzene forms 2 delocalized pi lobes.

(3) The carbon atoms in the double bond of an alkene have two attached groups, one above the double bond, one below the double bond. Consequently, some alkene molecules have geometric isomers. The carbon atoms in a benzene ring have one attached group other than the carbon atoms in the benzene ring. Consequently, aromatic molecules cannot have geometric isomers.

Section 12.7 Review:

(1) 1 and 2; (2) 1 and 3; (3) 1 and 4

Section 12.8 Review:

(1) They are nonpolar molecules that are less dense than water and do not dissolve in water. They dissolve in nonpolar solvents.
(2) Aromatic compounds react similarly to alkanes. They do not undergo addition reactions like alkenes and alkynes.
(3) Yes, the pi cloud in the aromatic compounds provides stability that a lone carbon-carbon double bond does not have. Consequently, aromatic compounds do not undergo addition reactions, but alkenes do.
(4) Benzopyrene contains only carbon and hydrogen. It is completely nonpolar. Riboflavin contains nitrogen and oxygen in addition to carbon and hydrogen. It is polar.
(5) Benzopyrene is a carcinogen and riboflavin is a vitamin. Riboflavin is important to maintaining the health of the human body, while benzopyrene can cause cancer within the human body.

Tying It All Together with a Laboratory Application:

benzene | 1-hexene | 1-hexyne | cyclohexene | cyclohexane

(1)

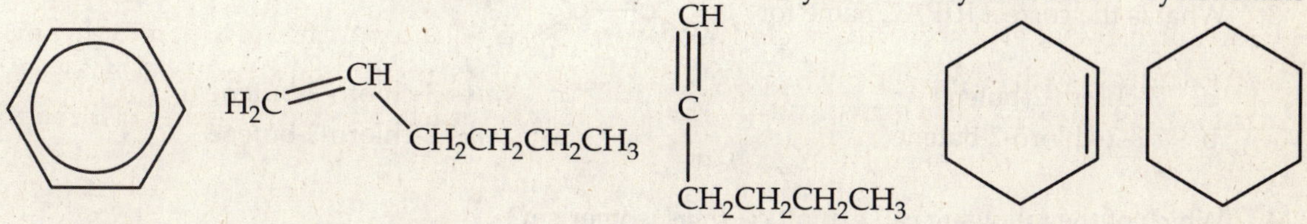

(2) C_6H_6 | C_6H_{12} | C_6H_{10} | C_6H_{10} | C_6H_{12}

(3) structural isomers = 1-hexene and cyclohexane (C_6H_{12}); 1-hexyne and cyclohexene (C_6H_{10})
(4) no geometric isomers (5) density (6) 1-hexene (7) 1-hexyne (8) cyclohexane (9) cyclohexene (10) benzene
(11) The compound with the lowest density only has a double bond, the next compound has a triple bond, the three highest density compounds are all cyclic. The lowest of these three does not contain any multiple bonds, the next compound contains one double bond, and the highest density compound is benzene which is aromatic. Based on this group of compounds, density increases with degree of unsaturation and with cyclic compounds.
(12) hydrocarbons (13) nonpolar (14) will not (15) less (16) will not
(17) $2\ C_6H_6 + 15\ O_2 \rightarrow 12\ CO_2 + 6\ H_2O$; $C_6H_{12} + 9\ O_2 \rightarrow 6\ CO_2 + 6\ H_2O$; $2\ C_6H_{10} + 17\ O_2 \rightarrow 12\ CO_2 + 10\ H_2O$
(18) aliphatic
(19)

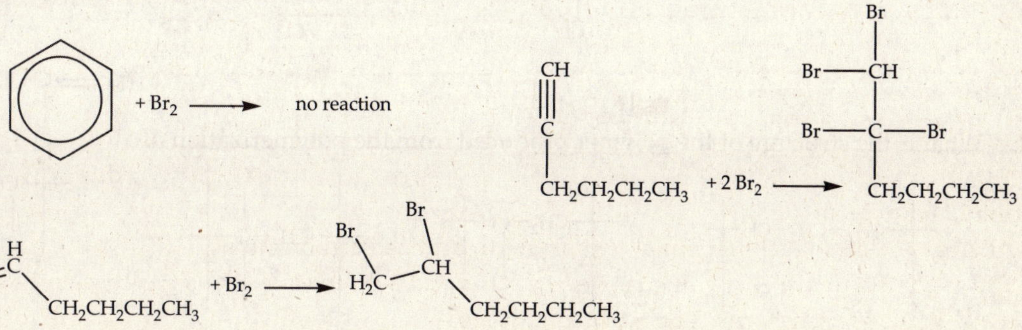

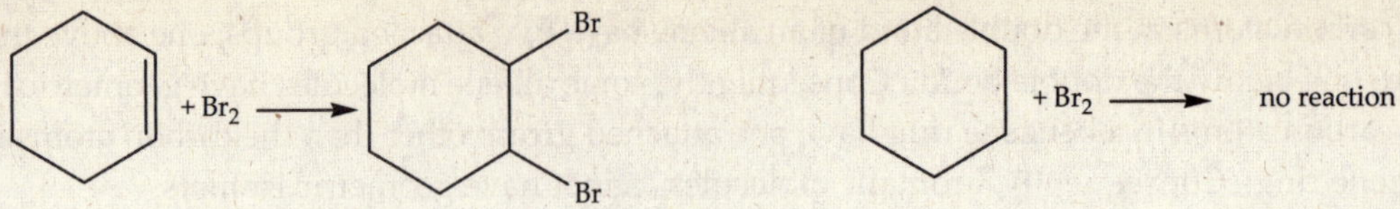

(20) did not
(21) benzene
(22) cyclohexane
(23) 1-hexyne
(24) it reacted with the most Br_2
(25) flammable
(26) cyclohexene
(27) 1-hexyne
(28) 1-hexene

SELF-TEST QUESTIONS

Multiple Choice

1. What is the correct IUPAC name for 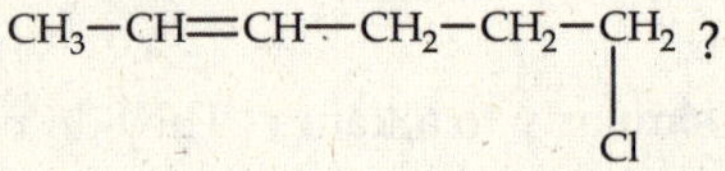 ?

   a. 1-chloro-2-hexene
   b. 6-chloro-2-hexene
   c. 1-chloro-4-hexene
   d. 1-chlorohexene

2. What is the correct IUPAC name for 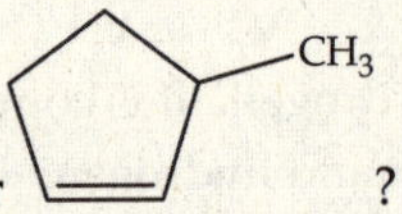 ?

   a. 1-methylcyclopentene
   b. 2-methylcyclopentene
   c. 3-methylcyclopentene
   d. 1-methyl-2-cyclopentene

3. What is the correct IUPAC name for 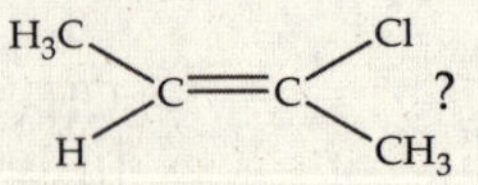 ?

   a. 3-chloro-2-butene
   b. *cis*-2-chloro-2-butene
   c. *trans*-3-chloro-2-butene
   d. *trans*-2-chloro-2-butene

4. Which of the following can exhibit *cis-trans* isomerism?

   a. 2-methyl-1-butene
   b. 2,3-dimethyl-2-butene
   c. 2-methyl-2-butene
   d. 2,3-dichloro-2-butene

5. Markovnikov's rule is useful in predicting the product of a reaction between an alkene and ____.

   a. H_2
   b. Br_2
   c. HBr
   d. O_2

6. What is the product of the following reaction?

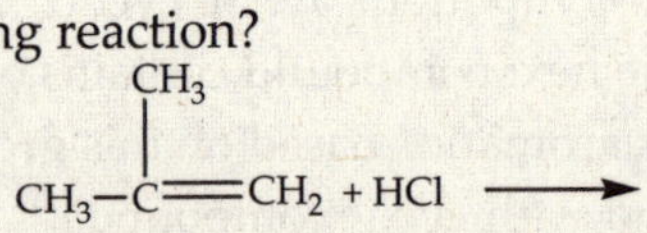

   a. 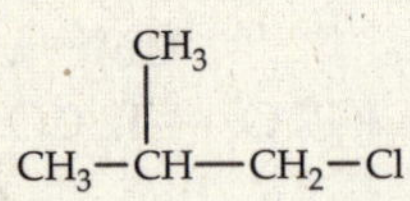
   b. 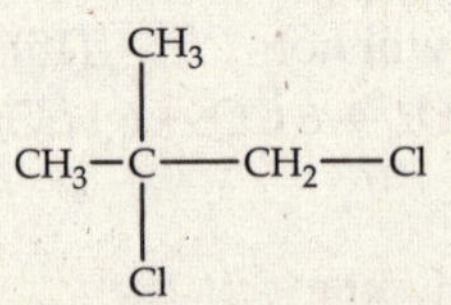
   c. 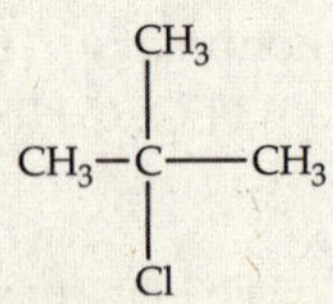
   d. 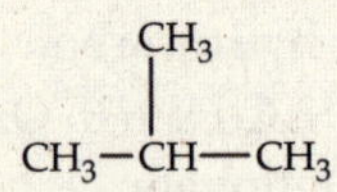

7. What is the structure of the polymer produced from the polymerization of 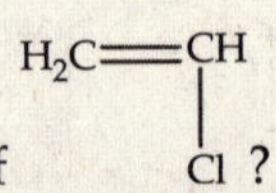 ?

   a. 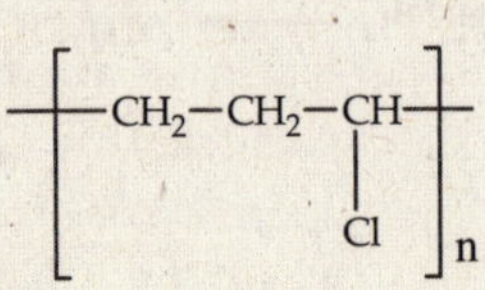
   b. 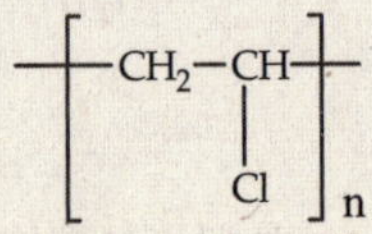
   c. 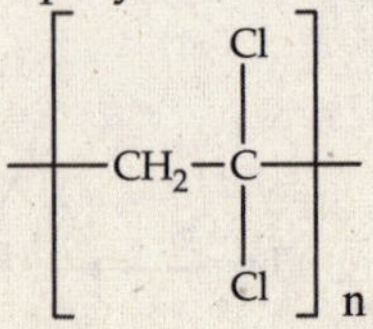
   d. 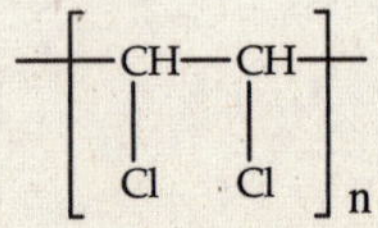

8. What is the correct name for 1,2,6-trichlorobenzene?
   a. 1,2,5-trichlorobenzene
   b. 1,3,5-trichlorobenzene
   c. 1,2,3-trichlorobenzene
   d. 1,2,4-trichlorobenzene

9. What is the correct IUPAC name for 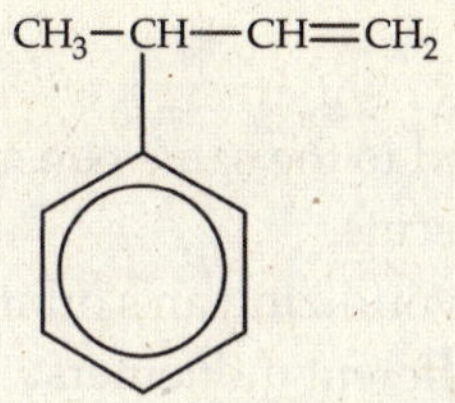 ?
   a. 2-phenyl-1-butene
   b. 3-phenyl-3-butene
   c. 2-phenyl-3-butene
   d. 3-phenyl-1-butene

10. What is produced when H_2O is added to 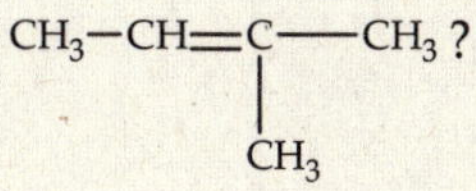 ?
   a. 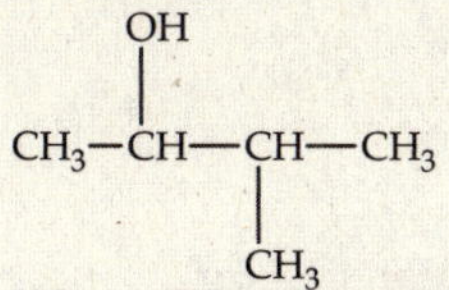
   b. 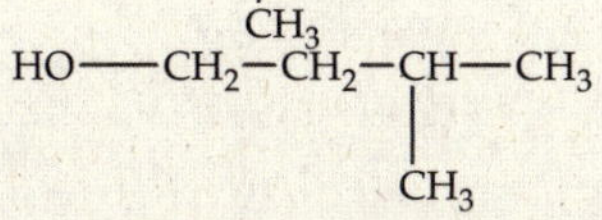
   c. 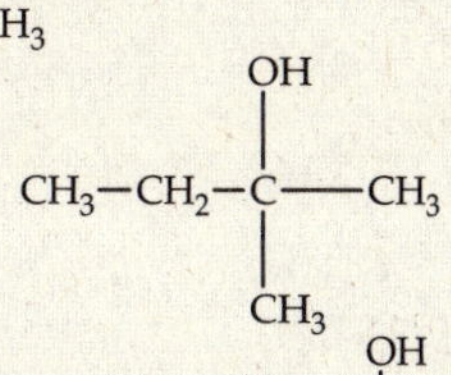
   d. 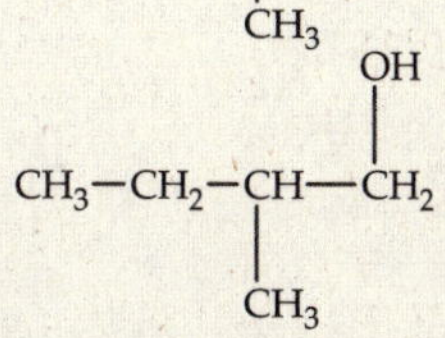

11. What reactant and catalyst are necessary to hydrogenate an alkene?
   a. H^+ and Pt
   b. HCl and Pt
   c. H_2 and Pt
   d. H_2O and Pt

12. What is the IUPAC name for 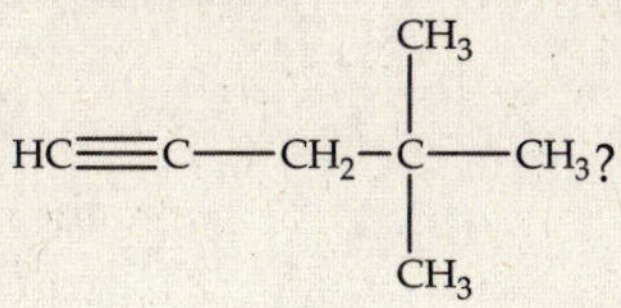?
   a. 2,2-dimethyl-2-pentyne
   b. 4,4-dimethyl-4-pentyne
   c. 2,2-dimethyl-4-pentyne
   d. 4,4-dimethyl-1-pentyne

Matching

For each aromatic compound on the left, select the correct use or derived product from the responses on the right.

13. phenol
14. toluene
15. aniline
16. riboflavin

a. a solvent
b. a vitamin
c. formica
d. dyes

For each description on the left, select a correct polymer from the responses on the right.

17. used for insulation
18. a copolymer
19. use in airplane windows

a. saran wrap
b. plexiglass
c. polystyrene
d. PVC

True-False

20. Carbon-carbon double bonds do not rotate as freely as carbon-carbon single bonds.
21. The compound $BrCH{=}CHCH_3$ can exhibit *cis-trans* isomerism.
22. Alkenes are polar substances.
23. All aromatic compounds are cyclic.
24. Two hydrogen atoms may be bonded to the same benzene carbon atom.
25. Benzene is a completely planar molecule.
26. A *sp* hybrid orbital is obtained by hybridizing an *s* orbital and two *p* orbitals.
27. A copolymer is formed from two different monomers.
28. All the atoms in ethene lie in the same plane.
29. Acetylene is the common name of the simplest alkene.
30. A triple bond contains 1 sigma bond and 2 pi bonds.

Chapter 13: Alcohols, Phenols, and Ethers

CHAPTER OUTLINE

13.1 The Nomenclature of Alcohols and Phenols
13.2 Classification of Alcohols
13.3 Physical Properties of Alcohols
13.4 Reactions of Alcohols
13.5 Important Alcohols
13.6 Characteristics and Uses of Phenols
13.7 Ethers
13.8 Properties of Ethers
13.9 Thiols
13.10 Polyfunctional Compounds

LEARNING OBJECTIVES/ASSESSMENT

When you have completed your study of this chapter, you should be able to:

1. Name and draw structural formulas for alcohols and phenols. (Section 13.1; Exercises 13.4 and 13.10)
2. Classify alcohols as primary, secondary, or tertiary on the basis of their structural formulas. (Section 13.2; Exercise 13.14)
3. Discuss how hydrogen bonding influences the physical properties of alcohols. (Section 13.3; Exercise 13.18)
4. Write equations for alcohol dehydration and oxidation reactions. (Section 13.4; Exercises 13.22 and 13.26)
5. Recognize uses for specific alcohols. (Section 13.5; Exercise 13.36)
6. Recognize uses for specific phenols. (Section 13.6; Exercise 13.38)
7. Name and draw structural formulas for ethers. (Section 13.7; Exercise 13.42)
8. Describe the key physical and chemical properties of ethers. (Section 13.8; Exercise 13.46)
9. Write equations for a thiol reaction with heavy metal ions and the production of disulfides that results when thiols are oxidized. (Section 13.9; Exercise 13.52)
10. Identify functional groups in polyfunctional compounds. (Section 13.10; Exercise 13.56)

SOLUTIONS FOR THE END OF CHAPTER EXERCISES

THE NOMENCLATURE OF ALCOHOLS AND PHENOLS (SECTION 13.1)

13.2 R—O—R′
where the R and the R′ are hydrocarbon groups

☑13.4 a. $CH_3CH_2CH_2CH_2—OH$
1-butanol

b. 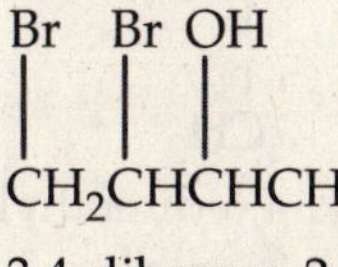

3,4-dibromo-2-butanol

c. 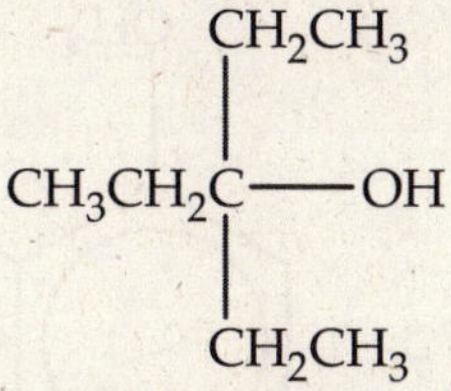

3-ethyl-3-pentanol

d. 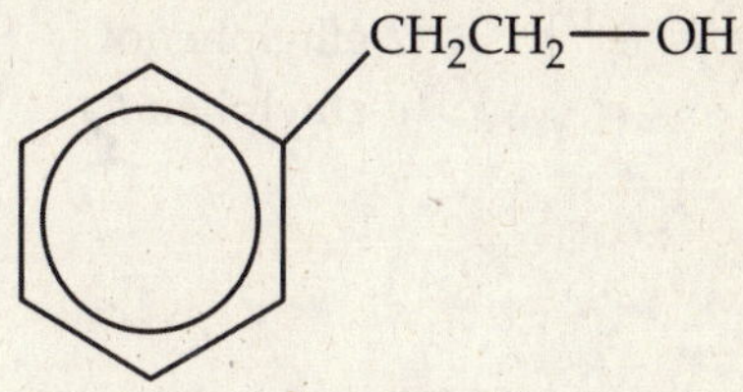

2-phenyl-1-ethanol

e.

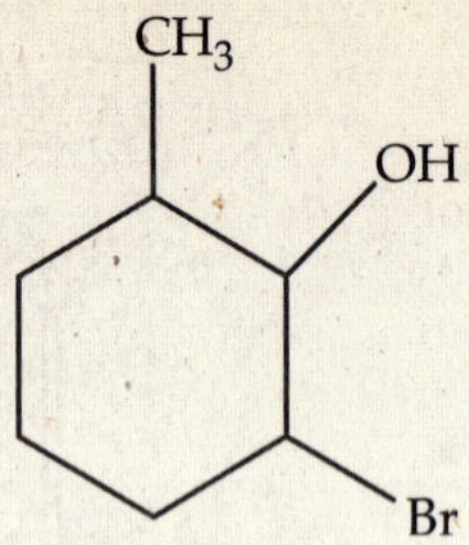

2-bromo-6-methylcyclohexanol

f.

OH OH
| |
$CH_3CHCH_2CHCH_3$

2,4-pentanediol

13.6 a. $CH_3{-}OH$

methanol

b.

OH
|
CH_3CHCH_3

2-propanol

c. $CH_3CH_2{-}OH$

ethanol

d.

OH OH
| |
$CH_2{-}CH_2$

1,2-ethanediol

e.

OH OH OH
| | |
$CH_2{-}CH{-}CH_2$

1,2,3-propanetriol

13.8 a. 2-methyl-2-pentanol

CH_3
|
$CH_3CCH_2CH_2CH_3$
|
OH

b. 1,3-butanediol

$CH_2CH_2CHCH_3$
| |
OH OH

c. 1-ethylcyclopentanol

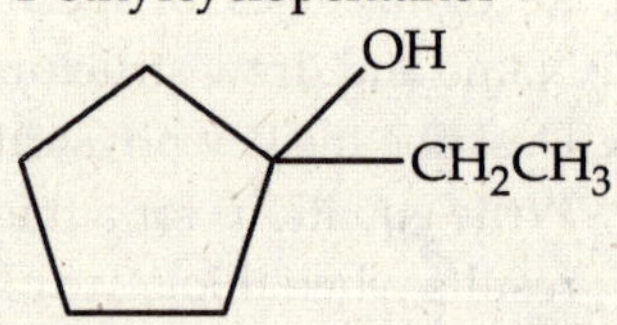

☑13.10 a.

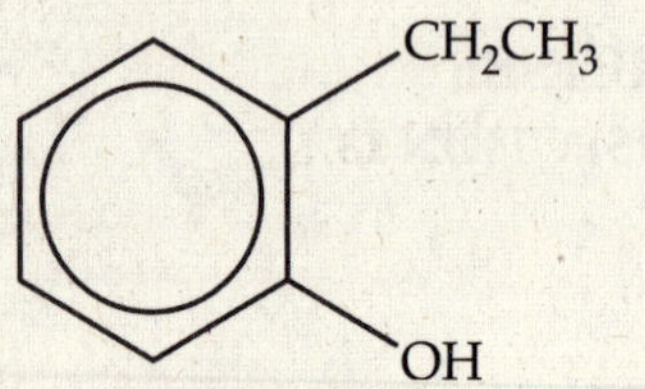

2-ethylphenol

o-ethylphenol

b.

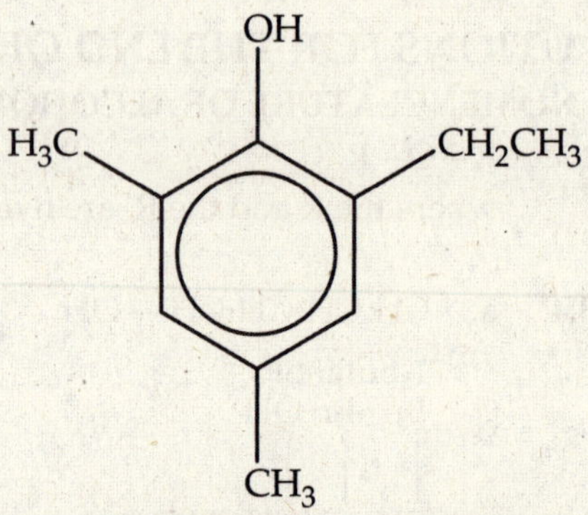

2-ethyl-4,6-dimethylphenol

13.12 a. *o*-bromophenol

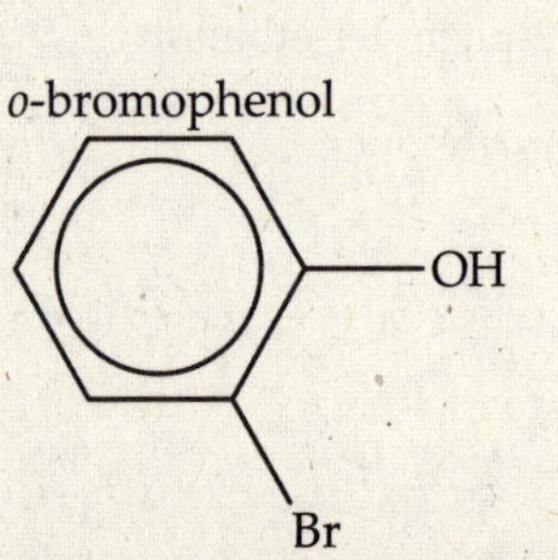

b. 2,3,5-triethylphenol

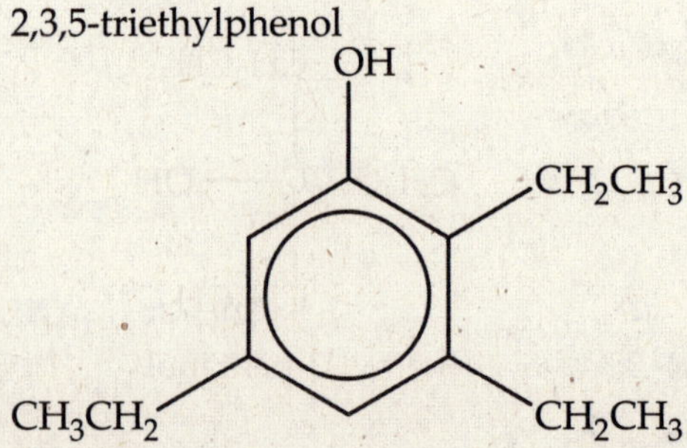

CLASSIFICATION OF ALCOHOLS (SECTION 13.2)

☑13.14 a. 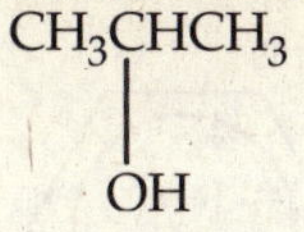
2°

b. $CH_3CH_2—OH$
1°

c. 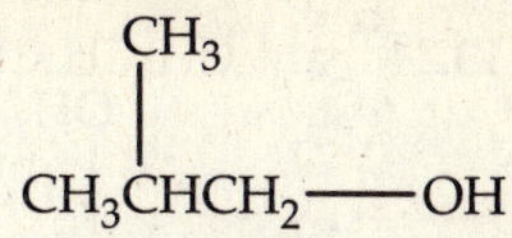
1°

13.16

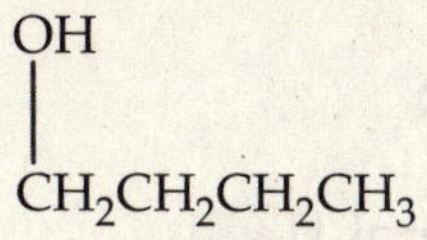
1-butanol/1°

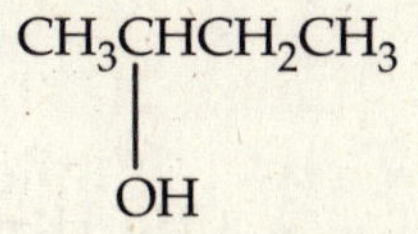
2-butanol/2°

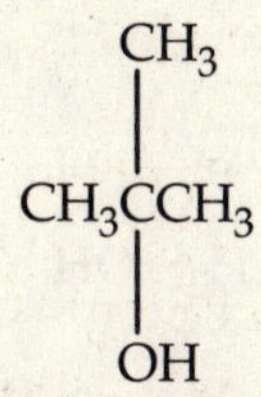
2-methyl-2-propanol/3°

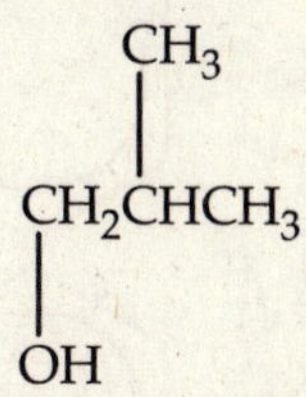
2-methyl-1-propanol/1°

PHYSICAL PROPERTIES OF ALCOHOLS (SECTION 13.3)

☑13.18 a. (lowest boiling point) methanol < ethanol < 1-propanol (highest boiling point)
All three molecules experience both dispersion forces and hydrogen bonding. Since they all hydrogen bond to the same extent, their boiling points differ according to the strength of their dispersion forces. Dispersion forces are stronger for larger molecules.

b. (lowest boiling point) butane < 1-propanol < ethylene glycol (highest boiling point)
These three compounds have comparable molar masses so their dispersion forces are approximately equal. Butane experiences only dispersion forces, and therefore, has the lowest boiling point. 1-Propanol has only one hydrogen bonding site per molecule, while ethylene glycol has two. Ethylene glycol can, therefore, hydrogen bond more strongly than 1-propanol and has the higher boiling point.

13.20 a. 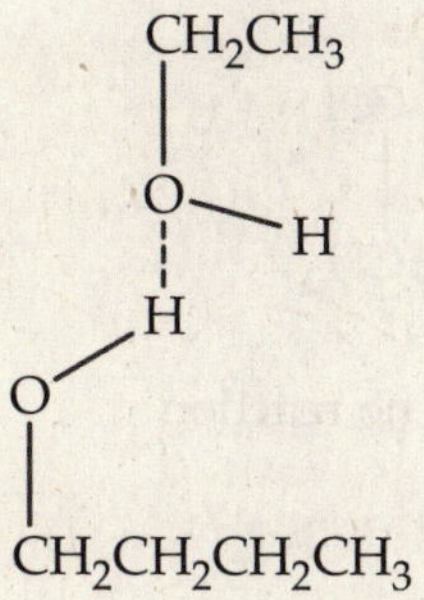

b. 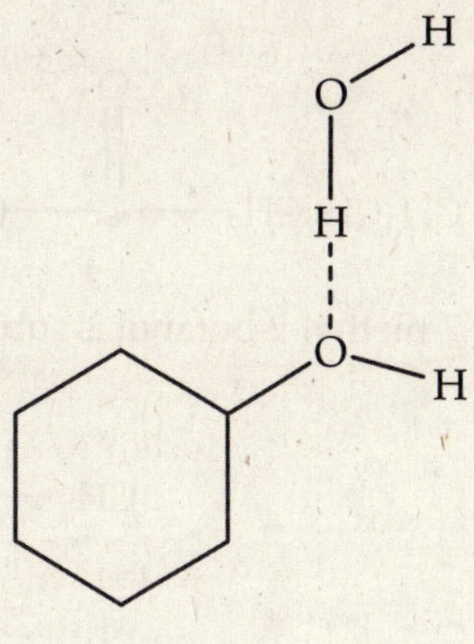

REACTIONS OF ALCOHOLS (SECTION 13.4)

	Reactant	Product
☑13.22 a.	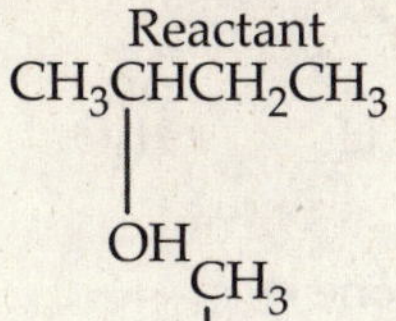	$CH_3CH{=}CHCH_3$
b.	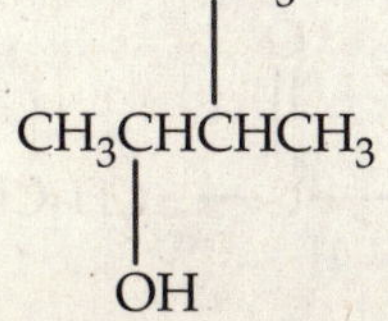	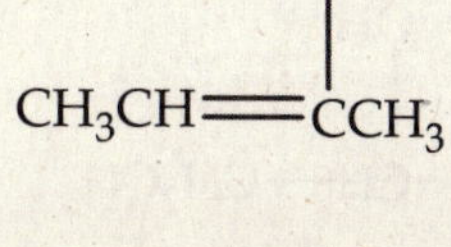

		Reactant	Product
13.24	a.	$CH_3CH_2CH_2-OH$	$CH_3CH_2CH_2-O-CH_2CH_2CH_3$
	b.	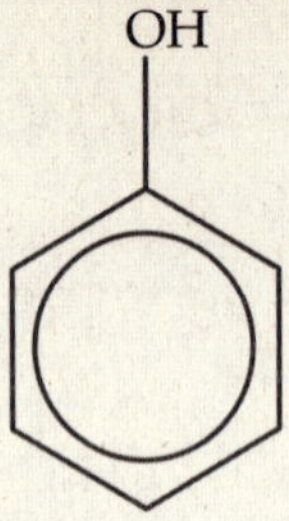	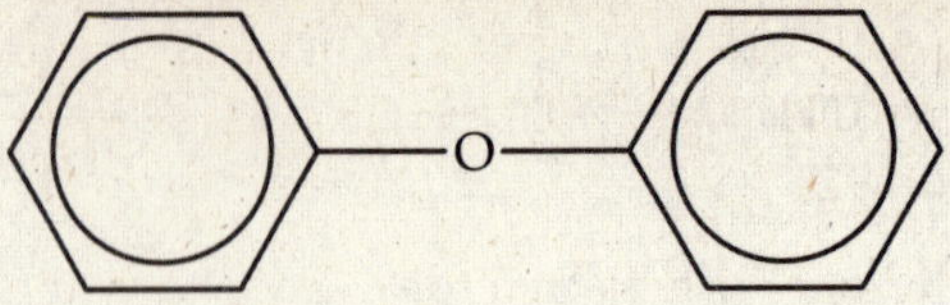
	c.	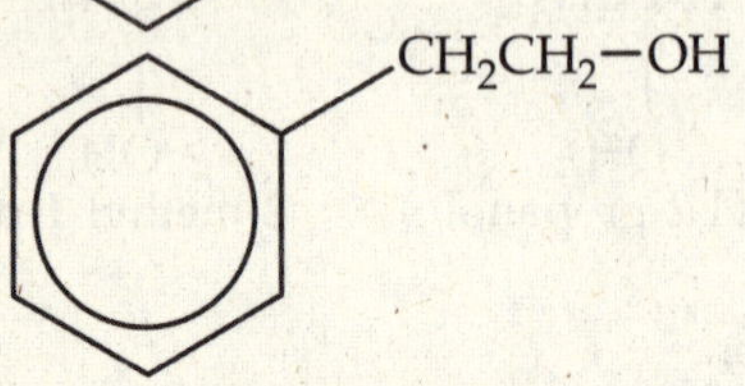	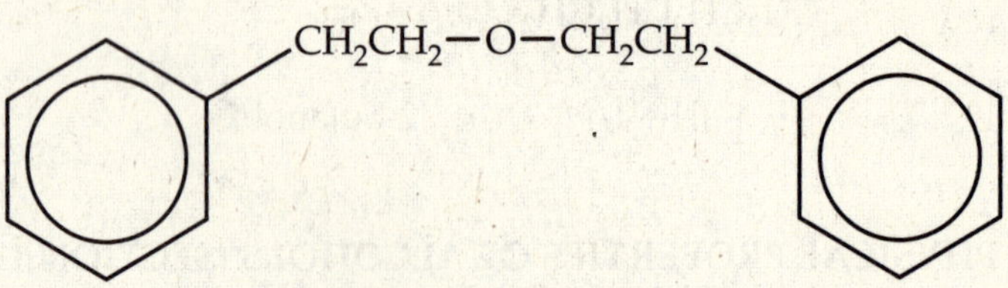

		Product	Reactant
☑13.26	a.	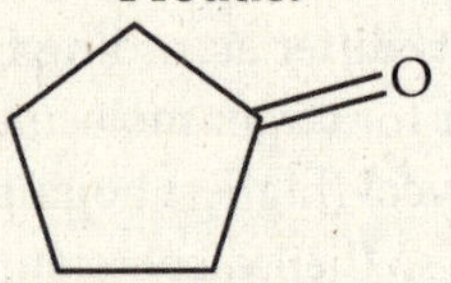	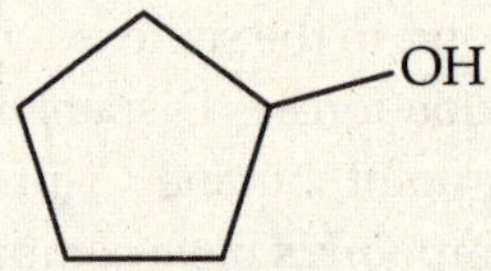
	b.	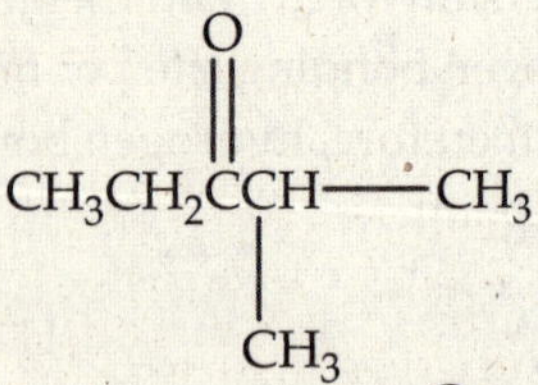	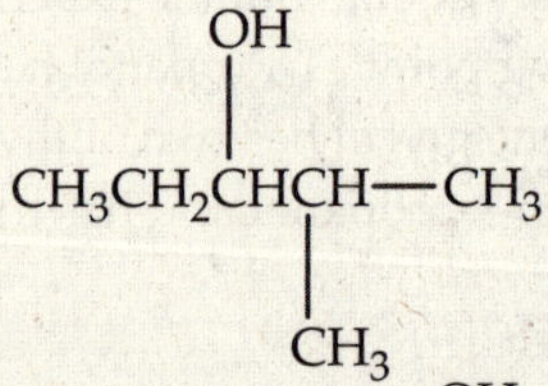
	c.	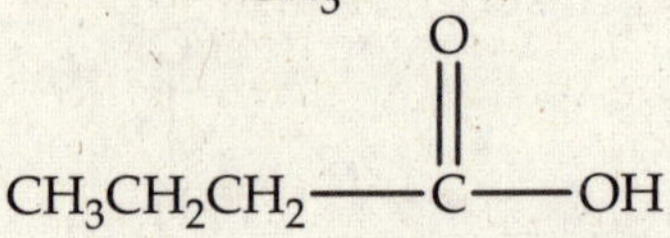	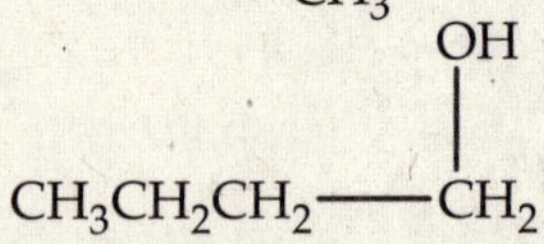

13.28 a. 2-methyl-2-butanol is subjected to controlled oxidation → no reaction

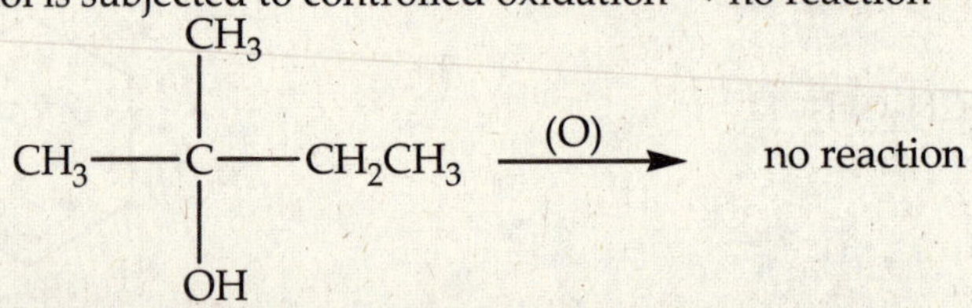

b. 1-propanol is heated to 140°C in the presence of sulfuric acid → dipropyl ether, water

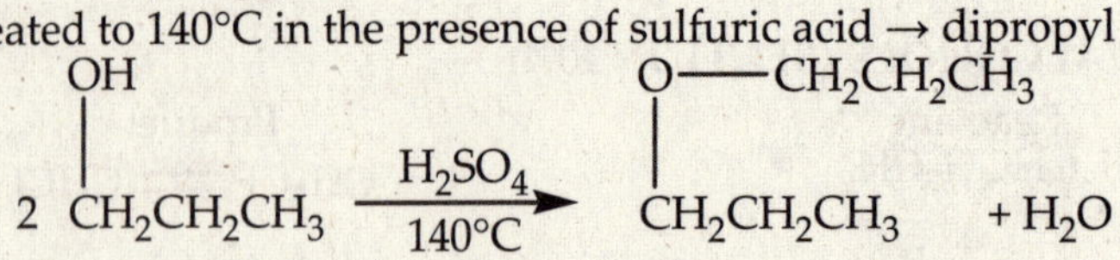

c. 3-pentanol is subjected to controlled oxidation → 3-pentanone

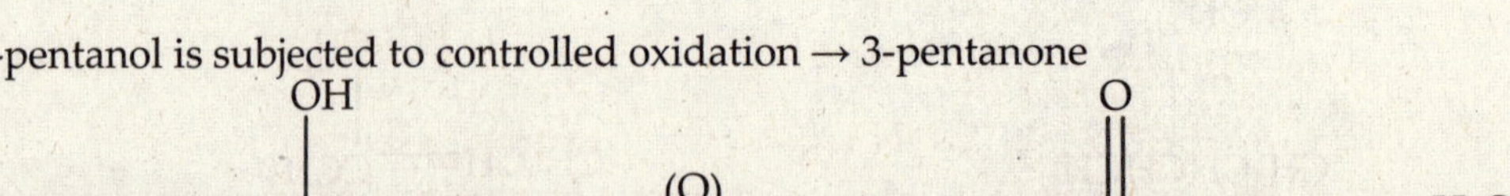

d. 3-pentanol is heated to 180°C in the presence of sulfuric acid → 2-pentene, water

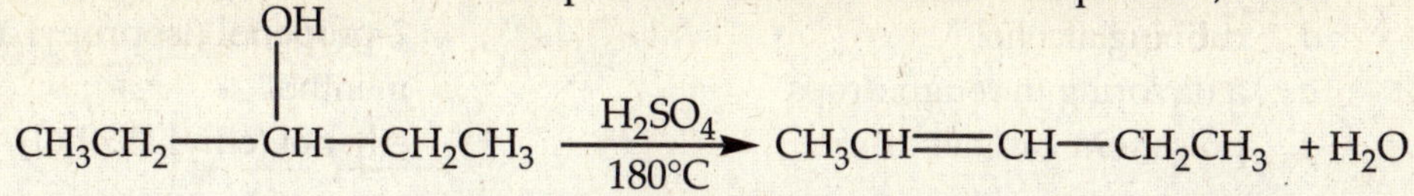

e. 1-hexanol is subjected to an *excess* of oxidizing agent → hexanoic acid, water

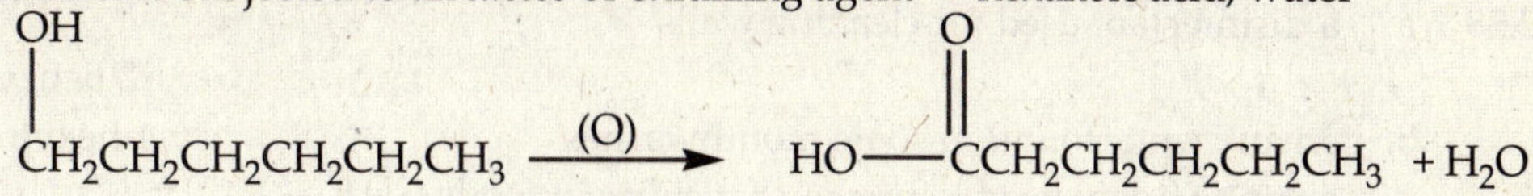

13.30 a.

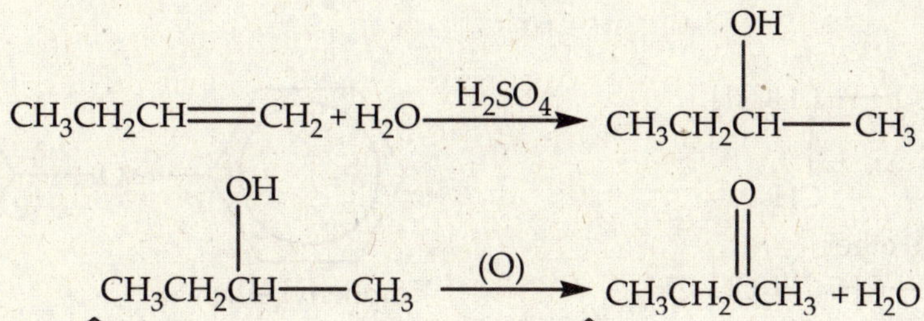

b.

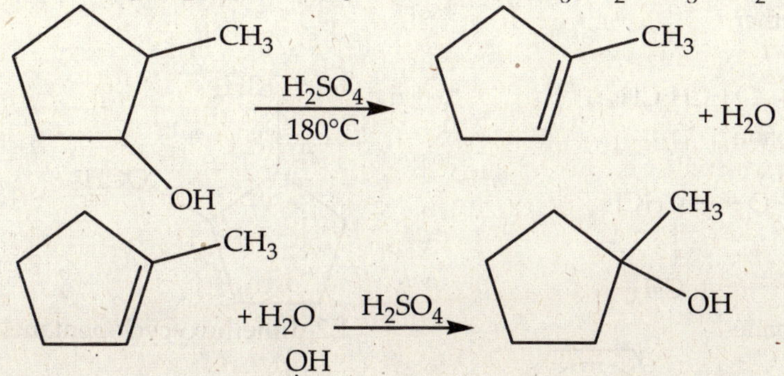

c.

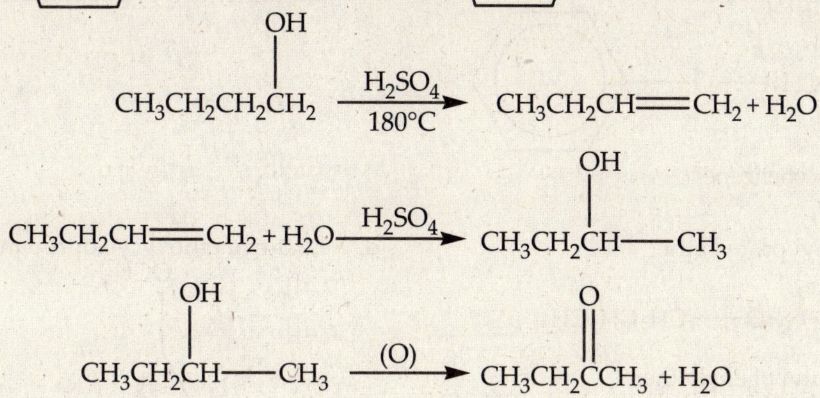

13.32 pyruvic acid

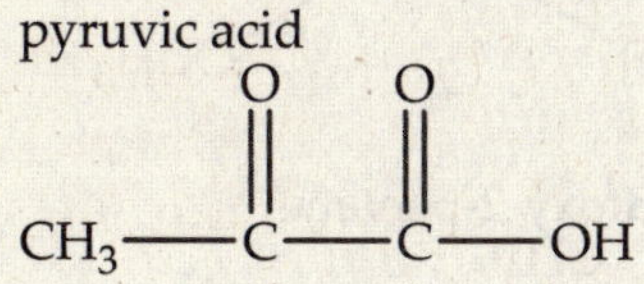

IMPORTANT ALCOHOLS (SECTION 13.5)

13.34 Methanol is an important industrial chemical because its oxidation product is formaldehyde, which is a major starting material for the production of plastics.

☑13.36 a. a moistening agent in many cosmetics — 1,2,3-propanetriol (glycerin, glycerol)
b. the solvent in solutions called tinctures — ethanol (ethyl alcohol)

c. automobile antifreeze — 1,2-ethanediol (ethylene glycol)
d. rubbing alcohol — 2-propanol (isopropyl alcohol)
e. a flavoring in cough drops — menthol
f. present in gasohol — ethanol (ethyl alcohol)

CHARACTERISTICS AND USES OF PHENOLS (SECTION 13.6)

☑13.38 a. a disinfectant used for cleaning walls — *o*-phenylphenol, 2-benzyl-4-chlorophenol
b. an antiseptic found in some mouthwashes — 4-hexylresorcinol, phenol
c. an antioxidant used to prevent rancidity in foods — BHA, BHT

ETHERS (SECTION 13.7)

13.40 a.
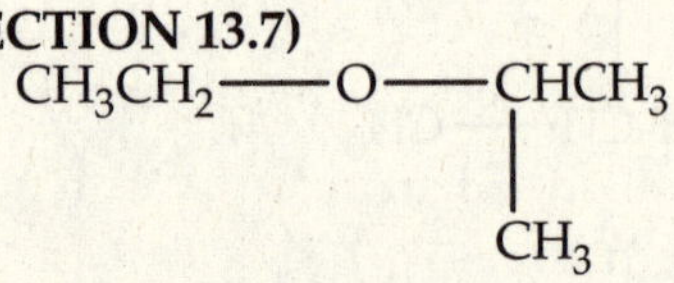
ethyl isopropyl ether

b. CH_3—O—$CH_2CH_2CH_2CH_3$
butyl methyl ether

c.
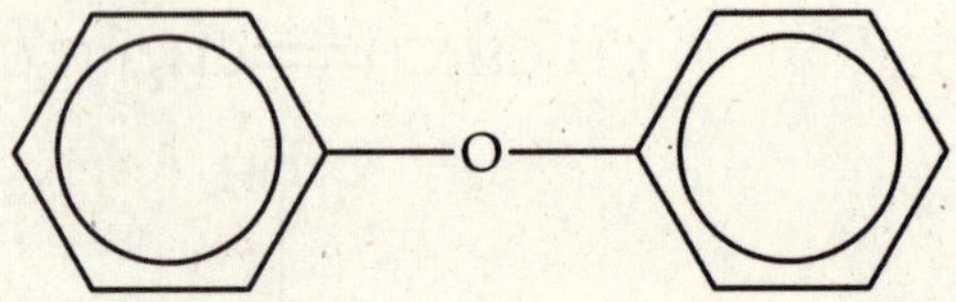
diphenyl ether

☑13.42 a. CH_3CH_2–O–$CH_2CH_2CH_3$
1-ethoxypropane

b.
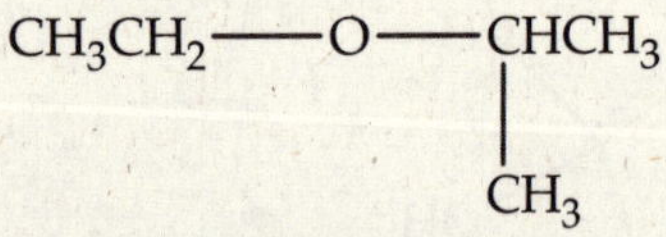
2-ethoxypropane

c.
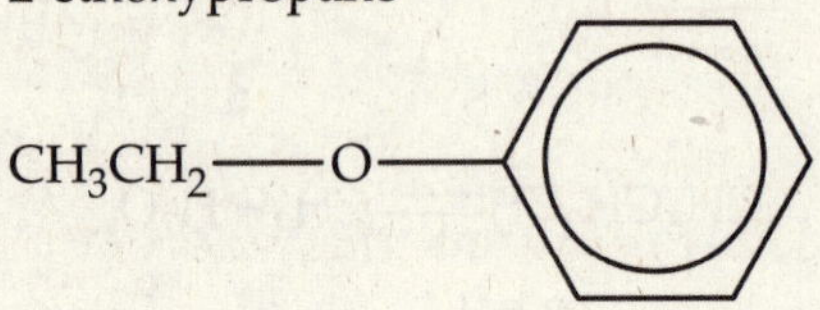
ethoxybenzene

d.
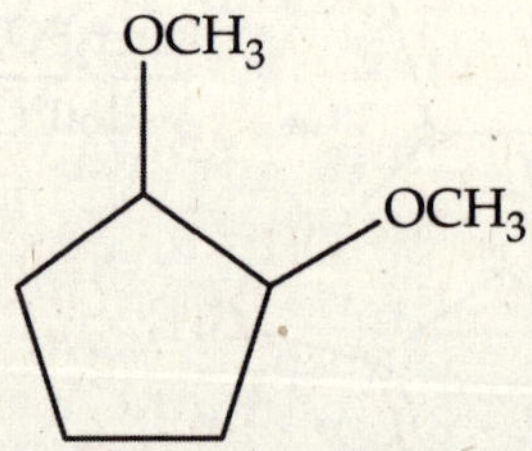
1,2-dimethoxycyclopentane

13.44 a. methyl propyl ether

CH_3—O—$CH_2CH_2CH_3$

b. isobutyl phenyl ether
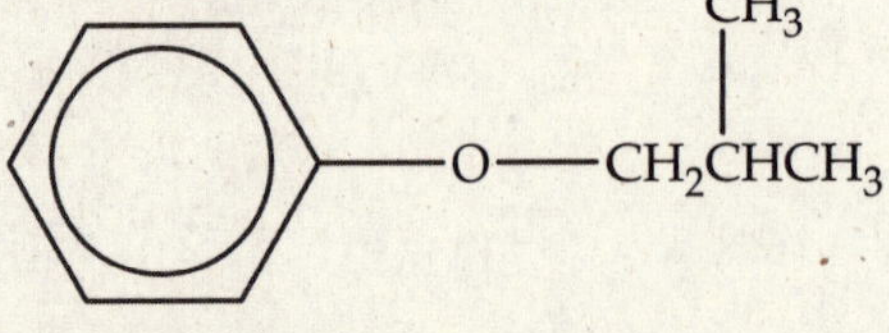

c. 2-methoxybutane
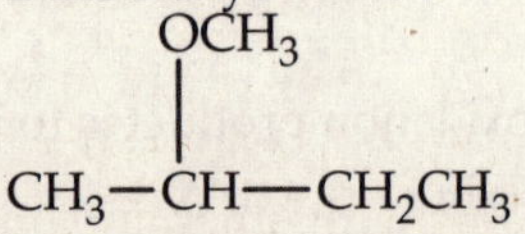

d. 1,2,3-trimethoxycyclopropane
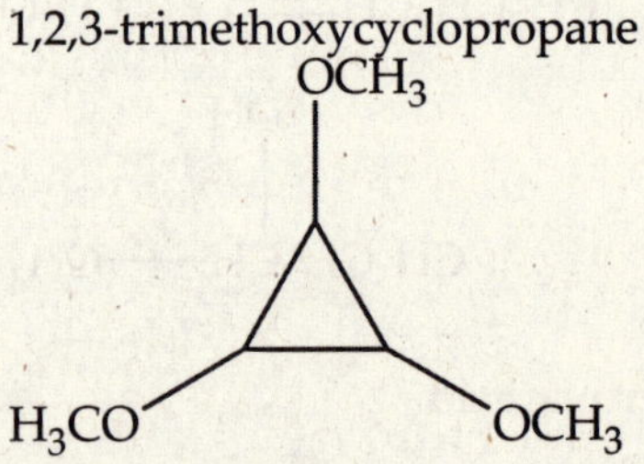

e. 2-*t*-butoxy-2-pentene
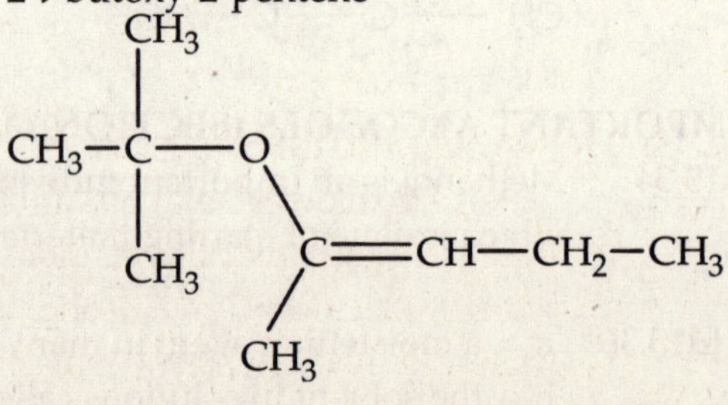

PROPERTIES OF ETHERS (SECTION 13.8)

☑13.46 The chief chemical property of ethers is they are inert (relatively unreactive) towards most reagents.

13.48 $CH_3CH_2CH_2{-}OH > CH_3CH_2{-}O{-}CH_3 > CH_3CH_2CH_2CH_3$

All of the compounds have approximately the same molecular mass. All compounds experience dispersion forces, but since these three have similar molecular masses, the strength of their dispersion forces is similar. The major differences between these molecules are their key functional groups and the strength of their interparticle forces. The 1-propanol will have the strongest hydrogen bonds to water because of its hydroxy group; therefore, 1-propanol is the most soluble compound. The ethyl methyl ether (methoxyethane) can also hydrogen bond to water; therefore, it is more soluble than the butane, but less soluble than the 1-propanol. The butane cannot form hydrogen bonds with water; therefore, it is the least soluble.

13.50

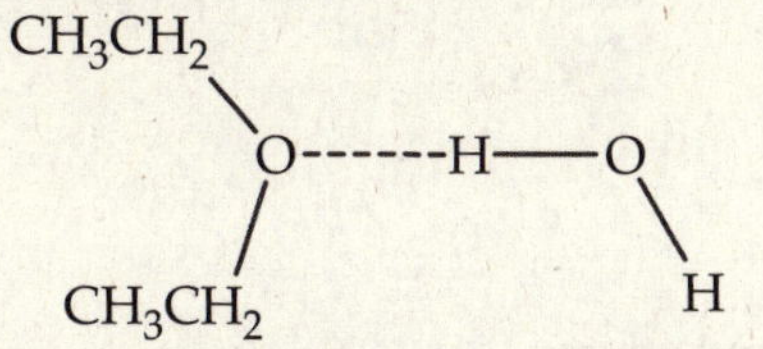

THIOLS (SECTION 13.9)

☑13.52 a. $2\ CH_3CH_2CH_2{-}SH + Hg^{2+} \rightarrow CH_3CH_2CH_2{-}S{-}Hg{-}S{-}CH_2CH_2CH_3 + 2\ H^+$

b. $2\ CH_3CH_2CH_2{-}SH + (O) \rightarrow CH_3CH_2CH_2{-}S{-}S{-}CH_2CH_2CH_3 + H_2O$

c.

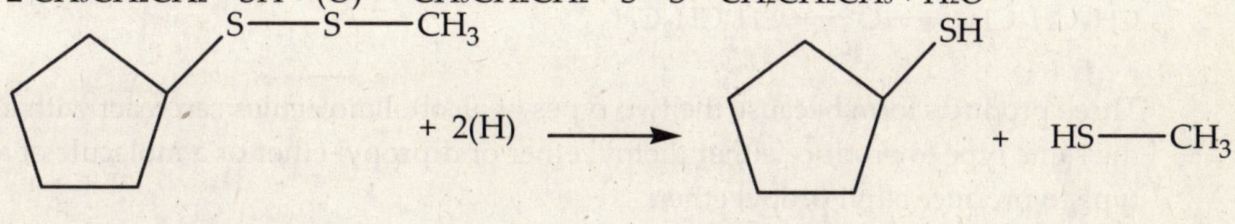

13.54 When alcohols are oxidized in a controlled way, they produce aldehydes or carboxylic acids (1° alcohols), ketones (2° alcohols), or no reaction (3° alcohols). When thiols are oxidized in a controlled way they produce disulfides.

POLYFUNCTIONAL COMPOUNDS (SECTION 13.10)

☑13.56

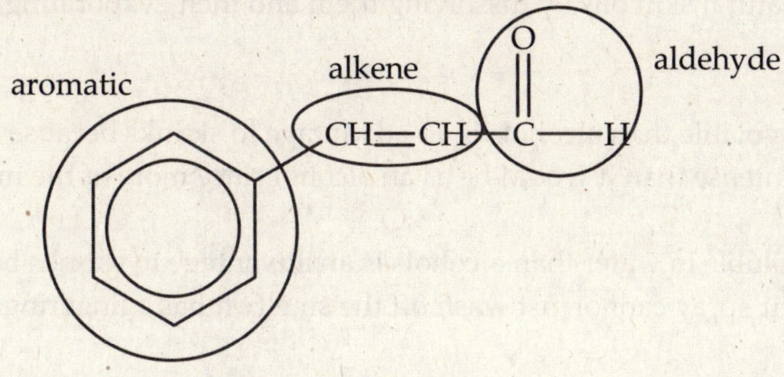

The key functional groups in this molecule include a benzene ring (aromatic), a carbon-carbon double bond (alkene), and a carbonyl group (aldehyde).

ADDITIONAL EXERCISES

13.58 Thiols have lower boiling points and are less water soluble than the corresponding alcohols because thiol molecules cannot experience hydrogen bonding with other thiol molecules or water molecules, while alcohol molecules experience hydrogen bonding with other alcohol molecules and water molecules. The weaker intermolecular forces that thiol molecules experience (dipolar forces) account for the lower boiling points and lower water solubility than the corresponding alcohols.

13.60 Phenol is a stronger acid than cyclohexanol because the aromatic ring in the phenol allows the conjugate base of phenol (the phenolate ion) to be stable, while cyclohexanol forms a less stable conjugate base because it does not have an aromatic ring. The aromatic ring in the phenolate ion delocalizes the negative charge throughout the ring through resonance, effectively stabilizing it. Consequently, phenol ionizes in water to produce hydronium ions more readily than cyclohexanol and is a stronger acid than cyclohexanol.

ALLIED HEALTH EXAM CONNECTION

13.62 (d) C_3H_7OH contains the alcohol functional group.

13.64 Alcoholic beverages contain (d) ethyl alcohol.

13.66 (b) $CH_3{-}O{-}CH_3$ is an ether.

CHEMISTRY FOR THOUGHT

13.68 The three products are:

diethyl ether

$CH_3CH_2{—}O{—}CH_2CH_3$

ethyl propyl ether

$CH_3CH_2{—}O{—}CH_2CH_2CH_3$

dipropyl ether

$CH_3CH_2CH_2{—}O{—}CH_2CH_2CH_3$

Three products form because the two types of alcohol molecules can react with a molecule of the same type to produce either diethyl ether or dipropyl ether or a molecule of a different type to produce ethyl propyl ether.

13.70 Ethanol poses greater risks for fetuses because they are much smaller than adults and their organ systems are still developing, whereas an adult's organ systems are fully developed.

13.72 Skin takes on a dry appearance after diethyl ether is spilled on the skin because diethyl ether, a volatile solvent, removes natural skin oils by dissolving them and then evaporating, taking these oils with it.

13.74 The fact that thiols are more volatile than alcohols is an advantage to skunks because their smell of their spray is more intense than it would be as an alcohol since more of the molecules are in the gas phase.
The fact that thiols are less soluble in water than alcohols is an advantage to skunks because the creatures that receive their spray cannot just wash off the smell. It has a lingering effect.

13.76 Glycerol, or glycerin is an excellent moisturizer because it contains three hydroxy groups. These attract water molecules by hydrogen bonding and will keep candy and shredded coconut moist.

ADDITIONAL ACTIVITIES

Section 13.1 Review:

(1) Is the *-ol* ending appropriate for compounds that contain a hydroxy group attached to an aliphatic hydrocarbon chain?

(2) What is the priority for numbering the longest carbon chain in a compound that contains a hydroxy group?

(3) What is the meaning of a compound name that ends in *-triol*?

(4) Which other class of compounds studied so far can have an ending similar to (3)?

(5) Phenols are studied here as their own compound class; however, this is not the first appearance of phenol in the textbook. Where else in the textbook were phenols introduced?

Section 13.2 Review:

(1) Draw and name the simplest secondary alcohol.

(2) Draw and name the simplest tertiary alcohol.

Section 13.3 Review:

Box the following properties that apply to low molecular weight alcohols. Correct the properties that do not apply to low molecular weight alcohols as they are currently written.

high boiling point   low melting point   polar   dispersion forces

soluble in water   less dense than water   hydrophobic

High molecular weight alcohols do not dissolve in water. Explain this observation.

Section 13.4 Review:

(1) Alcohol dehydration to produce an alkene is the opposite process of a previously studied chemical reaction. What is the process?

(2) How do the conditions differ between alcohol dehydration and the other process from (1)?

(3) Should alcohol dehydration to produce an alkene and the process from (1) be represented with equilibrium arrows?

(4) Can Markovnikov's rule be applied to predicting the major product of alcohol dehydration to an alkene?

(5) How do the conditions for alcohol dehydration to an alkene differ from alcohol dehydration to an ether?

(6) In general, what is the relationship between addition and elimination reactions?

(7) Write the reaction for the oxidation of isopropyl alcohol.

(8) Calculate the oxidation numbers for each atom in the reactant and product molecules of (7).

(9) Which atom(s) experience(s) a change in oxidation number? Is it (are they) oxidized or reduced?

(10) Classify primary, secondary, and tertiary alcohols in order of increasing reactivity.

Section 13.5 Review:

(1) Classify the reaction for the formation of methanol.

(2) How does the reaction to form glucose from ethylene differ from the reaction to form ethanol from glucose?

(3) What is different in the reaction conditions of the hydration of an alkene as studied in Chapter 12 and the industrial production of ethanol from ethylene?

(4) Classify the important alcohols in Section 13.5 as toxic or nontoxic.

Section 13.6 Review:

(1) Phenol is also known as carbolic acid. Is the use of the term "acid" appropriate?
(2) What are some of the uses of phenol derivatives?

Section 13.7 Review:

(1) Draw and label the following alkoxy groups: methoxy, ethoxy, proproxy, isoproproxy, butoxy, isobutoxy, *sec*-butoxy, *t*-butoxy.
(2) Which atom from the alkoxy groups in (1) would connect to a carbon chain in a compound?
(3) Propose an explanation for why the IUPAC system of nomenclature does not utilize an *-er* or *-ther* ending for compounds that belong to the compound class called ethers.
(4) Why are furan and pyran called heterocyclic rings?
(5) Give an example of a homocyclic ring.

Section 13.8 Review:

Box the following properties that apply to ethers. Correct the properties that do not apply to ethers as they are currently written.

high boiling point

polar

low melting point

nonflammable

slightly soluble in water

dispersion forces

hydrophobic

high volatility

Section 13.9 Review:

(1) Which element in the sulfhydryl group is responsible for the odors associated with thiols?
(2) Which of the following best corresponds to the odors associated with thiols: roses, baby powder, rotten eggs?
(3) Propose a reaction cycle to convert methanethiol to its corresponding disulfide and to return the disulfide to methanethiol.
(4) Are the reactions in (3) redox reactions?
(5) What happens to the methanethiol as it is converted to its corresponding disulfide?
(6) What happens to the corresponding disulfide as it is converted to methanethiol?
(7) Is the reaction of heavy metal ions with thiols as easily reversed as the reaction to convert methanethiol to its corresponding disulfide?

Section 13.10 Review:

(1) Does the term "polyfunctional compounds" adequately describe the molecules within this class?
(2) Most biological compounds are polyfunctional compounds. How can your current study of the individual functional groups benefit your future understanding of biological compounds?

Tying It All Together with a Laboratory Application:

A student must distinguish between five unknown compounds. The unknowns are labeled A, B, C, D, and E. The possible identities of the unknowns (not in order) are phenol, cyclohexanol, methoxycyclopentane, 2,3-dimethyl-2-butanol, and 1-hexanol.

(1) Draw the structural formulas for all of the unknowns.
(2) Write the molecular formulas for each of the unknowns.
(3) Are any of the compounds structural isomers? If so, which ones?

(4) Are any of the compounds geometric isomers? If so, which ones?
(5) Identify the compound(s) that is(are) alcohol(s).
(6) Classify the compound(s) from (5) as primary, secondary, or tertiary alcohol(s).
(7) Identify the compound(s) that is(are) ether(s).
(8) Identify the compound(s) that contain(s) a benzene ring.

The student is allowed to use "solutions" of the unknown compounds in water or cyclohexane solvents. All of the solutions are colorless. The unknown A in water mixture appears to have two layers; however, unknown A in cyclohexane is completely dissolved. This suggests that of the two regions within unknown A, the (9) _________ (polar or nonpolar) region of the molecule dominates. The student suspects that unknown A may be (10) _________ (phenol, cyclohexanol, methoxycyclopentane, 2,3-dimethyl-2-butanol, or 1-hexanol) because it has the (11) _________ (strongest or weakest) intermolecular forces of all of the compounds.

The student tests the pH of all of the aqueous solutions and determines unknown E has a pH of 6.2. The other solutions are closer to a pH of 7. The student suspects that unknown E is (12) _________ because it has a(n) (13) _________ pH which indicates that the compound (14) _________ in water. The student also notices that this compound has a (15) _________ odor, like Chloroseptic® throat spray.

The student reacts orange potassium dichromate with small samples of all of the unknowns dissolved in water and records the following observations:

Unknown	Observations
A	the reaction mixture turned orange
B	the reaction mixture turned gray-green immediately and felt hot to the touch
C	the reaction mixture turned gray-green immediately and felt hot to the touch
D	the reaction mixture turned orange
E	the reaction mixture turned orange

Unknowns (16) _________ reacted with potassium dichromate; therefore, they must be (17) _________ because they were (18) _________ by the potassium dichromate. The student filters the resulting mixtures from the potassium dichromate experiment and isolates the organic materials from the filtrate. The student then dissolves the organic materials from the filtrates in water and tests the pH of the resulting solutions. The pH of solution C and E are less than 7. The other compounds have pH values of approximately 7. This suggests to the student that unknown B is (19) _________ and unknown C is (20) _________. Therefore, unknown D is (21) _________. To confirm the identities of unknowns A and D, the student could check the boiling points of pure samples of the compounds. If the student is correct, unknown (22) _________ will have the higher boiling point.

SOLUTIONS FOR THE ADDITIONAL ACTIVITIES

Section 13.1 Review:

(1) The *-ol* ending is appropriate for compounds that contain a hydroxy group attached to an aliphatic hydrocarbon chain because the hydroxy group is the key functional group for the class of compounds known as alcohols. The compound names and compound class both have the same ending.
(2) When numbering the longest carbon chain in a compound that contains a hydroxy group, the chain is numbered in such a manner that the hydroxy group is attached to the carbon atom with the smallest possible number.
(3) A compound name that ends in *-triol* suggests that the compound contains three hydroxy groups.
(4) Alkenes with multiple double bonds (dienes, trienes, tetraenes) have endings showing multiple substitutions of the same type, as do the endings for diols, triols, and tetrols.

(5) Phenol was first introduced in the textbook as a benzene derivative. The structure of phenol contains a benzene ring.

Section 13.2 Review:

(1)

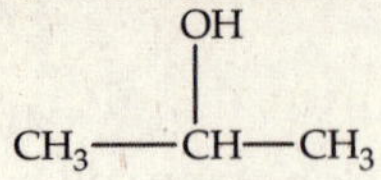

isopropyl alcohol
2-propanol

(2)

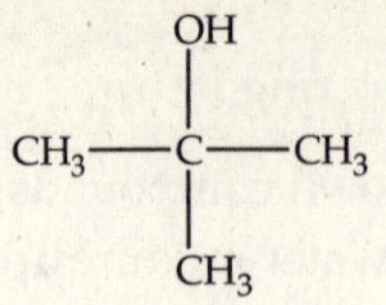

t-butyl alcohol
2-methyl-2-propanol

Section 13.3 Review:

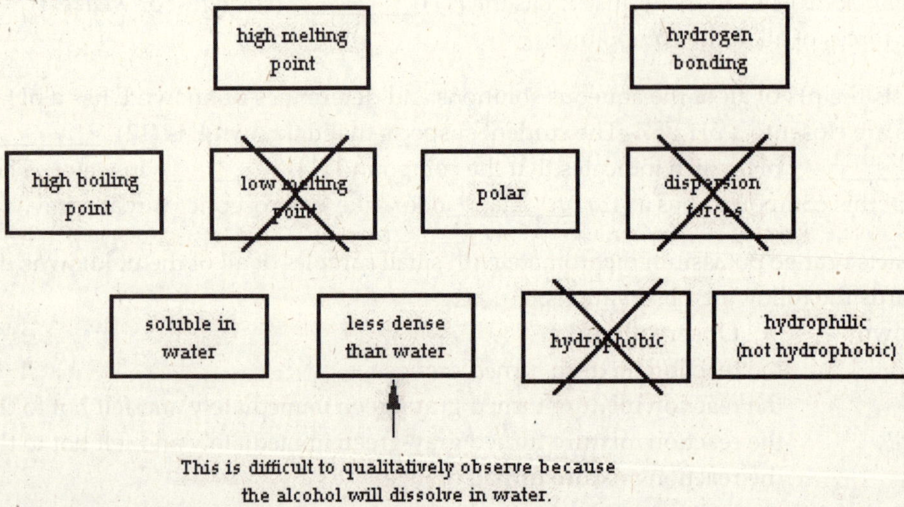

High molecular weight alcohols do not dissolve in water because the hydrogen bonds between the hydroxy group and the water is not sufficient to overcome the repulsion between the hydrophobic aliphatic tail and the polar water molecules. Water and high molecular weight alcohols are not enough like for dissolving to occur. High molecular weight alcohols behave more like hydrocarbons than like water.

Section 13.4 Review:

(1) hydration of an alkene
(2) Dehydration of an alcohol to form an alkene requires a sulfuric acid catalyst and temperatures of 180°C. Hydration of an alkene to form an alcohol requires water and a sulfuric acid catalyst.
(3) These processes do not require equilibrium arrows because of the difference in reaction conditions.
(4) The opposite of Markovnikov's rule can be applied to predicting the major product of alcohol dehydration to an alkene. The hydrogen atom will be removed from the carbon atom (adjacent to the carbon with the hydroxy group) that contains the fewest number of hydrogen atoms.
(5) Alcohol dehydration to an alkene requires a sulfuric acid catalyst and 180°C. Alcohol dehydration to an ether requires a sulfuric acid catalyst and 140°C. The temperatures of these reactions differ.
(6) Addition and elimination reactions are opposite processes. In an addition reaction, two reactants produce one product. In an elimination reaction, one reactant produces two products.
(7)

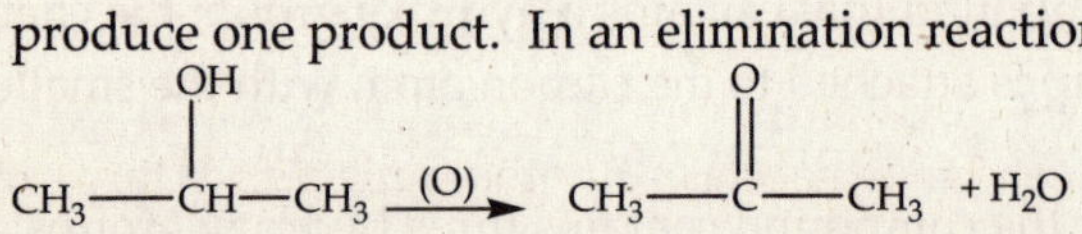

(8)

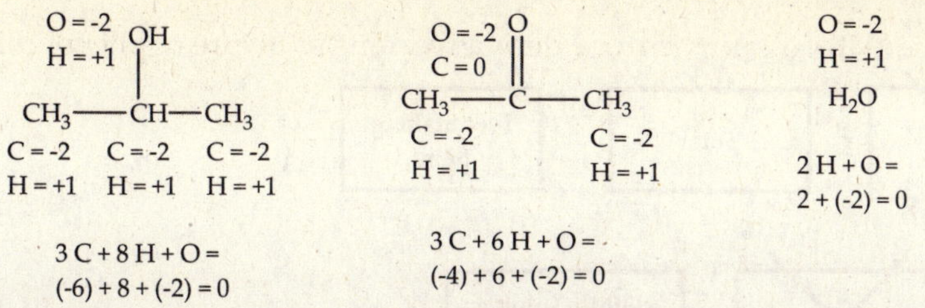

(9) The carbon atom, attached to the hydroxy group as a reactant and the double bonded oxygen as a product, begins with an oxidation number of -2 and ends with an oxidation number of 0. This carbon atom has been oxidized. (10) tertiary < secondary < primary

Section 13.5 Review:

(1) redox, addition; (2) ethylene = redox, addition; glucose = fermentation, redox, elimination

(3) The hydration of an alkene as studied in Chapter 12 required the presence of a sulfuric acid catalyst. The industrial production of ethanol from ethylene requires high pressure (70 atm) and high temperature (300°C).

(4) toxic: methanol, isopropyl alcohol
nontoxic: ethanol, glycerol
cannot be classified from the reading: ethylene glycol (toxic), propylene glycol (nontoxic)

Section 13.6 Review:

(1) Phenol reacts with water to produce hydronium and phenolate ions. This solution is acidic, thus the name carbolic acid is appropriate.

(2) Phenol derivatives are used as disinfectants, antiseptics, antioxidants, and many other industrial applications.

Section 13.7 Review:

(1)

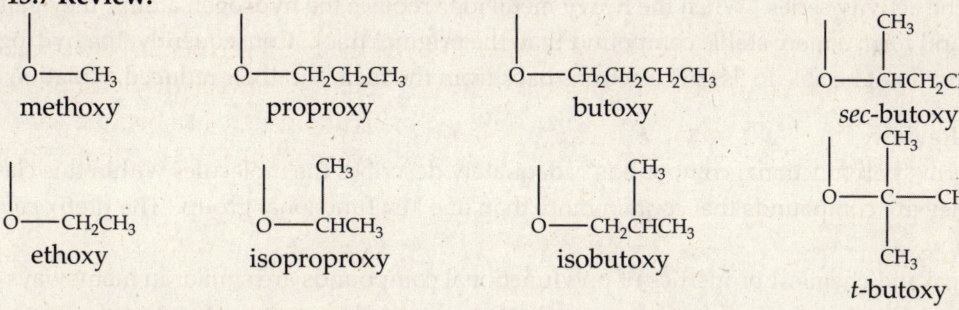

(2) The oxygen atom from the alkoxy groups in (1) connects to a carbon chain in a compound.

(3) The key functional group in ethers is unique from any of the other key functional groups studied in depth up to this point. The ether key functional group is a linkage of two carbon based groups through an oxygen atom. Numbering the longest carbon chain in such a way that the oxygen atom is attached to the smallest numbered carbon atom could be used identify the location of the oxygen atom, but not the identity of the other carbon group. An ending of *-er* or *-ther* would not contain enough information to be useful. Consequently, the IUPAC system names ethers using the hydrocarbon name for the longest chain connected to the oxygen atom and an alkoxy name to describe the location and structure of other atoms bonded to the oxygen atom.

(4) Furan and pyran are heterocyclic rings because they contain more than one type of atom in the ring structure.

(5) Any cycloalkane, cycloalkene, or benzene derivative is an example of a homocyclic ring. Only carbon atoms are bonded into the ring structure.

Section 13.8 Review:

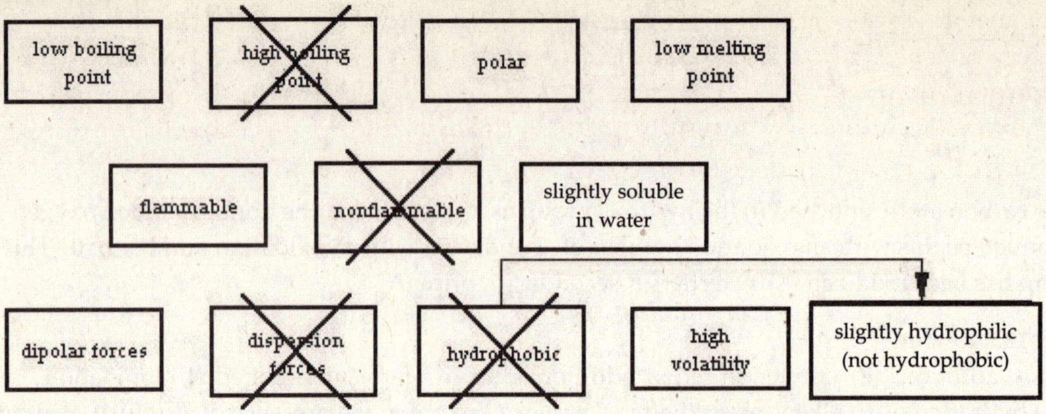

Section 13.9 Review:

(1) sulfur; (2) rotten eggs;

(4) yes; (5) The methanethiol is oxidized. (6) The corresponding disulfide is reduced.

(3)

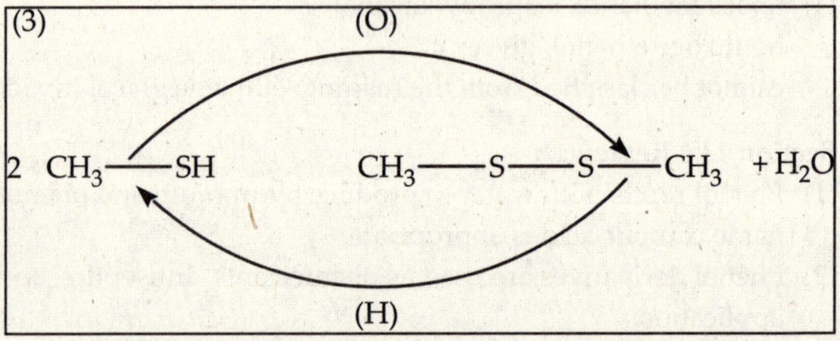

(7) No, the reaction of heavy metal ions with thiols is not as easily reversed as the reaction to convert methanethiol to its corresponding disulfide because this reaction is based on the activity series. When the heavy metal ions replace the hydrogen atoms, they gain electrons and form a more stable compound than the original thiol. Consequently, the hydrogen ions produced will not be able to "steal" electrons back from the metals in their reduced oxidation state.

Section 13.10 Review:

(1) Yes, the term "polyfunctional compounds" adequately describes the molecules within this class because they are compounds that contain more than one key functional group. The prefix *poly-* means many.

(2) The physical and chemical properties of polyfunctional compounds are similar in many ways to the physical and chemical properties of compounds containing only one of the functional groups in the polyfunctional compounds. Understanding the physical and chemical principles associated with the individual parts of a polyfunctional compound will enable better understanding of complex compounds.

Tying It All Together with a Laboratory Application:

(1)

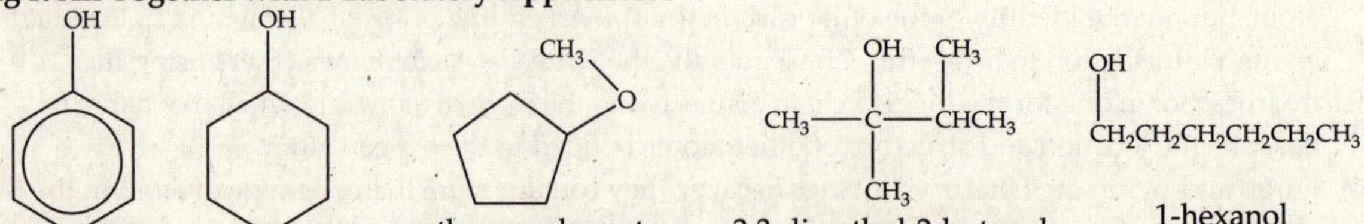

phenol | cyclohexanol | methoxycyclopentane | 2,3-dimethyl-2-butanol | 1-hexanol

(2) C_6H_6O | $C_6H_{12}O$ | $C_6H_{12}O$ | $C_6H_{14}O$ | $C_6H_{14}O$

(3) cyclohexanol and methoxycyclopentane ($C_6H_{12}O$); 2,3-dimethyl-2-butanol and 1-hexanol ($C_6H_{14}O$)

(4) no geometric isomers

(5) alcohols: cyclohexanol, 2,3-dimethyl-2-butanol, 1-hexanol

(6) 1° = 1-hexanol, 2° = cyclohexanol, 3° = 2,3-dimethyl-2-butanol
(7) ether: methoxycyclopentane
(8) benzene ring: phenol
(9) nonpolar
(10) methoxycyclopentane
(11) weakest
(12) phenol
(13) acidic
(14) dissociated
(15) medicinal
(16) B and C
(17) 1° or 2° alcohols
(18) oxidized
(19) cyclohexanol
(20) 1-hexanol
(21) 2,3-dimethyl-2-butanol
(22) D

SELF-TEST QUESTIONS

Multiple Choice

1. What is the structure of 1-propoxypropane?
   a. $CH_3-CH_2-O-CH_2-CH_2-CH_3$
   b. 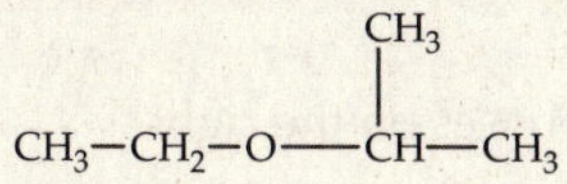
   c. $CH_3-CH_2-CH_2-O-CH_2-CH_2-CH_3$
   d. 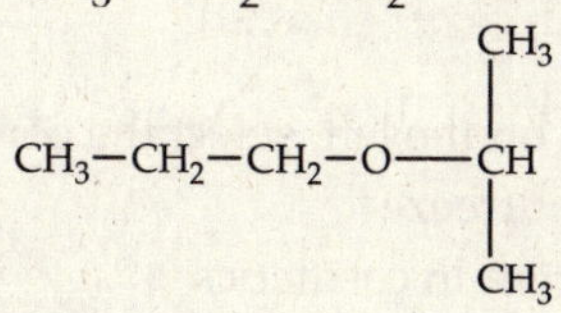

2. What is the correct IUPAC name for 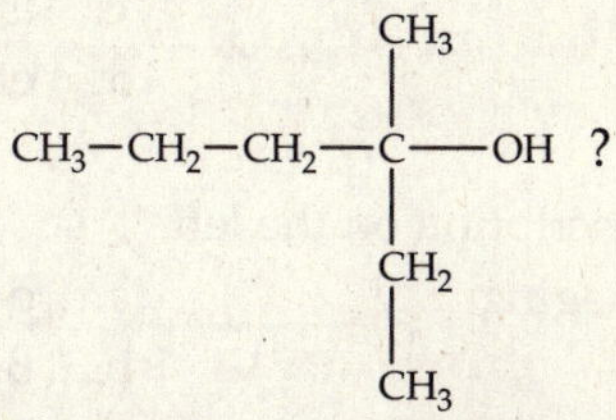 ?
   a. 1-ethyl-1-methylbutanol
   b. 2-ethyl-2-pentanol
   c. 3-methyl-3-hexanol
   d. 3-methylheptanol

3. What is the correct IUPAC name for 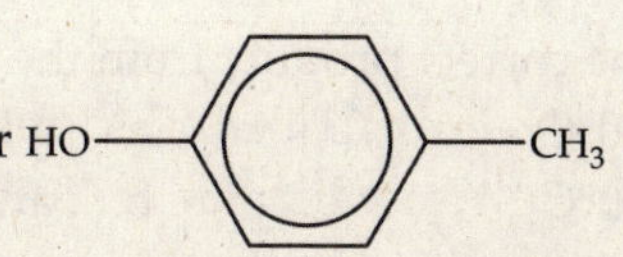 ?
   a. 4-methylphenol
   b. 1-methylphenol
   c. 2-methyl-4-phenol
   d. 4-methyl-2-phenol

4. 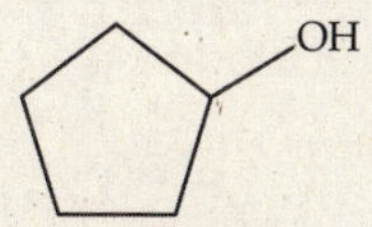 is a ____________ alcohol.
   a. primary
   b. secondary
   c. tertiary
   d. quaternary

5. Which of the following is a primary alcohol?
   a. 1-butanol
   b. 2-butanol
   c. 2-propanol
   d. 2-methyl-2-propanol

6. Which of the following is the most soluble in water?
   a. $CH_3CH_2CH_2-O-CH_3$
   b. $CH_3CH_2CH_2CH_2-OH$
   c. $CH_3CH_2-O-CH_2CH_3$
   d. $CH_3CH_2CH_2CH_2CH_2-OH$

7. What reagent is required to carry out the following reaction?
   $$2\ CH_3-SH + ? \rightarrow CH_3-S-S-CH_3 + H_2O$$
   a. NaOH
   b. [O]
   c. H^+
   d. H_2SO_4

8. The chemical reactivity of ethers is closest to the:
   a. alkanes.
   b. alkenes.
   c. aromatics.
   d. alcohols.

9. What is a common name for ethoxybenzene?
   a. ethyl benzene ether
   b. ethyl benzyl ether
   c. ethyl phenyl ether
   d. ethoxy benzyl ether

10. What is the IUPAC name of 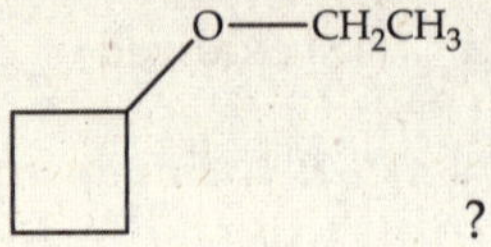 ?
   a. ethoxycyclobutane
   b. 1-ethyloxycyclobutane
   c. 1-ethylcyclobutane
   d. ethylcyclobutoxy

Matching

For each description on the left, select the correct alcohol from the responses on the right.

11. automobile antifreeze
12. moistening agent in cosmetics
13. rubbing alcohol
14. present in alcoholic beverages

a. glycerol
b. ethylene glycol
c. isopropyl alcohol
d. ethyl alcohol

Select the correct phenol for each description on the left.

15. antioxidant used in food packaging
16. used in throat lozenges
17. present in Lysol

a. o-phenylphenol
b. BHA
c. 4-hexylresorcinol
d. eugenol

For each reaction on the left, select the correct product from the responses on the right.

18. an alcohol is dehydrated at 140°C
19. an alcohol is dehydrated at 180°C
20. a secondary alcohol is oxidized

a. a ketone
b. an ether
c. an alkene
d. an aldehyde

True-False

21. Methyl alcohol is a product of the fermentation of sugars and starch.
22. Vitamin E is a natural antioxidant.
23. Tertiary alcohols undergo oxidation to produce ketones.
24. Hydrogen bonding accounts for the water solubility of certain alcohols.
25. Alcohols have a higher boiling point than ethers of similar molecular weight.
26. Oxidation of a thiol produces a disulfide.
27. Two Hg^{2+} ions are required for the reaction with one thiol.
28. Ethers can form hydrogen bonds with water molecules.
29. Thiols are responsible for the pleasant fragrance of many flowers.
30. Ethylene glycol is used as a moisturizer in foods.

Chapter 14: Aldehydes and Ketones

CHAPTER OUTLINE

14.1 The Nomenclature of Aldehydes and Ketones
14.2 Physical Properties
14.3 Chemical Properties
14.4 Important Aldehydes and Ketones

LEARNING OBJECTIVES/ASSESSMENT

When you have completed your study of this chapter, you should be able to:

1. Recognize the carbonyl group in compounds and classify the compounds as aldehydes or ketones. (Section 14.1; Exercise 14.4)
2. Assign IUPAC names to aldehydes and ketones. (Section 14.1; Exercise 14.6)
3. Compare the physical properties of aldehydes and ketones with those of compounds in other classes. (Section 14.2; Exercise 14.16)
4. Write key reactions for aldehydes and ketones. (Section 14.3; Exercise 14.42)
5. Give specific uses for aldehydes and ketones. (Section 14.4; Exercise 14.52)

SOLUTIONS FOR THE END OF CHAPTER EXERCISES

THE NOMENCLATURE OF ALDEHYDES AND KETONES (SECTION 14.1)

14.2

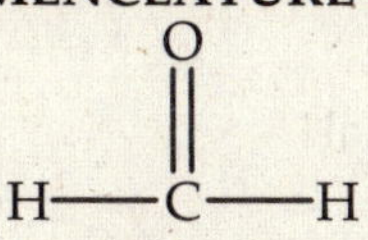
aldehyde

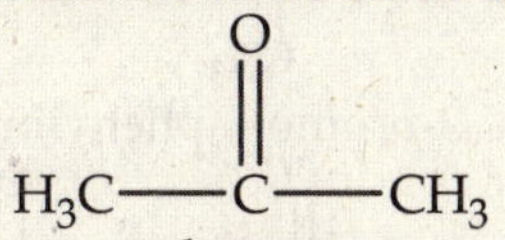
ketone

☑14.4 a.

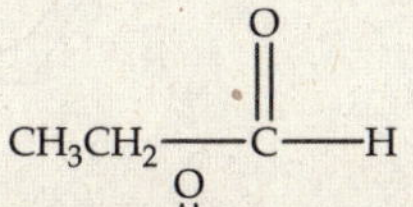
aldehyde

b.

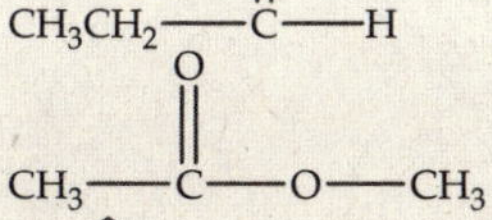
neither (ester)

c.

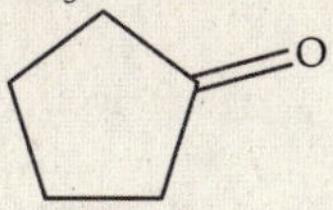
ketone

d.

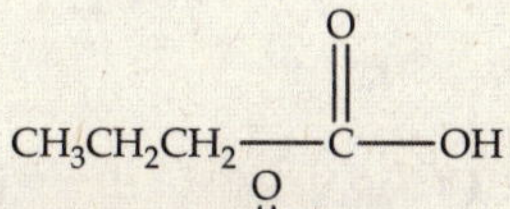
neither (carboxylic acid)

e.

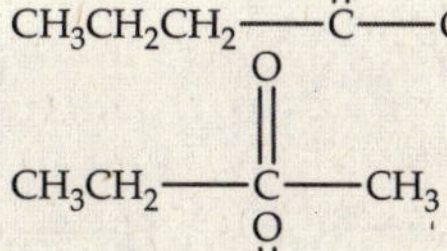
ketone

f.

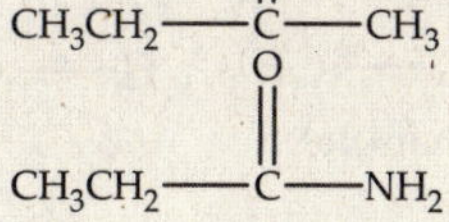
neither (amide)

☑14.6 a.

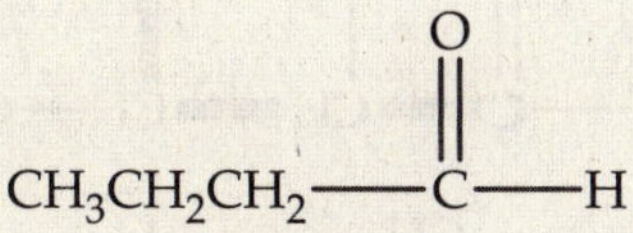
butanal

b.

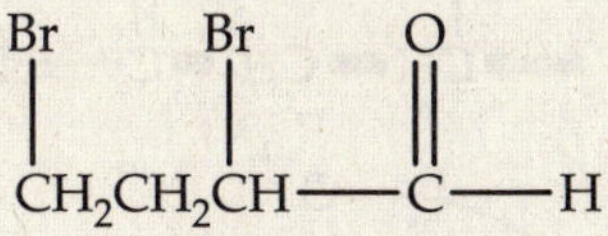
2,4-dibromobutanal

c.

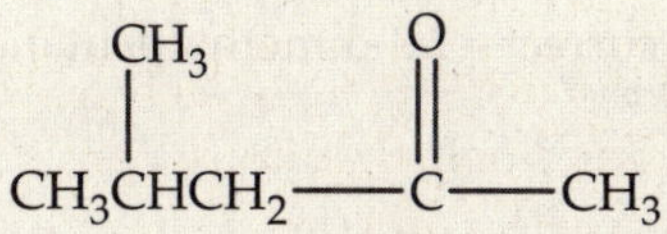
4-methyl-2-pentanone

d.

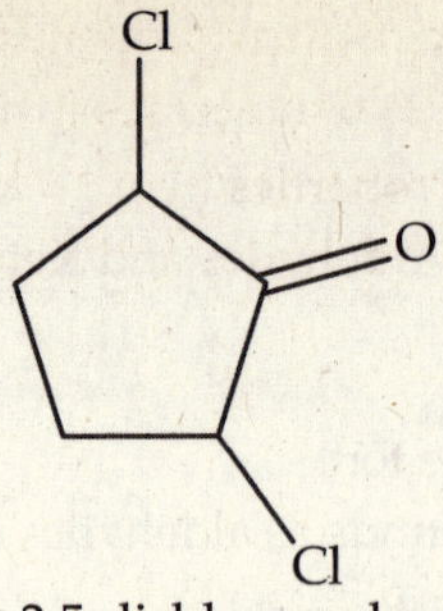

2,5-dichlorocyclopentanone

e.

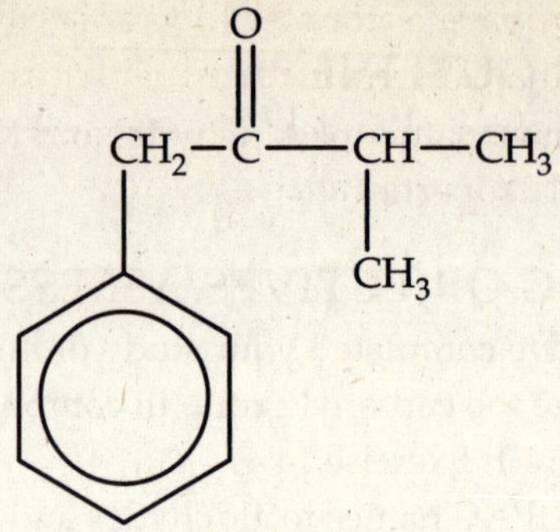

3-methyl-1-phenyl-2-butanone

14.8 a. propanal

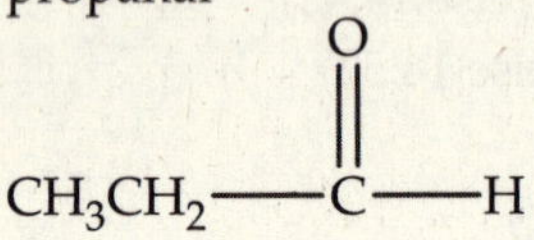

c. 2,2-dimethylcyclopentanone

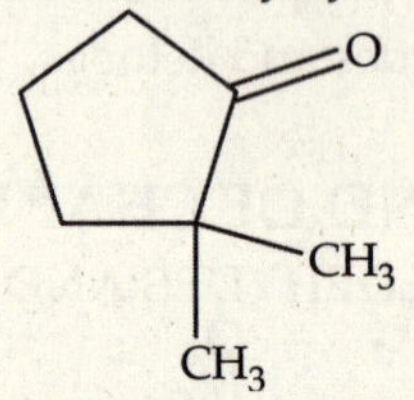

b. 3-methyl-2-butanone

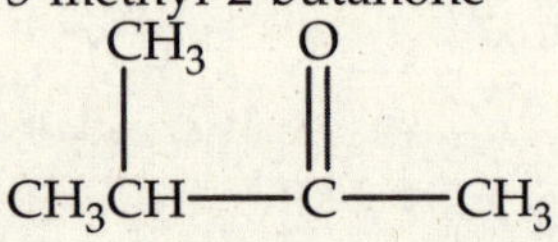

d. 3-bromo-4-phenylbutanal

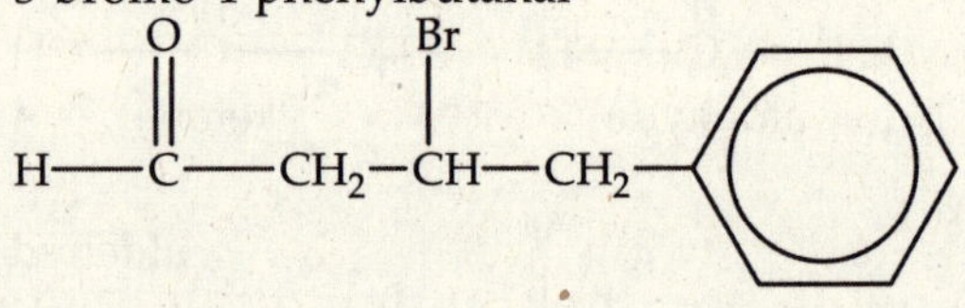

14.10

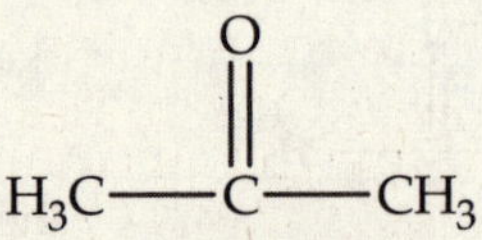

2-propanone

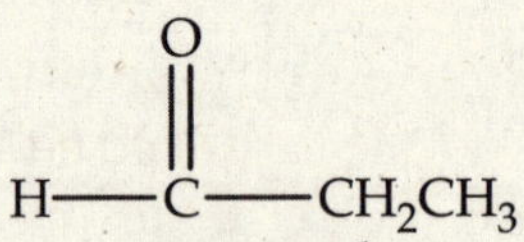

propanal

14.12 a. incorrect = 3-ethyl-2-methylbutanal
correct = 2,3-dimethylpentanal

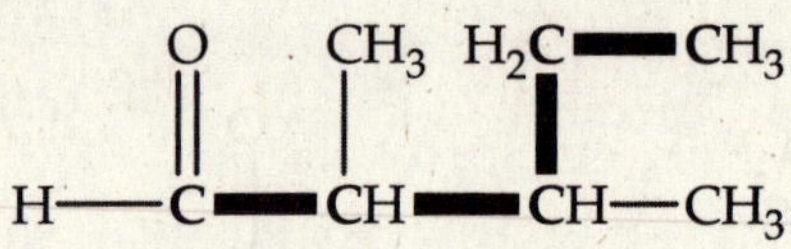

b. incorrect = 2-methyl-4-butanone
correct = 3-methylbutanal

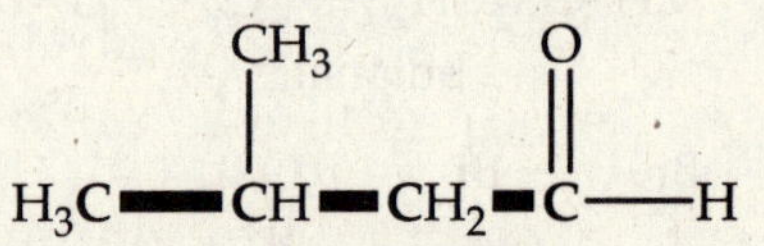

c. incorrect = 4,5-dibromocyclopentanone
correct = 2,3-dibromocyclopentanone

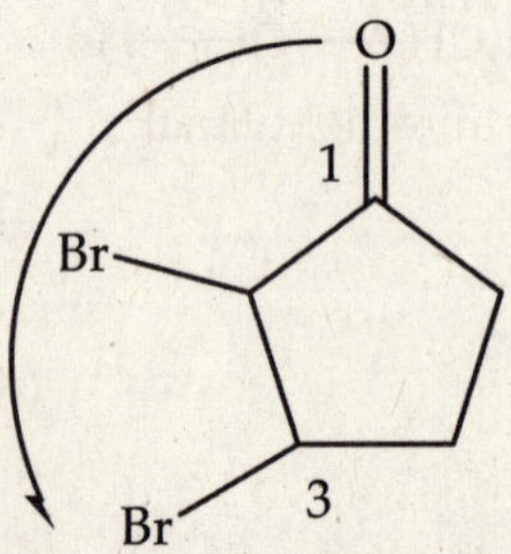

PHYSICAL PROPERTIES (SECTION 14.2)

14.14 The acetone dissolves the remaining water in the glassware and the bulk of the mixture is discarded. The remaining traces of acetone evaporate quickly because it is much more volatile than water.

☑14.16 Propane is nonpolar and only experiences dispersion forces with other propane molecules. Ethanal is polar and experiences dipolar forces with other ethanal molecules in addition to dispersion forces. The stronger interparticle forces in ethanal cause it to have a higher boiling point than propane.

14.18 a.

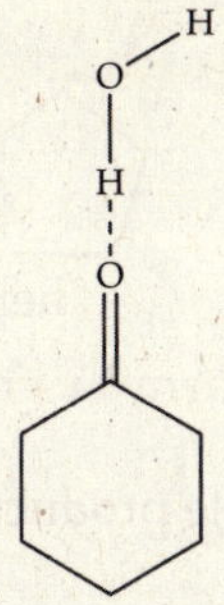

b.

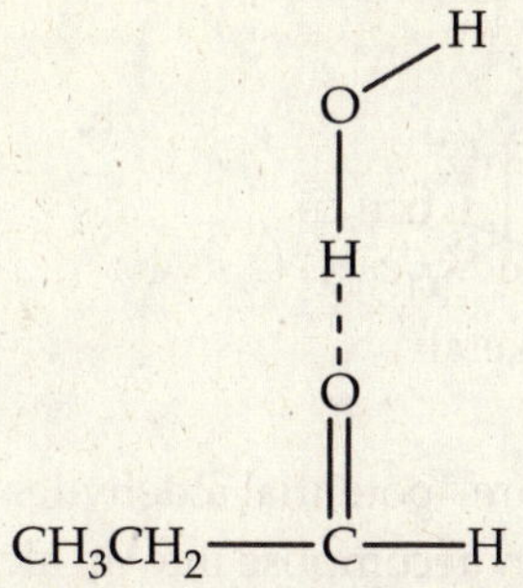

14.20

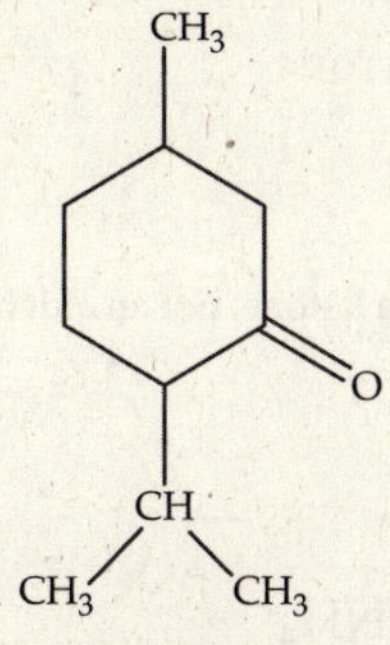

menthone

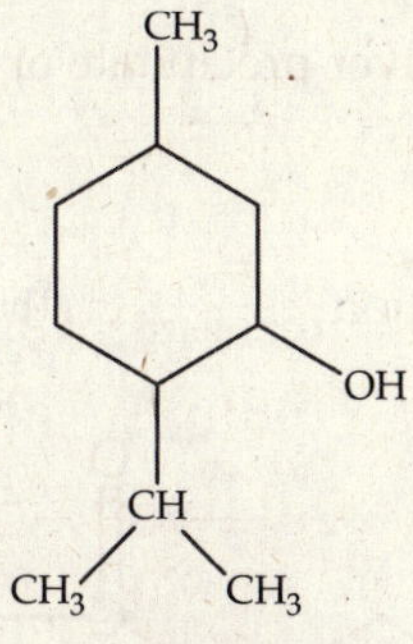

menthol

Both molecules experience dispersion forces, but menthol also experiences hydrogen bonding, while menthone experiences dipolar interparticle forces. Menthol is more likely to be the solid because hydrogen bonding is a stronger interparticle force than dipolar forces. Menthone is more likely to be the liquid.

CHEMICAL PROPERTIES (SECTION 14.3)

14.22 a.

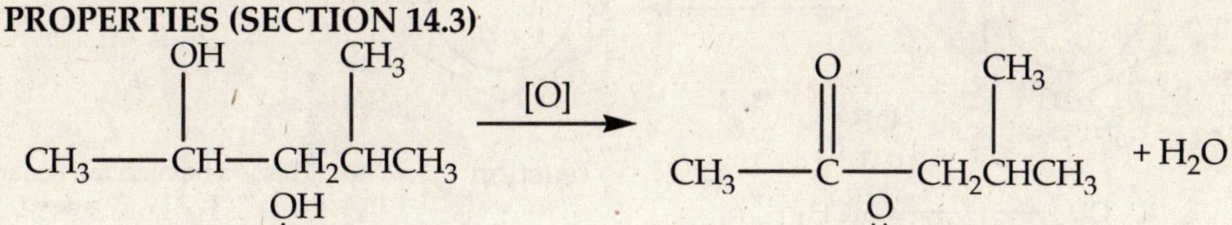

b.

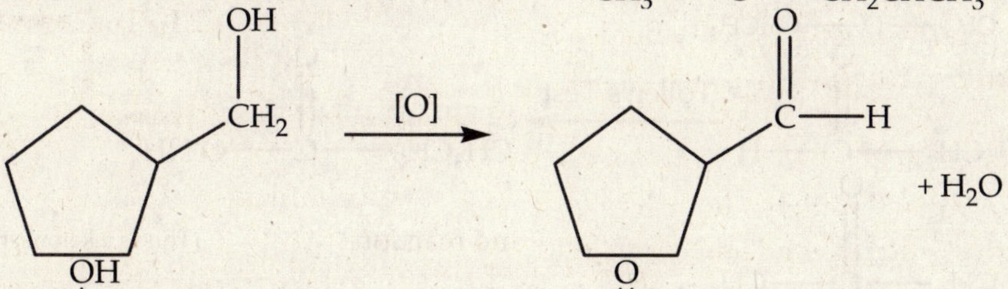

c.

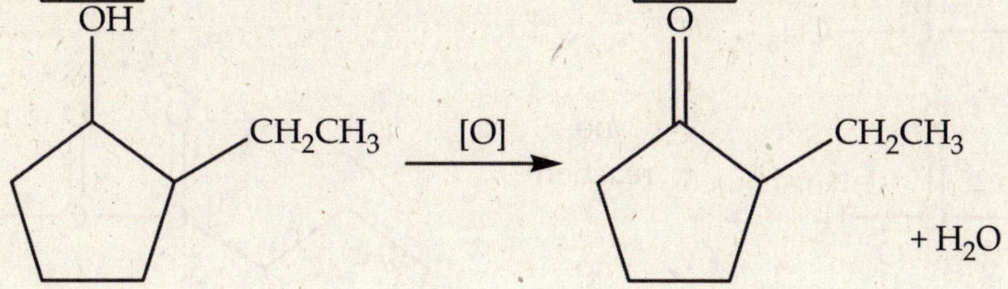

14.24 a.

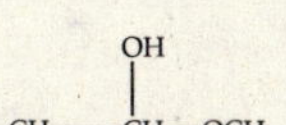

hemiacetal

b.

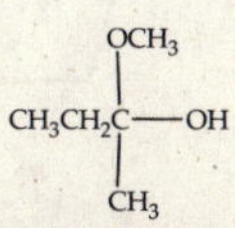

hemiketal

c.

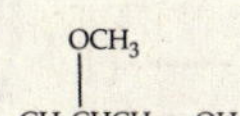

neither

(alcohol and ether functional groups)

d.

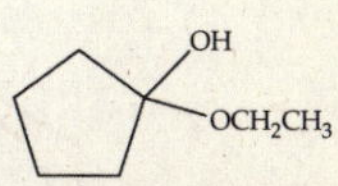

hemiketal

14.26 a. 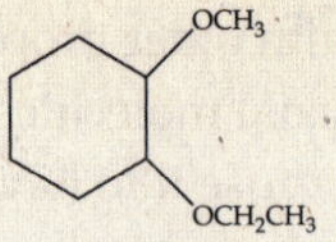

neither
(2 ether functional groups)

b. 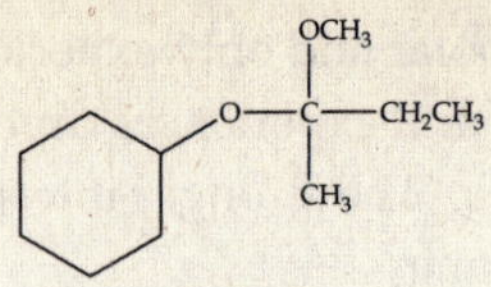

ketal

c. 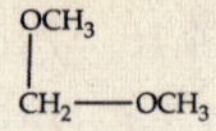

acetal

14.28 a. 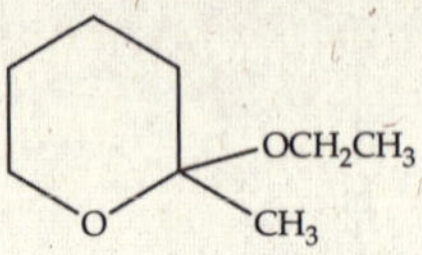

ketal

b. 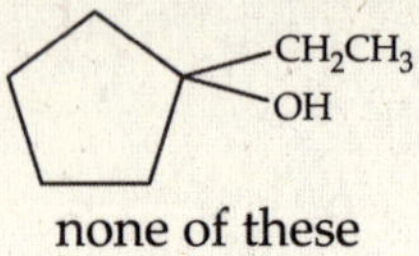

none of these
(alcohol)

c. 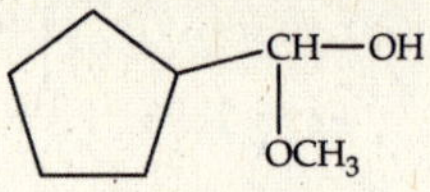

hemiacetal

14.30 Hemiacetals are "potential aldehydes" because hemiacetals are not stable products. The hemiacetal can decompose into an alcohol and an aldehyde.

14.32 A positive Tollen's test is the appearance of a silver precipitate or mirror.

14.34 a. 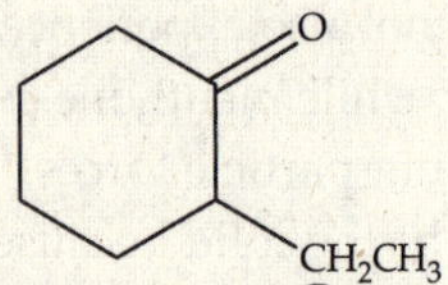

no reaction (This is a ketone, not an aldehyde.)

b. 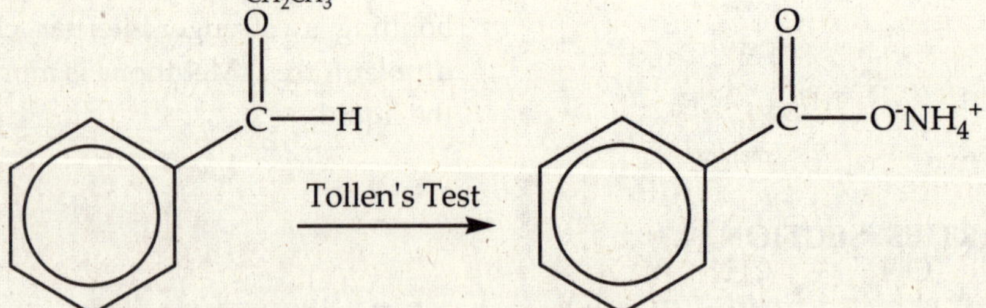

c. 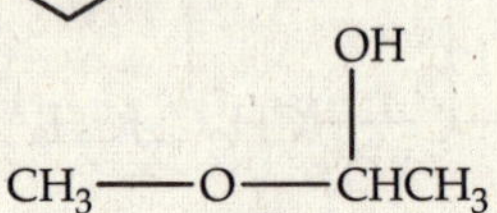

no reaction (This is a hemiacetal, not an aldehyde.)

d. 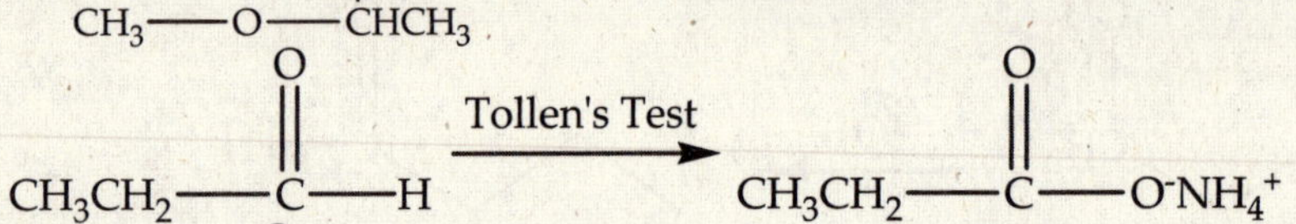

e. 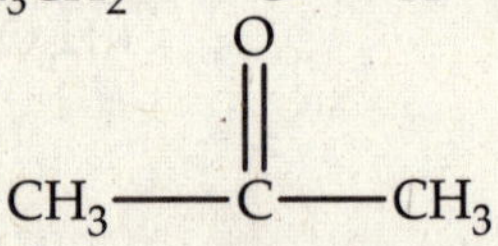

no reaction (This is a ketone, not an aldehyde.)

14.36 a. 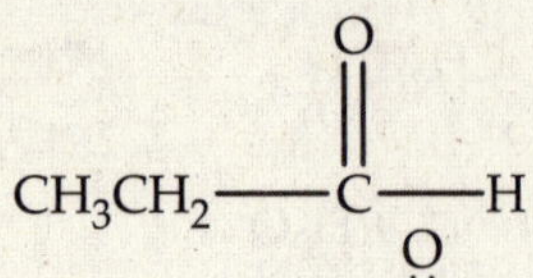

no reaction

b. 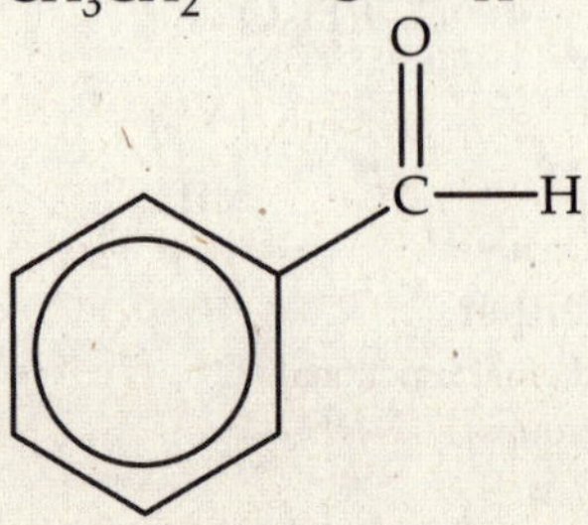

no reaction

c. 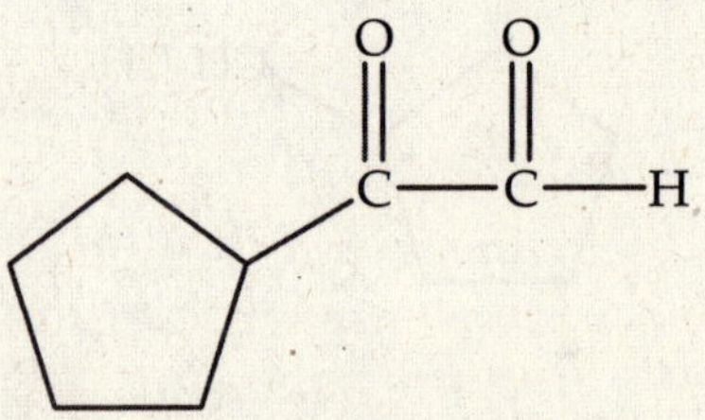

reacts

d. 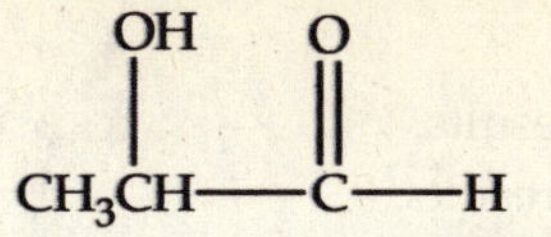 reacts

e. 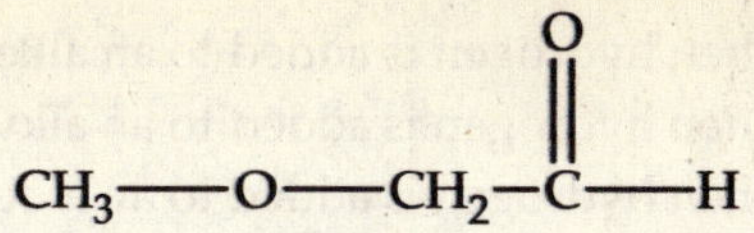 no reaction

14.38 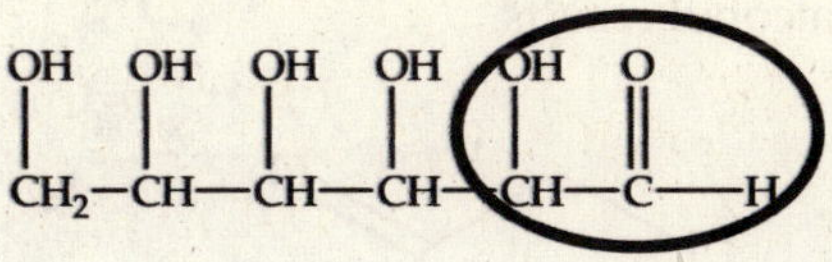 (an aldehyde with an adjacent alcohol group)

14.40

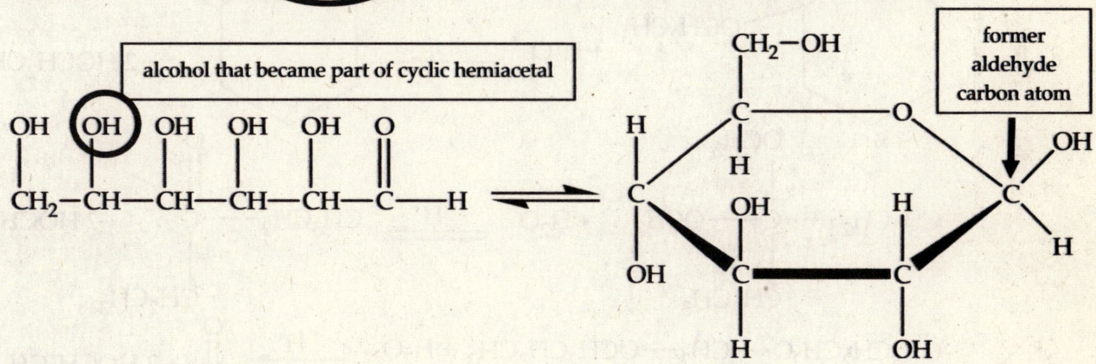

☑14.42 a.

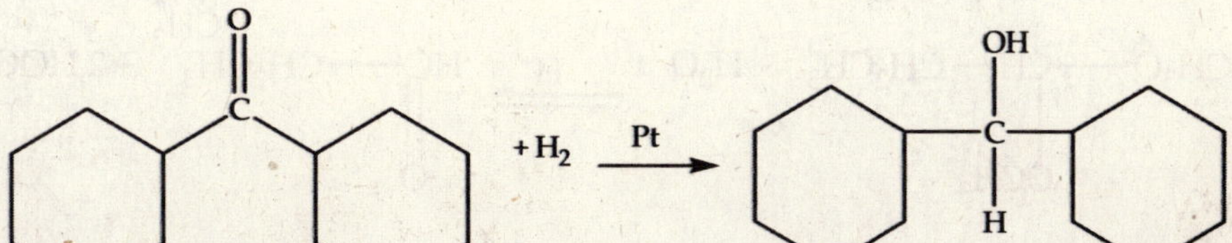

b.

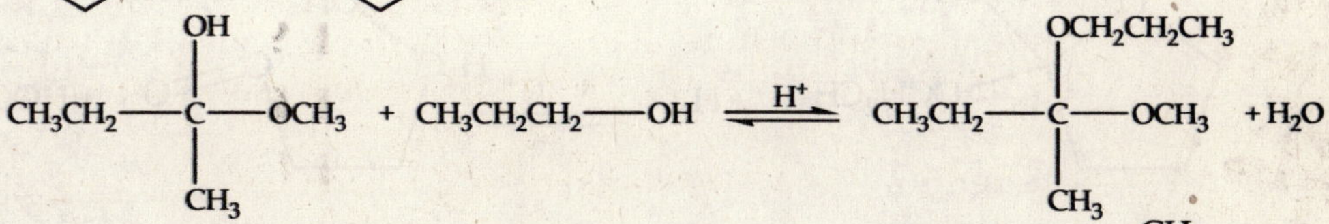

c.

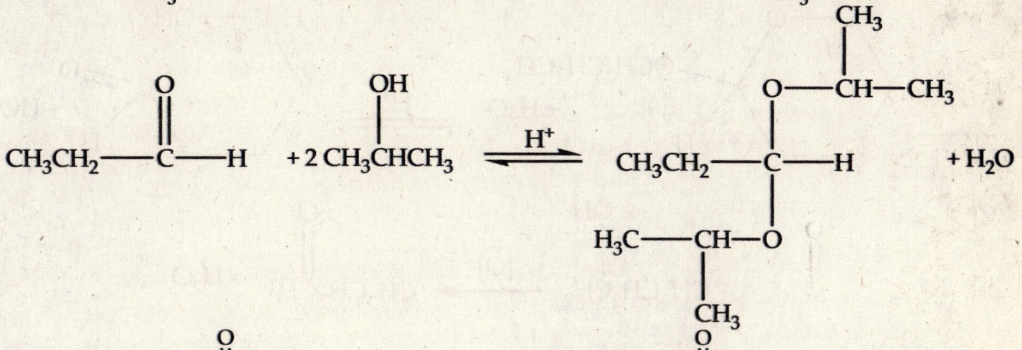

d.

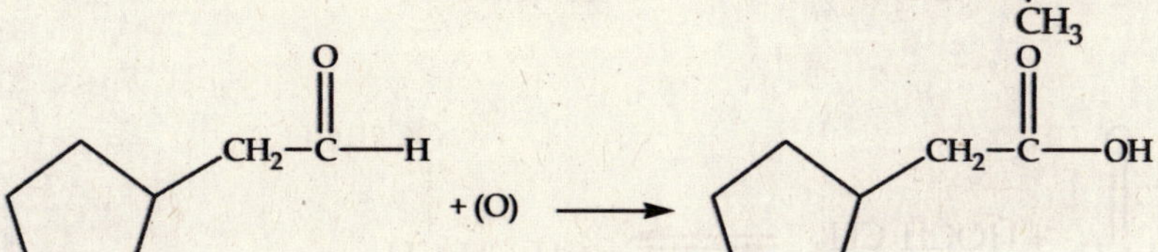

e.

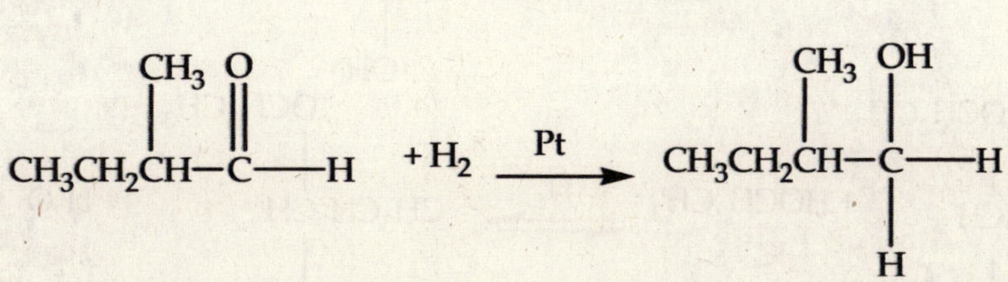

14.44 When hydrogen is added to an alkene, an alkane results.
When hydrogen is added to an alkyne, an alkene or alkane results.
When hydrogen is added to an aldehyde, a primary alcohol results.
When hydrogen is added to a ketone, a secondary alcohol results.

14.46 a.

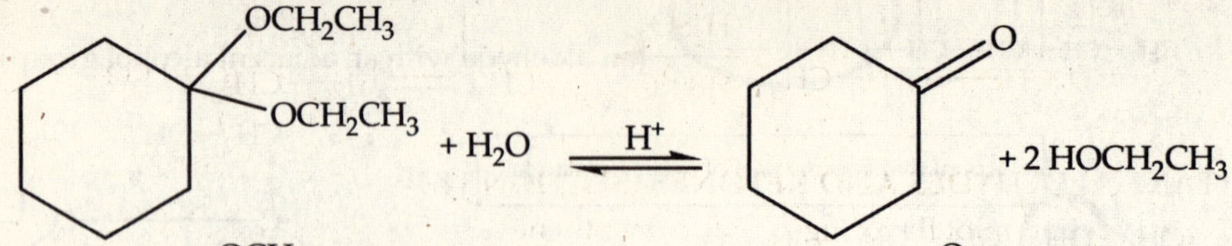

b.

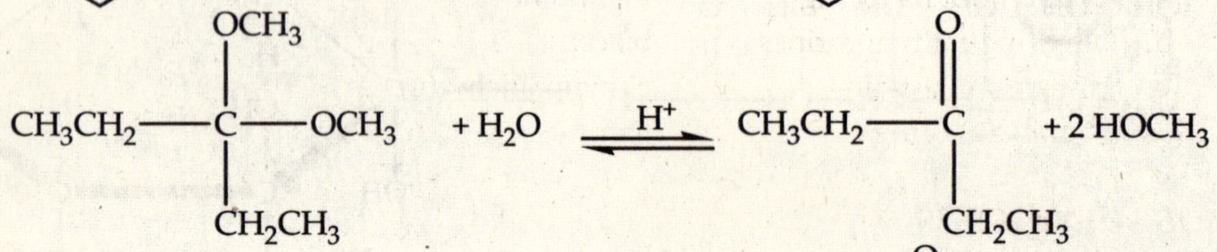

c.

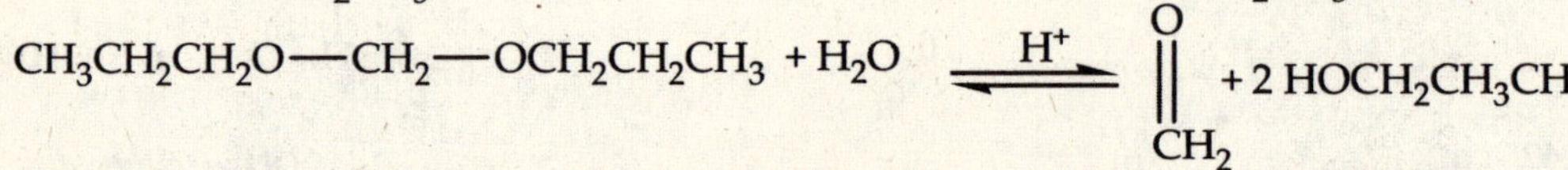

d.

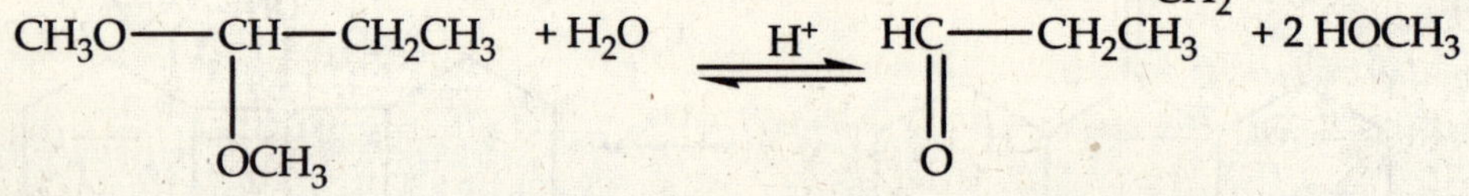

14.48 a.

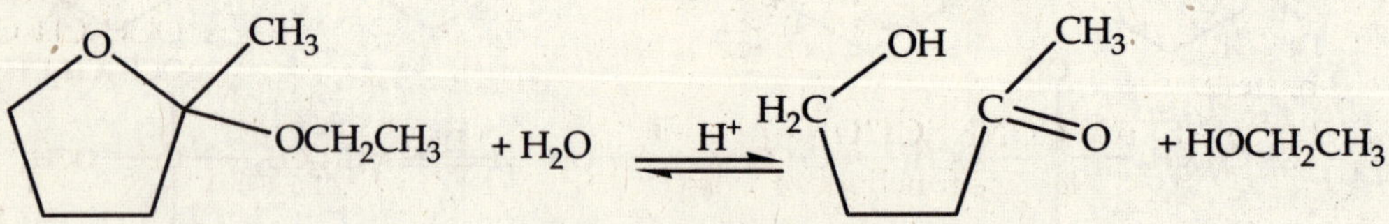

b.

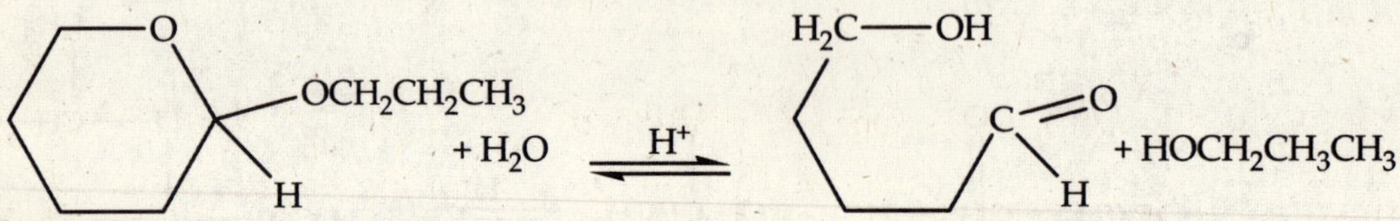

14.50 a.

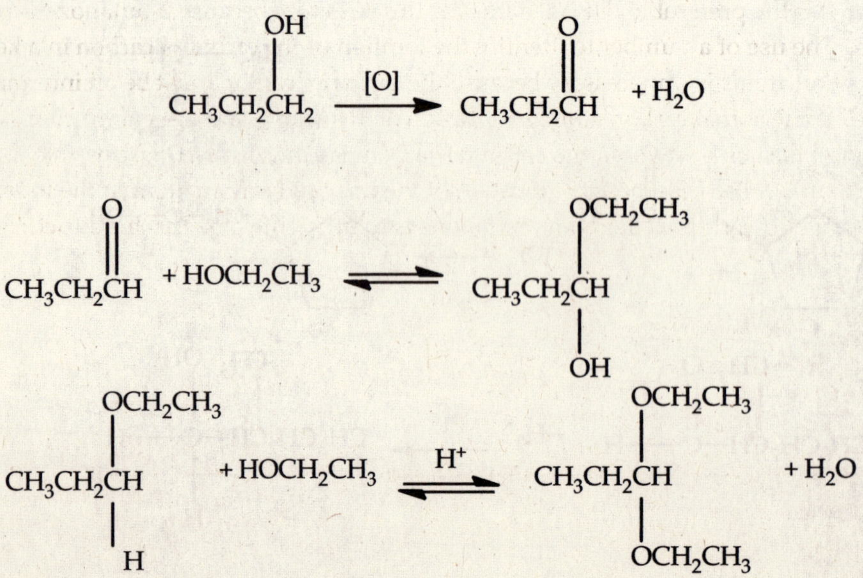

b.
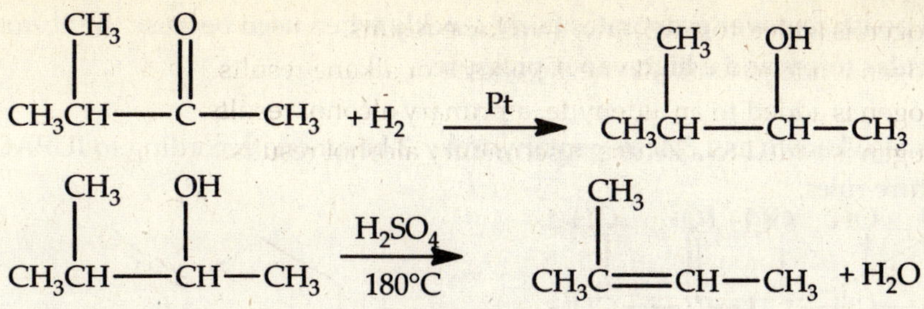

IMPORTANT ALDEHYDES AND KETONES (SECTION 14.4)

☑14.52	a.	peppermint flavoring	menthone
	b.	flavoring for margarines	biacetyl
	c.	cinnamon flavoring	cinnamaldehyde
	d.	vanilla flavoring	vanillin

ADDITIONAL EXERCISES

14.54 $$500\text{ g }C_2H_4O_2\left(\frac{1\text{ mole }C_2H_4O_2}{60.0\text{ g }C_2H_4O_2}\right)\left(\frac{1\text{ mole }C_2H_4O}{1\text{ mole }C_2H_4O_2}\right)\left(\frac{44.0\text{ g }C_2H_4O}{1\text{ mole }C_2H_4O}\right)\left(\frac{100\text{ g }C_2H_4O\text{ added}}{79.6\text{ g }C_2H_4O\text{ react}}\right) = 461\text{ g }C_2H_4O$$

14.56
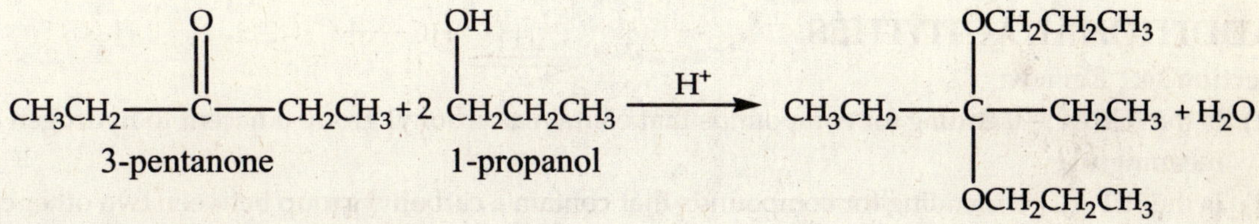

ALLIED HEALTH EXAM CONNECTION

14.58
a. RCHO = aldehyde functional group
b. ROH = alcohol functional group
c.
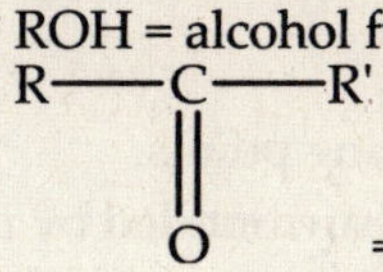
= ketone functional group

CHEMISTRY FOR THOUGHT

14.60 Butanone is the preferable IUPAC name for this structure because 2-butanone is repetitive in nature. The use of a number to identify the location of the carbonyl carbon in a ketone with four carbon atoms is unnecessary because the carbonyl carbon must be an interior carbon atom (not a terminal carbon atom, like carbon atoms 1 or 4) and the parent chain will be numbered in such a way that the carbonyl carbon has the lower of the possible numbers (2 rather than 3). The only possible location of the carbonyl carbon atom in the four carbon parent chain is carbon 2; therefore, butanone is a sufficient name for this structure.

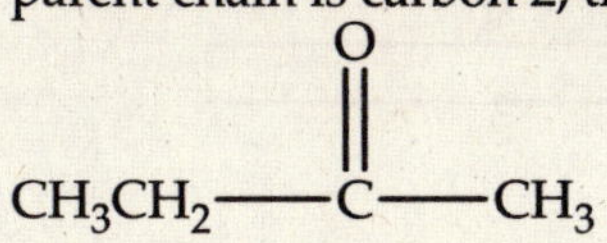

14.62
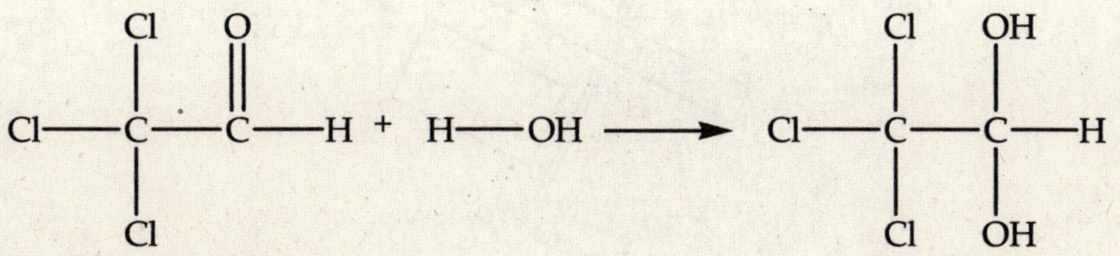

trichloroacetaldehyde — chloral hydrate

14.64 Fingernail polish remover evaporates fairly quickly when used because the acetone has weak intermolecular forces and a high vapor pressure.

14.66 Fructose is also known as 1,3,4,5,6-pentahydroxy-2-hexanone according to IUPAC nomenclature rules.

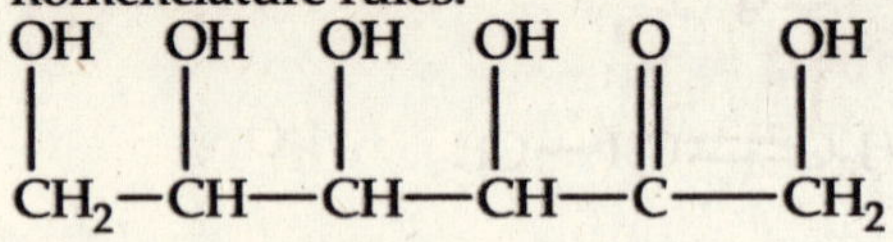

14.68 The acetyl group of maltose is circled below:

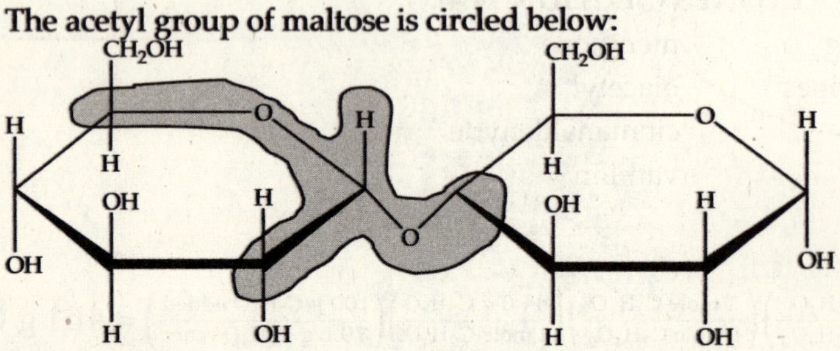

ADDITIONAL ACTIVITIES

Section 14.1 Review:

(1) Is the IUPAC *–al* ending for compounds that contain a carbonyl group adjacent to hydrogen atom meaningful?
(2) Is the IUPAC *–one* ending for compounds that contain a carbonyl group between two other carbon atoms meaningful?
(3) How is the longest carbon chain numbered for aldehydes?
(4) How is the longest carbon chain numbered for ketones?

Section 14.2 Review:

(1) Rank alcohols, ethers, and aldehydes/ketones in order of increasing boiling points.
(2) What intermolecular forces do aldehydes and ketones experience when surrounded by molecules of the same type?

Section 14.3 Review: Complete the following reaction diagrams.

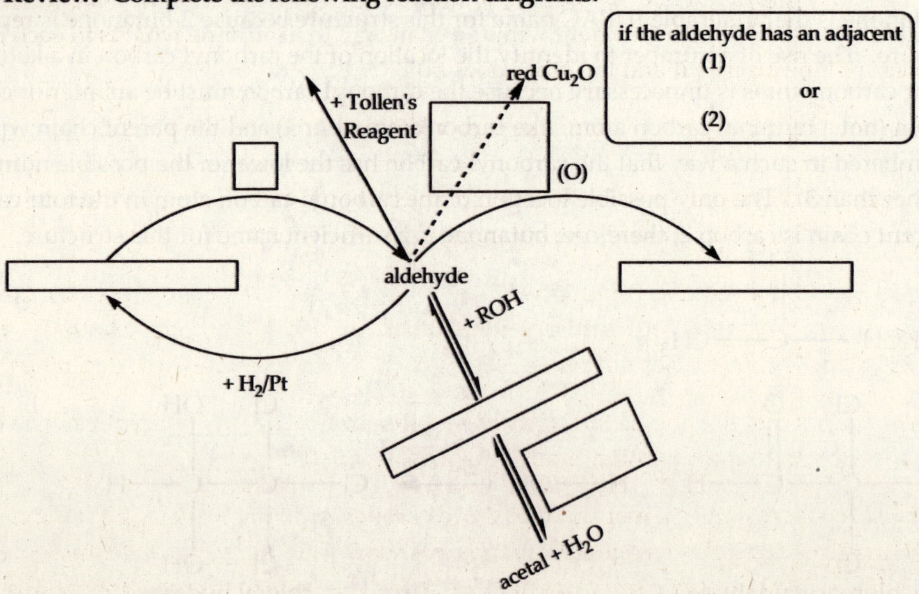

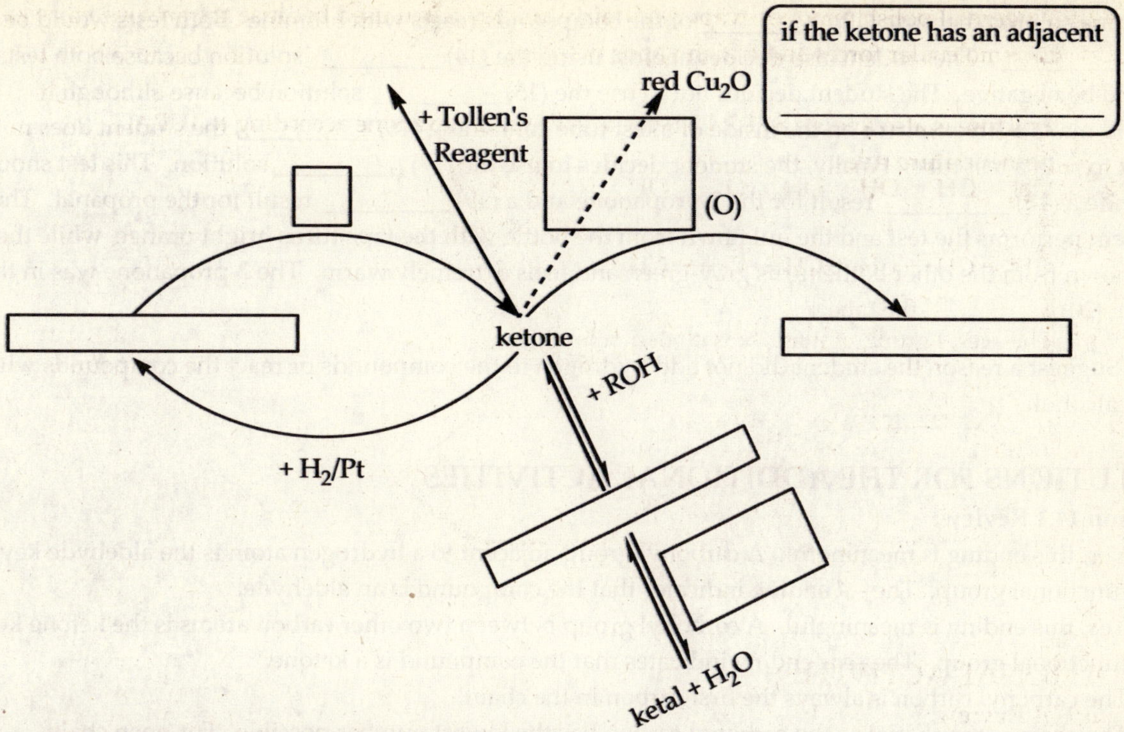

(1) What do hemiacetals and hemiketals have in common?
(2) What do hemiacetals and acetals have in common?
(3) What do hemiketals and ketals have in common?
(4) What do acetals and ketals have in common?
(5) How does hydrolysis differ from hydration?

Section 14.4 Review:
(1) What is the IUPAC name for formaldehyde?
(2) What is the IUPAC name for acetone?
(3) Do some aldehydes and ketones have an associated smell?

Tying It All Together with a Laboratory Application:
An instructor transfers approximately 50 mL of 2-propanone and propanal into two separate unlabeled containers and hands the bottles to a student with the challenge to determine what is in each bottle. One bottle has a piece of tape around it and the other does not.

The compound with the (1) __________ name 2-propanone has the common name (2) __________. 2-propanone is frequently used as (3) __________ remover. Propanal has an overwhelming fruity odor. The student knows that the two compounds have different (4) __________, but unfortunately, this student does not have the ability to smell the compounds.

The student also knows that the two compounds (5) __________ dissolve in water because they will experience (6) __________ intermolecular forces with water. These compounds are both (7) __________ molecules. Mixing these compounds with water (8) __________ differentiate these compounds.

The student also knows that the two compounds have different (9) __________ points; however, the student does not want to heat these compounds because they are (10) __________.

The student considers performing a reaction that could serve as a qualitative test. First the student draws and labels the structural formulas for the compounds (shown in (11) __________). Bromine, potassium dichromate, Tollen's, and Benedict's solutions are available. The student decides against using the bromine solution because (12) _________ of the compounds reacts with bromine. Both tests would be (13) _________. The student also decides against using the (14) _________ solution because both tests would be negative. The student decides not to use the (15) _________ solution because although it would make a silver mirror on the inside of a test tube containing (16) _________, the student does not want to ruin a test tube. Finally, the student decides to use the (17) _________ solution. This test should provide a (18) _________ result for the 2-propanone and a (19) _________ result for the propanal. The student performs the test and the unknown from the bottle with the tape turns bright orange, while the unknown from the other bottle turns gray-green and feels extremely warm. The 2-propanone was in the bottle (20) _________ the tape.

(21) Suggest a reason the student did not add hydrogen to the compounds or react the compounds with alcohol.

SOLUTIONS FOR THE ADDITIONAL ACTIVITIES

Section 14.1 Review:

(1) Yes, this ending is meaningful. A carbonyl group adjacent to a hydrogen atom is the aldehyde key functional group. The *–al* ending indicates that the compound is an aldehyde.
(2) Yes, this ending is meaningful. A carbonyl group between two other carbon atoms is the ketone key functional group. The *–one* ending indicates that the compound is a ketone.
(3) The carbonyl carbon is always the first carbon in the chain.
(4) The chain is numbered so the carbonyl carbon has the lowest number possible. For open chain compounds, the carbonyl carbon will always have a number 2 or higher. For closed chain (cyclic) compounds, the carbonyl carbon will be given the number 1.

Section 14.2 Review:

(1) ethers < aldehydes/ketones < alcohols
(2) dipolar forces

Section 14.3 Review:

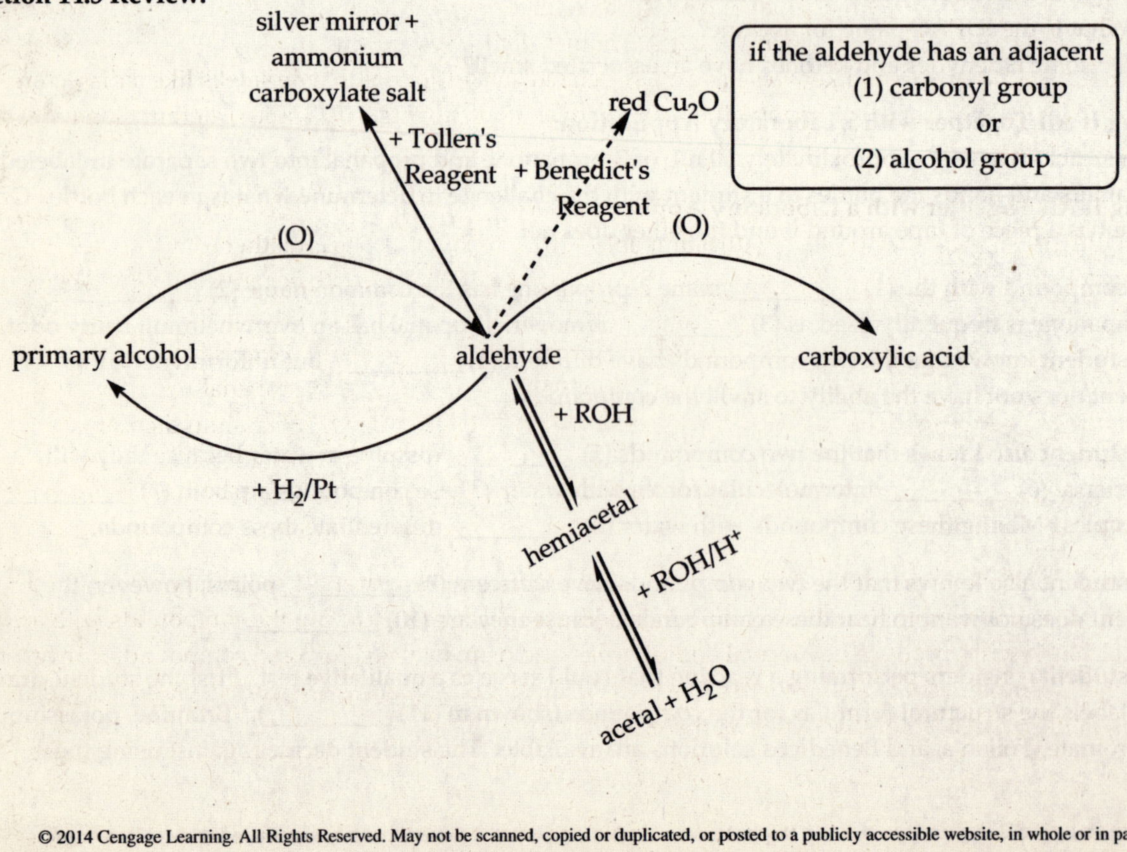

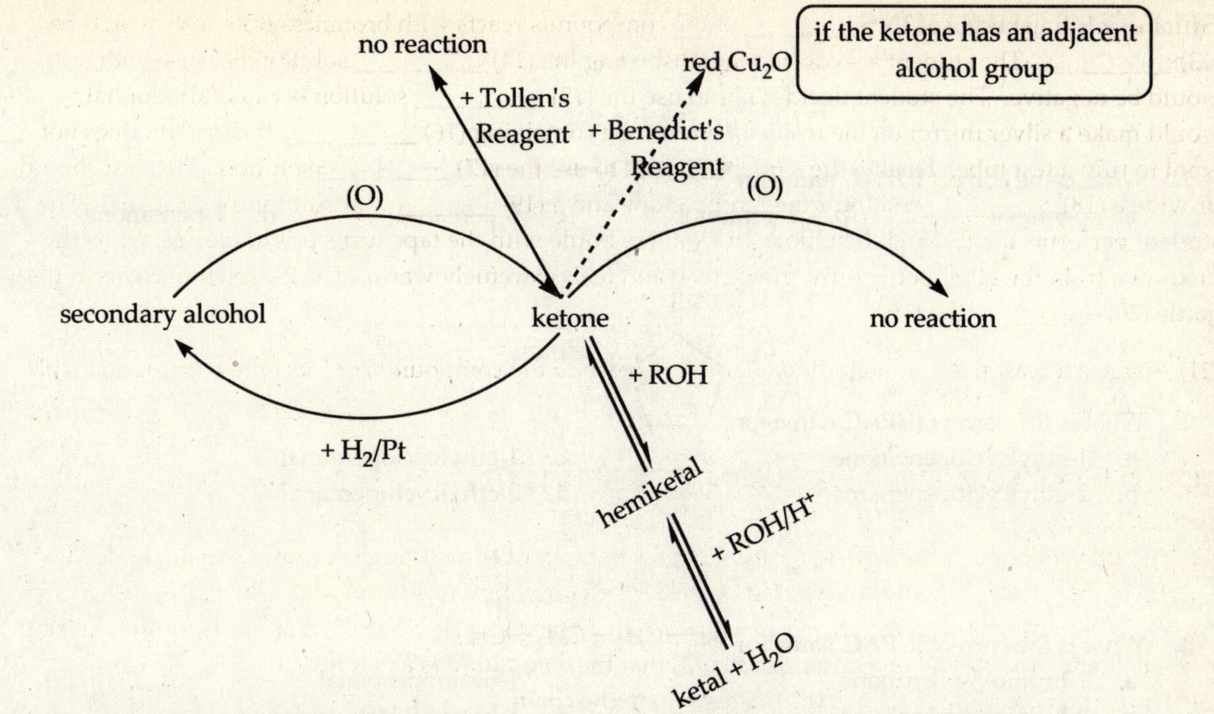

(1) Hemiacetal and hemiketals both have an –OH and an –OR bonded to the same carbon atom.
(2) Hemiacetals and acetals both have an –H and an –OR bonded to the same carbon atom.
(3) Hemiketals and ketals both have two –R groups and an –OR bonded to the same carbon atom.
(4) Acetals and ketals both have two –OR groups bonded to the same carbon atom.
(5) Hydrolysis involves breaking one of the bonds in a molecule using water. Hydration involves the addition of water to molecule (usually across a double bond).

Section 14.4 Review:
(1) Formaldehyde is called methanal in the IUPAC nomenclature system.
(2) Acetone is called 2-propanone in the IUPAC nomenclature system.
(3) Yes, some aldehydes and ketones have an associated smell. Formaldehyde smells like a dissecting lab, acetone is the smell of fingernail polish remover, and Table 14.3 lists other fragrant aldehydes and ketones.

Tying It All Together with a Laboratory Application:
(1) IUPAC
(2) acetone
(3) fingernail polish
(4) odors
(5) will
(6) hydrogen bonding
(7) polar
(8) will not
(9) boiling
(10) flammable
(11)

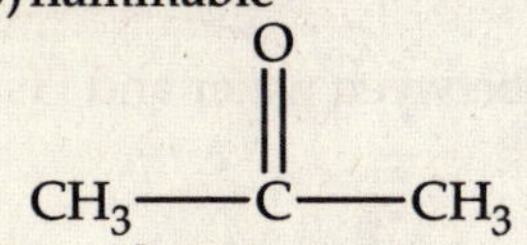

2-propanone

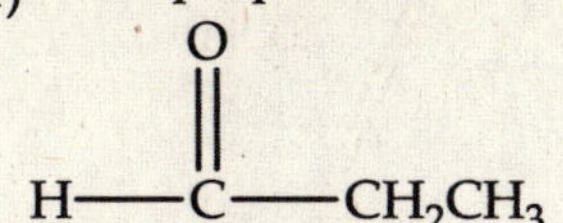

propanal

(12) neither
(13) negative
(14) Benedict's
(15) Tollen's
(16) propanal
(17) potassium dichromate
(18) negative
(19) positive
(20) with

(21) Reacting the compounds with hydrogen would produce a primary and a secondary alcohol; however, those compounds are colorless just like the reactants. Reacting the compounds with an alcohol would produce hemiacetal and/or acetal and hemiketal and /or ketal compounds; however, those compounds are also colorless like the reactants. Consequently, the student chose a colorful, qualitative test rather than either of these reactions.

SELF-TEST QUESTIONS

Multiple Choice

1. What is the correct IUPAC name for 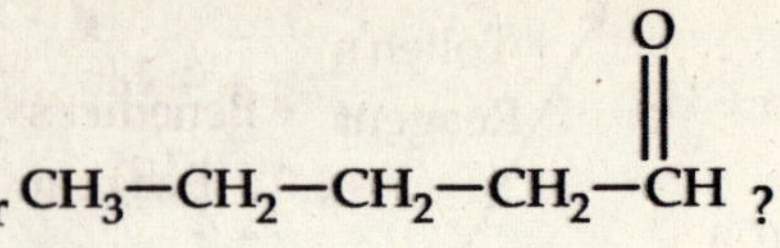 ?
   a. pentanal b. 1-pentanol c. 5-pentanal d. 1-pentanone

2. What is the correct IUPAC name for 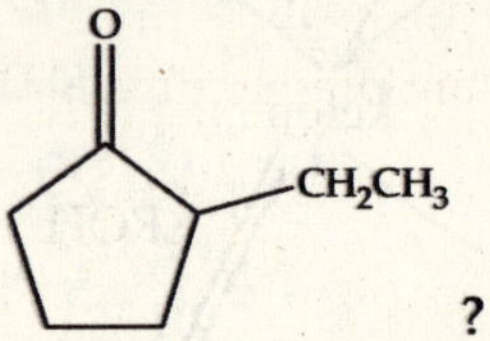 ?
   a. 1-ethylcyclopentanone
   b. 2-ethylcyclopentanone
   c. 1-ethylcyclopentanal
   d. 2-ethylcyclopentanal

3. What is the correct IUPAC name for 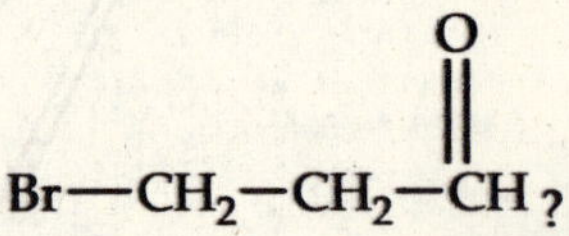 ?
   a. 1-bromo-3-propanone
   b. 3-bromopropanal
   c. 1-bromopropanal
   d. 3-bromo-1-propanone

4. Which of the following pure compounds can exhibit hydrogen bonding with other molecules of the same compound?
   a. 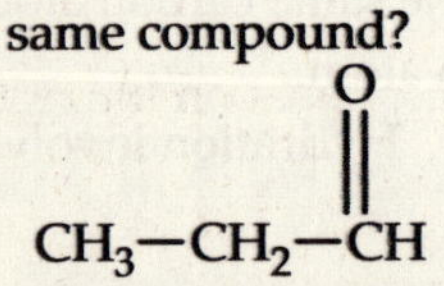
   b. $CH_3-CH_2-CH_2-OH$
   c. 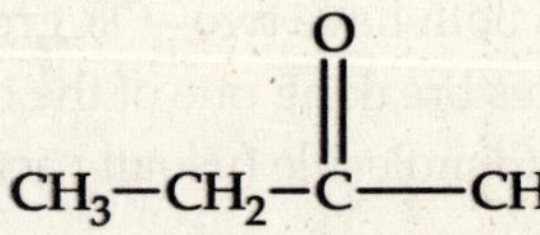
   d. more than one response is correct

5. Which of the chemical species below is a key component of a reagent mixture that may be used to test for the presence of an aldehyde?
   a. Ag^+ b. Br_2 c. Cu_2O d. NaOH

6. What are the products of the reaction between water and H^+ with 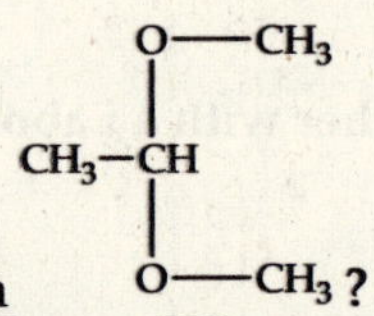 ?
   a. 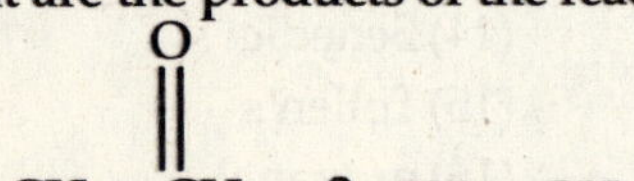
   b. 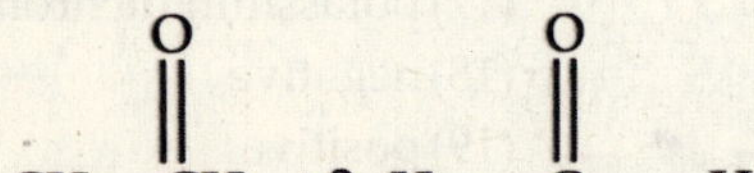
   c. 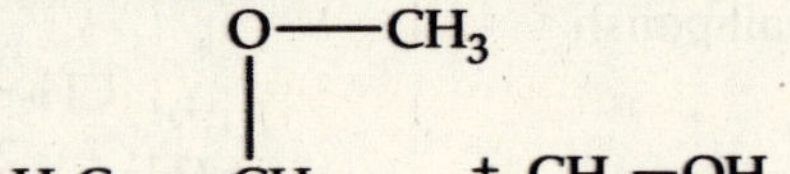
   d. 

7. Which of the following compounds could be oxidized by Benedict's reagent?

a. 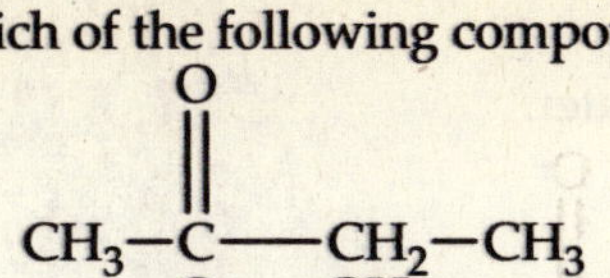

b. 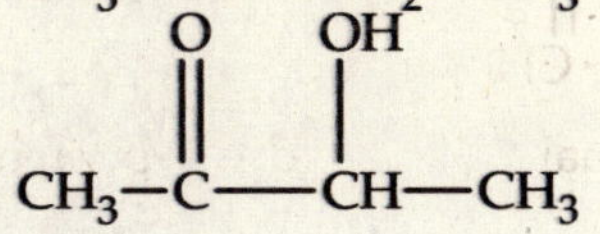

c. 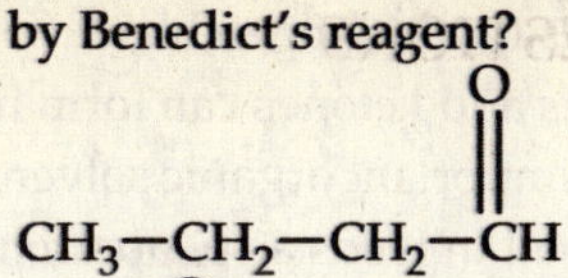

d. 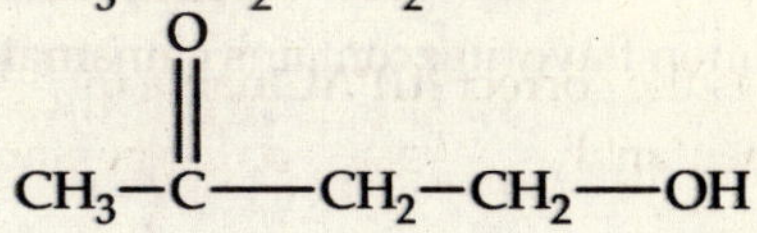

8. Which of the following is an important preservative for biological specimens?

a. 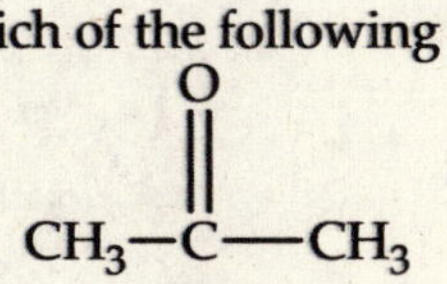

b. 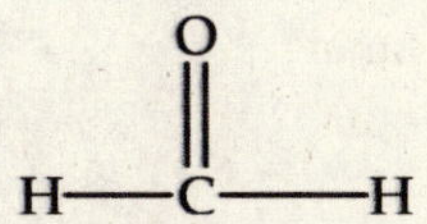

c. 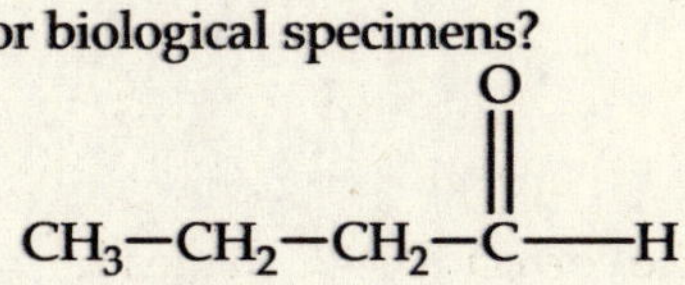

d. 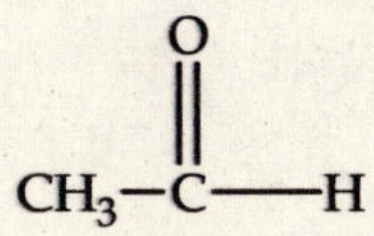

9. Which of following is a flavoring for margarine?

a. biacetyl b. benzaldehyde c. 2-butanone d. vanillin

10. What does the hydrogenation of an aldehyde produce?:

a. acetal b. primary alcohol c. secondary alcohol d. ketone

Matching

For each structure on the left, select the correct class of compound from the responses on the right.

11. 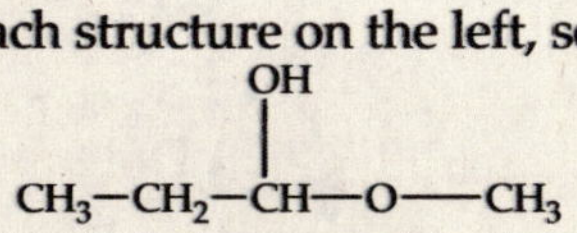

12. 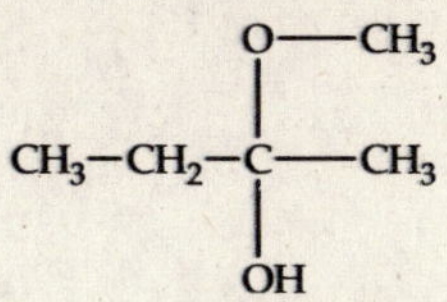

13. 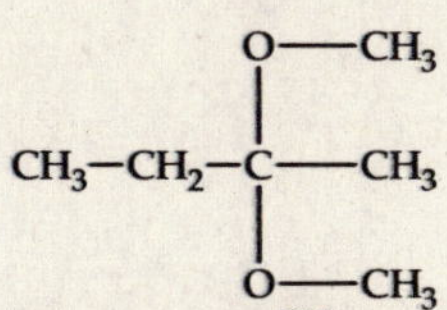

14. 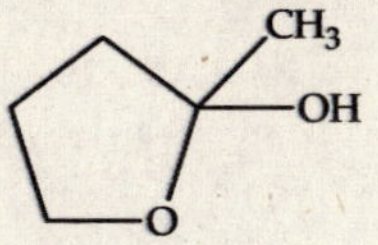

a. acetal
b. ketal
c. hemiacetal
d. hemiketal

Select the correct product for each of the reactions on the left.

15. aldehyde + [O] →
16. ketone + hydrogen →
17. aldehyde + alcohol ⇌
18. ketone + alcohol ⇌

a. hemiacetal
b. hemiketal
c. carboxylic acid
d. alcohol

True-False

19. Both aldehydes and ketones contain a carbonyl group.
20. Pure ketones can hydrogen bond.

21. Pure aldehydes can hydrogen bond.
22. Both aldehydes and ketones can form hydrogen bonds with water.
23. Acetone is an important organic solvent.
24. Camphor is used in foods as peppermint flavoring.
25. Cinnamon flavoring contains cinnamaldehyde.

Chapter 15: Carboxylic Acids and Esters

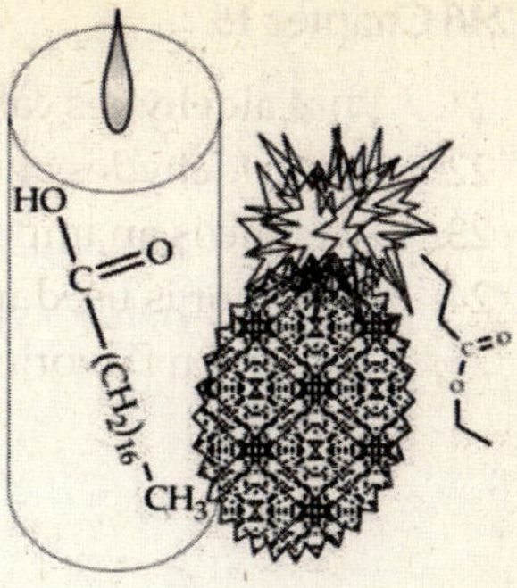

CHAPTER OUTLINE

15.1 The Nomenclature of Carboxylic Acids
15.2 Physical Properties of Carboxylic Acids
15.3 The Acidity of Carboxylic Acids
15.4 Salts of Carboxylic Acids
15.5 Carboxylic Esters
15.6 The Nomenclature of Esters
15.7 Reactions of Esters
15.8 Esters of Inorganic Acids

LEARNING OBJECTIVES/ASSESSMENT

When you have completed your study of this chapter, you should be able to:

1. Assign IUPAC names and draw structural formulas for carboxylic acids. (Section 15.1; Exercise 15.6)
2. Explain how hydrogen bonding affects the physical properties of carboxylic acids. (Section 15.2; Exercise 15.10)
3. Recognize and write key reactions of carboxylic acids. (Section 15.3; Exercise 15.26)
4. Assign common and IUPAC names to carboxylic acid salts. (Section 15.4; Exercise 15.28)
5. Describe uses for carboxylate salts. (Section 15.4; Exercise 15.32)
6. Recognize and write key reactions for ester formation. (Section 15.5; Exercise 15.36)
7. Assign common and IUPAC names to esters. (Section 15.6; Exercises 15.46)
8. Recognize and write key reactions of esters. (Section 15.7; Exercise 15.54)
9. Write reactions for the formation of phosphate esters. (Section 15.8; Exercise 15.56)

SOLUTIONS FOR THE END OF CHAPTER EXERCISES

THE NOMENCLATURE OF CARBOXYLIC ACIDS (SECTION 15.1)

15.2

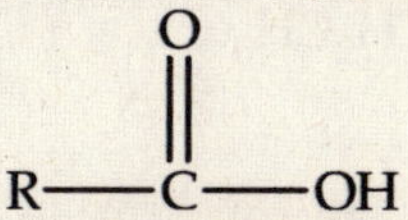

carboxylic acid

The structural features of a fatty acid are the carboxylic acid functional group and a long aliphatic tail. They are called fatty acids because they were originally isolated from fats.

15.4 The carboxylic acid present in sour milk and sauerkraut is lactic acid.

☑15.6 a.

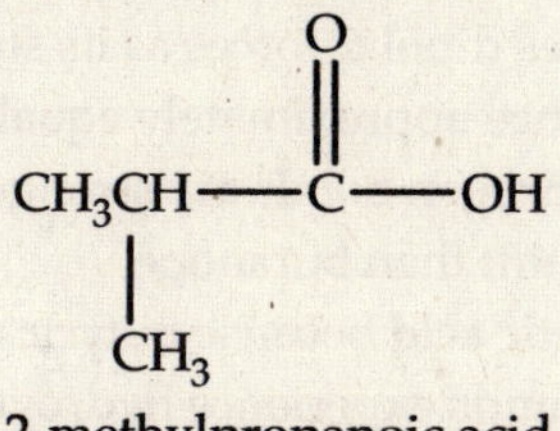

2-methylpropanoic acid

b.

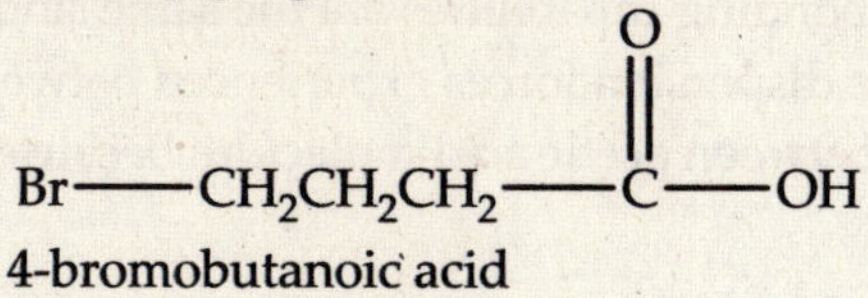

4-bromobutanoic acid

c.

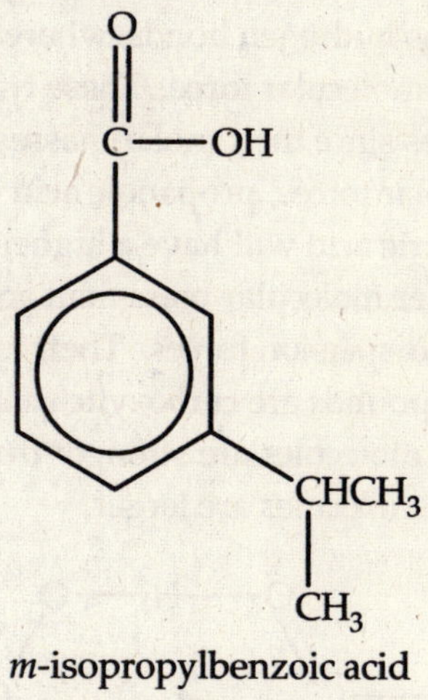

m-isopropylbenzoic acid
3-isopropylbenzoic acid

d.

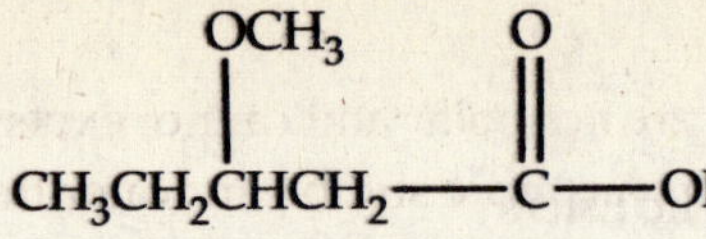

3-methoxypentanoic acid

e.

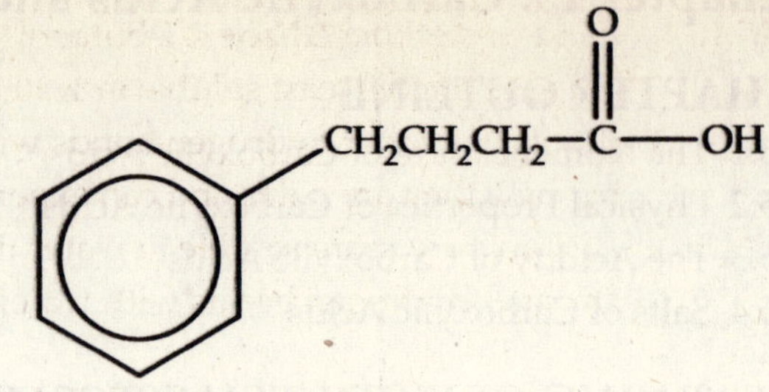

4-phenylbutanoic acid

15.8 a. pentanoic acid

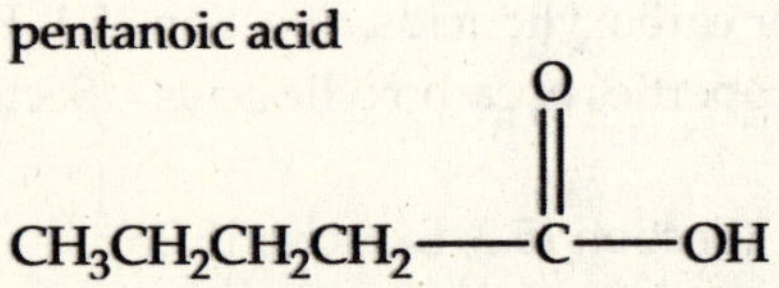

b. 2-bromo-3-methylhexanoic acid

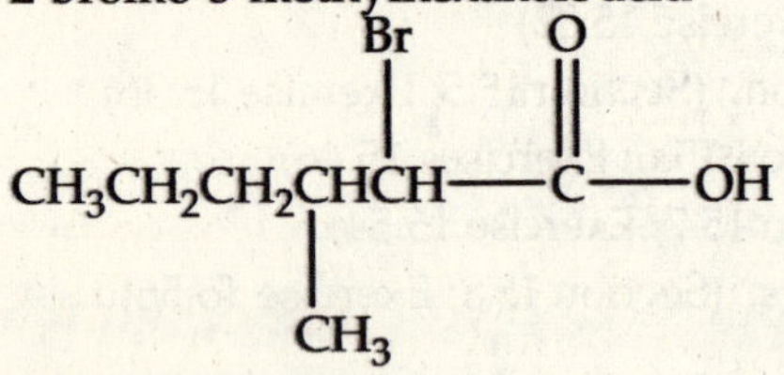

c. 4-propylbenzoic acid

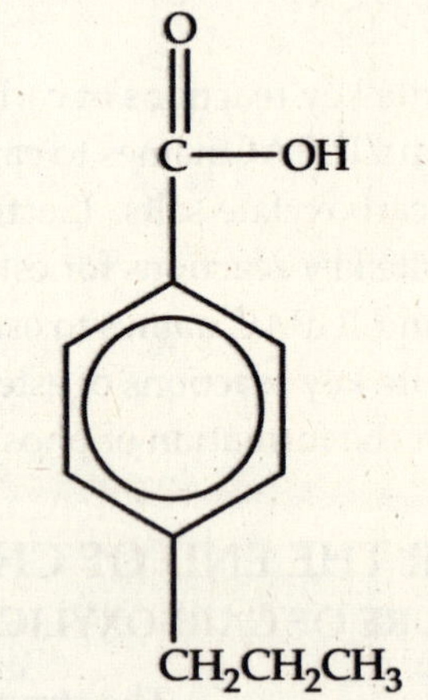

PHYSICAL PROPERTIES OF CARBOXYLIC ACIDS (SECTION 15.2)

☑15.10 a. Acetic acid will have a higher boiling point than 1-propanol because acetic acid forms stronger hydrogen bonds than 1-propanol. A carboxylic acid has two sites per molecule to form hydrogen bonds, whereas the alcohol has only one site.

b. Propanoic acid will have a higher boiling point than butanone because propanoic acid forms hydrogen bonds, whereas butanone experiences dipolar forces as its strongest intermolecular force. These two compounds experience approximately equal dispersion forces since their molar masses are similar. Since hydrogen bonds are stronger than dipolar forces, propanoic acid has a higher boiling point than butanone.

c. Butyric acid will have a higher boiling point than acetic acid because butyric acid has a higher molecular mass than acetic acid. Both compounds experience hydrogen bonding and dispersion forces. Their hydrogen bonding capabilities are the same since both compounds are carboxylic acids, but the dispersion forces experienced between butyric acid molecules are stronger than those between acetic acid molecules because butyric acid molecules are larger.

15.12

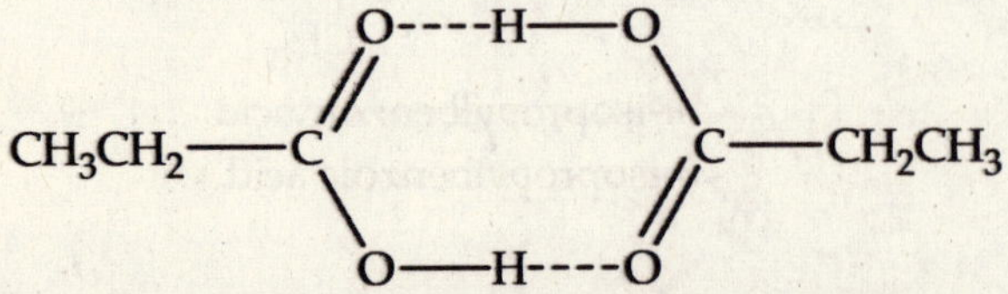

15.14 The carboxylic acid functional group allows caproic acid to be soluble in water. The solubility is limited by the aliphatic portion of the acid because it is hydrophobic.

15.16 pentane < ethoxyethane < 1-butanol < propanoic acid

Alkanes are the least soluble in water because they are nonpolar and cannot experience either dipolar forces or hydrogen bonds with water. Ethers are more soluble in water than alkanes because they are polar and can experience dipolar forces and hydrogen bonding with water. Alcohols are more soluble in water than ethers because they have a hydroxy group that has stronger hydrogen bonds with water than the oxygen atom in the ether. Carboxylic acids are more soluble in water than alcohols because they have two sites for hydrogen bonding per molecule, which results in a greater degree of hydrogen bonding to the water molecules.

THE ACIDITY OF CARBOXYLIC ACIDS (SECTION 15.3)

15.18

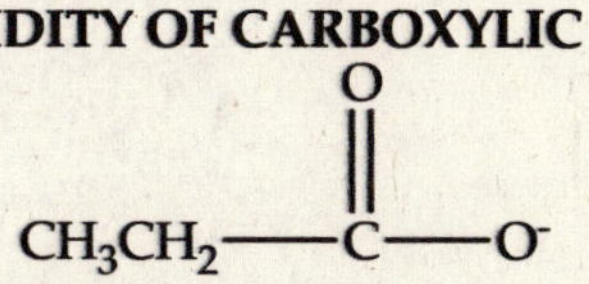

15.20 Within the cell, lactic acid will dissociate to form the lactate ion and H^+ because body fluids have a pH of 7.4 and lactic acid is a weak acid, which will dissociate in basic pH environments to form its conjugate base. The pK_a of lactic acid is 3.85, which is much lower than the pH of most body fluids, so at physiological pH (which ranges from slightly acidic to slightly basic), the acid will exist primarily as its conjugate base.

15.22

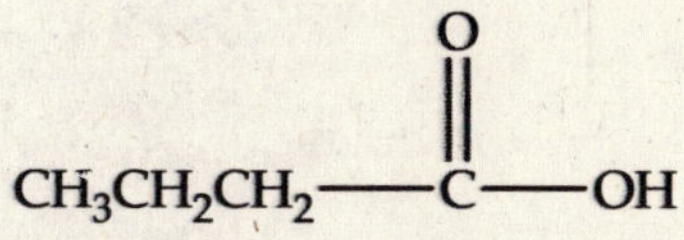

At a pH of 2, this weak acid will be primarily in undissociated form because the presence of hydronium ions in solution will push the equilibrium to favor the acidic form of butyric acid over its conjugate base.

15.24 a.

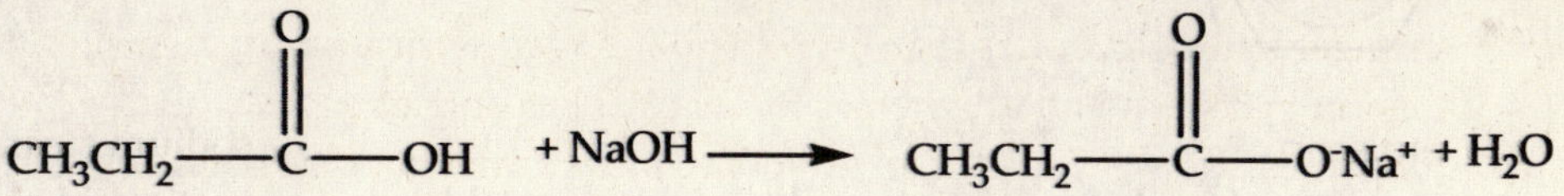

b.

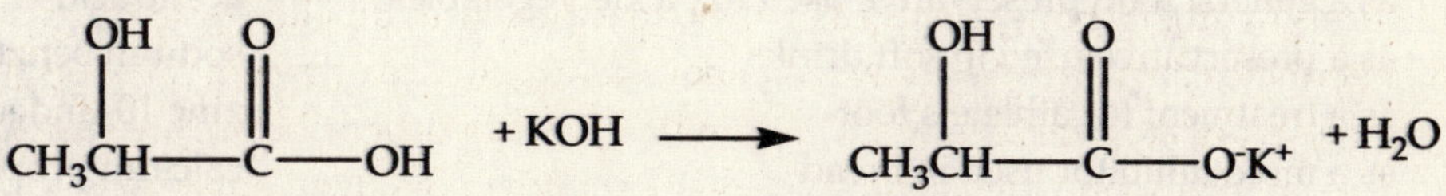

☑15.26 a.

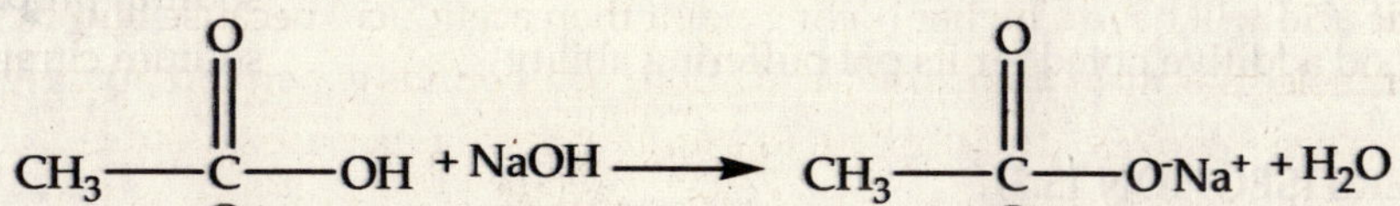

b.

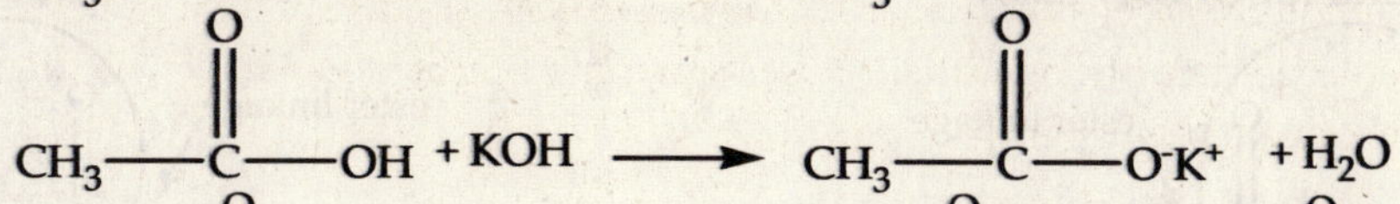

c.

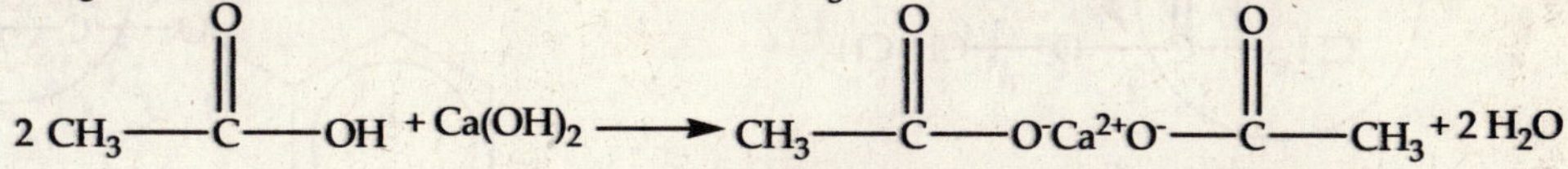

This can also be written as: $Ca(CH_3COO)_2$

SALTS OF CARBOXYLIC ACIDS (SECTION 15.4)

☑15.28 a.

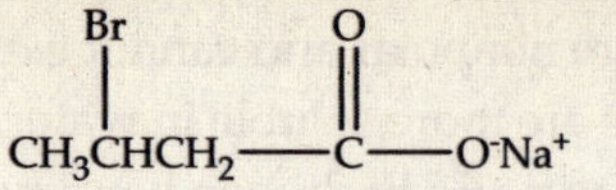

sodium 3-bromobutanoate

b.

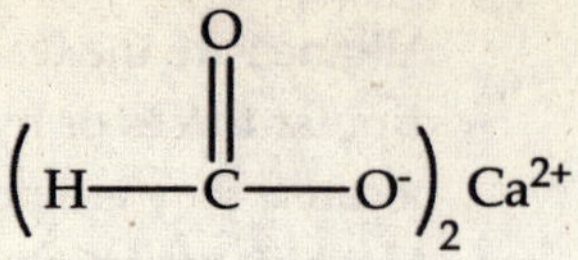

calcium methanoate

c.

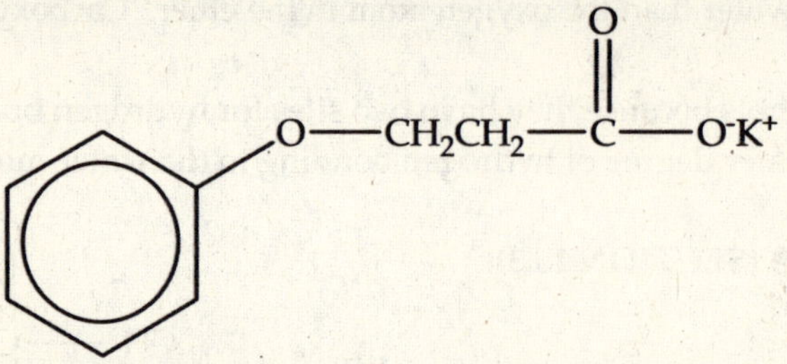

potassium 3-phenoxypropanoate

15.30 a. potassium ethanoate

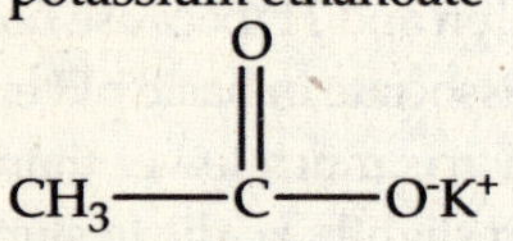

c. sodium 2-methylbutanoate

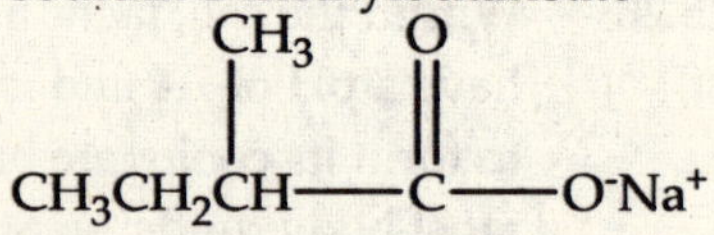

b. sodium *m*-methylbenzoate

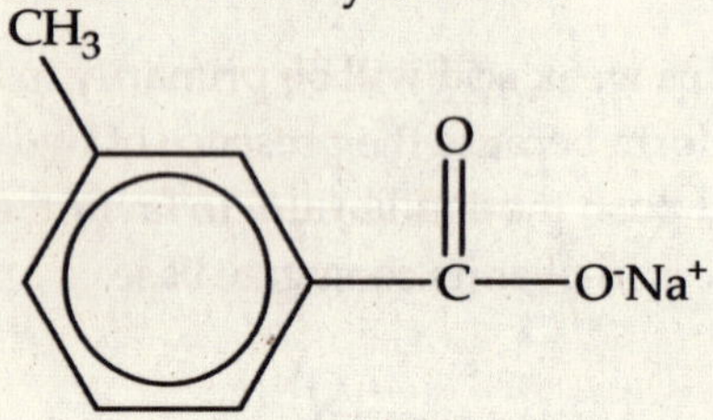

☑15.32

a. as a soap — sodium stearate
b. as a general food preservative used to pickle vegetables — acetic acid
c. as a preservative used in soft drinks — sodium benzoate
d. as a treatment for athlete's foot — zinc 10-undecylenate
e. as a mold inhibitor used in bread — calcium propanoate
sodium propanoate
f. as a food additive noted for its pH buffering ability — sodium citrate/citric acid

CARBOXYLIC ESTERS (SECTION 15.5)

15.34 b.

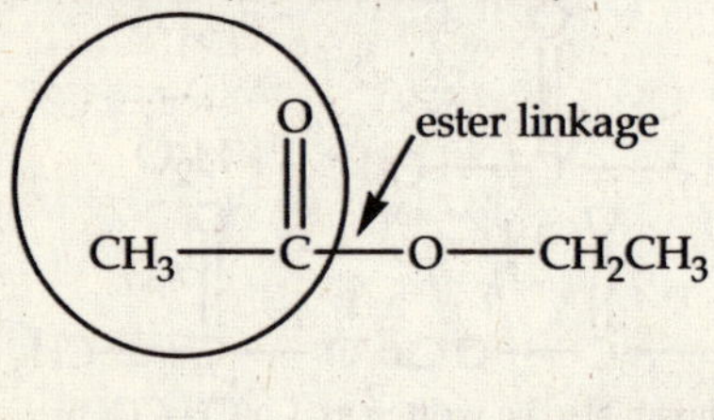

d.

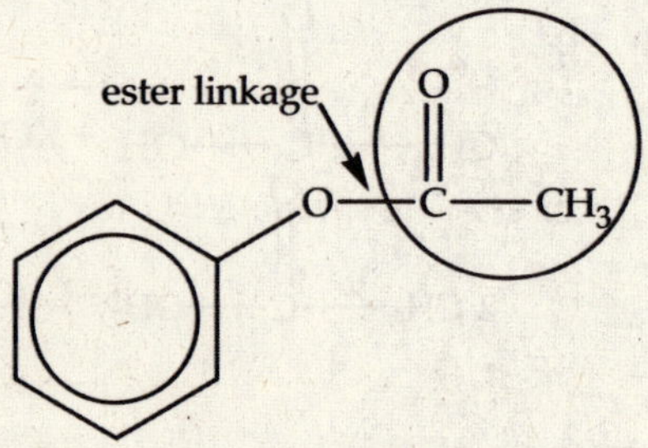

f.

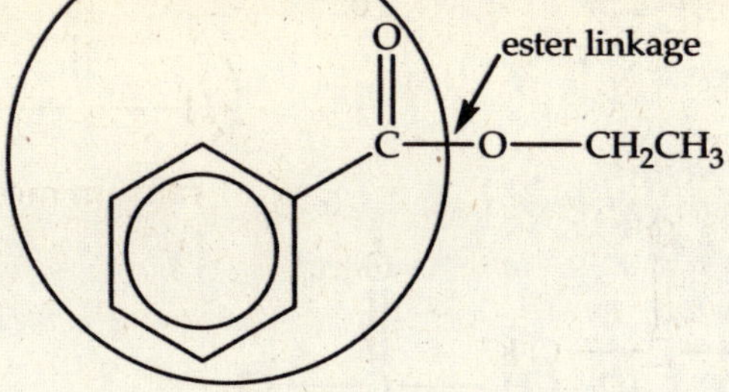

☑15.36 a.

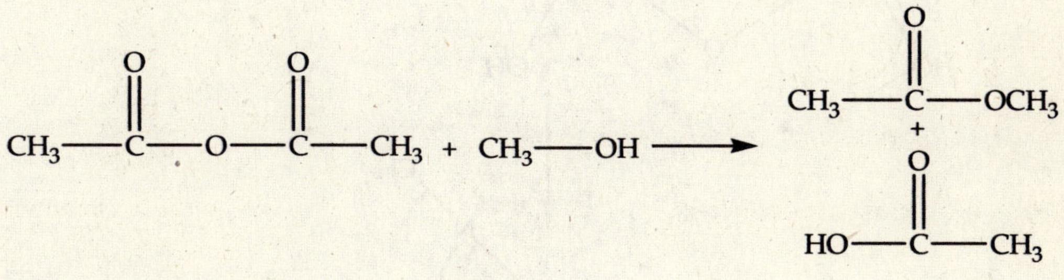

b.

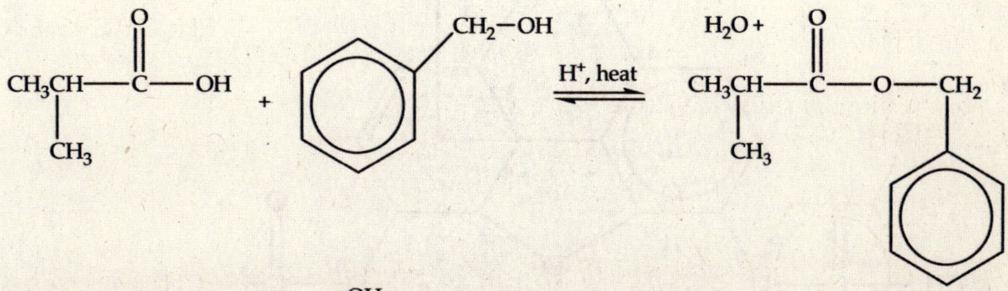

c.

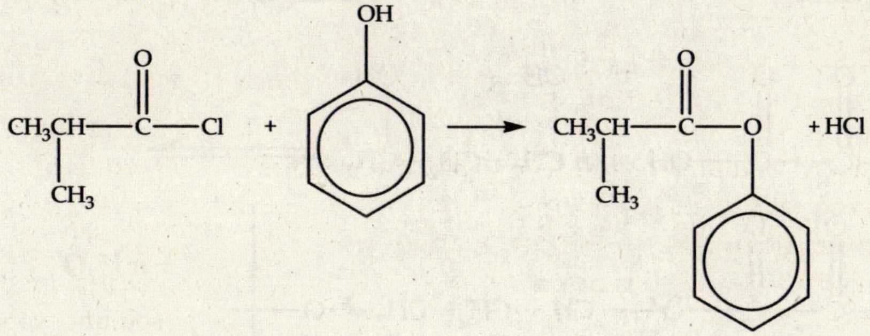

15.38 a. propanoic acid and methyl alcohol

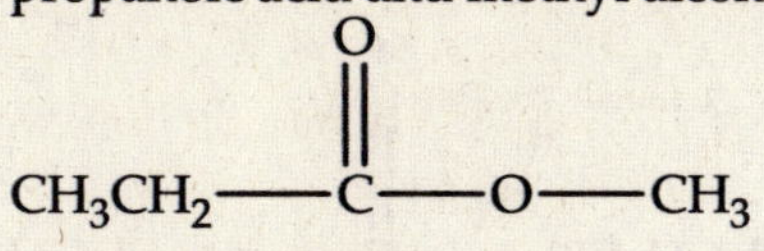

b. propanoic acid and phenol

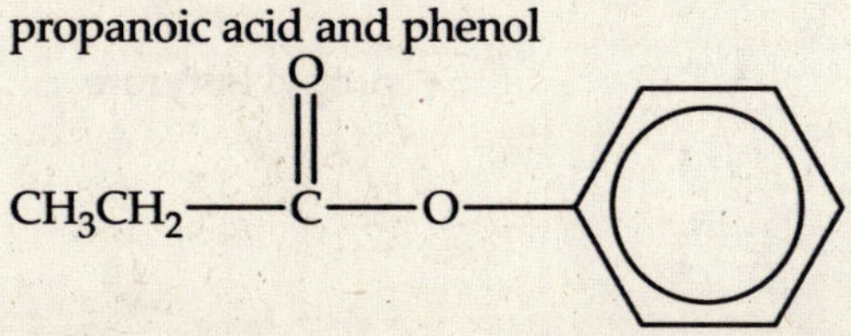

c. propanoic acid and 2-methyl-1-propanol

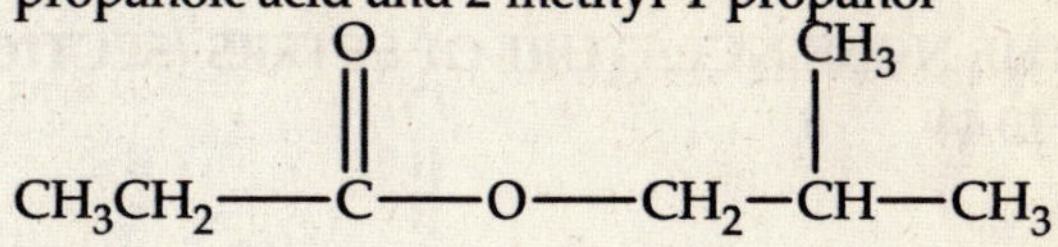

15.40

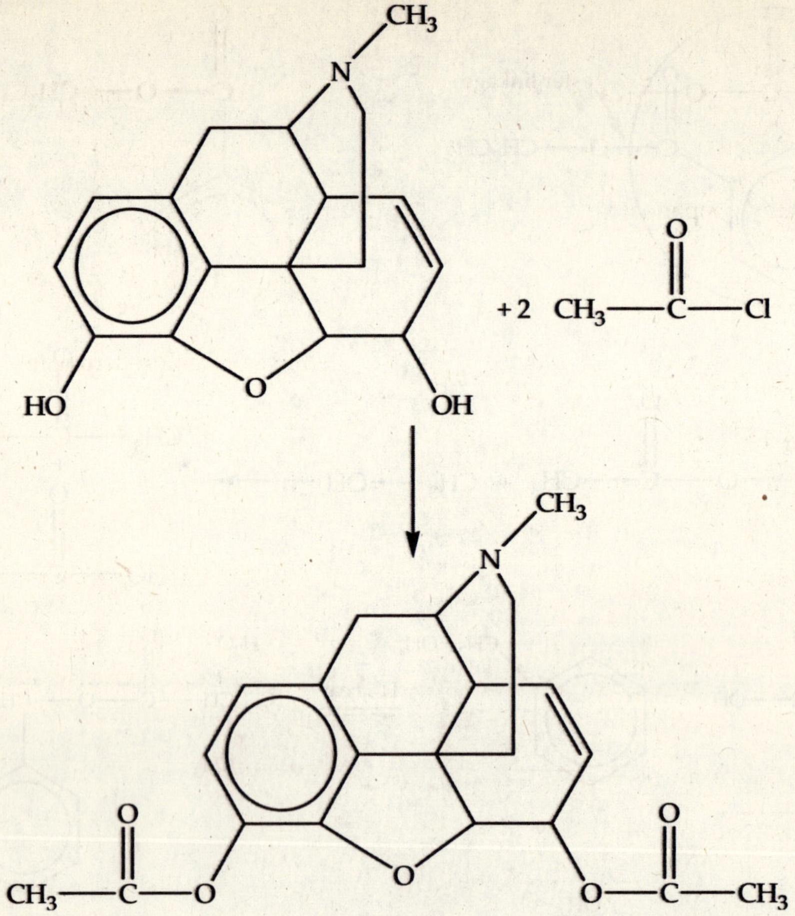

15.42

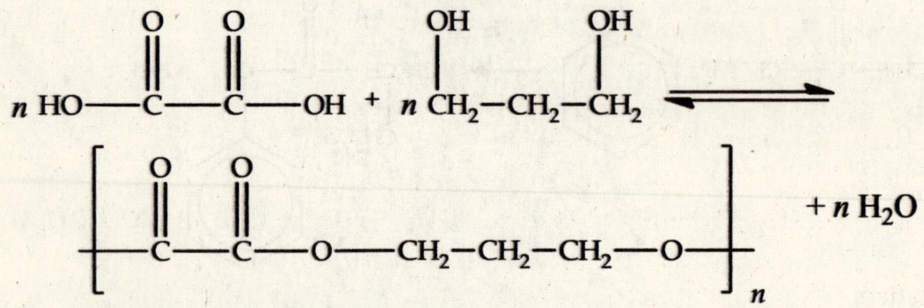

THE NOMENCLATURE OF ESTERS (SECTION 15.6)

15.44 a.

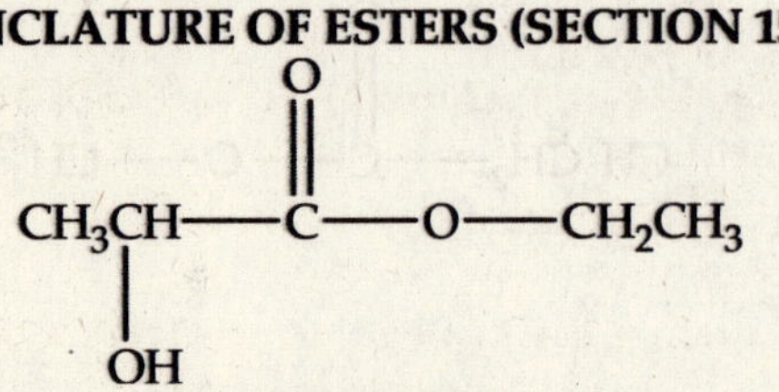

ethyl lactate

b.

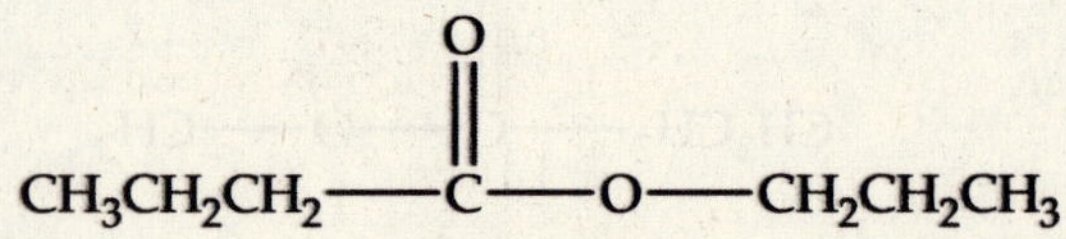

propyl butyrate

☑15.46 a.

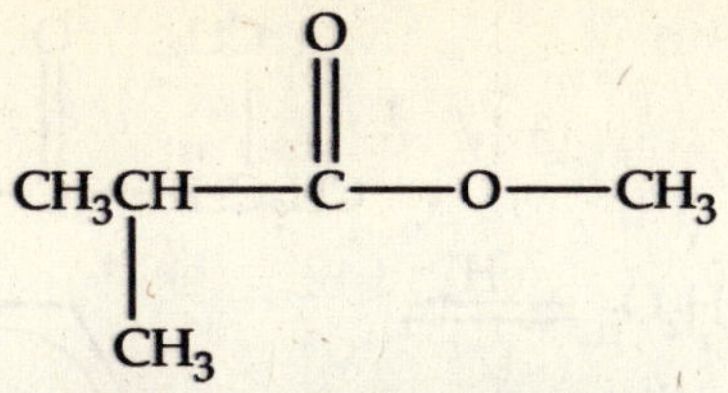

methyl 2-methylpropanoate

b.

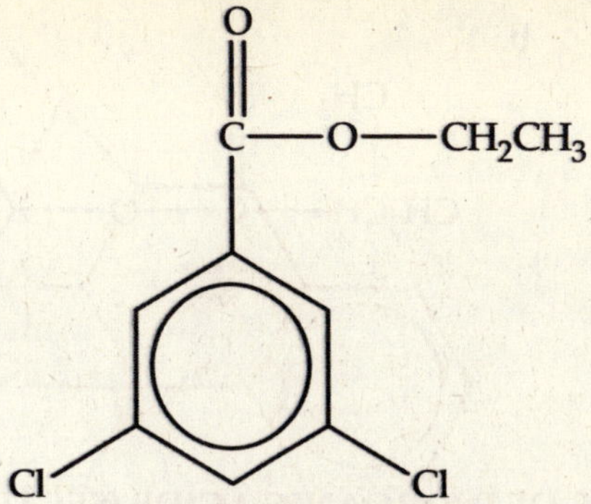

ethyl 3,5-dichlorobenzoate

15.48 a. methyl propionate
b. methyl butyrate
c. methyl lactate

15.50 a. phenyl formate

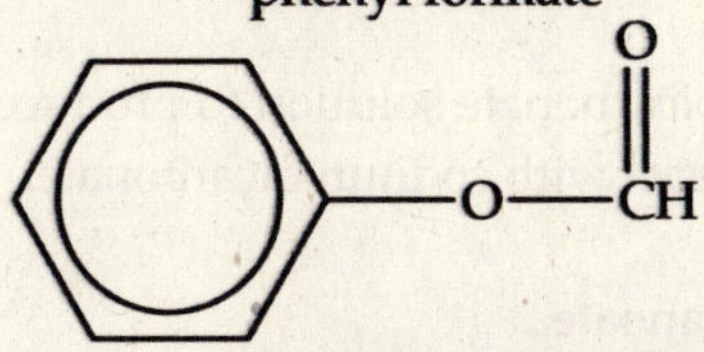

c. ethyl 2-chloropropanoate

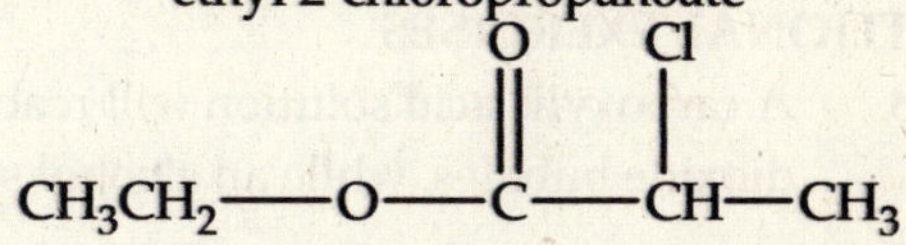

b. methyl 4-nitrobenzoate

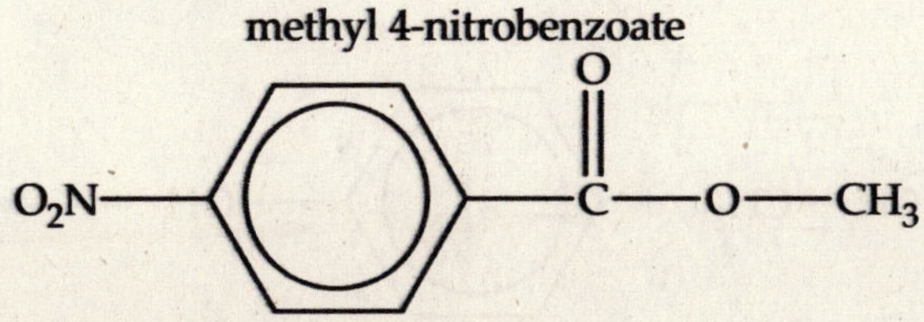

REACTIONS OF ESTERS (SECTION 15.7)

15.52 hydrolysis:

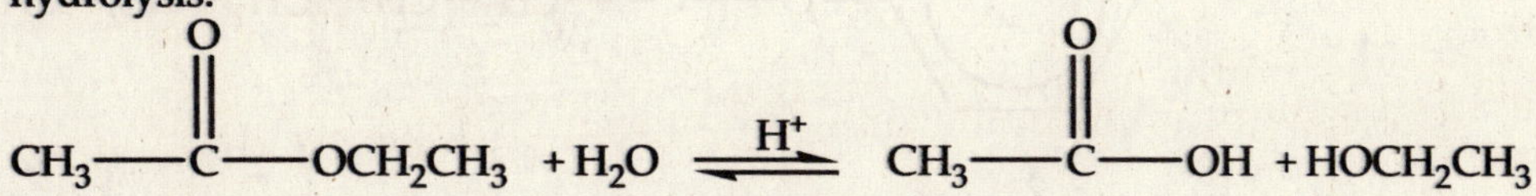

saponification:

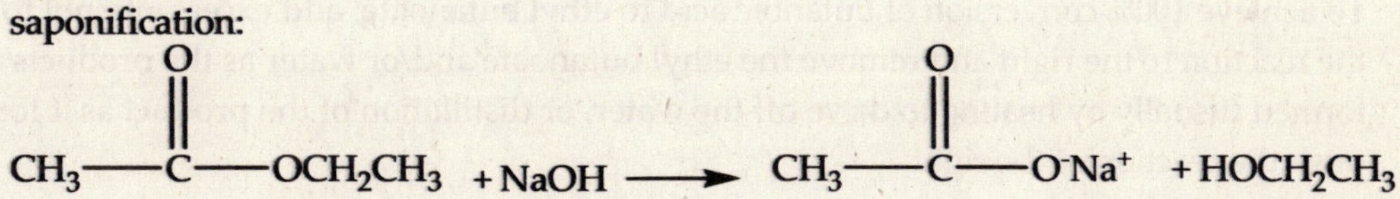

☑15.54 a.

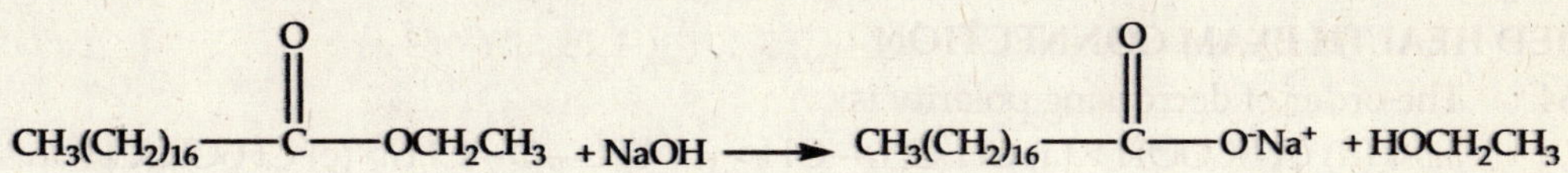

b.

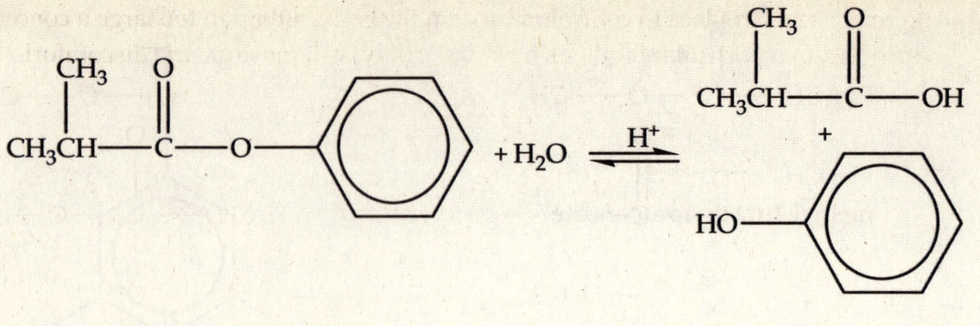

ESTERS OF INORGANIC ACIDS (SECTION 15.8)

☑15.56

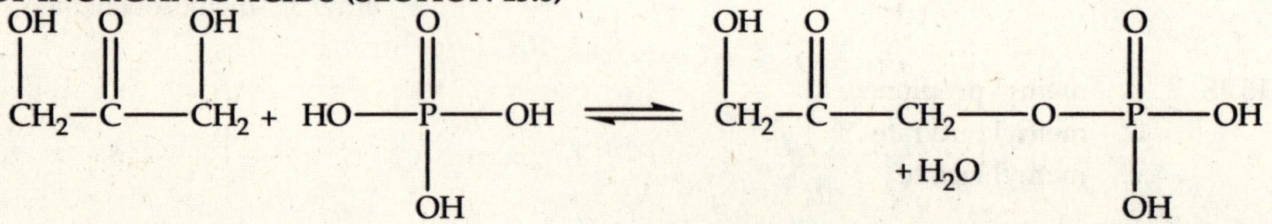

ADDITIONAL EXERCISES

15.58 A carboxylic acid solution will react with a sodium bicarbonate solution to produce carbon dioxide bubbles, while an alcohol solution will not react with sodium bicarbonate.

15.60 3-ethylhexanoic acid + phenol → phenyl 3-ethylhexanoate

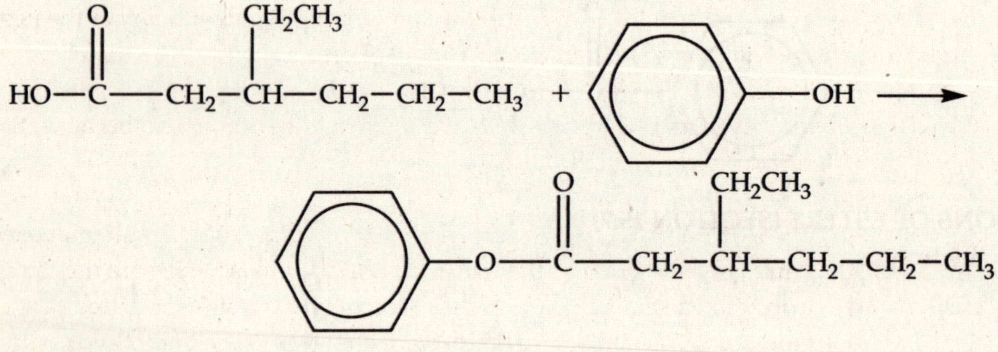

15.62 To achieve 100% conversion of butanoic acid to ethyl butanoate, add excess ethanol to push the reaction to the right and remove the ethyl butanoate and/or water as the products are formed (usually by heating to drive off the water, or distillation of the product as it forms) to push the reaction of the right.

ALLIED HEALTH EXAM CONNECTION

15.64 The order of decreasing polarity is:
most (d) CH_3COOH > (a) $CH_3CH_2{-}OH$ > (b) $CH_3{-}O{-}CH_3$ > (c) $CH_3CH_2CH_3$ *least*

The order of increasing boiling points is:
lowest (c) $CH_3CH_2CH_3$ < (b) $CH_3{-}O{-}CH_3$ < (a) $CH_3CH_2{-}OH$ < (d) CH_3COOH *highest*

CHEMISTRY FOR THOUGHT

15.66 Citric acid is viewed as a "safe" food additive because it is nontoxic. It is an acid that can be found in citrus fruit. The acidity of citric acid is not a concern because the body has many buffer systems in place to control pH within the body, although too large a concentration of citric acid in a particular food (such as sour candy) can cause gastric discomfort.

15.68

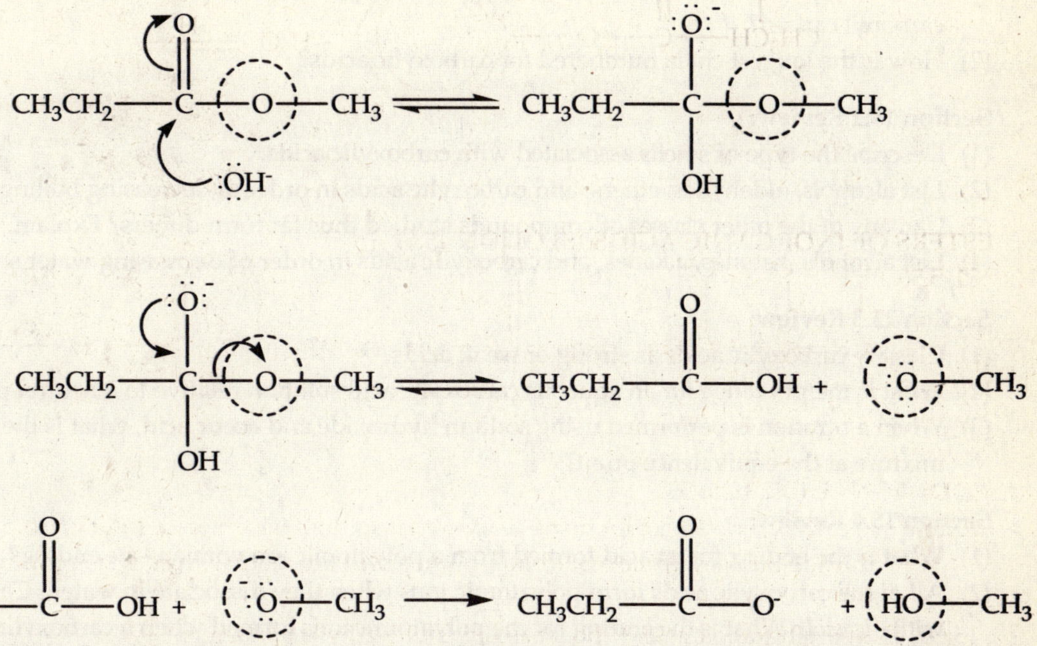

Note: The sodium ion is a spectator ion. It is the counterion present with any of the negatively charged ions in the equation to maintain a net charge of zero.

15.70 If oxalic acid were given an IUPAC name, its ending would be –dioic acid because it is a diacid that contains two carboxyl groups.

15.72 Polyesters like the Dacron patch resist hydrolysis reactions that would result in deterioration of the patch because they are polymers which are large molecules, which are much more difficult to hydrolyze than smaller single esters. Moreover, hydrolysis requires very acidic or very basic solutions, or the action of a specific esterase enzyme. The body fluids with which the Dacron comes into contact have a pH range of 6.8 to 7.2, which is close to neutral.

15.74

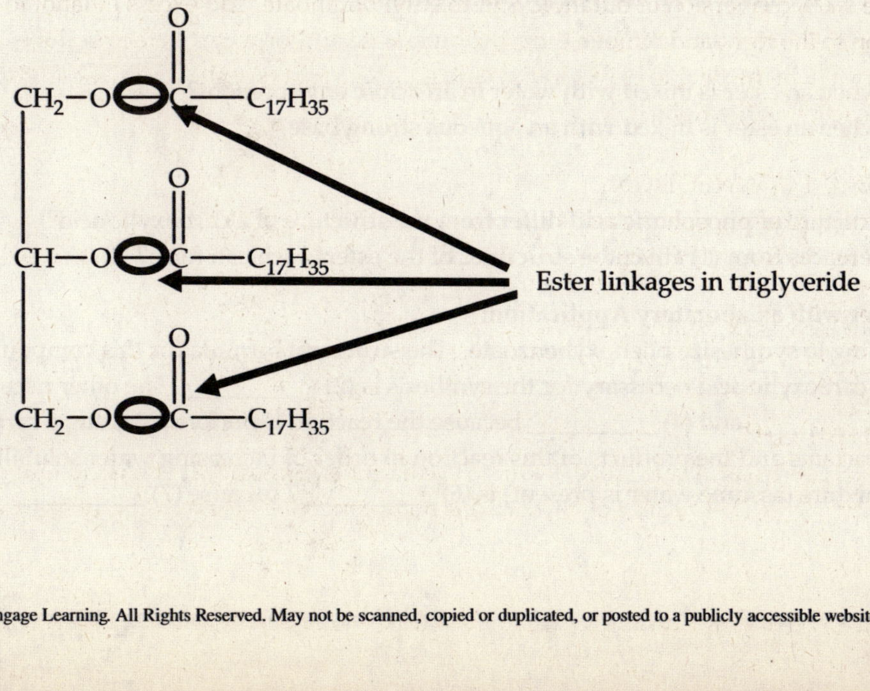

ADDITIONAL ACTIVITIES

Section 15.1 Review:

(1) What is importance of the *–oic acid* endings for compounds that contain hydroxy group bonded to a carbonyl carbon?

(2) How is the longest chain numbered for carboxylic acids?

Section 15.2 Review:

(1) Describe the type of smells associated with carboxylic acids.

(2) List alcohols, aldehydes, ethers, and carboxylic acids in order of decreasing boiling point.

(3) Can any of the other classes of compounds studied thus far form dimers? Explain.

(4) List alcohols, ketones, alkanes, and carboxylic acids in order of decreasing water solubility.

Section 15.3 Review:

(1) Classify carboxylic acids as strong or weak acids.

(2) What is the pH range for an aqueous carboxylic acid solution relative to a neutral pH of 7?

(3) When a titration is performed using sodium hydroxide and acetic acid, what is the pH range of the mixture at the equivalence point?

Section 15.4 Review:

(1) What is the ending for an acid formed from a polyatomic ion with an *–ate* ending?

(2) All of the carboxylic acids form polyatomic ions when they dissociate in water. Carboxylic acids end with *–ic acid*. What is the ending for the polyatomic ions formed when a carboxylic acid dissociates in water or reacts with a base?

(3) Are the carboxylate salts edible?

Section 15.5 Review:

(1) What is(are) joined by an ester linkage?

(2) When a carboxylic acid reacts with an alcohol to produce an ester, does the odor associated with the carboxylic acid become one of the characteristics of the ester?

(3) How does condensation polymerization differ from addition polymerization?

(4) What are three potential carbonyl containing starting materials for esters?

(5) What other reactant is needed to form an ester from the compounds in (4)?

Section 15.6 Review:

(1) Which carbon chain in an ester is named as a carboxylate group?

(2) Which carbon chain is named as an alkyl or aromatic group?

Section 15.7 Review:

(1) What happens when an ester is mixed with water in an acidic environment?

(2) What happens when an ester is mixed with an aqueous strong base?

Section 15.8 Review:

(1) How does the structure of phosphoric acid differ from the structure of a carboxylic acid?

(2) How do the differences from (1) affect the structures of the esters each can form?

Tying It All Together with a Laboratory Application:

A student is attempting to synthesize phenoxybenzoate. The structural formula for this compound is (1) __________. The carboxylic acid necessary for the synthesis is (2) __________. The other necessary components are (3) __________ and (4) __________ because the reaction is performed using a carboxylic acid. (5) Rank the reactants and the products of this reaction in order of increasing water solubility. The pH of the reactant mixture (assume water is present) is (6) __________ 7 because (7) __________. The pH of the ester in water is (8) __________ 7. This reaction system establishes equilibrium. The forward reaction is called (9) __________ and the reverse reaction is called (10) __________.

Phenoxybenzoate can also be synthesized from (11) __________ and the acid chloride with the following structure, (12) __________. The product obtained in addition to phenoxybenzoate is (13) __________. A third technique for synthesizing phenoxybenzoate uses (14) __________ and the carboxylic anhydride with the following structure, (15) __________. The other product produced in addition to phenoxybenzoate is (16) __________.

Once the phenoxybenzoate is isolated, the student reacts the ester with potassium hydroxide in water with heat. The products produced by this reaction are (17) __________ and __________. The same carboxylate salt produced by this reaction could also have been synthesized by reacting potassium hydroxide with (18) __________. The pH of an aqueous solution of the carboxylate salt is (19) __________ 7.

(20) Identify the major interparticle forces in esters, carboxylate salts, and carboxylic acids.
(21) Rank esters, carboxylate salts, and carboxylic acids in order of increasing boiling point.
(22) Rank esters, carboxylate salts, and carboxylic acids in order of increasing water solubility.

SOLUTIONS FOR THE ADDITIONAL ACTIVITIES

Section 15.1 Review:

(1) The *–oic acid* endings for compounds that contain a hydroxy group bonded to a carbonyl carbon indicates that the compound is a carboxylic acid.
(2) The carbonyl carbon is always the first carbon in the longest chain of a carboxylic acid.

Section 15.2 Review:

(1) Unpleasant odors (like unwashed bodies and dirty laundry) are associated with carboxylic acids.
(2) carboxylic acids > alcohols > aldehydes > ethers
(3) No, carboxylic acids are the only compounds studied that can form dimers. They are the first compound class to have a hydroxy group and a carboxyl group attached to the same carbon atom.
(4) carboxylic acids > alcohols > ketones > alkanes

Section 15.3 Review:

(1) Carboxylic acids are usually weak acids.
(2) The pH of an aqueous carboxylic acid solution is below 7 (acidic).
(3) When a titration is performed using sodium hydroxide and acetic acid, the pH range of the mixture at the equivalence point is above 7 (basic).

Section 15.4 Review:

(1) *–ic acid;* (2) *–ate;* (3) Some of the carboxylate salts are edible.

Section 15.5 Review:

(1) A carbonyl carbon with an aliphatic chain and an oxygen atom bonded to an aliphatic chain are joined by an ester linkage.
(2) No, the odors associated with esters are usually much more pleasant than the odors associated with carboxylic acids.
(3) A condensation polymer generally forms from a monomer or a set of monomers that contains two key functional groups. In addition to producing a polymer, a small molecule is also produced during this reaction. An addition polymer generally forms from a monomer or a set of monomers that contains a carbon-carbon double bond. The polymer is the only product produced by this reaction.

(4) Three potential carbonyl containing starting materials for esters are carboxylic acids, carboxylic acid chlorides, and carboxylic acid anhyrides.

(5) All of the reactants from (4) can react with an alcohol or phenol to produce an ester. The carboxylic acid also requires a catalyst.

Section 15.6 Review:

(1) The carbon chain that includes the carboxyl group is named as a carboxylate.

(2) The carbon chain connected to an oxygen atom by a single bond is named as an alkyl or aromatic group.

Section 15.7 Review:

(1) When an ester is mixed with water in an acidic environment, a carboxylic acid and an alcohol form.

(2) When an ester is mixed with an aqueous strong base, a carboxylate salt and an alcohol form.

Section 15.8 Review:

(1) Phosphoric acid contains a phosphorus atom double bonded to an oxygen atom and single bonded to three other oxygen atoms. A carboxylic acid contains a carbon atom double bonded to an oxygen atom and single bonded to one other oxygen atom. Phosphoric acid is triprotic. Carboxylic acids are monoprotic, although diacids contain two carboxylic acid functional groups and are diprotic.

(2) The esters formed from phosphoric acid can be monoesters, diesters, or triesters depending on how many alkyl or aromatic groups attach to the 3 available oxygen atoms. Esters formed from carboxylic acids can only be monoesters because the carboxylic acid only has one available oxygen atom to bond with an alkyl or aromatic group. Diacids can form diesters, if they react with an alcohol, or polyesters, if they react with a diol.

Tying It All Together with a Laboratory Application:

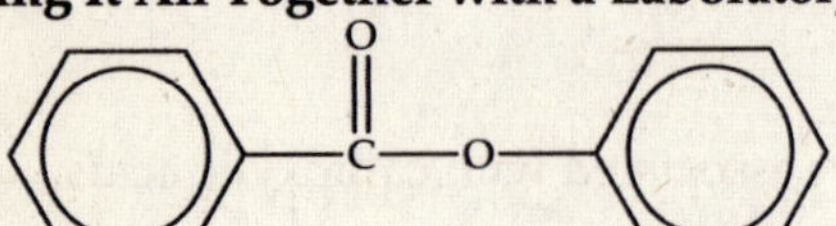

(1)

(2) benzoic acid

(3) phenol

(4) a catalyst

(5) phenoxyphenol < phenol < benzoic acid

(6) less than

(7) phenol and benzoic acid are acidic

(8) approximately

(9) esterification

(10) ester hydrolysis

(11) phenol

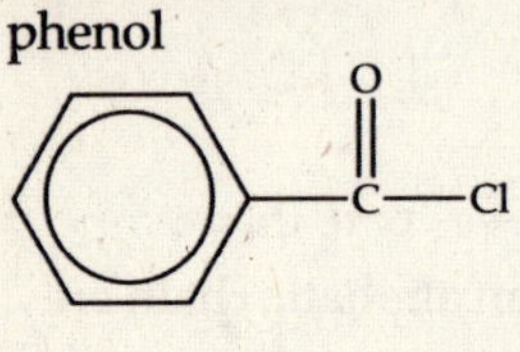

(12)

(13) HCl

(14) phenol

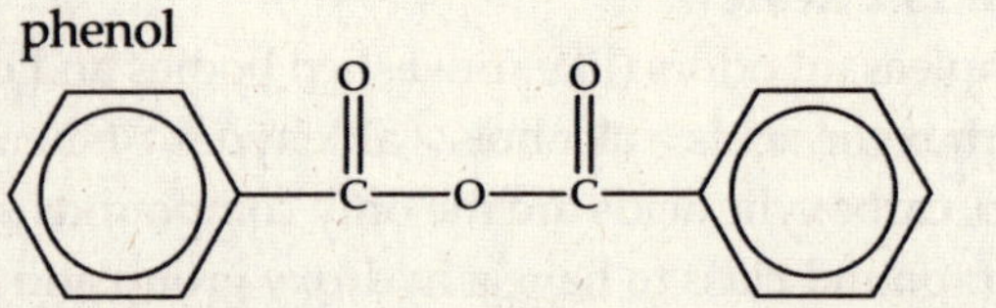

(15)

(16) benzoic acid

(17) potassium benzoate, phenol

(18) benzoic acid

(19) greater than

(20) esters – dipolar forces; carboxylate salts – ionic bonding; carboxylic acids – hydrogen bonding

(21) esters < carboxylic acids < carboxylate salts

(22) esters < carboxylic acids < carboxylate salts

SELF-TEST QUESTIONS

Multiple Choice

1. What is the correct IUPAC name for 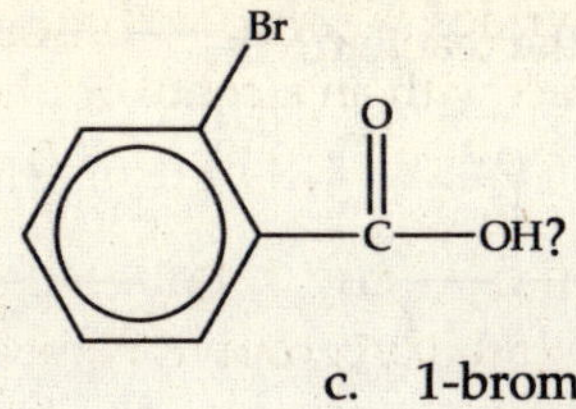

   a. bromobenzoic acid
   b. 2-bromobenzoic acid
   c. 1-bromobenzoic acid
   d. 2-bromo-1-benzoic acid

2. What is the correct IUPAC name for 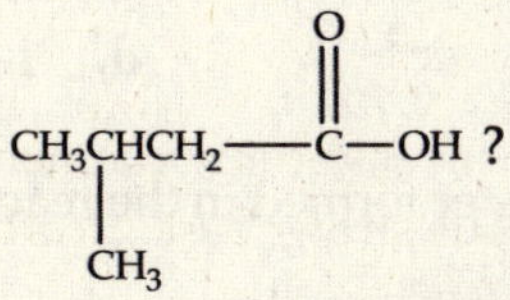

   a. 2-methylpentanoic acid
   b. 3-methylbutanoic acid
   c. 2-methylbutanoic acid
   d. 3-methylpentaoic acid

3. Which of the following pure substances would exhibit hydrogen bonding?

   a. aldehyde
   b. ketone
   c. ether
   d. carboxylic acid

4. Which of the following substances would you expect to be the most soluble in water?

   a. 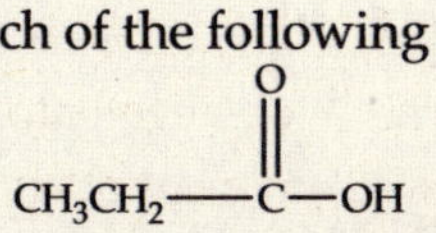
   b. 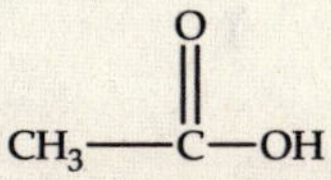
   c. 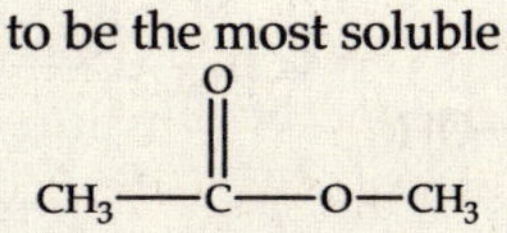
   d. 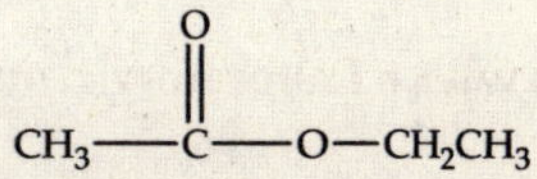

5. Which name is the most appropriate for an organic acid under body conditions of pH 7.4?

   a. lactic acid
   b. lactate acid
   c. lactic
   d. lactate

6. What is(are) the organic product(s) of the reaction between butanoic acid and NaOH?

   a. an ester
   b. a carboxylic acid and an alcohol
   c. a ketone
   d. a carboxylate salt

7. What reagent is needed to complete the following reaction?

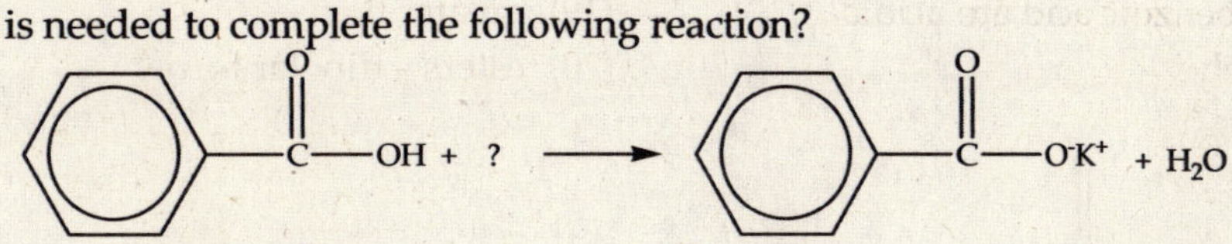

   a. K
   b. K^+
   c. KOH
   d. KO_2

8. Which of the molecules could be used as one of the reagents necessary to prepare this compound?

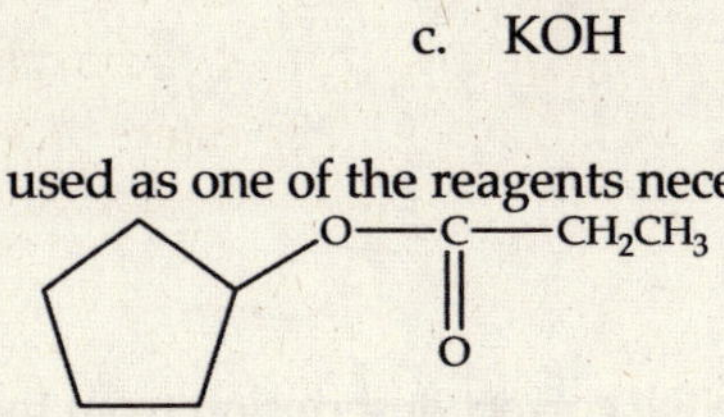

   a. CH_3CH_2—OH
   b. 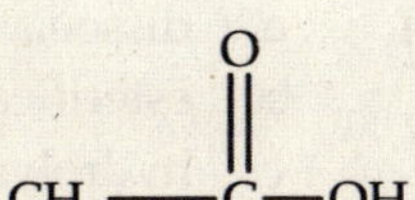
   c. 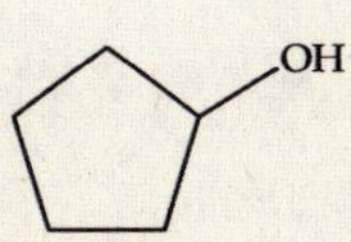
   d. 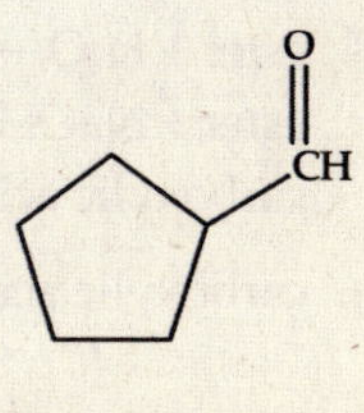

9. What is the organic product of the reaction 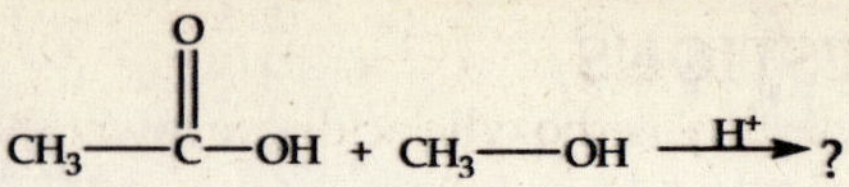

a. 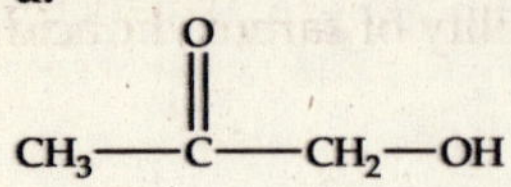
b. 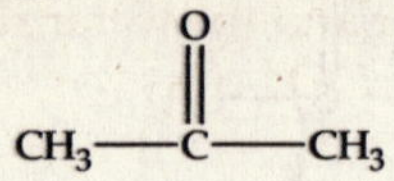
c. 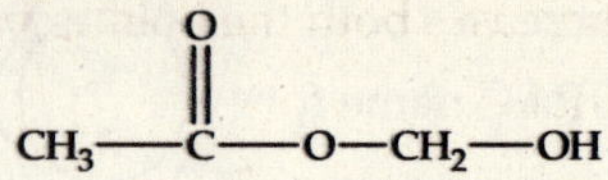
d. 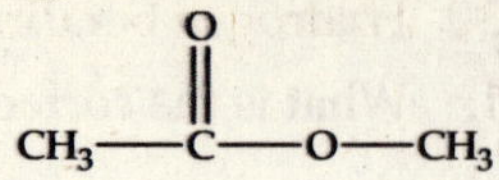

10. What is the name of the ester formed by reacting propanoic acid and isopropyl alcohol?
   a. propyl propanoate
   b. isopropyl propanoate
   c. isopropyl propanoic acid
   d. 1-propyl propanoate

11. What is the IUPAC name for the ester formed in the reaction of isopropyl alcohol and benzoic acid?
   a. benzyl isopropyl ester
   b. isopropyl benzoate
   c. benzyl isopropanoate
   d. isopropyl benzoic acid

12. Which of the following materials might be obtained as one of the products from the following reaction?

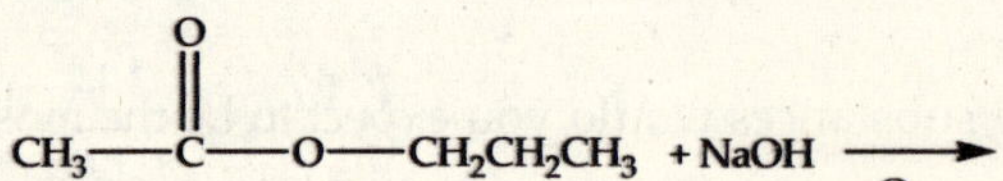

a. CH_3CH_2—OH

b. 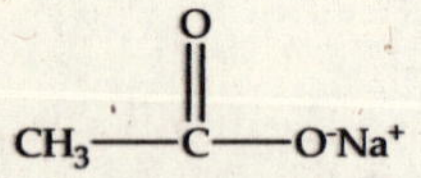

c. 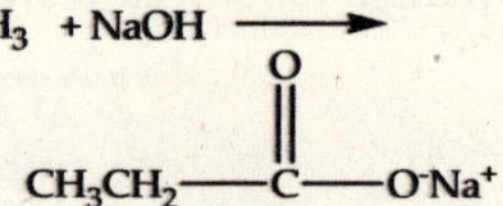

d. 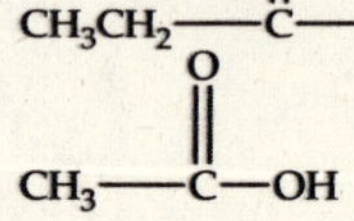

Matching

Select the best match for each of the following:

13. 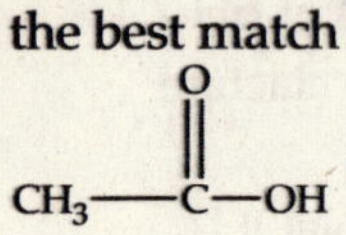

14. 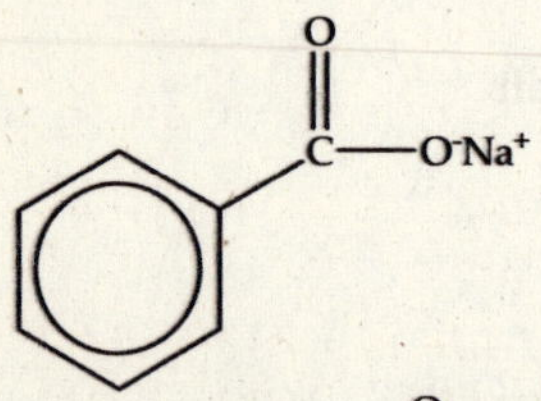

15. 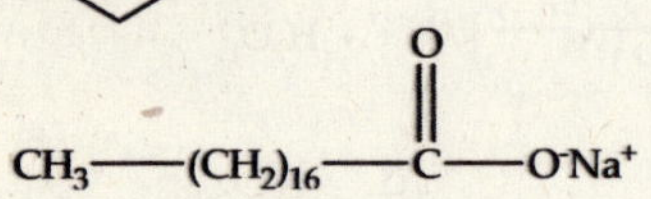

16. 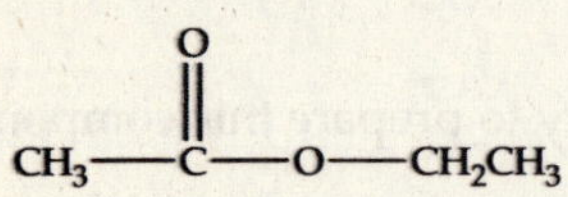

a. a preservative used in soda pop
b. a soap
c. present in vinegar
d. fingernail polish remover

For each reaction on the left, choose the correct description from the responses on the right.

17. ester + H_2O $\xrightarrow{H^+}$
18. ester + NaOH →
19. carboxylic acid + H_2O →
20. carboxylic acid + alcohol $\xrightarrow[\text{heat}]{H^+}$

a. dissociation
b. esterification
c. hydrolysis
d. saponification

True-False

21. The boiling points of carboxylic acids are lower than those of the corresponding alcohols.
22. Salts of carboxylic acids are not usually soluble in water.
23. Hydrogen bonding increases both the boiling point and the water solubility of carboxylic acids.
24. CH_3COOH has a higher boiling point than $CH_3{-}\overset{\overset{\displaystyle O}{\|}}{C}{-}H$.
25. Carboxylic acids are generally strong acids.
26. Both nitric acid and phosphoric acid can react with alcohols to form esters.
27. Certain phosphate esters are present in the body.
28. Esters are responsible for the pleasant fragrance of many flowers and fruits.
29. Carboxylate salts are fatty acids.
30. Esters form dimers more easily than carboxylic acids.

Chapter 16: Amines and Amides

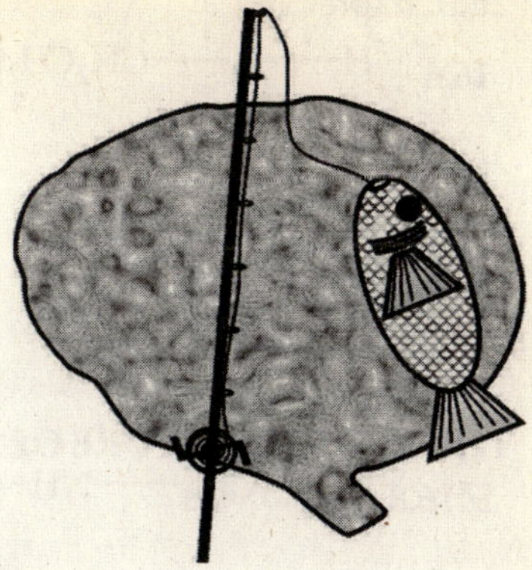

CHAPTER OUTLINE

16.1 Classification of Amines
16.2 The Nomenclature of Amines
16.3 Physical Properties of Amines
16.4 Chemical Properties of Amines
16.5 Amines as Neurotransmitters
16.6 Other Biologically Important Amines
16.7 The Nomenclature of Amides
16.8 Physical Properties of Amides
16.9 Chemical Properties of Amides

LEARNING OBJECTIVES/ASSESSMENT

When you have completed your study of this chapter, you should be able to:

1. Given structural formulas, classify amines as primary, secondary, or tertiary. (Section 16.1; Exercise 16.4)
2. Assign common and IUPAC names to simple amines. (Section 16.2; Exercises 16.8 and 16.10)
3. Discuss how hydrogen bonding influences the physical properties of amines. (Section 16.3; Exercises 16.16 and 16.18)
4. Recognize and write key reactions for amines. (Section 16.4; Exercise 16.26)
5. Name amines used as neurotransmitters. (Section 16.5; Exercise 16.38)
6. Give uses for specific biological amines. (Section 16.6; Exercises 16.40 and 16.44)
7. Assign IUPAC names for amides. (Section 16.7; Exercise 16.46)
8. Show the formation of hydrogen bonds with amides. (Section 16.8; Exercise 16.50)
9. Give the products of acidic and basic hydrolysis of amides. (Section 16.9; Exercise 16.52)

SOLUTIONS FOR THE END OF CHAPTER EXERCISES

CLASSIFICATION OF AMINES (SECTION 16.1)

16.2 Primary amines have the formula: $R-NH_2$

Secondary amines have the formula:

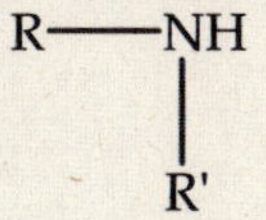

Tertiary amines have the formula:

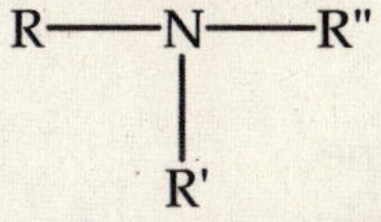

☑16.4 a. $CH_3CH_2-NH-CH_2CH_3$
secondary

b.

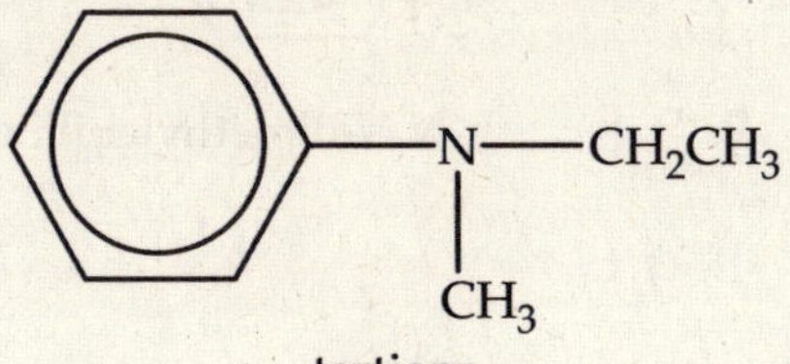

tertiary

c.

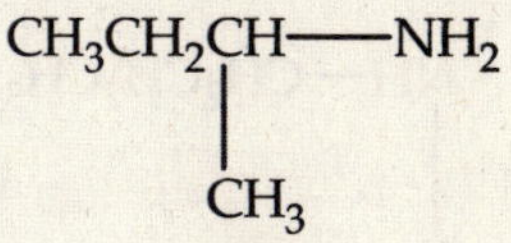

primary

d.

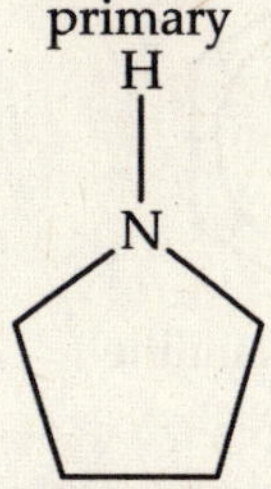

secondary

16.6 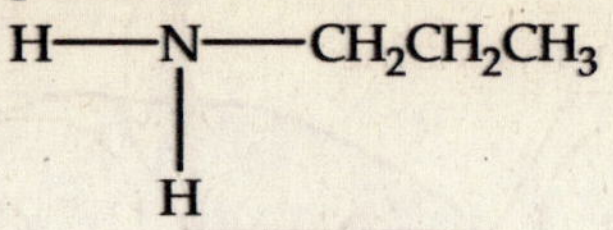
primary

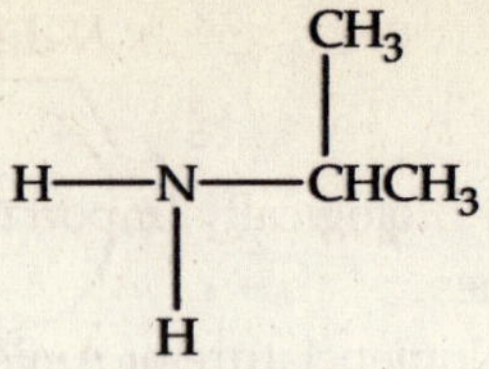
primary

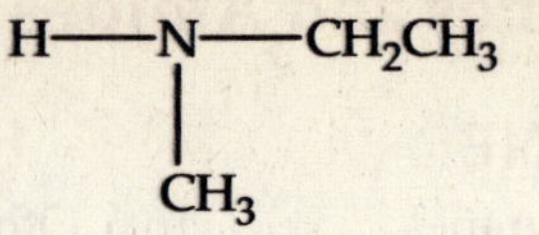
secondary

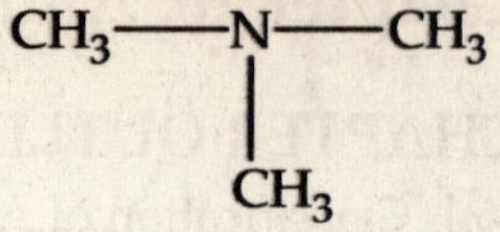
tertiary

THE NOMENCLATURE OF AMINES (SECTION 16.2)

☑16.8 a. 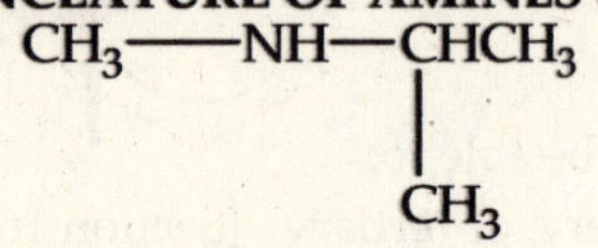
isopropylmethylamine

c. $CH_3CH_2CH_2CH_2—NH_2$
butylamine

b. 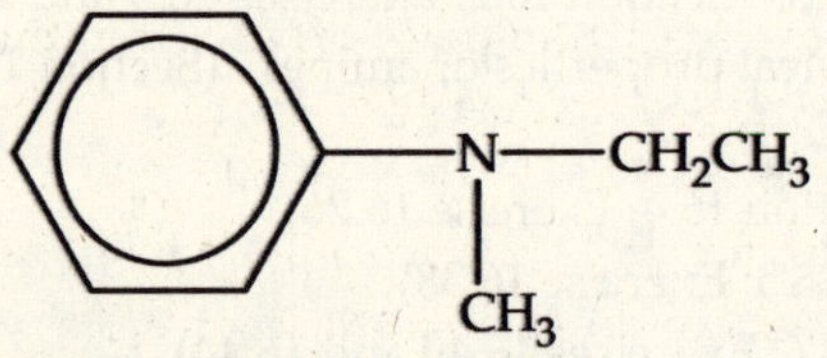
ethylmethylphenylamine

☑16.10 a. 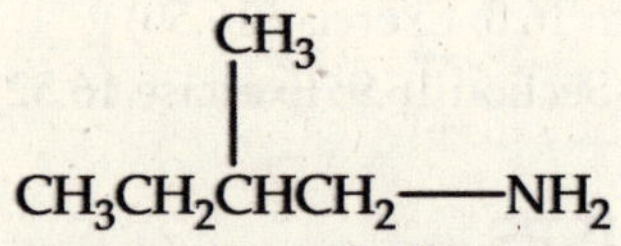
1-amino-2-methylbutane
or 2-methyl-1-butanamine

c. 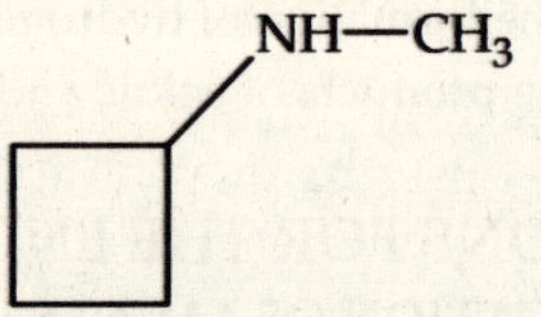
1-amino-*N*-methylcyclobutane
or *N*-methylcyclobutanamine

b. 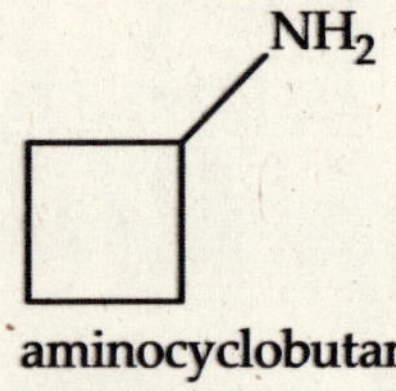
aminocyclobutane
or cyclobutanamine

16.12 a. 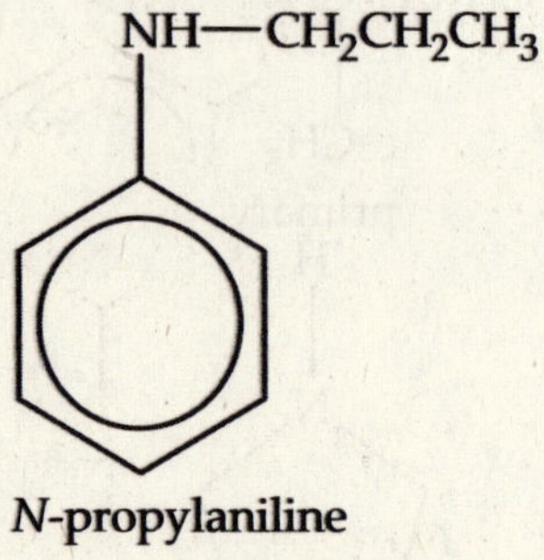
N-propylaniline

b. 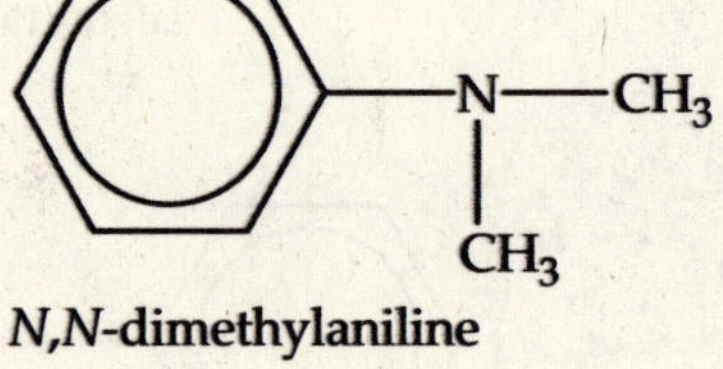
N,N-dimethylaniline

16.14 a. 3-ethyl-2-pentanamine
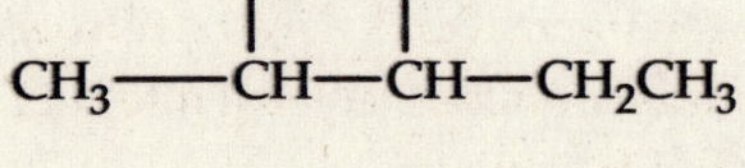

b. *m*-ethylaniline

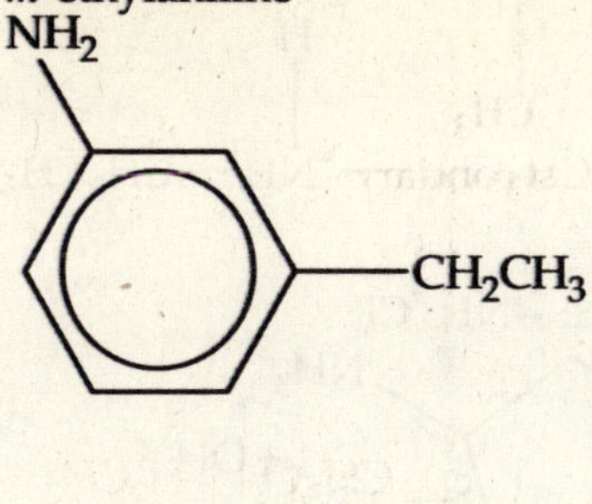

c. *N,N*-diphenylaniline

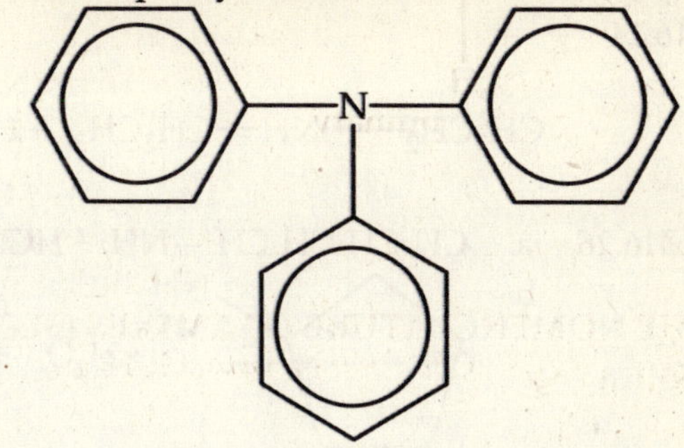

PHYSICAL PROPERTIES OF AMINES (SECTION 16.3)

☑16.16 All low molecular weight amines are water soluble because all amines can hydrogen bond with water. Low molecular weight amines have small aliphatic portions, thus the hydrogen bonds to water are strong enough to allow the amines to dissolve.

☑16.18 The boiling points of tertiary amines are lower than the boiling points of primary and secondary amines because tertiary amines cannot hydrogen bond to each other, while primary and secondary amines can. The weaker intermolecular forces in tertiary amines allow them to boil at a lower temperature.

16.20 a.

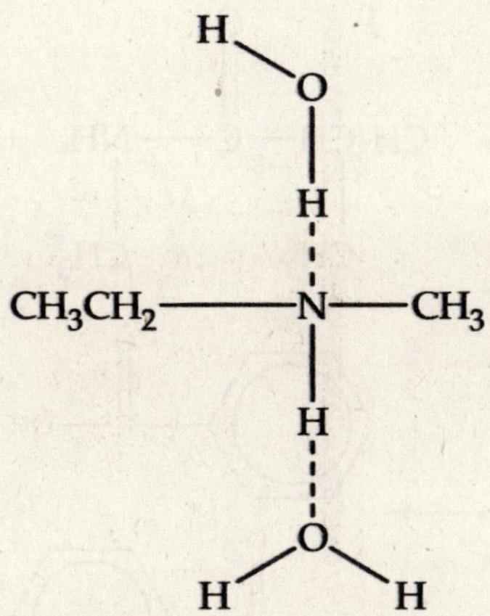

b.

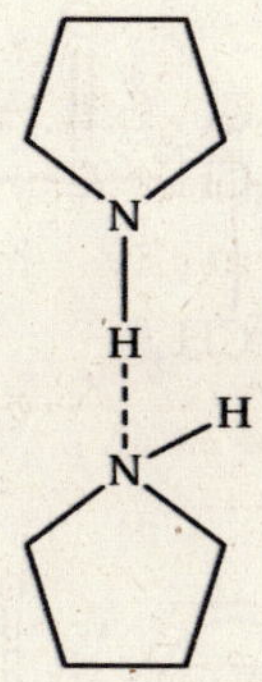

16.22

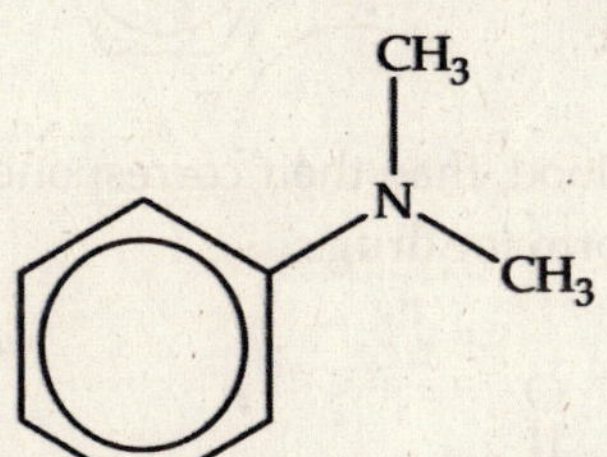

(b)

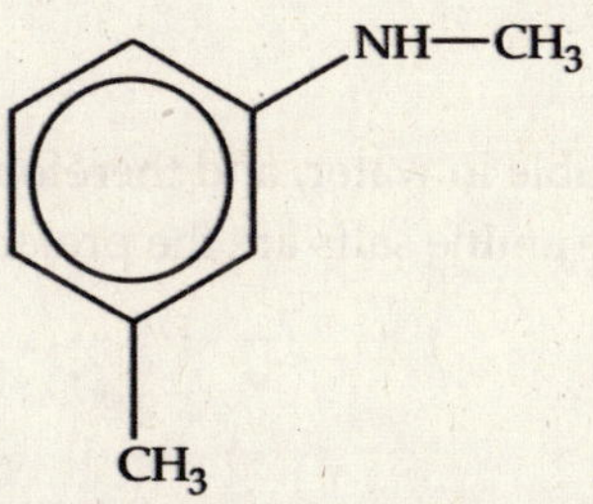

< (c)

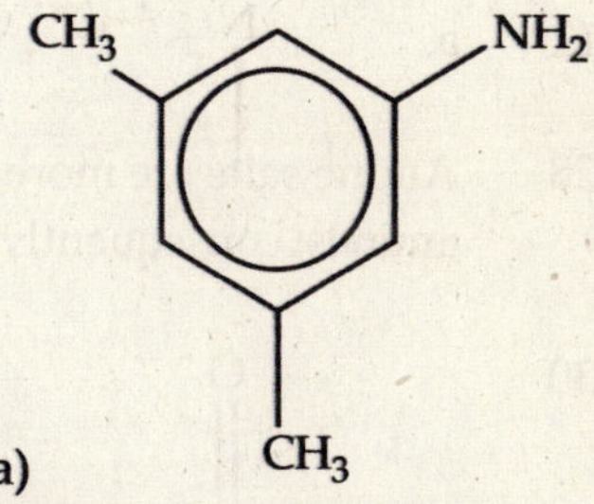

< (a)

Since the molecular weights of these three compounds are comparable, they experience comparable dispersion forces. The order of their boiling points is determined by the nature of the other intermolecular forces they experience. Tertiary amines have the lowest boiling point because they are unable to form hydrogen bonds with other tertiary amines. Secondary and primary amines are both capable of forming hydrogen bonds; however, the nitrogen atom is more accessible in primary amines. Therefore, secondary amines have an intermediate boiling point and primary amines have the highest boiling point.

CHEMICAL PROPERTIES OF AMINES (SECTION 16.4)

16.24

$$CH_3CH_2—NH—CH_2CH_3 + H_2O \rightleftharpoons CH_3CH_2—{}^{+}NH(H)—CH_2CH_3 + OH^-$$

☑16.26 a. $CH_3CH_2CH_2CH_2—NH_2 + HCl \rightarrow CH_3CH_2CH_2CH_2—NH_3^+Cl^-$

b.

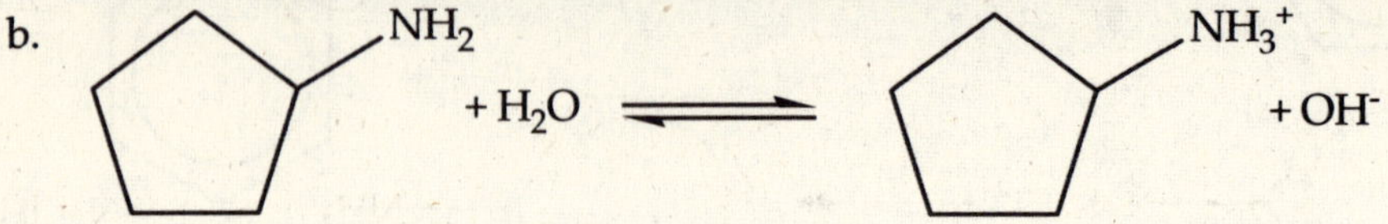

c.

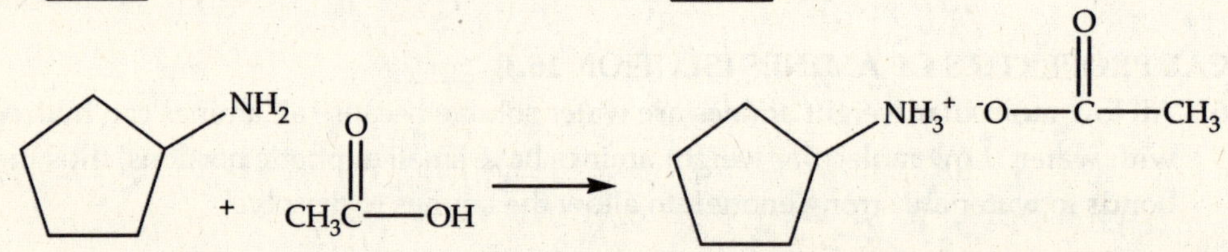

d.

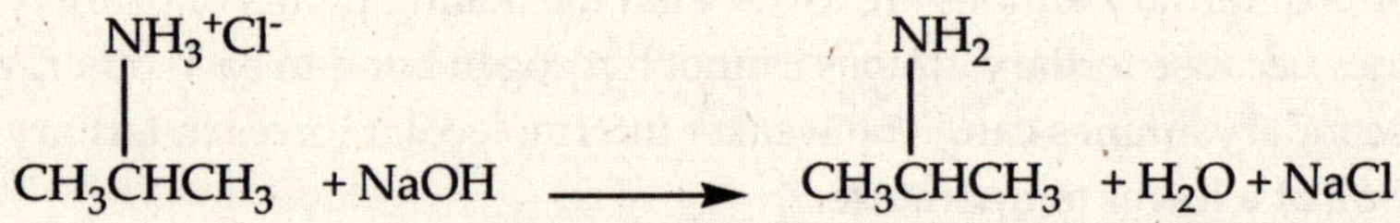

e.

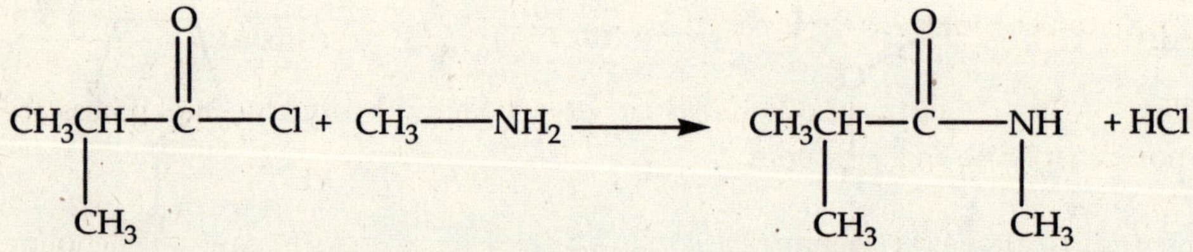

f.

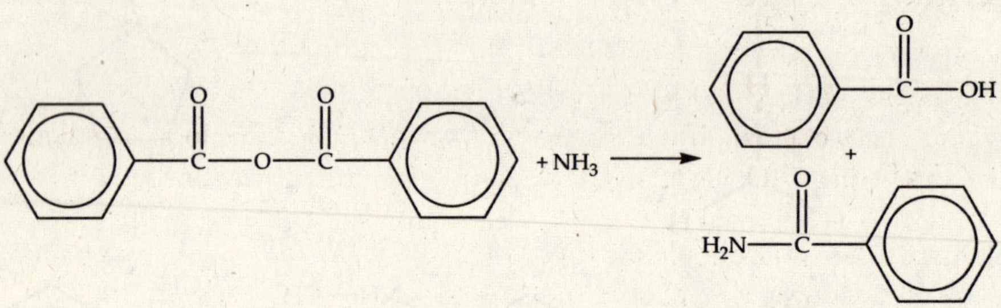

16.28 Amine salts are more soluble in water, and therefore in blood, than their corresponding amines; consequently, the amine salts are the preferred form for drugs.

16.30

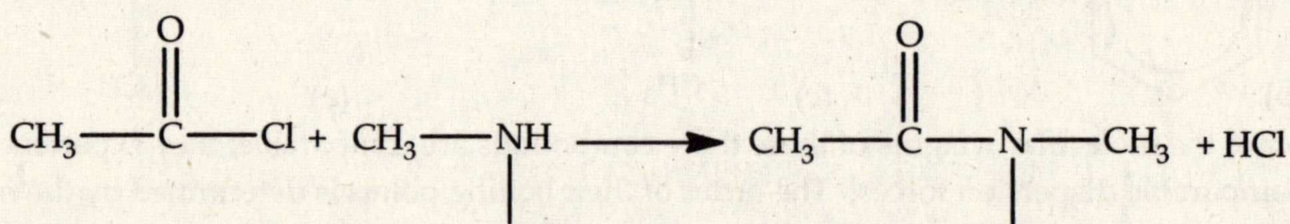

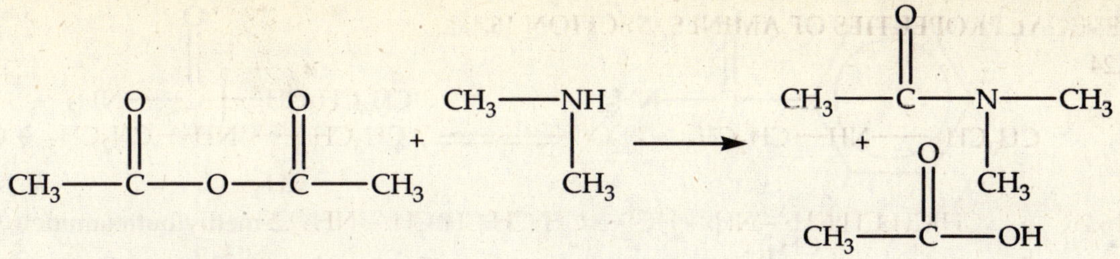

16.32 a.

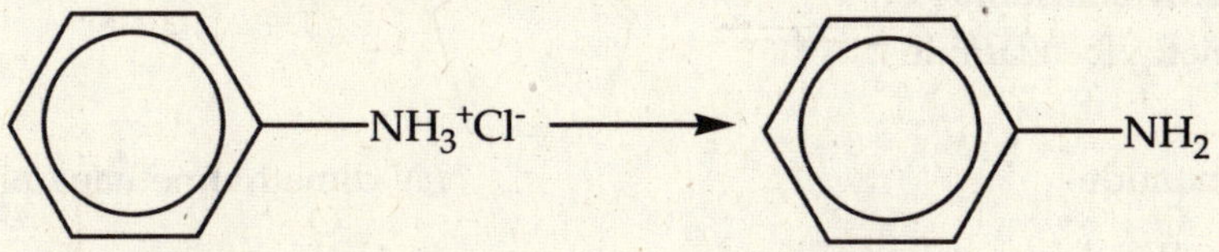

NaOH

b.

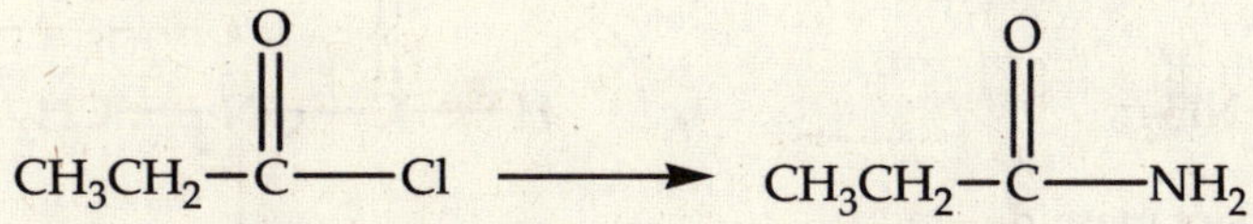

NH3

c.

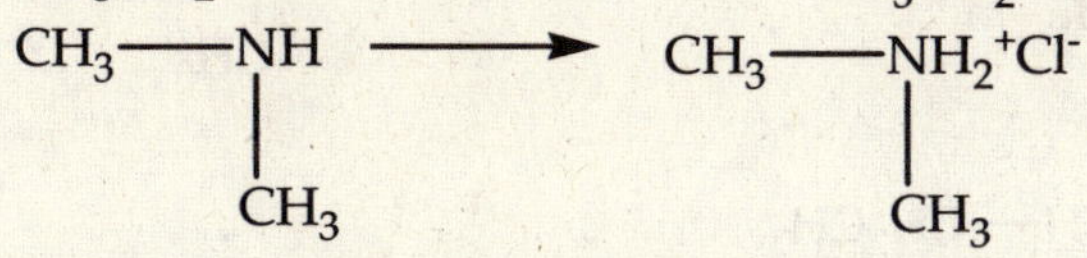

HCl

AMINES AS NEUROTRANSMITTERS (SECTION 16.5)

16.34 The gap between neurons is called a synapse.

16.36 The two amines often associated with the biochemical theory of mental illness are norepinephrine and serotonin.

☑16.38 Four neurotransmitters important in the central nervous system are acetylcholine, norepinephrine, dopamine, and serotonin.

OTHER BIOLOGICALLY IMPORTANT AMINES (SECTION 16.6)

☑16.40 The clinical uses of epinephrine are as a component of injectable local anesthetics as well as to reduce hemorrhage, treat asthma attacks, and combat anaphylactic shock.

16.42 Alkaloids are derived from plants.

☑16.44

a.	found in cola drinks	caffeine
b.	used to reduce saliva flow during surgery	atropine
c.	present in tobacco	nicotine
d.	a cough suppressant	codeine
e.	used to treat malaria	quinine
f.	an effective painkiller	morphine

THE NOMENCLATURE OF AMIDES (SECTION 16.7)

☑16.46 a.

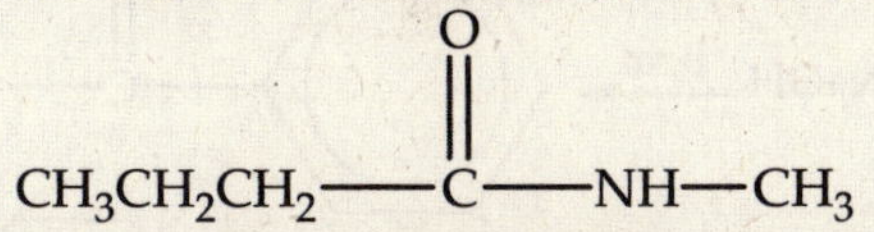

N-methylbutanamide

b.

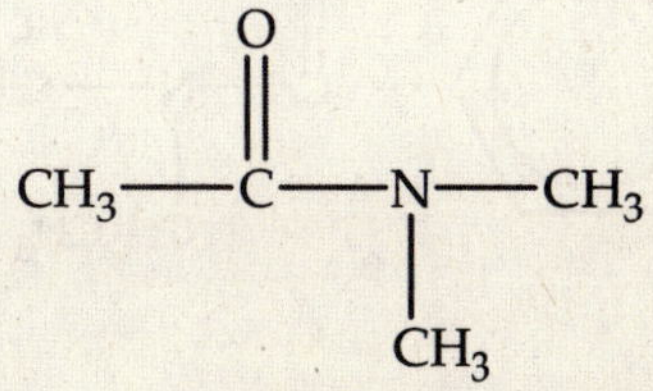

N,*N*-dimethylethanamide

c.

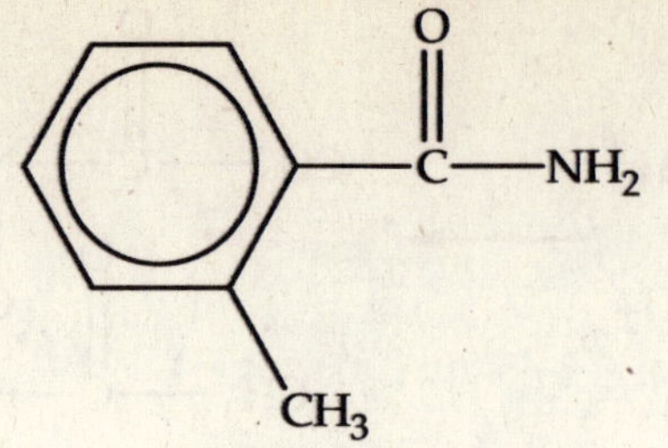

o-methylbenzamide or
2-methylbenzamide

d.

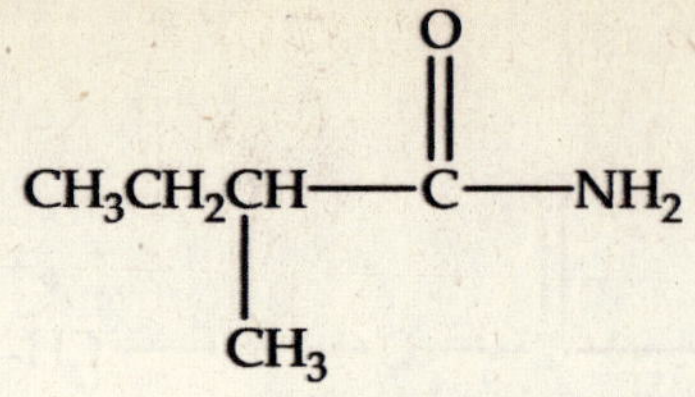

2-methylbutanamide

16.48 a. propanamide

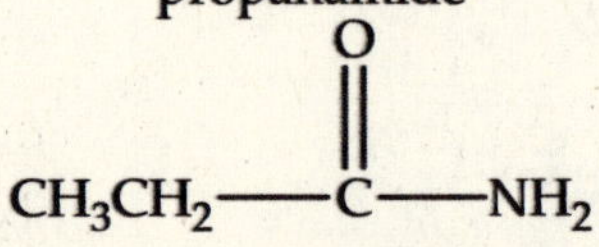

b. *N*-ethylpentanamide

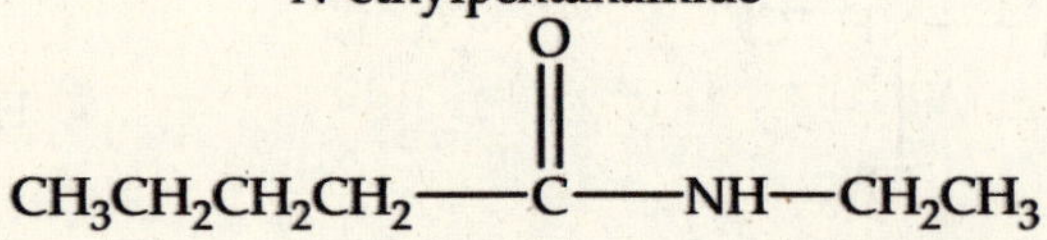

c. *N,N*-dimethylmethanamide

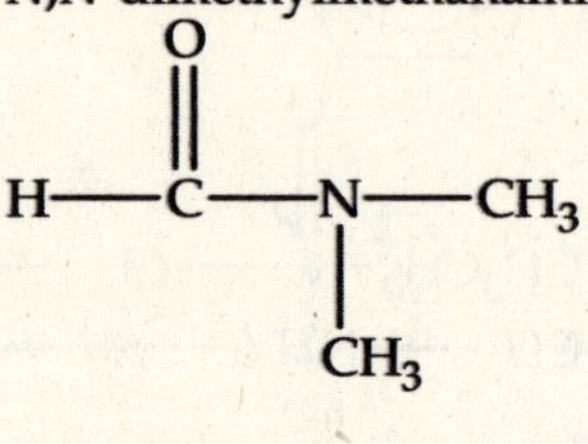

PHYSICAL PROPERTIES OF AMIDES (SECTION 16.8)

☑16.50 a.

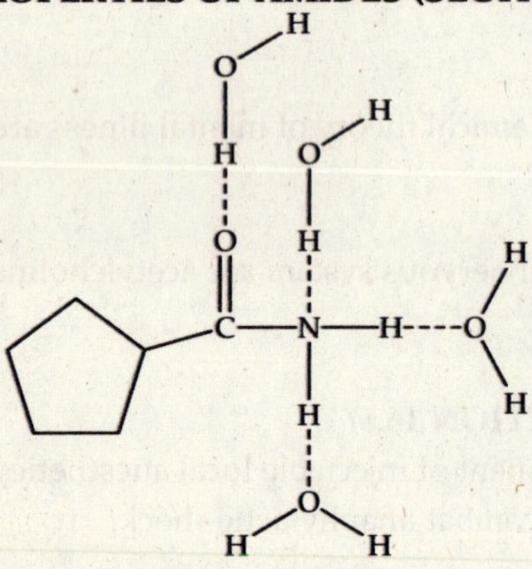

b.

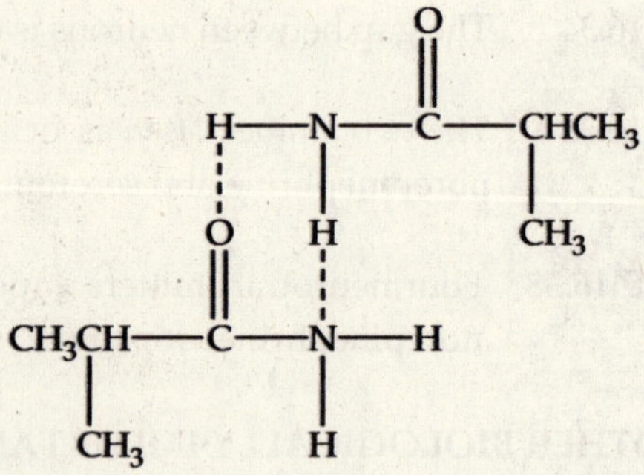

CHEMICAL PROPERTIES OF AMIDES (SECTION 16.9)

☑16.52 a.

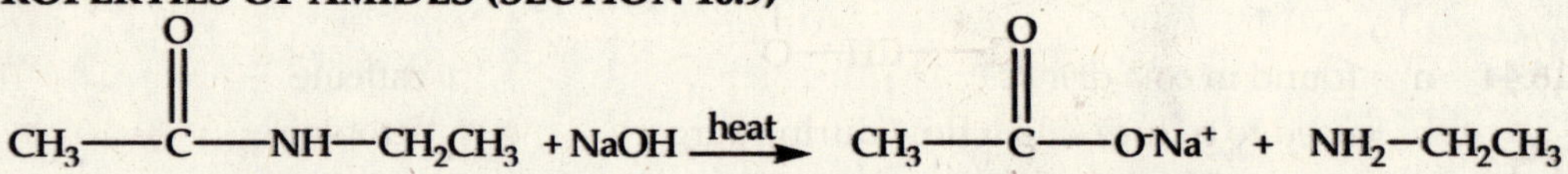

b.

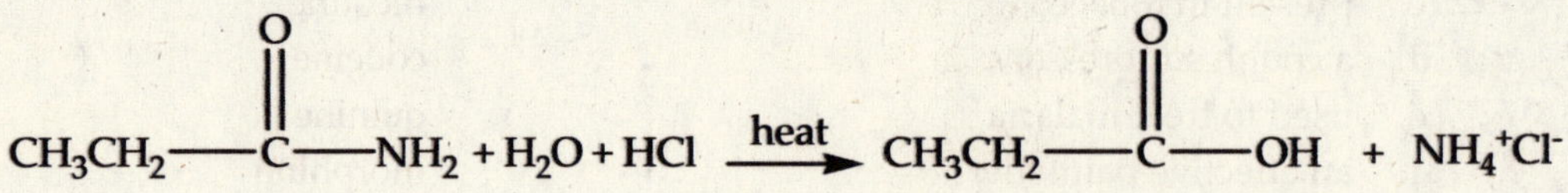

16.54

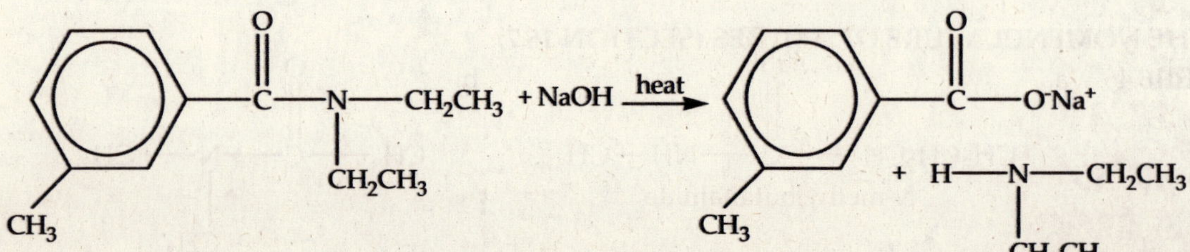

ADDITIONAL EXERCISES

16.56 a. $C_4H_{11}N$ is a saturated amine (butanamine, 2-butanamine, 2-methyl-1-propanamine, 2-methyl-2-propanamine, or diethylamine).

b. $C_5H_{12}N_2$ is an unsaturated amine with one carbon-carbon double bond.

c. C_4H_7N is an unsaturated amine with two carbon-carbon double bonds.

16.58 The hydrolysis of nylon 66 with NaOH would produce a sodium dicarboxylate salt and a diamine as shown below:

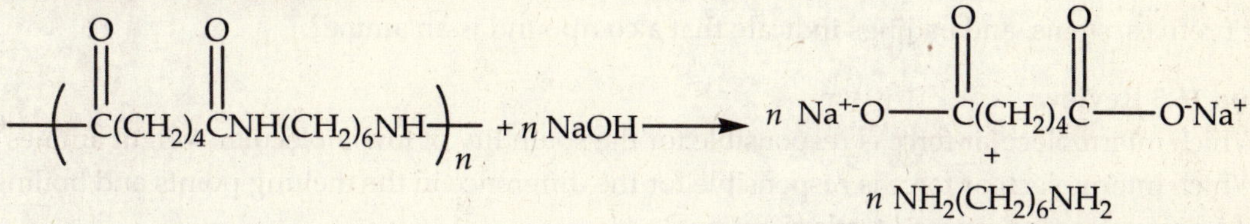

16.60

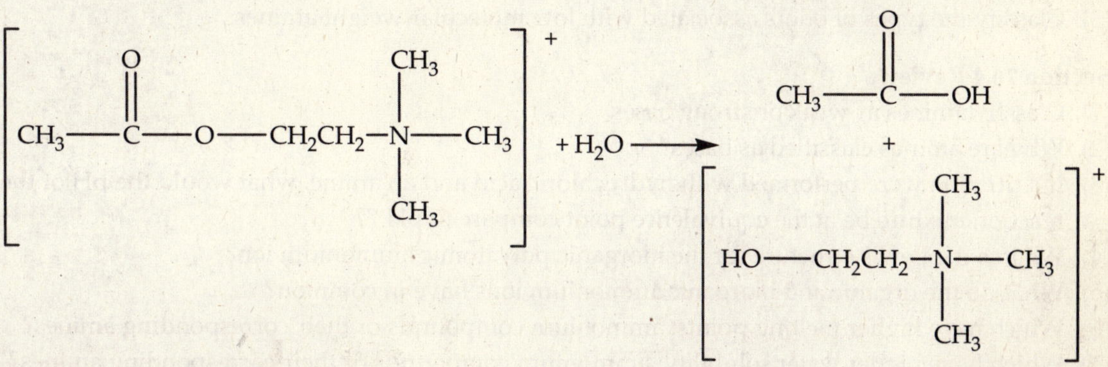

ALLIED HEALTH EXAM CONNECTION

16.62 Primary amines have the formula: $R-NH_2$

Secondary amines have the formula:

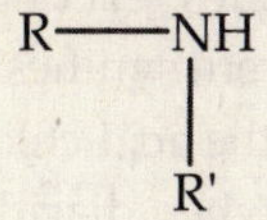

Tertiary amines have the formula:

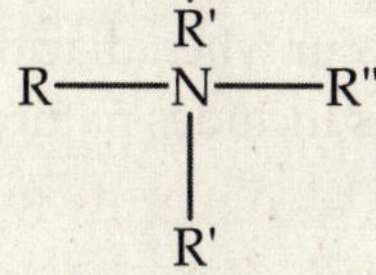

16.64 The most likely products of a reaction between $CH_3CH_2NH_2$ and H_2O are (a) $CH_3CH_2NH_3^+$ and OH^-.

CHEMISTRY FOR THOUGHT

16.66

$NH_2(CH_2)_4NH_2$ 4 carbon diamine

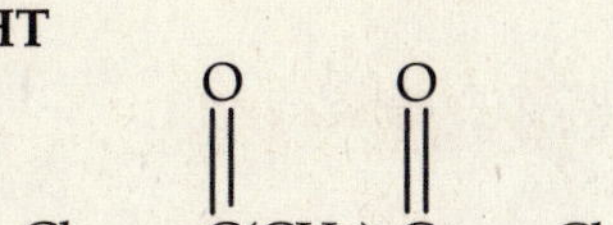

4 carbon diacid chloride

16.68 The ending of the word *maleate* is "-ate," which indicates that this compound is a salt or ester of maleic acid.

16.70 Hydrochloric acid is used to prepare an amine hydrochloride salt.

16.72 A catalyst can be used to hydrolyze amides under milder conditions than strong acids or bases and heat.

ADDITIONAL ACTIVITIES

Section 16.1 Review:

How does the classification of amines differ from the classification of alcohols?

Section 16.2 Review:

What prefixes, stems, and endings indicate that a compound is an amine?

Section 16.3 Review:

(1) Which intermolecular force is responsible for the solubility of low molecular weight amines in water?
(2) Which intermolecular force is responsible for the difference in the melting points and boiling points of primary, secondary, and tertiary amines?
(3) Classify the types of odors associated with low molecular weight amines.

Section 16.4 Review:

(1) Classify amines as weak or strong bases.
(2) Why are amines classified as bases?
(3) If a titration were performed with hydrochloric acid and an amine, what would the pH of the reaction mixture be at the equivalence point compare to pH 7?
(4) What is the Lewis structure for the inorganic polyatomic ammonium ion?
(5) What do the organic and inorganic ammonium ions have in common?
(6) Which have higher melting points: ammonium compounds or their corresponding amines?
(7) Which have higher water solubility: ammonium compounds or their corresponding amines?
(8) How can an amine salt be returned to an amine?
(9) Why is it possible to have a quaternary ammonium salt, but not a quaternary amine?
(10) Acid chlorides and acid anhydrides react with amines to form amides. Which other class of previously-studied molecules were synthesized from acid chlorides and acid anhydrides? Which class of molecules reacted with the acid chlorides or acid anhydrides?
(11) Why can polymerization occur when diacid chlorides react with diamines?
(12) Which class of biomolecules includes natural polyamides?

Section 16.5 Review:

(1) Where are neurotransmitters stored?
(2) How far do neurotransmitters travel?
(3) What are beta blockers?
(4) How do serotonin levels affect the body?

Section 16.6 Review:

(1) What are other names for epinephrine?
(2) What are some other names for amphetamines?
(3) Why are drugs from plant sources known as alkaloids?
(4) Why do alkaloids affect the central nervous system?

Section 16.7 Review:

(1) Draw the structures for *N*-methylpropanamide and methyl propanoate.
(2) Label the linkages in these structures.
(3) Circle the portion of the structures related to a carboxylic acid.
(4) Look at the structures. To which class of compounds do the uncircled portions of the structures belong?

(5) Is a reaction between a carboxylic acid and the compound class identified in (4) the most efficient way to make an amide or an ester? Explain.

Section 16.8 Review:

(1) Rank unsubstituted, monosubstituted, and disubstituted amides of similar molecular weight in order of increasing melting point.

(2) Which intermolecular force is responsible for the solubility of some amides in water?

Section 16.9 Review:

(1) Amides can be thought of as a combination of two compound classes as shown in the Section 16.7 Review. Based on their components, suggest a reason amides are neutral.

(2) When amides react with strong acids, what are the resulting products?

(3) When amides react with strong bases, what are the resulting products?

Tying It All Together with a Laboratory Application:

A student is attempting to identify three unknown amines labeled A, B, and C. Help the student complete the following chart (1).

IUPAC Name	N,N-dimethylaniline	(Hint: Use *amino*-.)	
Common Name(s)		phenethylamine	ethylphenylamine
Structure		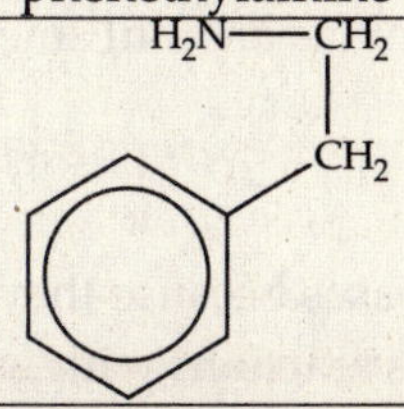	
Molecular Formula			
Melting Point	2.45°C	-60°C	-64°C
Boiling Point	194°C	200°C	204.7°C
Solubility in water	insoluble	soluble	insoluble
Amine Classification (1°, 2°, 3°)			

(2) Which of the compounds isomers? (3) Which type of isomers? If these compounds were derived from plant sources they would be called (4) ______. If these compounds were chemical messengers of the nervous system, they would be called (5) ______.

Working in a fume hood, the student places a small amount of each of the compounds into labeled, individual test tubes, and then places the test tubes into an ice bath. The student observes that compound B crystallizes after a few minutes. The identity of unknown B is (6) ______. Unknown B freezes at a higher temperature than the other two unknowns because (7) ______.

The student decides to mix unknowns A and C with water. Compound C dissolves in water, but compound A does not. The identity of unknown A is (8) ______ and the identity of unknown C is (9) ______. The difference in water solubility is caused by (10) ______. When the student checks the pH of the solution of compound C in water, the pH value is (11) ______ 7.

Unknown C reacts with hydrochloric acid to produce the compound (12) _______. The structure for this compound is (13) ______. In order to produce an amide, unknown C would have to react with a(n) (14) ______ or ______. The pH in water of the amide produced in the reaction would be (15) ______ 7. In order to regenerate unknown C from the amide, (16) ______ hydrolysis could be performed. The other product from that reaction is the (17) ______. (18) ______ hydrolysis of the amide would produce a(n)

(19) ______ and an amine salt. The amine salt could be converted back to unknown C by a reaction with (20) ______.

SOLUTIONS FOR THE ADDITIONAL ACTIVITIES

Section 16.1 Review:

Amines are classified based on the number of carbon atoms bonded to the nitrogen atom. Alcohols are classified based on the number of carbon atoms bonded to the carbon atom with the attached hydroxy group.

Section 16.2 Review:

An amine can be identified by the prefix *amino-*, the stem *aniline*, and the ending *–amine*.

Section 16.3 Review:

(1) Hydrogen bonding with water is the intermolecular force responsible for the solubility of low molecular weight amines.

(2) Hydrogen bonding (or lack thereof) is the intermolecular force is responsible for the difference in the melting points and boiling points of primary, secondary, and tertiary amines. Primary and secondary amines can hydrogen bond with molecules of the same type; however, tertiary amines cannot. Consequently, primary and secondary amines have higher melting points and boiling points than tertiary amines.

(3) Low molecular weight amines are foul smelling compounds with fishy or meat-like odors.

Section 16.4 Review:

(1) Amines are weak bases.

(2) Amines are classified as bases because they will produce hydroxide ions in an aqueous solution because of their reversible reactions with water.

(3) If a titration were performed with hydrochloric acid and an amine, the pH of the reaction mixture would be less than 7 at the equivalence point because the titration involves a strong acid and a weak base.

(4) 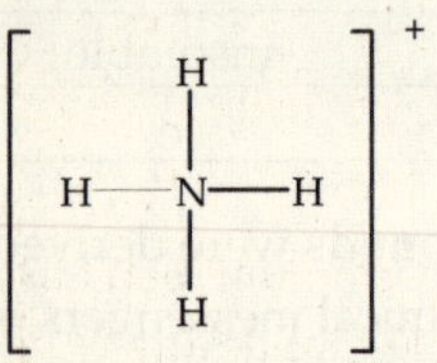

(5) Organic and inorganic ammonium ions both contain a nitrogen atom with four covalent bonds and a charge of +1.

(6) Ammonium compounds have higher melting points than their corresponding amines.

(7) Ammonium compounds have higher water solubility than their corresponding amines.

(8) Reacting an amine salt with a strong base will produce an amine.

(9) Nitrogen atoms typically form three bonds as seen in amines. When a nitrogen makes four bonds, it has a positive charge and forms an ammonium ion as seen in ammonium salts.

(10) Acid chlorides and acid anhydrides react with alcohols to form esters.

(11) Diacid chlorides and diamines both have two functional groups. Consequently, a polymer can form if both of the reactive ends of these molecules react.

(12) Proteins are natural polyamides.

Section 16.5 Review:

(1) Neurotransmitters are stored in small pockets in the axon near the synapse.

(2) Neurotransmitters do not travel very far. They are released from the axon, travel across the synapse, and bind to receptors on the dendrites of the next neuron.
(3) Beta blockers are drugs that reduce the stimulant action of epinephrine and norepinephrine on cells.
(4) Serotonin levels influence sleeping, the regulation of body temperature, and sensory perception.

Section 16.6 Review:
(1) Epinephrine is also known as adrenaline and the fight-or-flight hormone.
(2) Some amphetamines are also known as Benzedrine®, Methedrine®, speed, STP, and mescaline.
(3) Alkaloids are nitrogen containing compounds derived from plants. They are weak bases. The word alkaloid is derived from alkali, which means basic, and the *–oid* ending, which means like or similar to. These alkaloids are similar to bases.
(4) Alkaloids have similar chemical structures to neurotransmitters. These compounds also affect the central nervous system.

Section 16.7 Review:

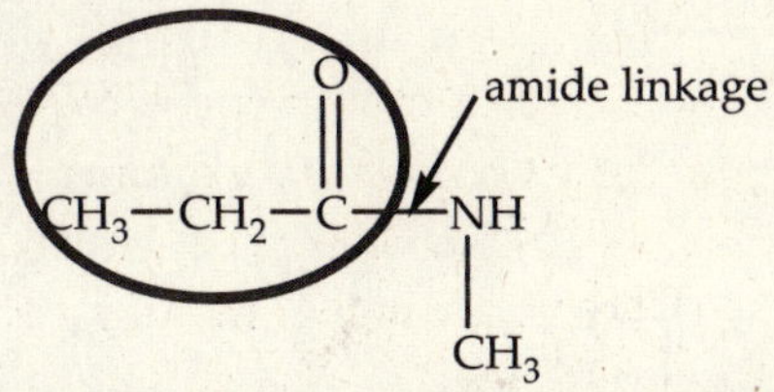

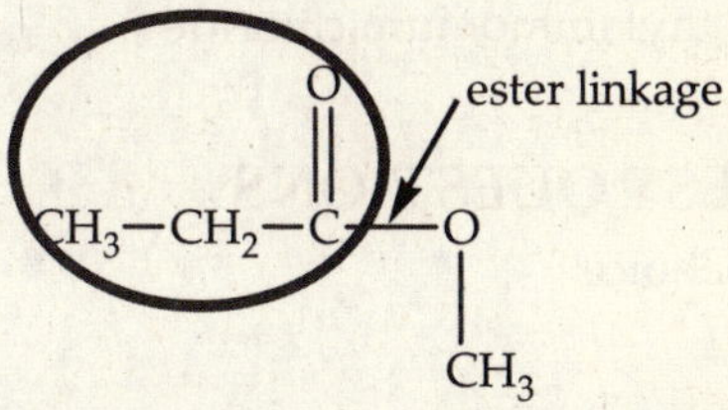

(1), (2), (3)
(4) The uncircled portion of the amide belongs to the amine class. The uncircled portion of the ester belongs to the alcohol class.
(5) The reaction between a carboxylic acid and an amine results in an ammonium carboxylate salt. The reaction between a carboxylic acid and an alcohol requires a catalyst and heat. The most efficient way to make an amide is to react an amine and an acid chloride or an amine and an acid anhydride. The most efficient way to make an ester is to react an alcohol and an acid chloride or an alcohol and an acid anhydride.

Section 16.8 Review:
(1) melting point: disubstituted < monosubstituted < unsubstituted
(2) Hydrogen bonding is the intermolecular force responsible for the solubility of some amides in water.

Section 16.9 Review:
(1) Carboxylic acids are weak acids and amines are weak bases. Consequently, an amide which has a carboxylic acid portion and an amine portion are neutral. It is important to note that amide formation does not occur when carboxylic acids react with amines. This is not a neutralization reaction.
(2) When amides react with strong acids, the products are carboxylic acids and amine (ammonium) salts.
(3) When amides react with strong bases, the products are carboxylate salts and amines.

Tying It All Together with a Laboratory Application:

IUPAC Name	*N,N-dimethylaniline*	**1-amino-2-phenylethane**	**N-ethylaniline**
Common Name(s)	**dimethylphenylamine**	*phenethylamine*	*ethylphenylamine*
Structure	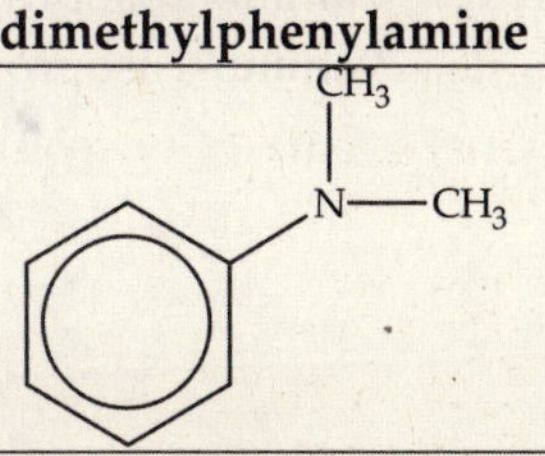	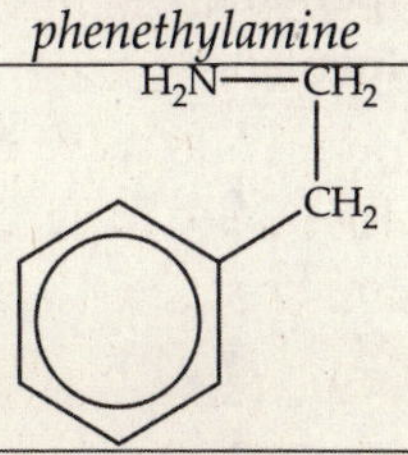	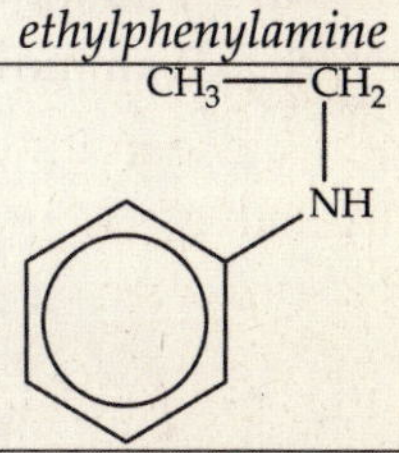
Molecular Formula	**$C_8H_{11}N$**	**$C_8H_{11}N$**	**$C_8H_{11}N$**
Amine Classification	**3°**	**1°**	**2°**

(2) All of the compounds are isomers.
(3) They are structural isomers.
(4) alkaloids
(5) neurotransmitters
(6) N,N-dimethylaniline
(7) it is a tertiary amine; consequently, it has weaker intermolecular forces than the other compounds.
(8) ethylphenylamine
(9) phenethylamine
(10) the greater hydrogen bonding that occurs with the primary amine with water.
(11) greater than
(12) phenethylammonium chloride

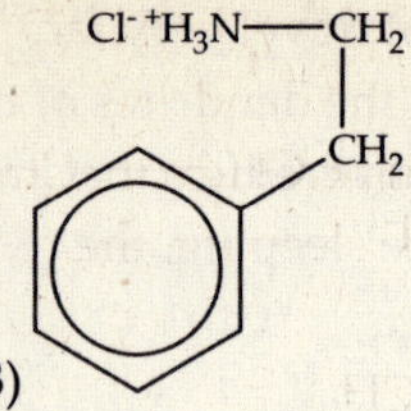

(13)
(14) an acid chloride or an acid anhydride
(15) approximately
(16) basic
(17) salt of a carboxylic acid
(18) Acidic
(19) carboxylic acid
(20) a strong base (like NaOH)

SELF-TEST QUESTIONS

Multiple Choice

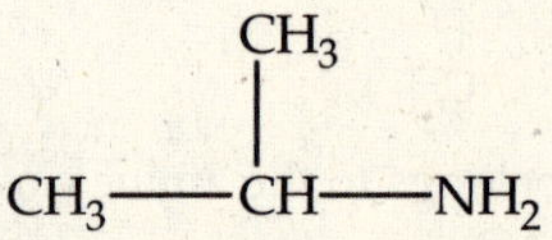

1. This is a _____ amine.
   a. primary   b. secondary   c. tertiary   d. quaternary

2. Which of the following is a secondary amine?
   a. 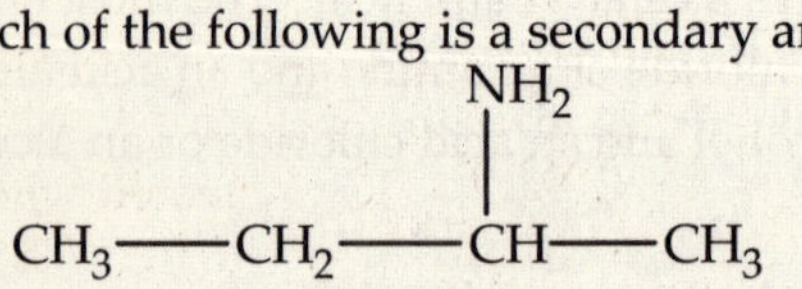

   c. 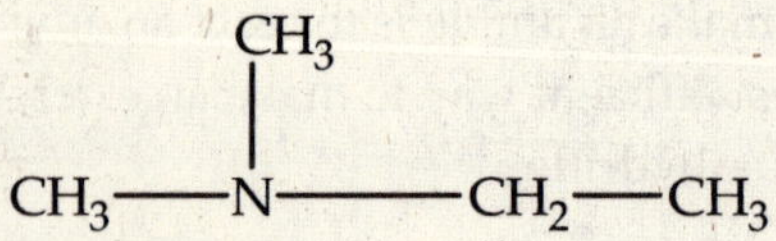

   b. CH_3—CH_2—NH_2
   d. 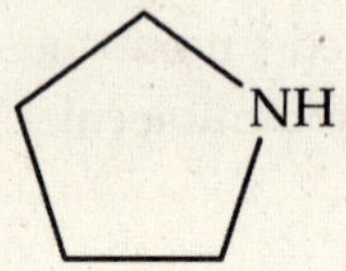

3. A common name for $CH_3CH_2NH-CH_2CH_3$ is:
   a. ethylaminoethane.   c. aminoethylethane.
   b. diethylamine.   d. diethylammonia.

4. What is the IUPAC name for 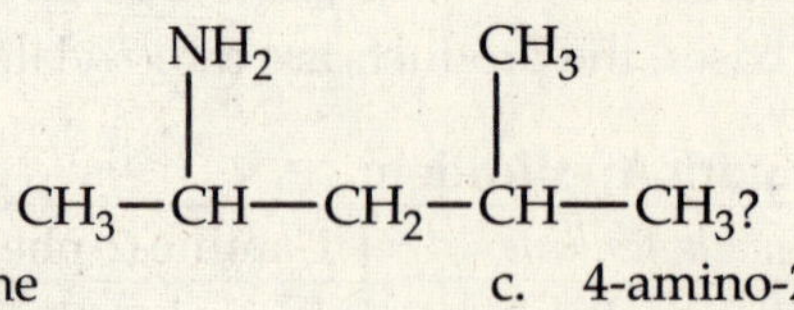

   a. 4-amino-2-methylpentane   c. 4-amino-2-isopropylpropane
   b. 2-amino-4-methylpentane   d. 2-amino-1-isopropylpropane

5. What does the reaction of $CH_3—CH(NH_2)—CH_3$ with HCl produce?

a. $CH_3—CH(NH_3^+Cl^-)—CH_3$

b. $CH_3—C(=NH)—CH_3$

c. $CH_3—CH(HN—Cl)—CH_3$

d. $CH_3—CH(H_2N—Cl)—CH_3$

6. The reaction of a primary amine with a carboxylic acid chloride produces a(n):

a. secondary amine. b. tertiary amine. c. amide. d. carboxylate salt.

7. What is the IUPAC name for $CH_3—CH_2—C(=O)—NH_2$?

a. 1-aminopropanamide
b. 1-aminobutanamide
c. propanamide
d. butanamide

8. What is the IUPAC name for $CH_3—CH_2—CH_2—C(=O)—NH—CH_3$?

a. pentanamide
b. 1-methylbutanmide
c. 1-methylpentanamide
d. *N*-methylbutanamide

9. What is one of the products produced when $CH_3CH_2CH_2—C(=O)—NH—CH_3$ is treated with NaOH?

a. $CH_3—NH_2$

c. $CH_3CH_2CH_2—C(=O)—NH_3^+CH_3$

b. $CH_3—CH_2—CH_2—NH_2$

d. $CH_3CH_2—C(=O)—NH_2$

10. Which of the following is one of the products produced when $CH_3—C(=O)—NH—CH_2—CH_3$ is treated with HCl and H₂O?

a. $CH_3—C(=O)—O^-$

c. $CH_3—C(=O)—NH_2$

b. $CH_3—CH_2—NH_3^+Cl^-$

d. $CH_3—CH_2—NH_2$

Matching

For each description on the left, select an amide from the list on the right.

11. a tranquilizer
12. an aspirin substitute
13. an antibiotic

a. penicillin
b. benzamide
c. acetaminophen
d. valium

For each description on the left, select the correct alkaloid.

14.	present in coffee	a.	atropine
15.	present in tobacco	b.	nicotine
16.	used as a cough suppressant	c.	codeine
17.	used to dilate the pupil of the eye	d.	caffeine

True-False

18. Triethylamine is a tertiary amine.
19. This structure is 1-methylaniline.

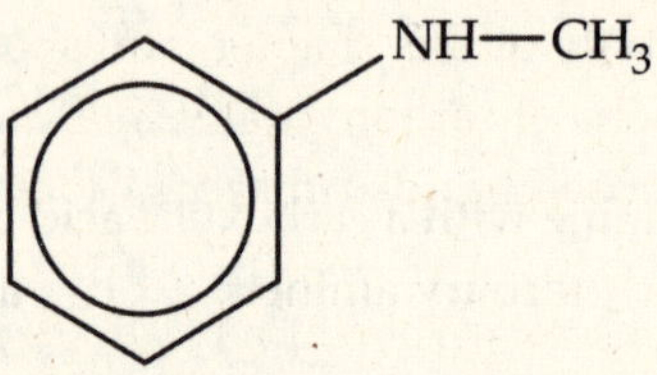

20. Tertiary amines have higher boiling points than primary and secondary amines.
21. Low molecular weight amines have a characteristic pleasant odor.
22. Both primary and tertiary amines can hydrogen bond with water molecules.
23. Amines under basic conditions exist as amine salts.
24. Disubstituted amides have lower boiling points than unsubstituted amides.
25. Amides with fewer than six carbons are water soluble.
26. Amide molecules are neither basic nor acidic.
27. Serotonin is an important neurotransmitter.
28. Amphetamine is a powerful nervous system stimulant.
29. Amine salts are frequently used in medication.
30. Alkaloids are acidic.

Chapter 17: Carbohydrates

CHAPTER OUTLINE

17.1 Classes of Carbohydrates
17.2 The Stereochemistry of Carbohydrates
17.3 Fischer Projections
17.4 Monosaccharides
17.5 Properties of Monosaccharides
17.6 Important Monosaccharides
17.7 Disaccharides
17.8 Polysaccharides

LEARNING OBJECTIVES/ASSESSMENT

When you have completed your study of this chapter, you should be able to:

1. Describe the four major functions of carbohydrates in living organisms. (Section 17.1; Exercise 17.2)
2. Classify carbohydrates as monosaccharides, disaccharides, or polysaccharides. (Section 17.1; Exercise 17.4)
3. Identify molecules possessing chiral carbon atoms. (Section 17.2; Exercise 17.8)
4. Use Fischer projections to represent D and L compounds. (Section 17.3; Exercise 17.12)
5. Classify monosaccharides as aldoses or ketoses, and classify them according to the number of carbon atoms they contain. (Section 17.4; Exercise 17.22)
6. Write reactions for monosaccharide oxidation and glycoside formation. (Section 17.5; Exercise 17.34)
7. Describe uses for important monosaccharides. (Section 17.6; Exercise 17.37)
8. Draw the structures and list sources and uses for important disaccharides. (Section 17.7; Exercise 17.44)
9. Write reactions for the hydrolysis of disaccharides. (Section 17.7; Exercise 17.52)
10. Describe the structures and list sources and uses for important polysaccharides. (Section 17.8; Exercise 17.54)

SOLUTIONS FOR THE END OF CHAPTER EXERCISES

CLASSES OF CARBOHYDRATES (SECTION 17.1)

☑17.2
a. cellulose – structural material in plants
b. sucrose, table sugar – an energy source in our diet
c. glycogen – form of stored energy in animals
d. starch – form of stored energy in plants

☑17.4
a. table sugar – carbohydrate, disaccharide
b. 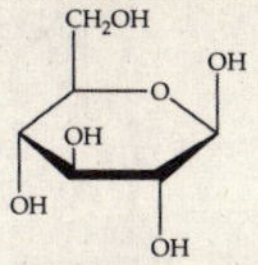 – carbohydrate, monosaccharide
c. starch – carbohydrate, polysaccharide
d. fructose – carbohydrate, monosaccharide
e. cellulose – carbohydrate, polysaccharide
f. 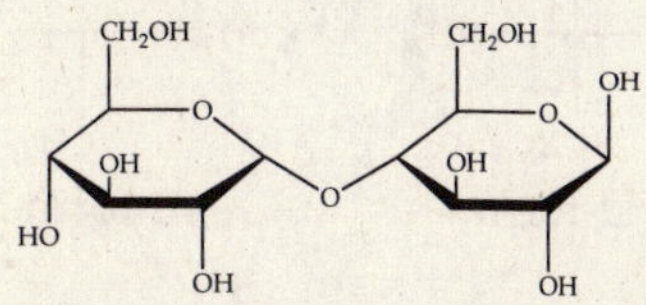 – carbohydrate, disaccharide
g. glycogen – carbohydrate, polysaccharide
h. amylose – carbohydrate, polysaccharide

THE STEREOCHEMISTRY OF CARBOHYDRATES (SECTION 17.2)

17.6 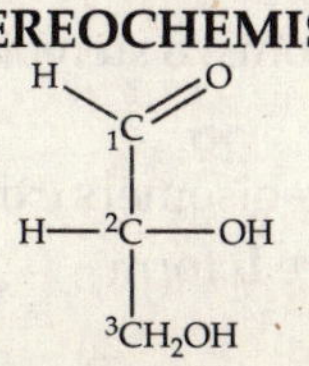

Carbon atom 1 is not chiral because it is only bonded to three other atoms.
Carbon atom 3 is not chiral because it is bonded to two hydrogen atoms.
Chiral carbon atoms are bonded to four unique groups.

☑17.8 a. 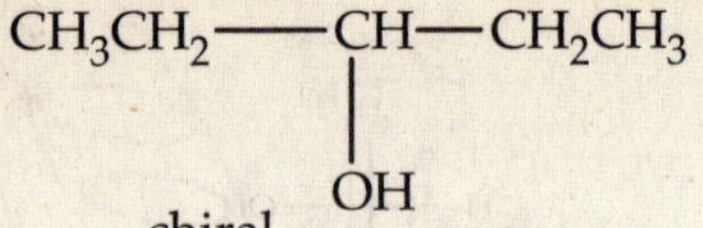

no chiral carbon atoms; no enantiomers

b. 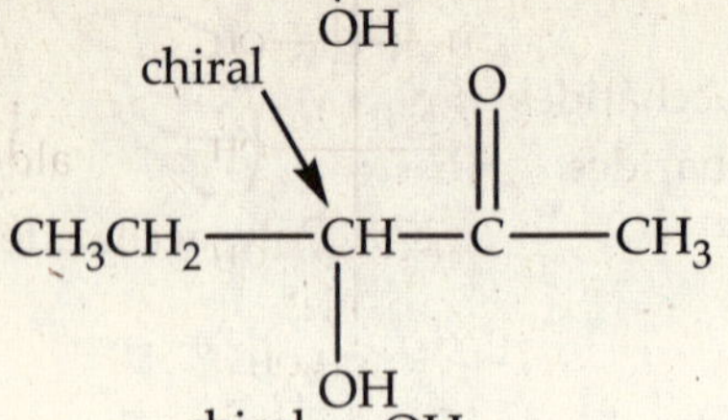

1 chiral carbon atom;
2 stereoisomers; a pair of enantiomers

c. 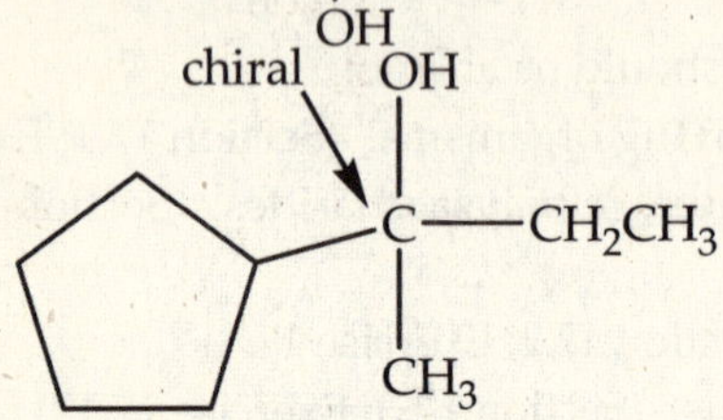

1 chiral carbon atom; 2 stereoisomers; a pair of enantiomers

FISCHER PROJECTIONS (SECTION 17.3)

17.10 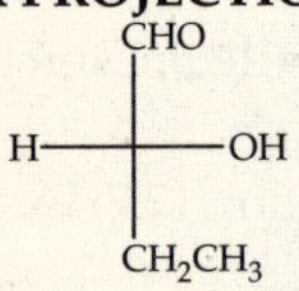

The intersection of the lines represents a chiral carbon atom. The H and the OH groups (groups on horizontal bonds) stick out in front of the plane of the paper and the CHO and CH_2CH_3 groups (groups on vertical bonds) stick out behind the plane of the paper.

☑17.12 a. D-form b. L-form

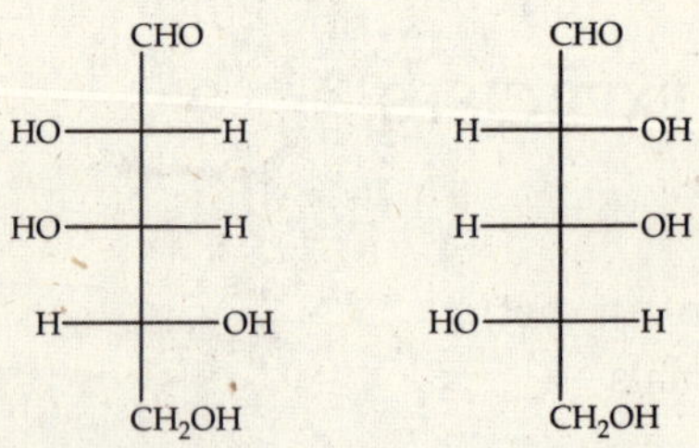

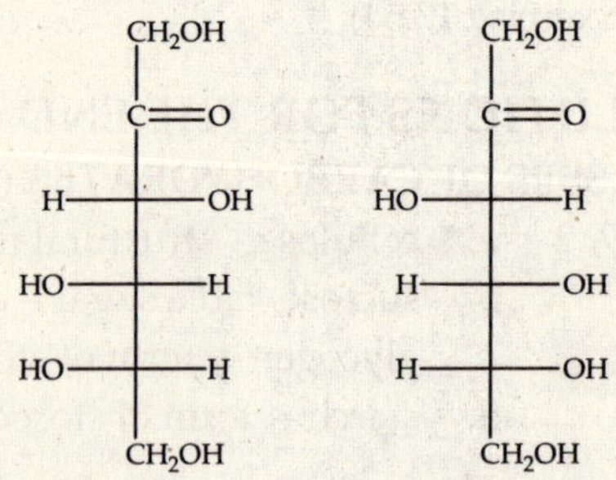

17.14 a. 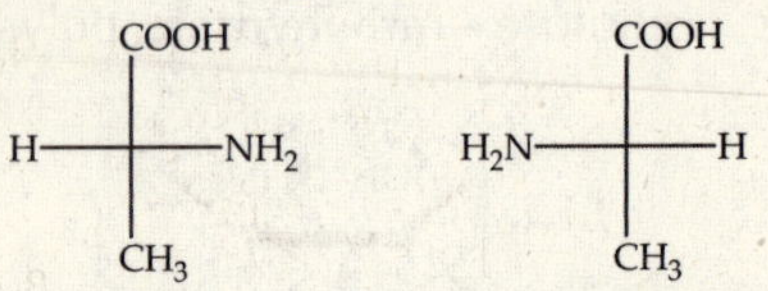

D-form L-form

b. 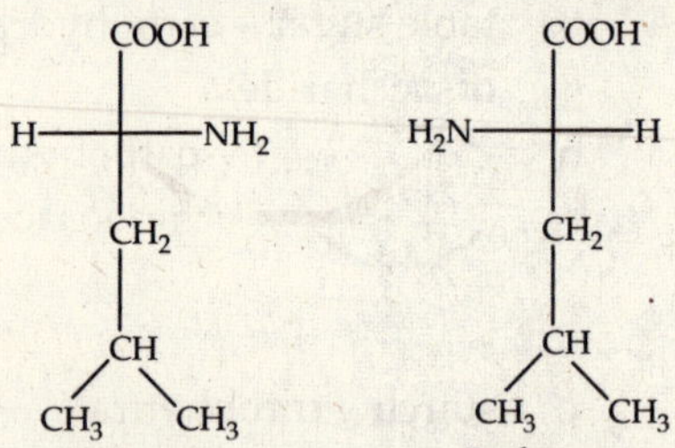

D-form L-form

17.16 a. 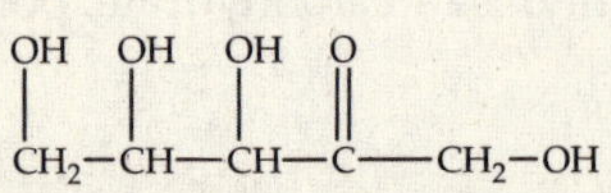

2 chiral carbon atoms; 4 stereoisomers

b. 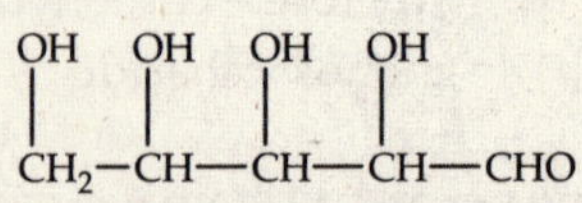

3 chiral carbon atoms; 8 stereoisomers

17.18 Aldopentoses contain 3 chiral carbon atoms; therefore, eight possible stereoisomers exist. Four of these stereoisomers are D form and four of these stereoisomers are L form.

17.20 Optically active molecules contain at least one chiral carbon atom, have nonsuperimposable mirror images, and rotate the plane of polarized light.

MONOSACCHARIDES (SECTION 17.4)

☑17.22 a. 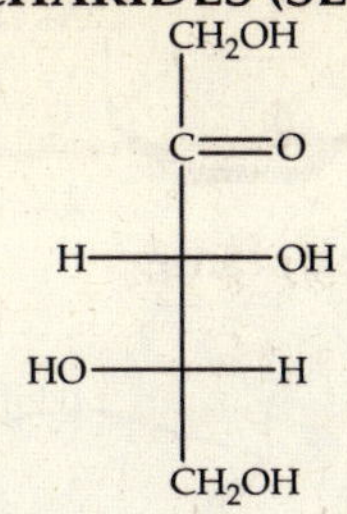 ketopentose

b. 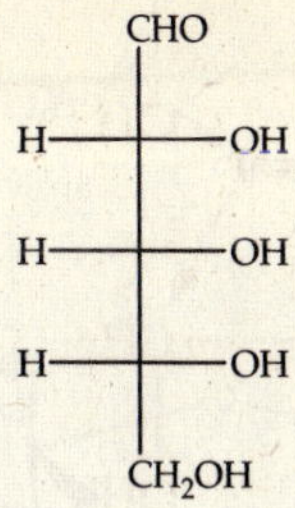 aldopentose

PROPERTIES OF MONOSACCHARIDES (SECTION 17.5)

17.24 Certain carbohydrates are called sugars because they taste sweet.

17.26 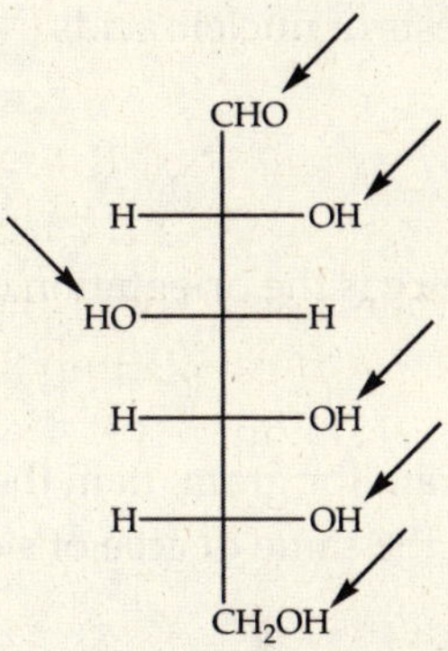

The arrows indicate the sites on the molecule where water can hydrogen bond to glucose.

17.28 a. 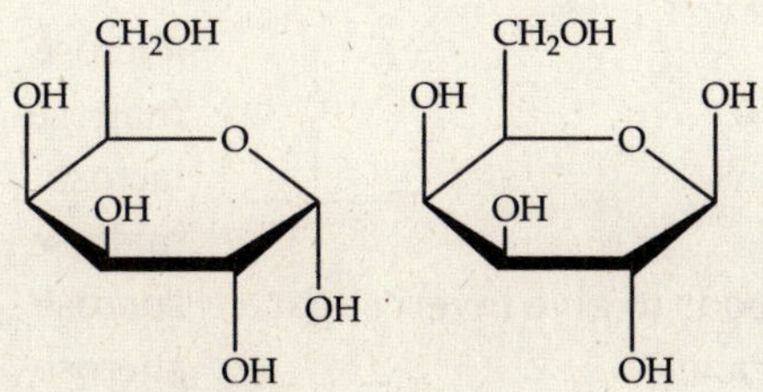

α form

b. 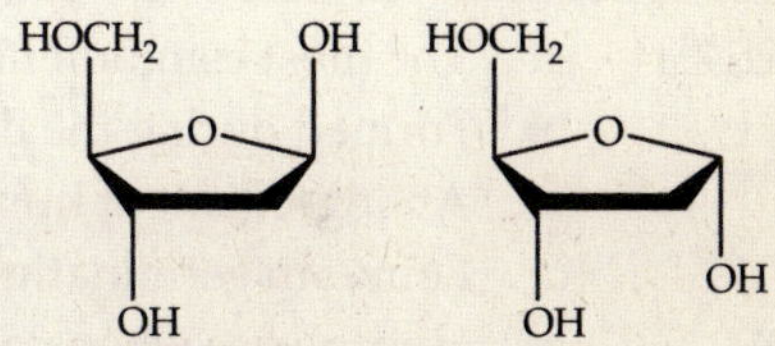

β form

17.30 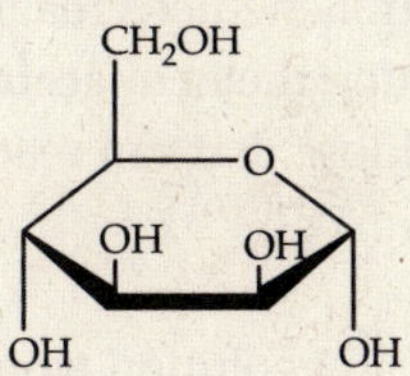 α-mannose

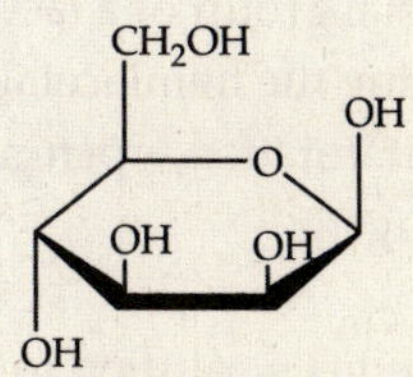 β-mannose

17.32 When a sugar fails to react with Cu^{2+}, it is a nonreducing sugar.

☑17.34 a. 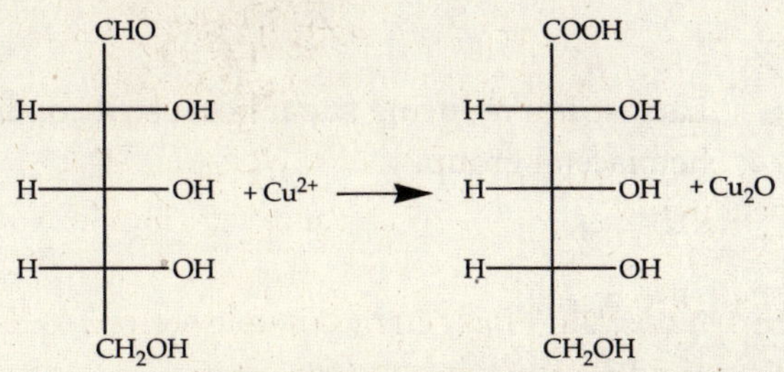

b. 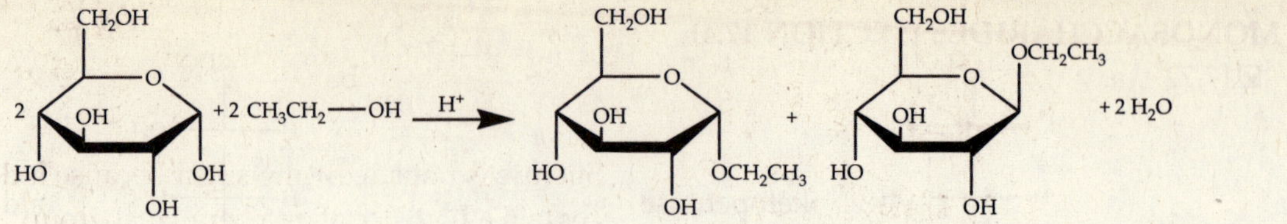

c. 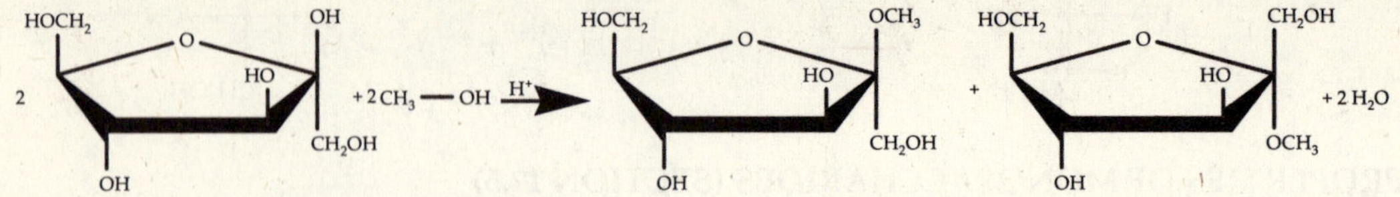

IMPORTANT MONOSACCHARIDES (SECTION 17.6)

17.36 Ribose and deoxyribose are monosaccharides used in the synthesis of nucleic acids.

17.38 Dextrose and blood sugar are two other names for D-glucose.

17.40 Glucose and galactose have similar structures. The only difference is the orientation of the hydroxy group attached to carbon 4.

17.42 Fructose can be a low-calorie sweetener because it is sweeter, gram for gram, than the other common sugars; consequently, less fructose is needed to obtain the same degree of sweetness.

DISACCHARIDES (SECTION 17.7)

☑17.44
a. The most common household sugar — sucrose
b. Formed during the digestion of starch — maltose
c. An ingredient of human milk — lactose
d. Found in germinating grain — maltose
e. Hydrolyzes when cooked with acidic foods to give invert sugar — sucrose
f. Found in high concentrations in sugar cane — sucrose

17.46 The hemiacetal group of a lactose molecule is able to react with Benedict's reagent because the ring containing the hemiacetal group is not "locked." In solution the hemiacetal group can undergo mutarotation, opening the ring into an open-chain aldose that can react with Bendict's reagent.

17.48 Adding Benedict's solution to a sample of the sweet-tasting water solution would provide a qualitative test for the presence of sucrose or honey. Honey contains invert sugar that will react with Benedict's solution. Sucrose will not react with Benedict's solution.

17.50 a. 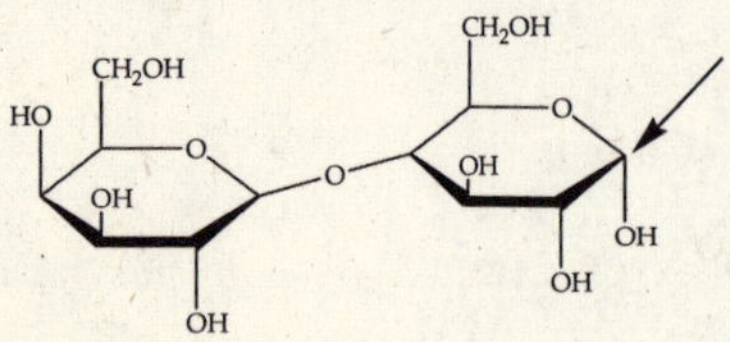

Lactose is a reducing sugar because it contains a hemiacetal group.

b.

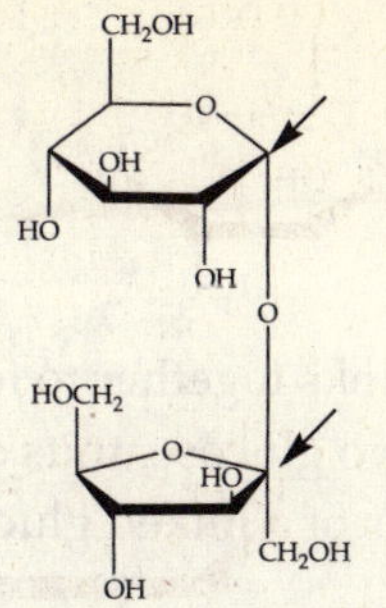

Sucrose is not a reducing sugar because it does not contain a hemiacetal or a hemiketal group.

☑17.52 a. When melibiose is hydrolyzed, glucose and galactose are the monosaccharides produced.
b. Melibiose is a reducing sugar because the glucose ring contains a hemiacetal group.
c. The glycosidic linkage in melibiose is an α $(1\rightarrow6)$ linkage.

POLYSACCHARIDES (SECTION 17.8)

☑17.54

a.	The unbranched polysaccharide in starch	amylose
b.	A polysaccharide widely used as a textile fiber	cellulose
c.	The most abundant polysaccharide in starch	amylopectin
d.	The primary constituent of paper	cellulose
e.	A storage form of carbohydrates in animals	glycogen

ADDITIONAL EXERCISES

17.56 Maltose ($C_{12}H_{22}O_{11}$) has approximately twice the mass per molecule of glucose ($C_6H_{12}O_6$); therefore if two 10% (w/v) solutions are made with maltose and glucose as solutes, the maltose solution would contain roughly half the number of molecules contained in the glucose solution. Neither of the solutes dissociates in water since both are molecular. The boiling point of a solution increases with the number of solute particles in solution; therefore, the glucose solution would contain more molecules of solute and would have the higher boiling point.

17.58

$$100\text{ g sucrose}\left(\frac{1\text{ mole sucrose}}{342\text{ g sucrose}}\right)\left(\frac{1\text{ mole glucose}}{1\text{ mole sucrose}}\right)\left(\frac{180\text{ g glucose}}{1\text{ mole glucose}}\right)$$
$$= 52.6\text{ g glucose}$$

17.60 If maltose was used in the "Osmosis through carrot membranes" demonstration and the solution level rose 5 cm above the carrot in the tube and then maltase was added to the solution, the solution level in the tube would further increase. Adding maltase would cause each maltose molecule to split into two molecules of glucose, which would double the osmolarity of the original maltose solution and cause the water level in the tube to further increase.

ALLIED HEALTH EXAM CONNECTION

17.62 (a) Cellulose, (b) starch, and (d) glycogen are composed of long chains of glucose molecules. (c) Cholesterol is not composed of glucose molecules.

17.64 Hydrolysis is the general type of reaction used to decompose sucrose to glucose and fructose.

17.66 (c) Glycogen is the polysaccharide that is a branched polymer of glucose found in the liver and muscle cells.

CHEMISTRY FOR THOUGHT

17.68 $$\frac{2.80\times10^5\ \text{u}}{1\ \text{molecule starch}}\left(\frac{1\ \text{molecule linked glucose}}{171\ \text{u linked glucose}}\right)$$

$$=1.64\times10^3\ \text{glucose molecules}$$

Note: The molecular mass of glucose is 180 u, but when glucose links together to form a molecule of starch, 1 molecule of water (18.0 u) is lost for every two glucose units or 0.5 molecules of water per glucose unit; therefore, the molecular mass of a linked glucose unit is 180-9.0 = 171 u.

17.70 Aspartame (Nutrasweet) contains calories and yet is used in diet drinks. A drink can contain aspartame as a sweetener and yet be low in calories because the amount needed to sweeten a diet drink is so small that very few calories are added to the drink.

17.72 Foods that would be expected to give a positive starch test include bread, crackers, pasta, and rice.

17.74 Amylose is a straight-chain glucose polymer similar to cellulose. Paper manufactured with amylose instead of cellulose would have less longevity because amylose forms loose spiral structures unlike cellulose which forms extended straight chains that can be aligned side by side to form well-organized, water-insoluble fibers in which the hydroxy groups form numerous hydrogen bonds with the neighboring chains that confer rigidity and strength to the overall structure of the paper.

ADDITIONAL ACTIVITIES

Section 17.1 Review:

(1) Carbohydrates are polyfunctional compounds. Which key functional groups do they contain?

(2) Complete the following table.

Prefix	Meaning	Classes of Carbohydrates	Definition
mono-		monosaccharide	
di-		disaccharide	
oligo-		oligosaccharide	
poly-		polysaccharide	

Section 17.2 Review:

(1) What are the major types of isomers?

(2) Which class of isomers from (1) includes enantiomers?

(3) Which other subclass of isomers belongs to the same class as enantiomers?

(4) Are all mirror images stereoisomers? Explain.

(5) What is the requirement for a chiral carbon atom?

(6) Are all stereoisomers enantiomers? Explain.

Section 17.3 Review:

(1) Draw the Fischer Projection of glyceraldehyde or look at the drawing on page 525. Imagine that this is a stick figure of a person facing you.

(2) What is the "head" of the glyceraldehyde?

(3) What is the "body" of the glyceraldehyde?

(4) What are the "hands" of the glyceraldehyde?

(5) What is the "feet" of the glyceraldehyde?

(6) What are the relative positions of the "head" and the "body"?

(7) What are the relative positions of the "body" and the "hands"?
(8) What are the relative positions of the "body" and the "feet"?
(9) What are the relative positions of the "head" and the "feet"?
(10) What are the relative positions of the "head" and the "hands"?
(11) What are the relative positions of the "hands" and the "feet"?
(12) Could the glyceraldehyde be described as "doing a back bend"? Explain.
(13) Could the glyceraldehyde be described as "giving a hug"? Explain.
(14) What is the difference between D-glyceraldehyde and L-glyceraldehyde?
(15) Does D-glyceraldehyde rotate light to the right?
(16) Which isomers of monosaccharides are used by humans?
(17) Which isomers of amino acids are used by humans?

Section 17.4 Review:

(1) How many chiral carbon atoms are found in an L-ketohexose?
(2) How many stereoisomers are classified as L-ketohexoses?
(3) Are L-ketohexoses abundant in nature? Explain.

Section 17.5 Review:

(1) How do the many hydroxy groups on monosaccharides influence the state of matter of monosaccharides at room temperature?
(2) When monosaccharides are in cyclic form, do they contain aldehyde or ketone groups? Explain.
(3) How might the expression, "Birds fly. Fish swim.", help you to remember the difference between α- and β-anomers?
(4) What is the difference between a Fischer projection and a Haworth structure?
(5) What are the rules for drawing Haworth structures?
(6) Are all reducing sugars monosaccharides? Explain.
(7) What is a glycoside?

Section 17.6 Review:

(1) Do the cyclic monosaccharides ribose and deoxyribose contain aldehyde, ketone, hemiacetal, hemiketal, acetal, or ketal key functional groups?
(2) Which internal organ converts other sugars into glucose?
(3) Where is lactose synthesized in the body?
(4) What is the major difference between fructose and the other monosaccharides in this section?

Section 17.7 Review:

(1) In glycosidic linkages of disaccharides, which ring determines whether the linkage is an α or β?
(2) In glycosidic linkages of disaccharides, which carbon atom is identified first?
(3) In glycosidic linkages of disaccharides, which carbon atom is identified second?
(4) Are all disaccharides reducing sugars? Explain.

Section 17.8 Review:

(1) Polysaccharides are condensation polymers. What is the small molecule produced in addition to the polysaccharide during polymerization?
(2) When polysaccharides are mixed with water and a colloidal dispersion forms, what is the dispersing medium?
(3) When polysaccharides are mixed with water and a colloidal dispersion forms, what is the dispersed phase?
(4) Are polysaccharide-water colloidal dispersions transparent? Explain.
(5) Which disaccharide has the same type of glycosidic linkage as amylose?
(6) What is the major difference between amylose and the disaccharide from (5)?

(7) When peeled potatoes are boiled, the water become cloudy. How could you check to see if the "potato water" contains amylose?
(8) Will "potato water" contain glycogen? Explain.
(9) How does the molecular formula of an amylose molecule with 1500 monosaccharide units compare to the molecular formula of a cellulose molecule with 1500 monosaccharide units?
(10) How does the structural formula for an amylose molecule with 1500 monosaccharide units compare to the structural formula of a cellulose molecule with 1500 monosaccharide units?

Tying It All Together with a Laboratory Application:

While cooking sweet-and-sour sauce at home, a chemistry student realizes the ingredients are similar to the compounds studied in class. Pineapple juice, lemon juice, granulated sugar, and corn starch all contain (1) ______. The pineapple and lemon juices as well as the red wine vinegar contain (2) ______ in the form of acetic acid and citric acid.

The recipe calls for mixing the pineapple juice and water in a sauce pan over low heat and slowly adding the table sugar while stirring. The solubility of sugar (3) ______ when heat is applied and the (4) ______ of dissolving increases as a result of the stirring. After the table sugar is dissolved, the vinegar and lemon juice are added to the sauce pan. When table sugar reacts with water in an (5) ______ environment, it undergoes (6) ______ to produce (7) ______ and ______. The resulting mixture is (8) ______ than sugar-water.

The corn starch is mixed in enough cold water to make a slurry. The student remembers that (9) ______ and ______ are the molecules contained in starch. They are both (10) ______ that contain many glucose groups. No matter how much water is added the mixture will remain cloudy because starch and water form a(n) (11) ______. Starch is the (12) ______ and water is the (13) ______. The corn starch slurry is then added to the sauce pan and stirred constantly. The corn starch slurry is added to the mixture to (14) ______.

The student decides to take the ingredients and final product to chemistry class and obtain permission to test the materials for the presence of reducing sugars. The student selects (15) ______ (Tollen's or Benedict's) reagent and makes the following predictions:

The reagent has a (16) ______ color. The product of a negative test will have a (17) ______ color. The product of a positive test will be (18) ______.
Pineapple juice should test positive for reducing sugars because it contains (19) ______, fruit sugar.
Lemon juice will be similar to pineapple juice.
Table sugar and water should test (20) ______ for reducing sugars because the table sugar is a(n) (21) ______ (monosaccharide, disaccharide, polysaccharide) with locked (22) ______ carbon atoms and no acid catalyst is present.
After heating, table sugar, acetic acid, and water should test (23) ______ for reducing sugars because (24) ______ will occur.
Corn starch should test (25) ______ for reducing sugars because it contains (26) ______ (monosaccharides, disaccharides, polysaccharides) with (27) ______ linkages.
The red wine vinegar should test (28) ______ for reducing sugars because the sugar in the wine was converted to ethanol which was then (29) ______ to form acetic acid.
The sweet-and-sour sauce should test (30) ______ for reducing sugars.

The student also asks for permission to use a brown iodine solution. Corn starch should be the only ingredient tested that has a (31) ______ test with iodine. The corn starch-iodine mixture should have a (32) ______ color because it contains (33) ______. The student does not know whether the iodine will have a (34) ______ test with the sweet-and-sour sauce. Although the sauce contains starch, it also contains acids that can (35) ______ starch and form (36) ______, ______, and ______.

SOLUTIONS FOR THE ADDITIONAL ACTIVITIES

Section 17.1 Review:

(1) Carbohydrates are polyfunctional compounds that contain aldehyde or ketone key functional groups (or can form aldehyde or ketone key functional groups) as well as several hydroxy groups.

(2)

Prefix	Meaning	Classes of Carbohydrates	Definition
mono-	1	monosaccharide	a simple carbohydrate
di-	2	disaccharide	a carbohydrate formed by combining two monosaccharides
oligo-	small	oligosaccharide	a carbohydrate formed by combining 3-10 monosaccharides
poly-	many	polysaccharide	a carbohydrate formed by combining many monosaccharides

Section 17.2 Review:

(1) structural isomers, stereoisomers
(2) stereoisomers
(3) geometric isomers (*cis-trans*)
(4) Only nonsuperimposable mirror images are stereoisomers. Superimposable mirror images are the same molecule.
(5) A chiral carbon atom must be bonded to four unique groups.
(6) No, not all stereoisomers are enantiomers. If a molecule contains 2 chiral carbon atoms, it is one of four stereoisomers of that structure. It has a mirror image (enantiomer) and two other stereoisomers that are enantiomers of each other, but not of the original molecule.

Section 17.3 Review:

(1)

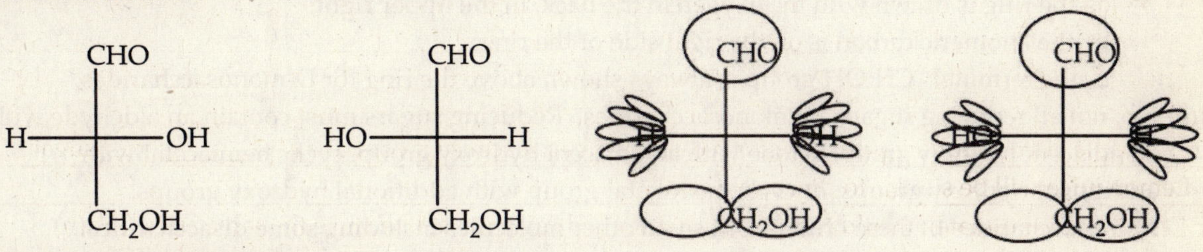

(2) The head is the CHO group.
(3) The body is the chiral carbon atom.
(4) The H and OH are the hands.
(5) The CH_2OH is the feet.
(6) The head is behind the body.
(7) The body is behind the hands.
(8) The body is in front of the feet.
(9) The head is above the feet, but both are in the same plane.
(10) The head is behind the hands.
(11) The hands are in front of the feet.

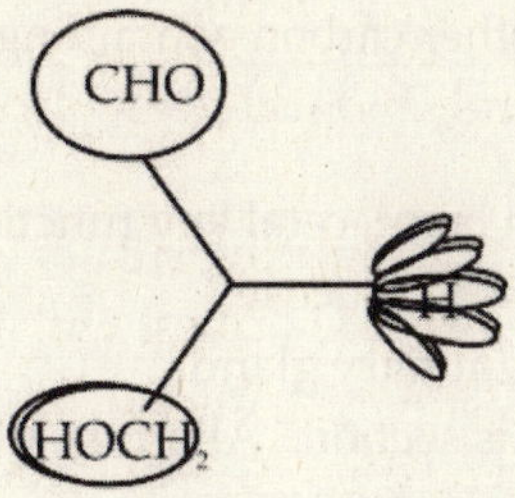

Side view - back bend
Question 12

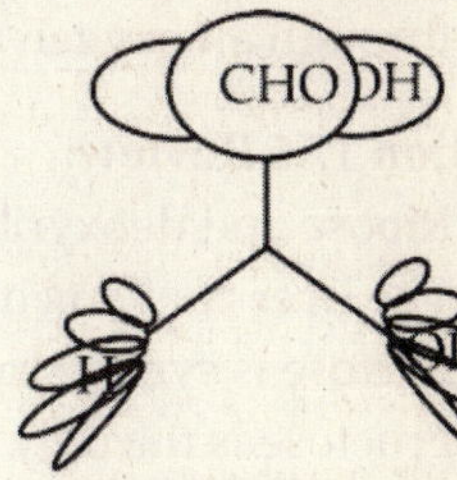

Top view - hug
Question 13

(12) Yes, the glyceraldehyde could be described as "doing a back bend" because the head and feet are behind the body.

(13) Yes, the glyceraldehyde could be described as "giving a hug" because the hands are outstretched in front of the body.

(14) The hydroxy group is on the right side of the chiral carbon in D-glyceraldehyde and the hydroxy group is on the left side of the chiral carbon in L-glyceraldehyde. (Note: This is from the perspective of the observer, not the glyceraldehyde stick person.)

(15) D-glyceraldehyde may rotate light to the right or to the left because it is optically active. The name and structure alone are not enough to determine which direction it will rotate the light.

(16) Humans use D-isomers of monosaccharides.

(17) Humans use L- isomers of amino acids.

Section 17.4 Review:

(1) An L-ketohexose contains 3 chiral carbon atoms.

(2) Because the stereoisomer contains 3 chiral carbon atoms, eight stereoisomers are ketohexoses because 2^3 equals 8. Of these eight stereoisomers, four stereoisomers are classified as L-ketohexoses.

(3) L-ketohexoses are not abundant in nature because naturally occurring monosaccharides are primarily D-form aldoses.

Section 17.5 Review:

(1) The hydroxy groups on monosaccharides allow them to hydrogen bond to other monosaccharide molecules. These strong intermolecular forces between monosaccharide molecules cause them to be solids at room temperature.

(2) When monsaccharides are in cyclic form, they do not contain aldehyde or ketone groups, but rather hemiacetal or hemiketal groups.

(3) The β-anomers have an –OH group on the anomeric carbon that points up. Birds fly up in the sky like the –OH group on the anomeric carbon of a β-anomer. The α-anomers have an –OH group on the anomeric carbon that points down. The α symbol looks like a fish. Fish swim down in the water like the –OH group on the anomeric carbon of an α-anomer.

(4) A Fischer projection is used to draw open chain structure of molecules, including carbohydrates. A Haworth structure is used to draw a closed ring structure of carbohydrates.

(5) The rules for drawing Haworth structures are:
   a. the ring is drawn with the oxygen in the back, in the upper right;
   b. the anomeric carbon is on the right side of the ring;
   c. the terminal $-CH_2OH$ group is always shown above the ring for D-monosaccharides.

(6) No, not all reducing sugars are monosaccharides. Reducing sugars must contain an aldehyde with an adjacent hydroxy group, ketone with an adjacent hydroxy group, cyclic hemiacetal with additional hydroxy groups, or cyclic hemiketal group with additional hydroxy groups. Monosaccharides fit these criteria, but so do other molecules including some disaccharides.

(7) A glycoside is a molecule that contains a cyclic monosaccharide bonded at the anomeric carbon atom through an oxygen atom to another carbon containing group.

Section 17.6 Review:

(1) Ribose and deoxyribose contain hemiacetal key functional groups.

(2) The liver converts other sugars into glucose.

(3) Lactose is synthesized in the mammary glands.

(4) Fructose is the only ketose in this section. All of the other important monosaccharides are aldoses.

Section 17.7 Review:

(1) In glycosidic linkages of disaccharides, the ring containing an anomeric carbon in the glycosidic linkage determines whether the linkage is classified as α or β.

(2) In glycosidic linkages of disaccharides, the anomeric carbon atom involved in the glycosidic linkage is identified first. If two anomeric carbon atoms are involved in the glycosidic linkage, then the anomeric carbon atom with the smaller number is identified first.

(3) In glycosidic linkages of disaccharides, the carbon atom that is not anomeric is identified second. If two anomeric carbon atoms are involved in the glycosidic linkage, then the anomeric carbon atom with the larger number is identified second.

(4) No, not all disaccharides reducing sugars because a reducing sugar must be able to form an open-chain. Disaccharides that have two anomeric carbon atoms in the glycosidic linkage cannot form an open-chain.

Section 17.8 Review:

(1) Water is the small molecule formed during the condensation reaction to form a polysaccharide.

(2) When polysaccharides are mixed with water and a colloidal dispersion forms, water is the dispersing medium.

(3) When polysaccharides are mixed with water and a colloidal dispersion forms, the polysaccharide is the dispersed phase.

(4) Polysaccharide-water colloidal dispersions are not transparent because they are colloids. Colloids appear cloudy because the dispersed phase scatters light.

(5) Maltose has the same type of glycosidic linkage as amylose.

(6) Amylose contains 1000-2000 glucose units. Maltose contains 2 glucose units.

(7) Add iodine to a sample of the "potato water." If the brown iodine turns dark bluish-purple, amylose is present.

(8) No, "potato water" will not contain glycogen because glycogen is the storage carbohydrate for animals, not plants.

(9) The molecular formula of an amylose molecule with 1500 monosaccharide units is the same as the molecular formula of a cellulose molecule with 1500 monosaccharide units.

(10) An amylose molecule with 1500 monosaccharide units is joined by $\alpha(1\rightarrow4)$ linkages that create a helical structure. A cellulose molecule with 1500 monosaccharide units is joined by $\beta(1\rightarrow4)$ linkages that create an extended straight chain.

Tying It All Together with a Laboratory Application:

(1) carbohydrates
(2) carboxylic acids
(3) increases
(4) rate
(5) acidic
(6) hydrolysis
(7) D-glucose, D-fructose
(8) sweeter
(9) amylose, amylopectin
(10) polymers
(11) colloidal dispersion
(12) dispersed phase
(13) dispersing medium
(14) thicken it
(15) Benedict's
(16) blue
(17) blue
(18) a brick red solid
(19) fructose
(20) negative
(21) disaccharide
(22) anomeric
(23) positive
(24) hydrolysis
(25) negative
(26) polysaccharides
(27) glycosidic
(28) negative
(29) oxidized
(30) positive
(31) positive
(32) deep purple
(33) amylose
(34) positive
(35) hydrolyze
(36) dextrins, maltose, glucose

SELF-TEST QUESTIONS

Multiple Choice

1. Which of the following is a monosaccharide?
   a. amylose b. ribose c. cellulose d. lactose

2. Which of the following is a polysaccharide?
   a. amylose b. lactose c. maltose d. galactose

3. How many chiral carbon atoms are in 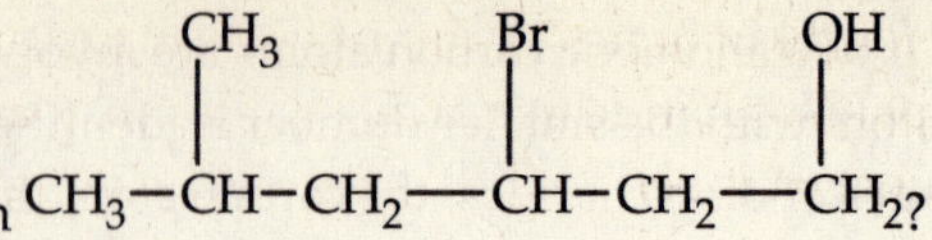

a. 0 b. 1 c. 2 d. 3

4. How many chiral carbon atoms are in 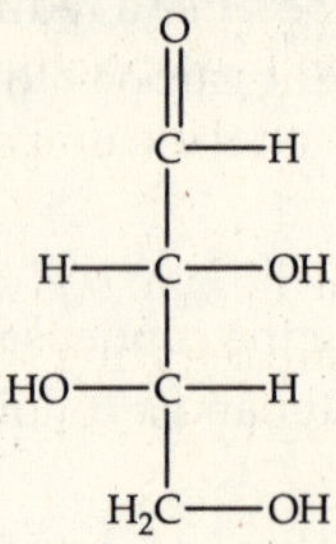 ?

a. 0 b. 1 c. 2 d. 3

5. How many stereoisomers are possible for 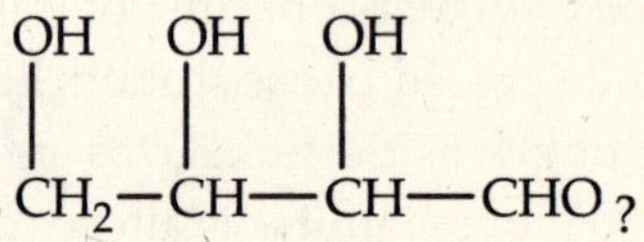

a. 0 b. 2 c. 4 d. 8

6. Glucose is a(n):

a. ketopentose. b. ketohexose. c. aldopentose. d. aldohexose.

7. Fructose is a(n):

a. ketopentose. b. ketohexose. c. aldopentose. d. aldohexose.

8. A positive Benedict's test is indicated by the formation of:

a. Ag. b. CuOH. c. Cu_2O. d. Cu^{2+}.

Matching

For each monosaccharide described on the left, select the best response from the right.

9. given intravenously	a. fructose
10. present with glucose in invert sugar	b. galactose
11. combines with glucose to form lactose	c. glucose
12. found in genetic material	d. ribose

For each disaccharide described on the left, select the best match from the responses on the right.

13. used as household sugar	a. glycogen
14. found in milk	b. sucrose
15. formed in germinating grain	c. maltose
	d. lactose

For each disaccharide on the left, select the correct hydrolysis products from the right.

16. sucrose	a. glucose and galactose
17. maltose	b. glucose and fructose
18. lactose	c. only glucose
	d. hydrolysis does not occur

Select the correct polysaccharide for each description on the left.

19. a storage form of carbohydrates in animals
20. most abundant polysaccharide in starch
21. primary constituent of paper

a. amylopectin
b. amylose
c. glycogen
d. cellulose

True-False

22. A D-enantiomer is the mirror image of an L-enantiomer.
23. In a D-carbohydrate, the hydroxy group on the chiral carbon farthest from the carbonyl group points to the left.
24. The L-carbohydrates are preferred by the human body.
25. Sugars that contain a hemiacetal group are reducing sugars.
26. In β-galactose, the hydroxy group at carbon 1 points up.
27. Maltose contains a glycosidic linkage.
28. The glucose ring of lactose can exist in an open-chain form.
29. Sucrose contains a hemiacetal group.
30. Linen is prepared from cellulose.

Chapter 18: Lipids

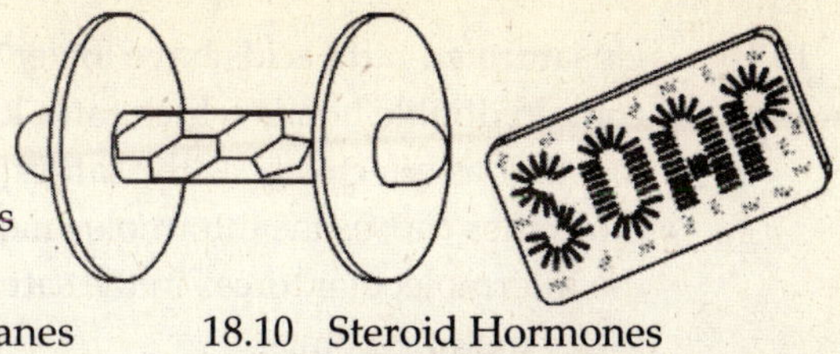

CHAPTER OUTLINE

18.1 Classification of Lipids
18.2 Fatty Acids
18.3 The Structure of Fats and Oils
18.4 Chemical Properties of Fats and Oils
18.5 Waxes
18.6 Phosphoglycerides
18.7 Sphingolipids
18.8 Biological Membranes
18.9 Steroids
18.10 Steroid Hormones
18.11 Prostaglandins

LEARNING OBJECTIVES/ASSESSMENT

When you have completed your study of this chapter, you should be able to:

1. Classify lipids as saponifiable or nonsaponifiable and list five major functions of lipids. (Section 18.1; Exercises 18.2 and 18.4)
2. Describe four general characteristics of fatty acids. (Section 18.2; Exercise 18.6)
3. Draw structural formulas of triglycerides given the formulas of the component parts. (Section 18.3; Exercise 18.14)
4. Describe the structural similarities and differences of fats and oils. (Section 18.3; Exercise 18.12)
5. Write key reactions for fats and oils. (Section 18.4; Exercise 18.18)
6. Compare the structures of fats and waxes. (Section 18.5; Exercise 18.24)
7. Draw structural formulas and describe uses for phosphoglycerides. (Section 18.6; Exercises 18.28 and 18.30)
8. Draw structural formulas and describe uses for sphingolipids. (Section 18.7; Exercise 18.34)
9. Describe the major features of cell membrane structure. (Section 18.8; Exercise 18.42)
10. Identify the structural characteristic typical of steroids and list important groups of steroids in the body. (Section 18.9; Exercises 18.44 and 18.46)
11. Name the major categories of steroid hormones. (Section 18.10; Exercise 18.50)
12. Describe the biological importance and therapeutic uses of the prostaglandins. (Section 18.11; Exercise 18.58)

SOLUTIONS FOR THE END OF CHAPTER EXERCISES

INTRODUCTION AND CLASSIFICATION OF LIPIDS (SECTION 18.1)

☑18.2 Lipids are a form of stored energy and a structural component of the human body. Some hormones in the human body are lipids.

☑18.4

a.	A steroid	nonsaponifiable	d.	A phosphoglyceride	saponifiable
b.	A wax	saponifiable	e.	A glycolipid	saponifiable
c.	A triglyceride	saponifiable	f.	A prostaglandin	nonsaponifiable

FATTY ACIDS (SECTION 18.2)

☑18.6 Four structural characteristics exhibited by most fatty acids are:

(1) they are usually straight-chain carboxylic acids;
(2) the sizes of most common fatty acids range from 10 to 20 carbon atoms;
(3) fatty acids usually have an even number of carbon atoms;
(4) fatty acids can be saturated or unsaturated.

18.8 Two essential fatty acids are linolenic acid and linoleic acid. The human body cannot synthesize these acids; therefore, they must be obtained through diet.

18.10 Unsaturated fatty acids have lower melting points than saturated fatty acids because they contain double bonds which form kinks and prevent the unsaturated fatty acids from packing together as effectively as the saturated fatty acids. The greater separation between the molecules causes the intermolecular forces in the unsaturated fatty acids to be weaker than the intermolecular forces in the saturated fatty acids. Weaker intermolecular forces result in lower melting points.

THE STRUCTURE OF FATS AND OILS (SECTION 18.3)

☑18.12 Fats and oils are both triglycerides that contain a glycerol backbone and three fatty acid chains. Fats contain more saturated than unsaturated fatty acids and oils contain more unsaturated than saturated fatty acids. This causes fats to be solid at room temperature, while oils are liquid under the same conditions.

18.14

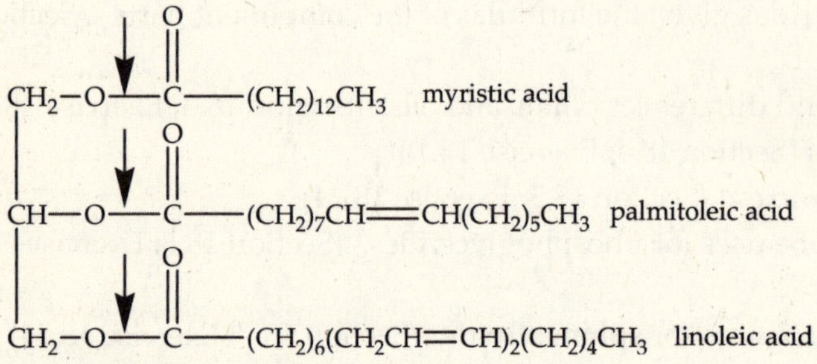

18.16 Triglyceride B should have the lower melting point because it contains a higher percentage of unsaturated fatty acids than Triglyceride A.

CHEMICAL PROPERTIES OF FATS AND OILS (SECTION 18.4)

☑18.18 The process used to prepare a number of useful products such as margarines and cooking shortenings from vegetable oils is hydrogenation. In this process some of the double bonds in the unsaturated fatty acids are hydrogenated by reaction with hydrogen gas in the presence of a catalyst. Since fewer double bonds are present, the melting point of the mixture increases and the vegetable oils become more solid in consistency.

18.20 Hydrogenation of vegetable oils is of great commercial importance because it increases the melting point of the fat or oil and allows the product to be used as margarine and cooking shortenings.

18.22 a. Glycerol from beef fat:
Acid-catalyzed hydrolysis

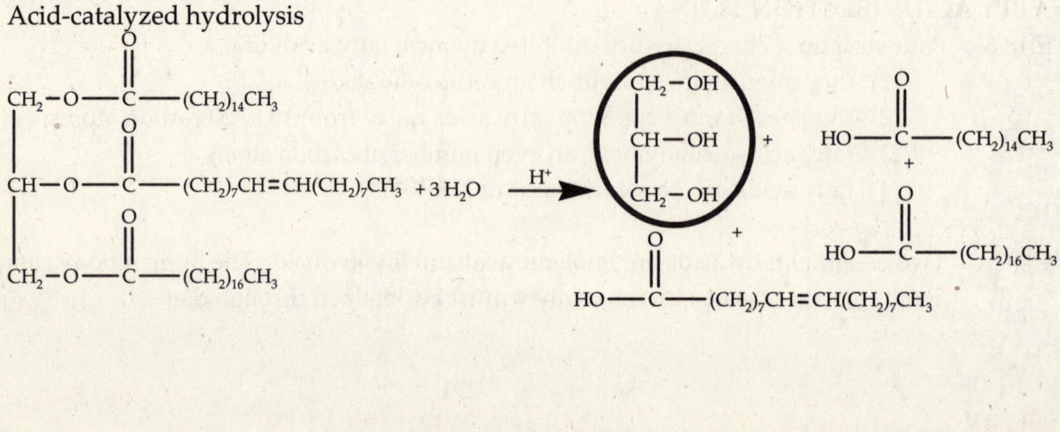

Saponification

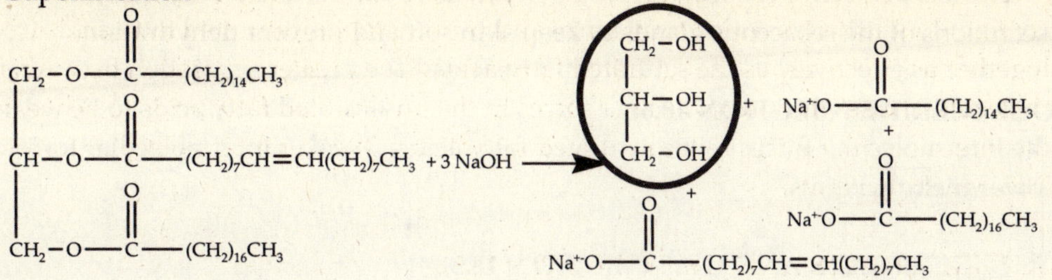

b. Stearic acid from beef fat:
Acid-catalyzed hydrolysis

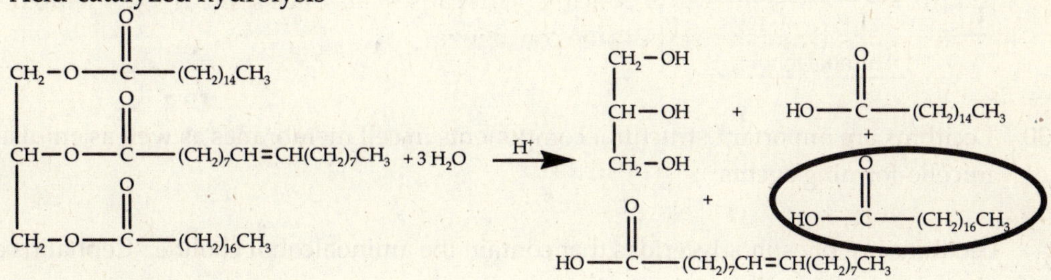

c. A margarine from corn oil:
Hydrogenation

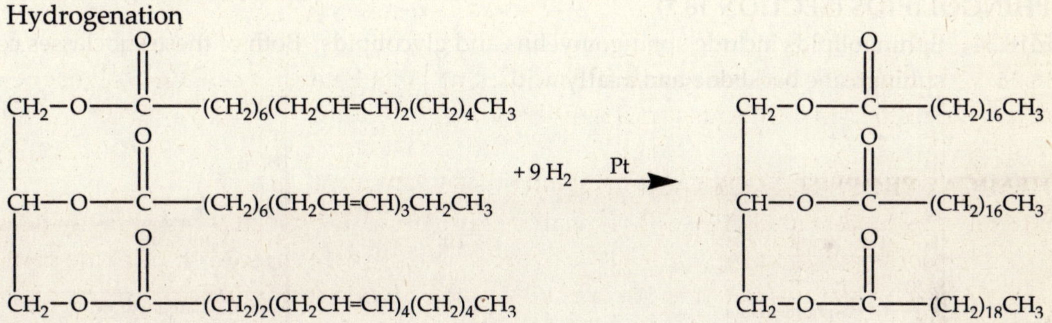

Note: To keep a margarine soft and spreadable, only a certain percentage of the double bonds should be hydrogenated. When all of the double bonds are hydrogenated, the resulting margarine will be hard, like a fat.

d. Soaps from lard:
Saponification

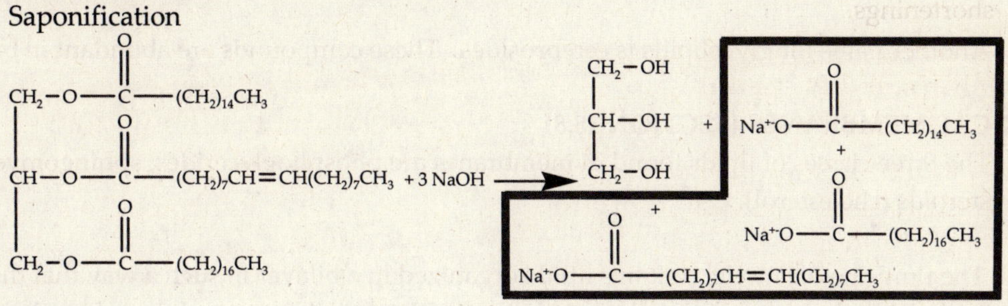

WAXES (SECTION 18.5)

☑18.24 Waxes and fats are esters of long-chain fatty acids; however, waxes do not contain glycerol backbones and fats do. Waxes contain one fatty acid chain, while fats contain three fatty acid chains.

18.26 Waxes are protective coatings on feathers, fur, skin, leaves, and fruits. They also occur in secretions of the sebaceous glands to keep skin soft and prevent dehydration.

PHOSPHOGLYCERIDES (SECTION 18.6)

☑18.28

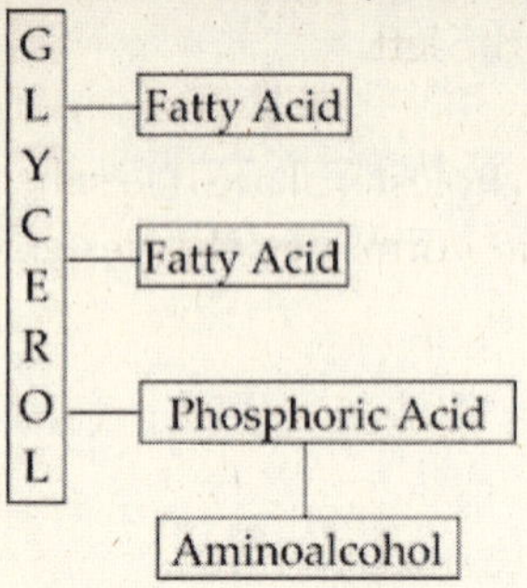

☑18.30 Lecithins are important structural components in cell membranes as well as emulsifying and micelle-forming agents.

18.32 Lecithins are phosphoglycerides that contain the aminoalcohol choline. Cephalins are phosphoglycerides that contain ethanolamine or serine.

SPHINGOLIPIDS (SECTION 18.7)

☑18.34 Sphingolipids include sphingomyelins and glycolipids. Both of these subclasses contain the sphingosine backbone and a fatty acid.

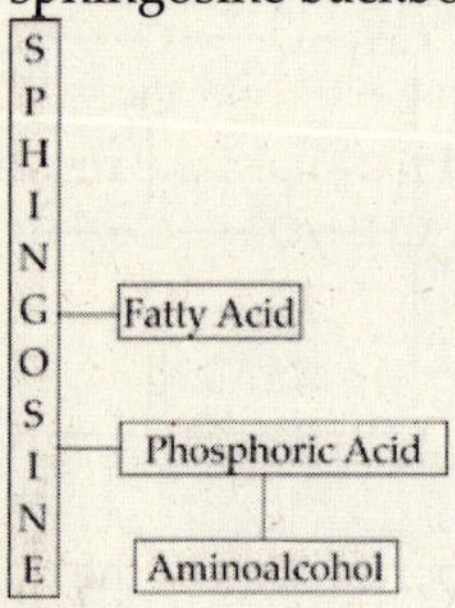

or

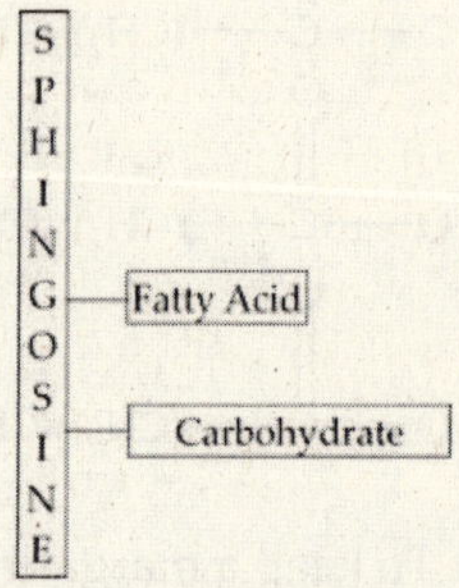

18.36 Three diseases caused by abnormal metabolism and accumulation of sphingolipids are Tay-Sachs, Gaucher's, and Niemann-Pick diseases.

18.38 Another name for glycolipids is cerebrosides. These compounds are abundant in brain tissue.

BIOLOGICAL MEMBRANES (SECTION 18.8)

18.40 The three classes of lipids found in membranes are phosphoglycerides, sphingomyelins, and steroids (cholesterol).

☑18.42 The fluid-mosaic model contains lipids organized in a bilayer in such a way that the hydrophilic heads are pointed toward the outside of the bilayer and the hydrophobic tails are pointed toward the inside of the bilayer. Some proteins float in the lipid bilayer and other proteins extend completely through the bilayer. The lipid molecules move freely laterally within the bilayer.

STEROIDS (SECTION 18.9)

☑18.44 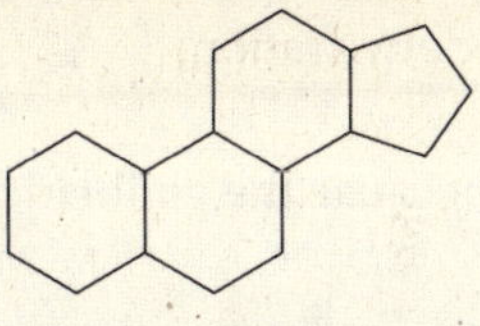 All steroids contain a 4 fused ring system composed of 3 six membered rings and 1 five membered ring arranged as shown in the structure to the left.

☑18.46 Important groups of compounds the body synthesizes from cholesterol are bile salts, male and female sex hormones, vitamin D, and the adrenocorticoid hormones. Cholesterol is an essential component of cell membranes.

18.48 The major component in most gallstones is cholesterol.

STEROID HORMONES (SECTION 18.10)

☑18.50 The two groups of adrenocorticoid hormones are mineralcorticoids and glucocorticoids. Mineralcorticoids regulate the concentration of ions in body fluids. An example of a mineralcorticoid is aldosterone which regulates the retention of sodium and chloride ions in urine formation. Glucocorticoids regulate carbohydrate metabolism. An example of a glucocorticoid is cortisol which helps to increase glucose and glycogen concentrations in the body.

18.52 The primary male sex hormone is testosterone. The three principal female sex hormones are estradiol, estrone, and progesterone.

18.54 The estrogens are involved in egg development in the ovaries. Progesterone causes changes in the wall of the uterus to prepare it to accept a fertilized egg and maintain the resulting pregnancy.

PROSTAGLANDINS (SECTION 18.11)

18.56 The starting material for prostaglandin synthesis is arachidonic acid.

☑18.58 Therapeutic uses of prostaglandins include inducing labor, therapeutic abortion, treating asthma, inhibiting gastric secretions, and treating peptic ulcers.

ADDITIONAL EXERCISES

18.60 If a red-brown solution of bromine is added to a lipid, and the characteristic red-brown bromine color disappears, then the lipid contains at least one carbon-carbon double bond because the disappearance of the characteristic red-brown bromine color is a positive test for carbon-carbon double bonds.

18.62 Complex lipids are more predominant in cell membranes than simple lipids because the cell membranes have a lipid bilayer which is composed of lipids with polar head groups and nonpolar tails. Phospholipid molecules have a structure that is more conducive to forming a lipid bilayer than a fat, oil, or wax. Simple lipids do not form as effective a barrier as the complex lipids.

ALLIED HEALTH EXAM CONNECTION

18.64 The functional group present in fats is an ester (RCOOR'). The ester functional group is common to all saponifiable lipids.

18.66 a. Estrogens and progesterone are produced in the ovaries.
b. Testosterone is produced in the testes.

18.68 (c) Fats are limited in digestibility in diseases of the gallbladder because the gallbladder stores the bile that helps digest fat.

18.70 Accumulation of cholesterol leads to the hardening of the arteries. This is called (c) atherosclerosis.

18.72 Bile is manufactured in the (b) liver.

18.74 The mineralocorticoid (d) aldosterone is a product of the adrenal cortex.

18.76 The cells of a human body are (d) eukaryotic cells.

CHEMISTRY FOR THOUGHT

18.78 The structure of cellular membranes is such that ruptures are closed naturally. The molecular forces that cause the closing to occur are dispersion forces between the nonpolar "tails" in the lipid bilayer as well as dipolar forces between the polar "heads" in the lipid bilayer and hydrogen bonding between the polar "heads" in the lipid bilayer and the water in the surrounding fluids.

18.80 Oils used in automobiles are typically petroleum-based products, while vegetable oil is derived from plants. The vegetable oil does not contain sulfur atoms, but petroleum based oil does contain sulfur atoms.

18.82 The micelle structure with the nonpolar tails in the center of the structure is more stable than a structure in which the polar heads are together because the dispersion forces between the nonpolar tails is stronger when they are facing into the center of the spherical micelle. The nonpolar tails are much closer together and many more fatty acid molecules can be part of the micelle structure shown in Figure 18.5 than in an "inverse" micelle that has the nonpolar tails facing the outside of the micelle.

18.84 A soap has a fatty acid tail and a polar head group that has a negative charge, which has a positive counterion. A lecithin has two fatty acid tails and a polar head group that contains both a negative charge (phosphate group) and a positive charge (choline). These compounds are similar in their structure because they both contain fatty acids as well as a highly charged polar head group.

ADDITIONAL ACTIVITIES

Section 18.1 Review:

Complete the following diagram by adding the subclasses of lipids into the appropriate locations.

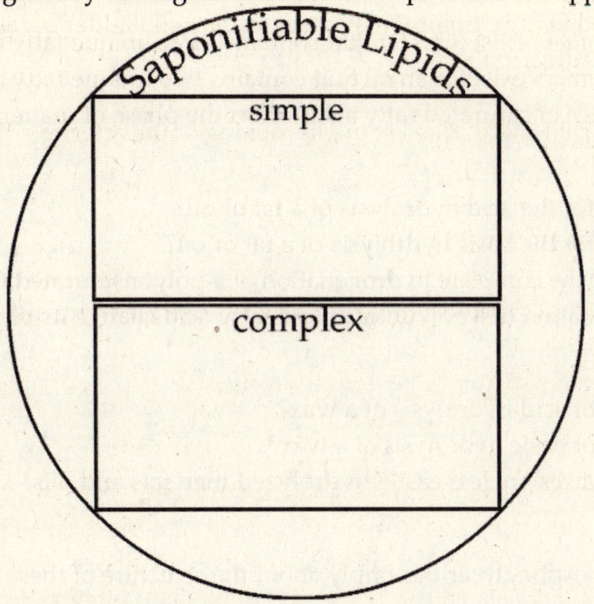

Section 18.2 Review:

(1) Draw a micelle of fatty acid anions in an oil based environment with small amounts of aqueous materials.

(2) What is the angle between the *cis*-carbon groups attached to the carbon atoms in a double bond? The following drawing may help along with the reminder that a straight line has an angle of 180° and the sum of the angles of a triangle is 180°.

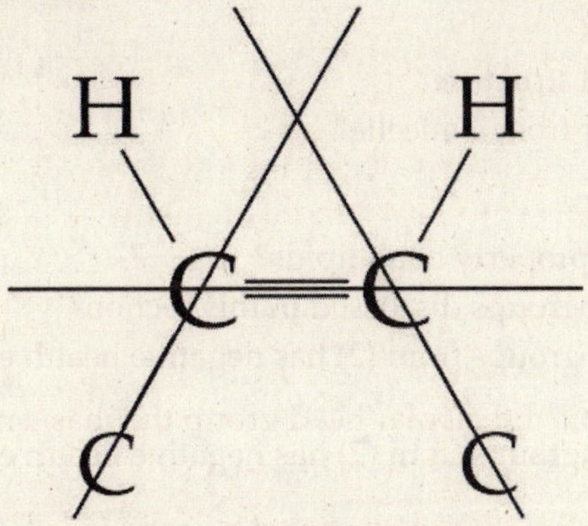

(3) What is the angle of the kink in a *cis*-unsaturated fatty acid?

(4) What is the angle between *trans*-carbon groups attached to the carbon atoms in a double bond?

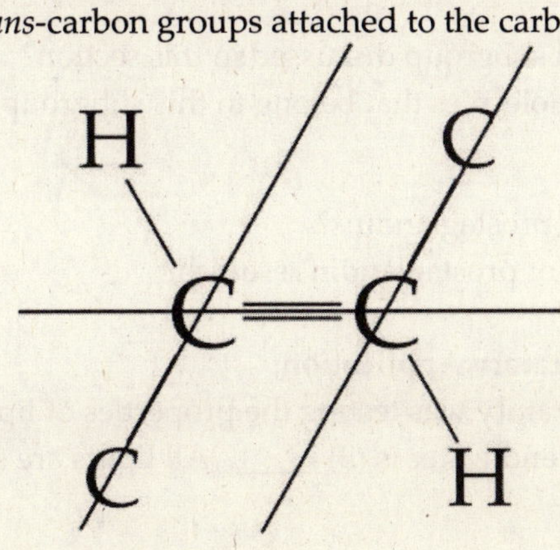

(5) Would *trans*-unsaturated fatty acids have a kink? Explain.
(6) Are the two essential fatty acids saturated, monounsaturated, or polyunsaturated? Explain.

Section 18.3 Review:
(1) How many possible isomers exist for a fat that contains three unique fatty acids?
(2) How many possible isomers exist for an oil that contains two unique fatty acids?
(3) How does the percentage of saturated fatty acids affect the phase of matter of the compound?

Section 18.4 Review:
(1) What are the products for the acid hydrolysis of a fat or oil?
(2) What are the products for the basic hydrolysis of a fat or oil?
(3) What is the product for the complete hydrogenation of a polyunsaturated fatty acid?
(4) How does the hydrogenation of a polyunsaturated fatty acid change its physical properties?

Section 18.5 Review:
(1) What are the products of acid hydrolysis of a wax?
(2) What are the products of basic hydrolysis of a wax?
(3) Suggest a reason that waxes are less easily hydrolyzed than fats and oils.

Section 18.6 Review:
(1) What does the name phosphoglycerides imply about the structure of these compounds?
(2) What parts of the phosphoglyceride structure is not implied by the name of this group?
(3) What are the two categories of phosphoglycerides?

Section 18.7 Review:
(1) What does the name sphingolipid imply about the structures of these compounds?
(2) What are the structural differences between a sphingomyelin and a glycolipid?

Section 18.8 Review:
(1) What role do lipids play in cell structure?
(2) How does a lipid bilayer differ from a micelle?

Section 18.9 Review:
(1) What is the common physical property of all lipids?
(2) What are the major steroid subgroups discussed in this section?
(3) Which of the major steroid subgroups from (2) has negative health effects when the levels in the body are high?
(4) Which of the major steroid subgroups from (2) has negative health effects when the levels in the body are low?

Section 18.10 Review:
(1) What is the name of the steroid subgroup discussed in this section?
(2) What are the major classes of molecules that belong to this subgroup?

Section 18.11 Review:
(1) What is the common feature of prostaglandins?
(2) What is a potential application of prostaglandin research?

Tying It All Together with a Laboratory Application:
A student in the biochemistry laboratory was testing the properties of lipids. All lipids are (1) _____ in water because lipids are (2) ______ and water is (3) _____. All lipids are (4) _____ in cyclohexane because

cyclohexane is (5) _____, (6) _____ lipids. The solubility concept of "(7) _____ dissolves like" applies even in biochemistry.

The student determined that a hydrolysis reaction under (8) _____ conditions would separate the class of lipids into two subclasses, (9) _____ and _____. The (10) _____ lipids contain two groups: (11) _____ and prostaglandins. Of these two groups only prostaglandins contain a(n) (12) _____ (carboxylic acid or ketone) key functional group. This might cause a difference in their (13) _____ (chemical or physical) properties because of the reactivity of this key functional group.

The other major lipid subclass is divided into two smaller groups, (14) _____ and _____. One similarity between all of these compounds is they contain a (15) _____ chain, which can be saturated or (16) _____. A bromine test would be (17) _____ for a saturated (18) _____ chain because it (19) _____ carbon-carbon double bonds.

When these compounds undergo saponification, (20) _____, _____, and _____ will produce glycerol. These compounds are classified in (21) _____ subclass(es) because an alcohol and (22) _____ are the only components of (23) _____ (simple or complex) lipids. (24) _____ contain a phosphate group in addition to the other components. These compounds contain an (25) _____ (amide or ester) linkage as do (26) _____, which are (27) _____ (simple or complex) lipids that do not contain glycerol. The saponification of (28) _____ will produce sphingosine as one of the products. These compounds are joined by an (29) _____ linkage to the part of the molecule that can become a soap. These compounds are (30) _____ (simple or complex) lipids.

SOLUTIONS FOR THE ADDITIONAL ACTIVITIES

Section 18.1 Review:

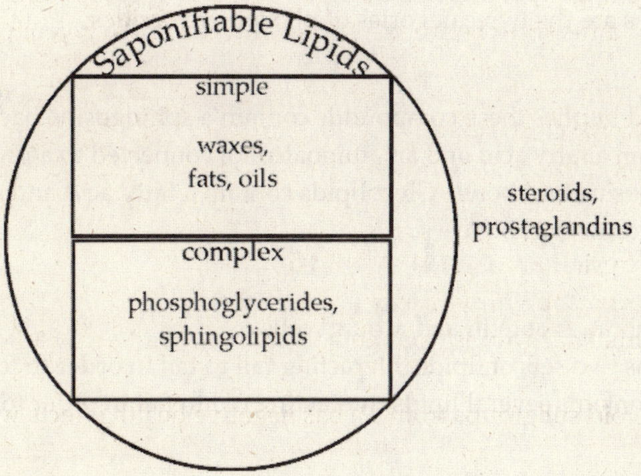

Section 18.2 Review:

(1)

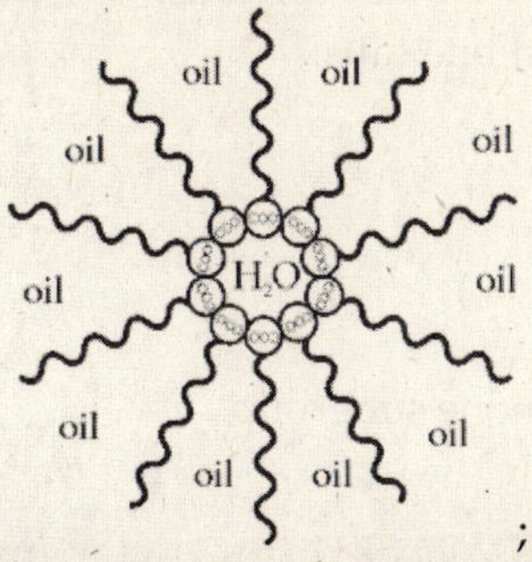

; (2) 60°;

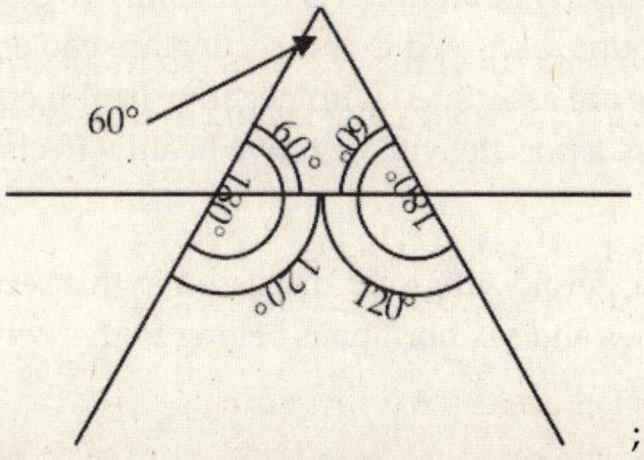

; (3) 60°; (4) 180°;

(5) The *trans*-unsaturated fatty acids do not have a kink because the angle between the carbon groups attached to the double bonded carbon atoms is 180°, which is a straight line.

(6) The two essential fatty acids are polyunsaturated because they contain more than one double bond.

Section 18.3 Review:

(1) The glycerol backbone of fats is symmetrical; therefore, only three isomers exist.

(2) The glycerol backbone of oils is symmetrical; therefore, only two isomers exist.

(3) The higher the percentage of saturated fatty acids in a compound, the more likely the compound is to be a solid. The lower the percentage of saturated fatty acids in a compound, the more likely the compound is to be a liquid.

Section 18.4 Review:

(1) The products for the acid hydrolysis of a fat or oil are glycerol and three fatty acid molecules.

(2) The products for the basic hydrolysis of a fat or oil are glycerol and three salts of fatty acids.

(3) The product for the complete hydrogenation of a polyunsaturated fatty acid is a saturated fatty acid.

(4) The hydrogenation of a polyunsaturated fatty acid will increase the melting and boiling points.

Section 18.5 Review:

(1) The products of acid hydrolysis of a wax are an alcohol and a fatty acid.

(2) The products of basic hydrolysis of a wax are an alcohol and a salt of a fatty acid.

(3) Waxes are less easily hydrolyzed than fats and oils because their ester linkages are less accessible than the ester linkages in fats and oils. Waxes have hydrophobic groups on either side of the ester linkage, while fats and oils have hydrophobic groups on only one side of the ester linkage.

Section 18.6 Review:

(1) The name phosphoglycerides implies these compounds contain a glycerol backbone and a phosphate group.

(2) The name phosphoglyceride does not imply that these compounds contain two fatty acids or that an aminoalcohol is attached to the phosphate group.

(3) Lecithins and cephalins are the two categories of phosphoglycerides.

Section 18.7 Review:

(1) The name sphingolipid implies these compounds contain a sphingosine backbone.

(2) Sphingomyelins contain a fatty acid and an aminoalcohol connected to a phosphate group in addition to the sphingosine backbone. Glycolipids contain a fatty acid and a carbohydrate in addition to the sphingosine backbone.

Section 18.8 Review:

(1) Lipids help form membranes both in and around cells.

(2) A lipid bilayer contains two sets of lipids interacting tail to tail in order to form an extended structure. A micelle contains several lipids interacting tail to tail in order to form a sphere.

Section 18.9 Review:

(1) The common physical property of all lipids is their ability to dissolve in nonpolar solvents.

(2) The major steroid subgroups discussed in this section are cholesterol and bile salts.

(3) High levels of cholesterol are associated with negative health effects.

(4) Low levels of bile salts are associate with negative health effects.

Section 18.10 Review:

(1) Steroid hormones are the steroid subgroup discussed in this section.

(2) Adrenocorticoid hormones and sex hormones belong to the steroid hormone subgroup.

Section 18.11 Review:

(1) Prostaglandins are synthesized from arachidonic acid, a 20-carbon unsaturated fatty acid.

(2) A potential application of prostaglandin research is new treatments for a variety of ailments.

Tying It All Together with a Laboratory Application:

(1) insoluble
(2) nonpolar
(3) polar
(4) soluble
(5) nonpolar
(6) like
(7) like
(8) basic
(9) saponifiable, nonsaponifiable
(10) nonsaponifiable
(11) steroids
(12) carboxylic acid
(13) chemical
(14) simple, complex
(15) fatty acid
(16) unsaturated
(17) negative
(18) fatty acid
(19) lacks (does not contain)
(20) fats, oils, phosphoglycerides
(21) different
(22) fatty acid chain
(23) simple
(24) Phosphoglycerides
(25) ester
(26) waxes
(27) simple
(28) sphingolipids
(29) amide
(30) complex

SELF-TEST QUESTIONS

Multiple Choice

1. All simple lipids are:
   a. salts of fatty acids.
   b. esters of fatty acids with various alcohols.
   c. esters of fatty acids with glycerol.
   d. esters of fatty acids with alcohol and other additional components.

2. The esters of fatty acids and long chain alcohols are known as:
   a. waxes. b. phospholipids. c. compound lipids. d. fats.

3. Which of the following is a glycerol-containing lipid?
   a. sphingolipid b. glycolipid c. phosphoglyceride d. prostaglandin

4. Which of the following is a complex lipid?
   a. steroid b. sphingomyelin c. prostaglandin d. triacylglycerol

5. Which fatty acid is most likely to be found in an oil?
   a. $CH_3(CH_2)_7CH{=}CH(CH_2)_7COOH$
   b. $CH_3(CH_2)_{16}COOH$
   c. $CH_3(CH_2)_{14}COOH$
   d. $CH_3(CH_2)_{18}COOH$

6. Which of the following food sources would most likely be highest in saturated fatty acids?
   a. cottonseed b. corn c. beef d. sunflower seeds

7. Generally, the structural difference between a fat and an oil is the:
   a. alcohol.
   b. chain length of fatty acid.
   c. degree of fatty-acid unsaturation.
   d. degree of fatty-acid chain-branching.

8. In triglycerides, fatty acids are joined to glycerol by:
   a. ester linkages. b. ether linkages. c. amide linkages. d. hydrogen bonds.

9. Which of the following is an essential fatty acid?
   a. stearic acid b. myristic acid c. linoleic acid d. palmitic acid

Matching

Match the following formulas to the correct lipid classification given as a response

a. steroid
b. phosphoglyceride
c. fat or oil
d. wax

10. 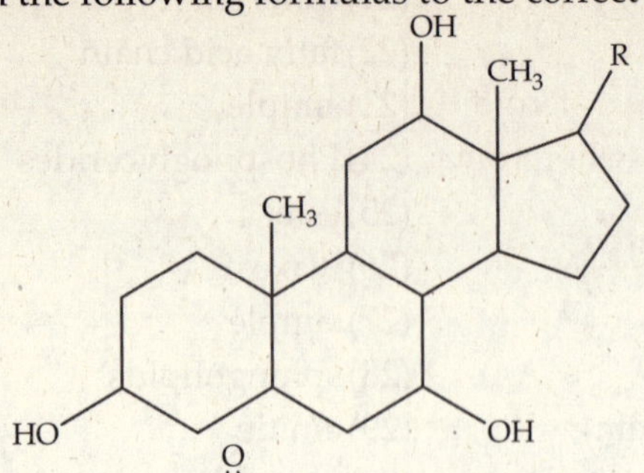

11. 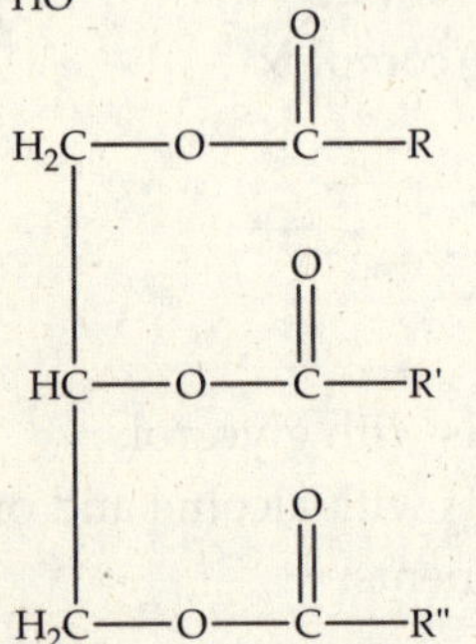

12. 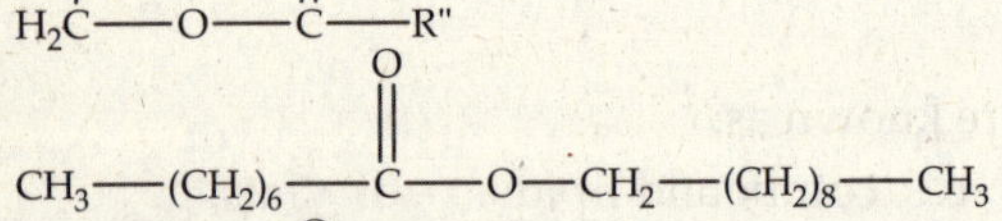

13. 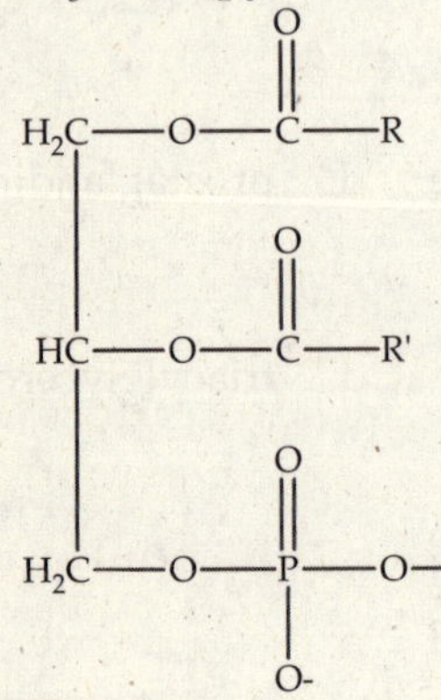

Materials can be obtained from lipids when the lipids changed by chemical processes. Choose the correct process to accomplish each change described below.

14. obtain a high molecular weight alcohol from a plant wax
15. obtain glycerol from an oil
16. obtain soaps from an oil
17. raise the melting point of an oil

a. hydrogenation
b. acid-catalyzed hydrolysis
c. saponification
d. more than one listed process would work

Match the lipids on the right with their possible hydrolysis products, W, X, Y, and Z.

18. W = fatty acids, X = glycerol, Y = phosphoric acid, Z = choline
19. W = fatty acids, X= glycerol, no other products
20. W = fatty acids, X = sphingosine, Y = phosphoric acid, Z = choline
21. W = fatty acids, X = sphingosine, Y = carbohydrate, no other product forms

a. simple lipid
b. glycolipid
c. phospholipid
d. sphingomyelin

True-False

22. Cell membranes contain about 60% lipid and 40% carbohydrate.
23. Cell membranes are thought to be relatively flexible.
24. A compound containing nine carbon atoms could not be a steroid.
25. In their physiological action, the prostaglandins resemble hormones.
26. All of the male and female sex hormones are steroids.
27. The hormone aldosterone exerts its influence at the pancreas.
28. Lipids are polar molecules.
29. Lipids are soaps.
30. Bile salts are steroids.

Chapter 19: Proteins

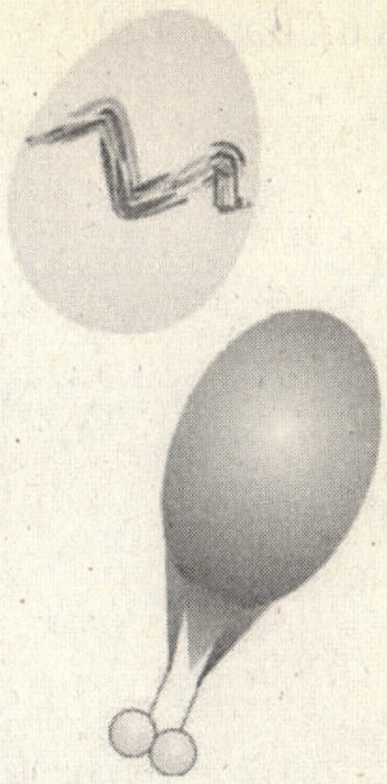

CHAPTER OUTLINE

19.1 The Amino Acids
19.2 Zwitterions
19.3 Reactions of Amino Acids
19.4 Important Peptides
19.5 Characteristics of Proteins
19.6 The Primary Structure of Proteins
19.7 The Secondary Structure of Proteins
19.8 The Tertiary Structure of Proteins
19.9 The Quaternary Structure of Proteins
19.10 Protein Hydrolysis and Denaturation

LEARNING OBJECTIVES/ASSESSMENT

When you have completed your study of this chapter, you should be able to:

1. Identify the characteristic parts of alpha-amino acids. (Section 19.1; Exercise 19.2)
2. Draw structural formulas to illustrate the various ionic forms assumed by amino acids. (Section 19.2; Exercise 19.12)
3. Write reactions to represent the formation of peptides and proteins. (Section 19.3; Exercise 19.16)
4. Describe uses for important peptides. (Section 19.4; Exercise 19.22)
5. Describe proteins in terms of the following characteristics: size, function, classification as fibrous or globular, and classification as simple or conjugated. (Section 19.5; Exercises 19.30 and 19.32)
6. Explain what is meant by the primary structure of proteins. (Section 19.6; Exercise 19.34)
7. Describe the role of hydrogen bonding in the secondary structure of proteins. (Section 19.7; Exercise 19.38)
8. Describe the role of side-chain interactions in the tertiary structure of proteins. (Section 19.8; Exercise 19.42)
9. Explain what is meant by the quaternary structure of proteins. (Section 19.9; Exercise 19.46)
10. Describe the conditions that can cause proteins to hydrolyze or become denatured. (Section 19.10; Exercise 19.50)

SOLUTIONS FOR THE END OF CHAPTER EXERCISES

THE AMINO ACIDS (SECTION 19.1)

☑19.2 All amino acids contain an amine and a carboxylic acid key functional group.

19.4 a. threonine

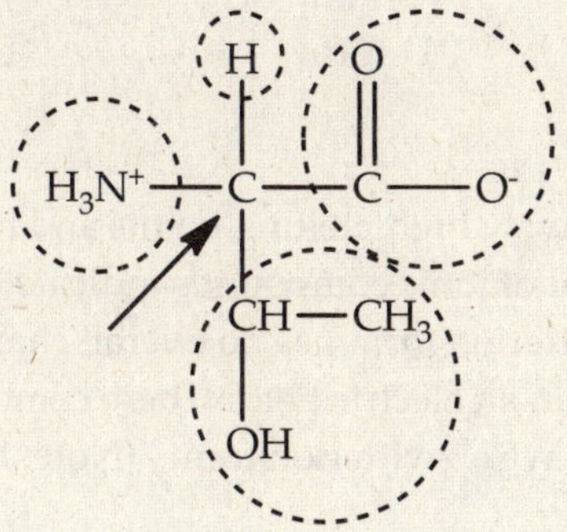

b. aspartate

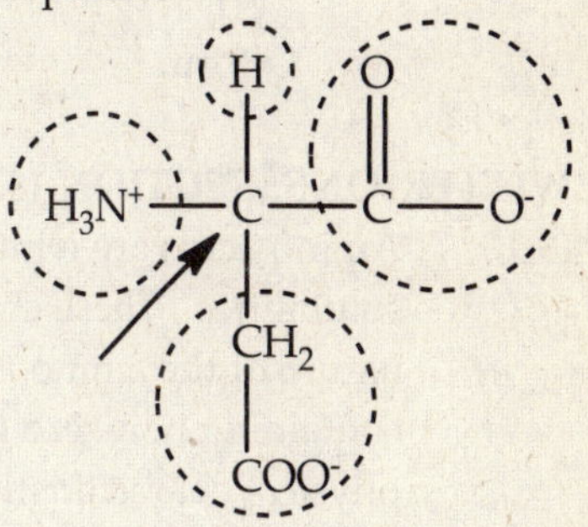

c. serine

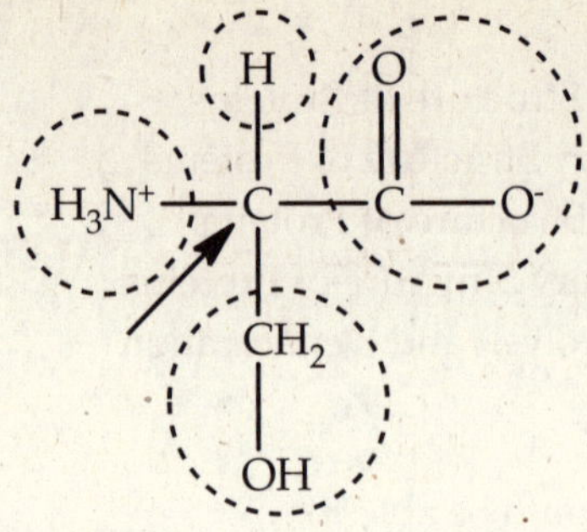

d. phenylalanine

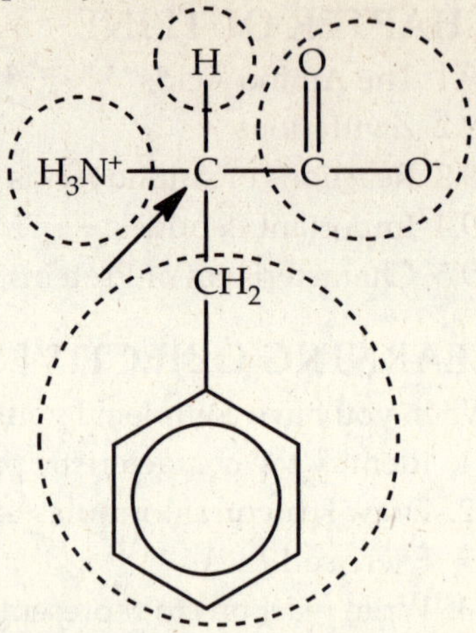

19.6

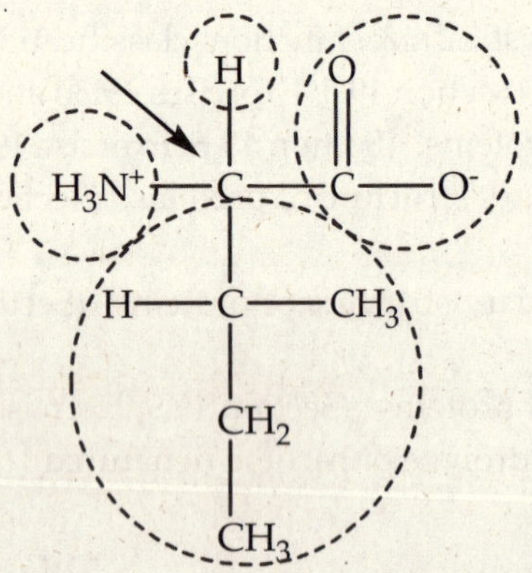

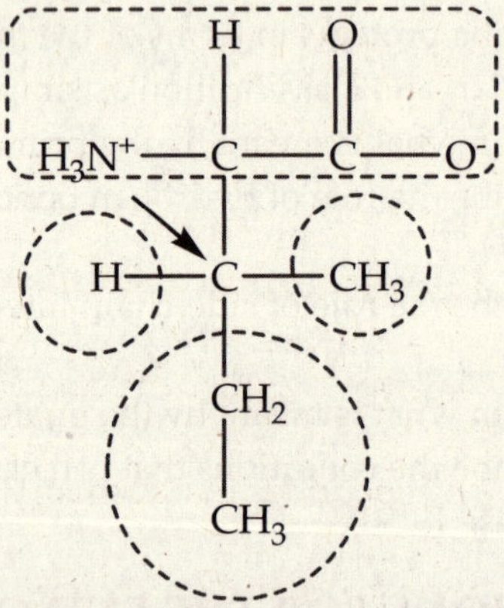

19.8 a.

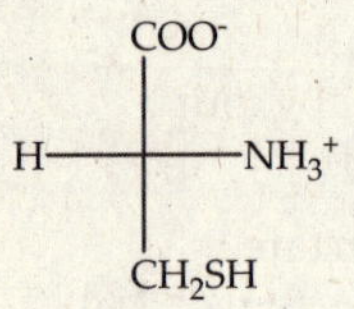

D-form

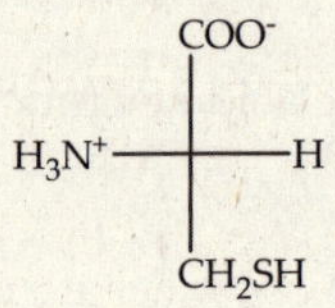

L-form

b.

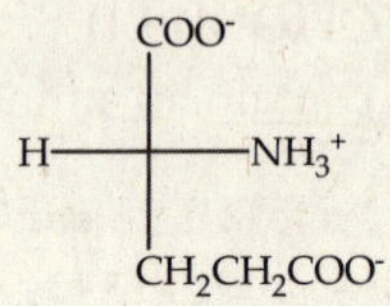

D-form

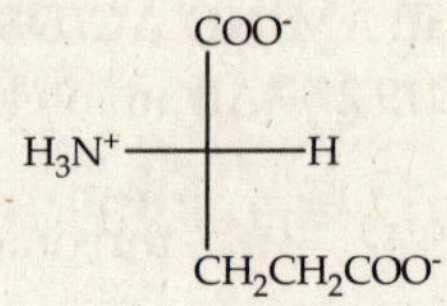

L-form

ZWITTERIONS (SECTION 19.2)

19.10 Amino acids are white crystalline solids with relatively high melting points and high water solubilities. These characteristics, which are typical of ionic compounds indicate the ionic nature of the amino acids. The molecule in its zwitterion form has no overall charge, despite containing non-zero charges, and will not migrate in an electric field. These combined physical characteristics indicate that amino acids have a zwitterion form. (Note: Ions with an overall charge do migrate in an electric field.)

☑19.12 a.

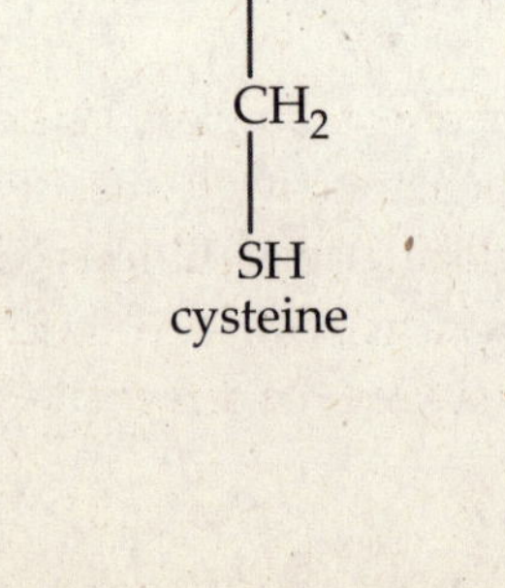

cysteine

b.

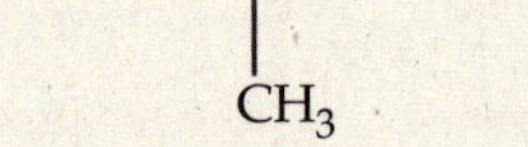

alanine

19.14 a., b.

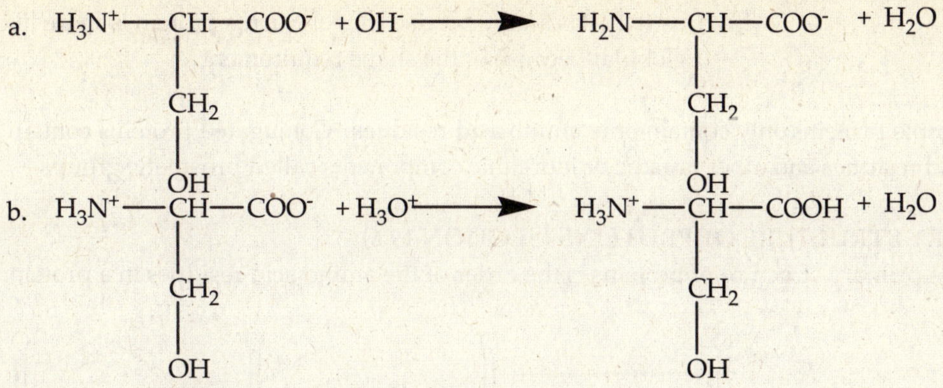

REACTIONS OF AMINO ACIDS (SECTION 19.3)

☑19.16

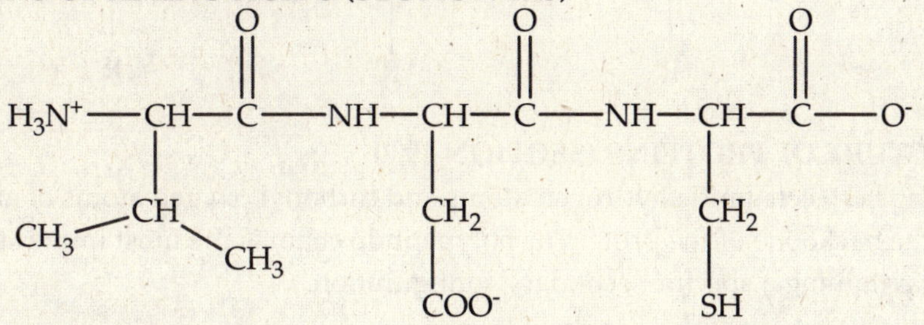

Val-Asp-Cys

19.18 Ala-Phe-Arg, Ala-Arg-Phe, Phe-Ala-Arg, Phe-Arg-Ala, Arg-Phe-Ala, Arg-Ala-Phe

19.20 All tripeptides that contain three unique amino acids can have 6 possible structural isomers.

IMPORTANT PEPTIDES (SECTION 19.4)

☑19.22 Cysteine forms disulfide bridges in the peptides vasopressin and oxytocin. This affects the overall structure of these peptides, altering their shape, and therefore, their reactivity.

CHARACTERISTICS OF PROTEINS (SECTION 19.5)

19.24 The presence of certain proteins in urine or blood can indicate that cellular damage has occurred in the body because most proteins are too large to leave the cell in which they were synthesized. The presence of these proteins outside of their original cells indicates that the cell walls have leaked the proteins into body fluids.

19.26 A protein is least soluble in an aqueous medium that has a pH equal to the isoelectric point of the protein because the protein does not have a net charge at its isoelectric point. The protein molecules will clump together at the isoelectric point because the individual molecules do not contain like charges that repel other molecules.

19.28 The eight principle functions of proteins are: (1) catalysis, (2) structure, (3) storage, (4) protection, (5) regulation, (6) nerve impulse transmission, (7) movement, and (8) transport.

☑19.30
a. collagen — This protein is more likely to be fibrous than globular because it is a structural protein found in fibrous connective tissue.
b. lactate dehydrogenase — This protein is more likely to be globular than fibrous because it is a catalyst that must work to oxidize lactic acid. A protein which acts as a catalyst has a specialized area called the active site, which

usually has a very precise shape. Such a protein is more likely to be globular because of the shape requirement.

☑19.32 Simple proteins only contain only amino acid residues. Conjugated proteins contain amino acid residues and other organic or inorganic components called prosthetic groups.

THE PRIMARY STRUCTURE OF PROTEINS (SECTION 19.6)

☑19.34 The primary structure of proteins is the order of the amino acid residues in a protein.

19.36

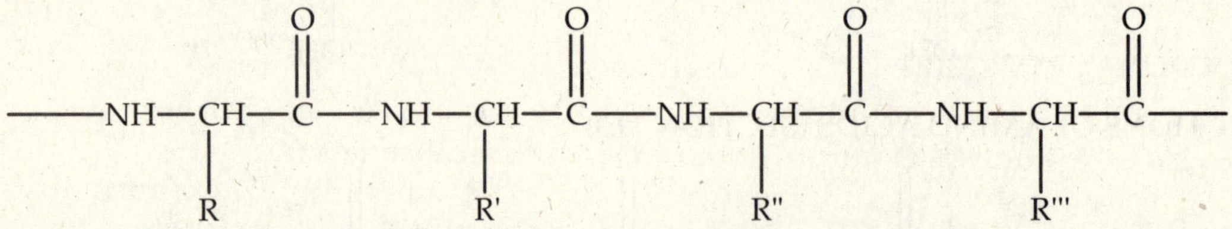

THE SECONDARY STRUCTURE OF PROTEINS (SECTION 19.7)

☑19.38 Hydrogen bonding between amide hydrogen atoms and carbonyl oxygen atoms of amino acid residues in the backbone of the protein or polypeptide chain is the most important type of bonding for maintaining a specific secondary configuration.

THE TERTIARY STRUCTURE OF PROTEINS (SECTION 19.8)

19.40 The hydrogen bonding in the tertiary structure of a protein occurs between the side chains on the amino acid residues. The hydrogen bonding in the secondary structure of a protein occurs between the atoms in the backbone of the protein.

☑19.42

a.	tyrosine and glutamine	hydrogen bonds
b.	aspartate and lysine	salt bridge
c.	leucine and isoleucine	hydrophobic interactions
d.	phenylalanine and valine	hydrophobic interactions

19.44 The alanine, phenylalanine, and methionine would be directed toward the inside of the protein structure. The lysine and glutamate would be directed toward the aqueous surroundings.

THE QUATERNARY STRUCTURE OF PROTEINS (SECTION 19.9)

☑19.46 Hydrophobic forces, hydrogen bonds, disulfide bridges, and salt bridges between subunits give rise to quaternary structure.

19.48 The term subunit describes a polypeptide chain that has its own primary, secondary, and tertiary structure that combines with other polypeptide chains to form the quaternary structure of a protein.

PROTEIN HYDROLYSIS AND DENATURATION (SECTION 19.10)

☑19.50 A protein that is completely hydrolyzed will result in amino acids. A protein that is denatured will maintain its primary structure, but will lose its secondary, tertiary, and quaternary structure.

19.52 The primary structure of the protein in a raw egg and a cooked egg are the same.

19.54 Cooking an egg results in a permanent change to its three dimensional structure. This process involves irreversible protein denaturation. Beating an egg white does not result in a completely permanent change to its three dimensional structure. This process involves reversible protein denaturation.

19.56 Egg white can serve as an emergency antidote for heavy-metal poisoning because the proteins in the egg bind to the metal ions and form a precipitate that resists further digestion in the stomach. The metal-containing precipitate is then removed from the stomach by pumping or is ejected by induced vomiting.

ADDITIONAL EXERCISES

19.58 Valine is an amino acid that could be referred to as 2-amino-3-methylbutanoic acid. Methionine and threonine are also derivitatives of butanoic acid.

19.60

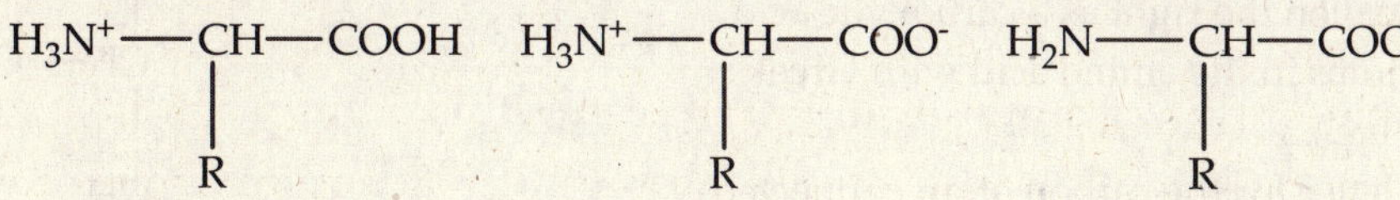

conjugate acid zwitterion conjugate base

ALLIED HEALTH EXAM CONNECTION

19.62 Oxytocin stimulates the smooth muscle of the uterine wall during the labor and delivery process. After delivery, it promotes the ejection of milk.

19.64 The bond that forms between the carboxyl (RCOOH) group of one amino acid and the amino ($RC\text{-}NH_2$) group of another amino acid is (a) called a peptide bond and (c) it is formed by a dehydration reaction.

19.66 The quaternary protein structure of hemoglobin involves four chains (2 alpha chains containing 141 amino acid residues each and 2 beta chains containing 146 amino acid residues each) and four heme groups (1 per chain) located in crevices near the exterior of the molecule.

The hemoglobin molecule is nearly spherical and the four subunits are held together rather tightly by hydrophobic forces.

19.68 Proteins do not (a) contain genetic information.

CHEMISTRY FOR THOUGHT

19.70 During exercise the levels of endorphins increase in the body. These neurotransmitters are manufactured in the brain and chemically similar to morphine. It is possible that the "high" experienced by runners is due to the brain's synthesis of these peptides which are chemically similar to a drug.

19.72 The ethyl ester of alanine melts 200 degrees below the melting point of alanine because the ethyl ester of alanine is unable to form a zwitterion, and consequently, it is unable to make ionic bonds to other molecules of the same type. The weak interparticle forces between the molecules of the ethyl ester of alanine correspond to a low melting point, while the strong interparticle forces between the alanine zwitterions correspond to a melting point more like an ionic compound than an organic compound.

19.74 Plasma proteins are likely to be globular so that they can be transported in the blood vessels.

19.76 The properties of fibrous proteins that make hair, fur, and spider webs useful are that these proteins, which are made up of long rod-shaped or string like molecules, can intertwine with one another and form strong fibers that are water-insoluble.

19.78 Two elements that are found in proteins but are not present in fats, oils, or carbohydrates are sulfur and nitrogen.

ADDITIONAL ACTIVITIES

Section 19.1 Review:

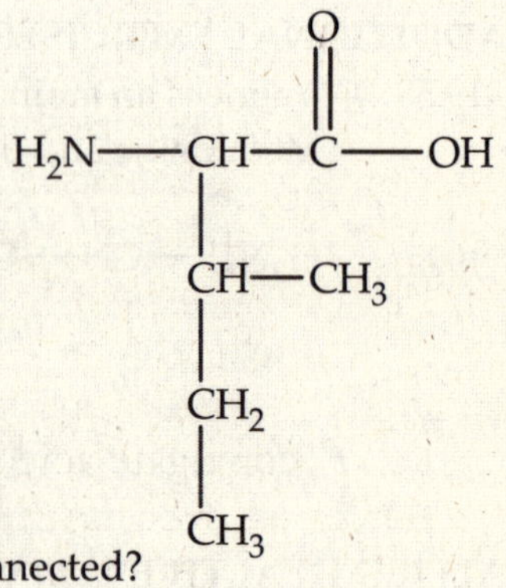

(1) Which are larger biomolecules: lipids or proteins?
(2) Number the carbon chain of the unionized amino acid on the right.
(3) Name the amino acid on the right as a carboxylic acid.
(4) Label the carbon atoms in the amino acid with Greek letters (α, β, γ, δ, ε, etc.).
(5) What is the Greek letter for the carbon atom connected to the $-NH_2$ group?
(6) Circle the side chain.
(7) What is the Greek letter for the carbon atom to which the side chain is connected?
(8) Is the side chain polar or nonpolar?
(9) Is the side chain acidic, basic, or neutral?
(10) What is the common name for this amino acid?
(11) What are the abbreviations for this amino acid?
(12) Draw this amino acid as a Fischer Projection of the L-form.

Section 19.2 Review:

(1) Label the acidic and basic areas of the amino acid from the Section 19.1 Review.
(2) Use an arrow to show the internal acid-base reaction within the amino acid from (1).
(3) What is a general name of structures like the one produced by the reaction in (2)?
(4) What is the net charge on the amino acid in (2)?
(5) Write an equation that shows what happens when hydronium ions are added to a solution of (2).
(6) What is the net charge on the amino acid after the addition of hydronium ions?
(7) Write an equation that shows what happens when hydroxide ions are added to a solution of (2).
(8) What is the net charge on the amino acid after the addition of hydroxide ions?
(9) What is the name of the pH at which the amino acid exists in the form produced in (2)?
(10) Is this amino acid able to act as a buffer? Explain.

Section 19.3 Review:

(1) Can the amino acid from the Section 19.1 Review produce a disulfide bond with another amino acid? Explain.
(2) Show the two possible amide-forming reactions between glycine and the amino acid from the Section 19.1 Review.
(3) What is the general term used to describe the larger products from (2)?
(4) Name the products from (2).
(5) Write the abbreviations for the products from (2).
(6) Label the peptide linkages in the structures from (2).
(7) What are the other names for a peptide linkage?

(8) Circle the N-terminal end of the products from (2).
(9) Box the C-terminal end of the products from (2).

Section 19.4 Review:
(1) What happens to the polypeptide structure when two cysteine residues form a disulfide bridge in a polypeptide?
(2) Does the order of the amino acid residues influence biological function? Explain.

Section 19.5 Review:
(1) Which class of carbohydrates contains molecules with similar molecular weights to proteins?
(2) What classification can be used for both proteins and the class of carbohydrates identified in (1)?
(3) Which of the following expressions best help to explain the solubility of proteins: "Like dissolves like." or "Opposites attract."? Explain.
(4) Differentiate between the following protein functions:
  a. catalytic and regulatory
  b. structural and protective
  c. storage and nerve impulse transmission
  d. movement and transport
(5) Can a protein be both fibrous and simple? Explain.
(6) Can a protein be both globular and conjugated? Explain.
(7) Which class of proteins contains prosthetic groups?

Section 19.6 Review:
(1) What do all proteins have in common with regard to primary structure?
(2) What is unique about the primary structure of different proteins?

Section 19.7 Review:
(1) Which intermolecular force is responsible for secondary structure?
(2) How is primary structure related to secondary structure?

Section 19.8 Review:
(1) Distinguish between disulfide bridges and salt bridges.
(2) Can the amino acid residues with neutral side chains be involved in tertiary structure hydrogen bonds? Explain.
(3) Which groups of amino acid residues are involved in hydrophobic interactions?

Section 19.9 Review:
(1) What is the difference between a subunit and a prosthetic group?
(2) Do all conjugated proteins have quaternary structure?

Section 19.10 Review:
(1) What was the result of the hydrolysis reactions involving simple lipids?
(2) What is the result of the hydrolysis reactions involving proteins?
(3) Which level(s) of structure is(are) changed for a protein that undergoes hydrolysis?
(4) What factors will denature a protein?
(5) Which level(s) of structure is(are) changed for a protein that undergoes denaturation?
(6) Which level(s) of structure is(are) important for a protein to perform its original biological function?

Tying It All Together with a Laboratory Application:
A pentapeptide contains the amino acid residues of valine, cysteine, asparagine, lysine, and glutamate. These residues are the (1) ______-form stereoisomers found in most living organisms. At its isoelectric point, the Val-Cys-Asn-Lys-Glu pentapeptide has the structure (2) ______. The N-terminal amino acid

residue in this pentapeptide is (3) ______. The C-terminal amino acid residue in this pentapeptide is (4) ______. This is one of (5) ______ possible pentapeptides formed from these five amino acid residues. The major difference between the pentapeptides is the (6) ______ of the amino acid residues, which is called the (7) ______ level of structure.

Hydrogen bonding between the carbonyl group of the glutamate amino acid residue and the amide group of the cysteine amino acid residue of the Val-Cys-Asn-Lys-Glu pentapeptide would be considered the (8) ______ level of structure. If this pattern of hydrogen bonding occurred again and again in a longer polypeptide or protein it would result in a(n) (9) ______.

If the side chains of these amino acids were to interact, the (10) ______ level of structure would result. (11) The following table lists the classification of side chains of each amino acid as well as the type of interactions that side chain is likely to experience:

Amino Acid Residue	Classification of Side Chain	Probable Side Chain Interactions
valine		
cysteine		
asparagine		
lysine		
glutamate		

Side chain interaction is (12) ______ (likely or unlikely) because this peptide chain is much (13) ______ than the polypeptides with this level of structure. The (14) ______ level of structure is unlikely to occur with this pentapeptide because (15) ______.

This pentapeptide will likely have its isoelectric point (16) ______ 7 because it contains (17) ______ neutral side chain(s), (18) ______ acidic side chain(s), and (19) ______ basic side chain(s). Increasing the pH of a solution containing this pentapeptide would initially (20) ______ the pentapeptide by changing its (21) ______ state; however, adding a strong base to this pentapeptide could result in (22) ______ which destroys the (23) ______ level of structure.

The amino acids that result from (24) ______ under basic conditions could be isolated. The techniques that could be employed include gel electrophoresis, electrodialysis, or ion-exchange chromatography. All of these techniques rely on the (25) ______ of the amino acids at different pH values. The two amino acids that will be most easily separated by these techniques are (26) ______ and ______. The amino acids with (27) ______ side chains will be more difficult to separate; however, their side chains differ in (28) ______ and reactivity.

(29) ______ could be isolated by mixing the aqueous solution with a nonpolar solvent like benzene in a separatory funnel because it is more soluble in benzene than water. The remaining residues could be mixed with an oxidizing agent or heavy metal ions. Only the (30) ______ will react with those reagents. The (31) ______ is the better choice because this reaction can be reversed by adding (32) ______; however, the (33) ______ will form insoluble salts that can be filtered from the aqueous solution, leaving (34) ______ as the only amino acid in the aqueous solution.

SOLUTIONS FOR THE ADDITIONAL ACTIVITIES

Section 19.1 Review:

(1) Proteins are larger biomolecules than lipids. Proteins are polymers, but lipids are not.

(2)

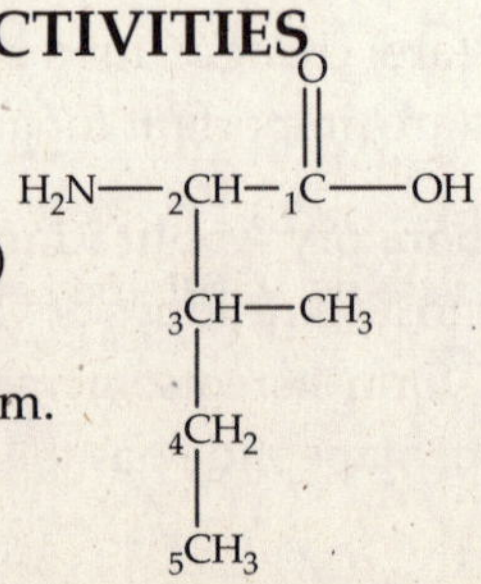

(3) 2-amino-3-methylpentanoic acid

(4)

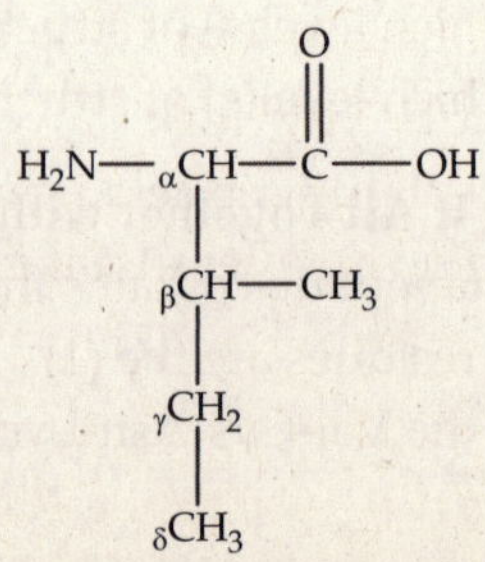

(5) The amino group is attached to the α-carbon atom.

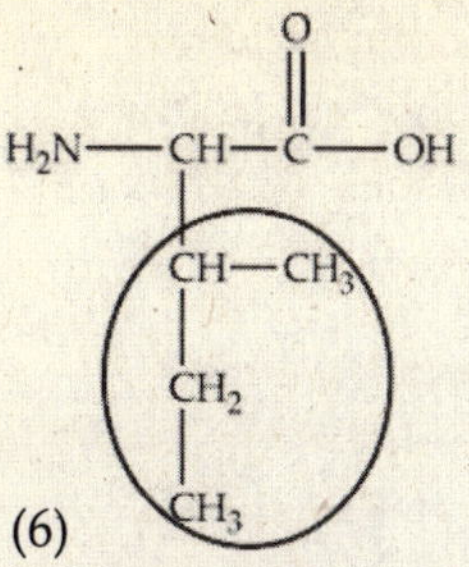

(6)

(7) The side chain is attached to the α-carbon atom.

(8) The side chain is nonpolar.

(9) The side chain is neutral.

(10) The common name for this amino acid is isoleucine.

(11) The three letter abbreviation for this amino acid is Ile.
The one letter abbreviation for this amino acid is I.

(12) 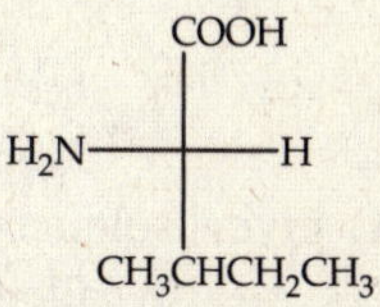

Section 19.2 Review:

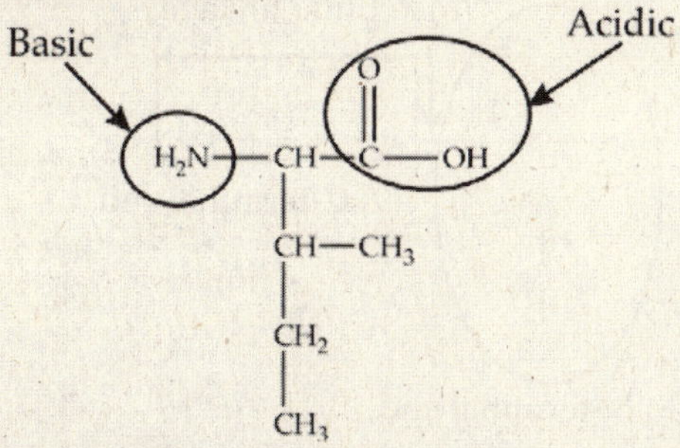

(1)

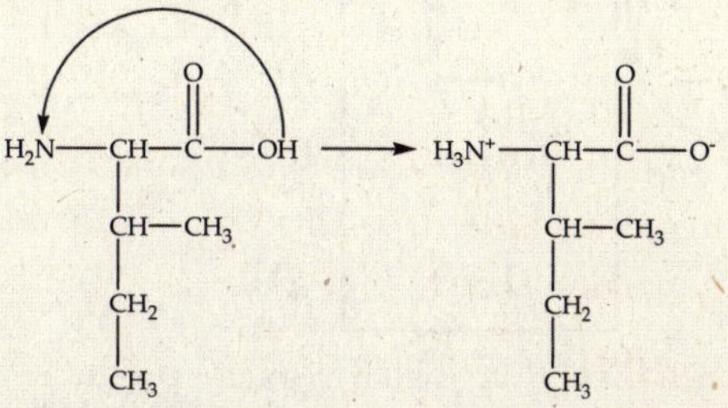

(2)

(3) zwitterion; (4) net charge = 0;

Note: This arrow does not show electron movement, but relocation of the hydrogen atom.

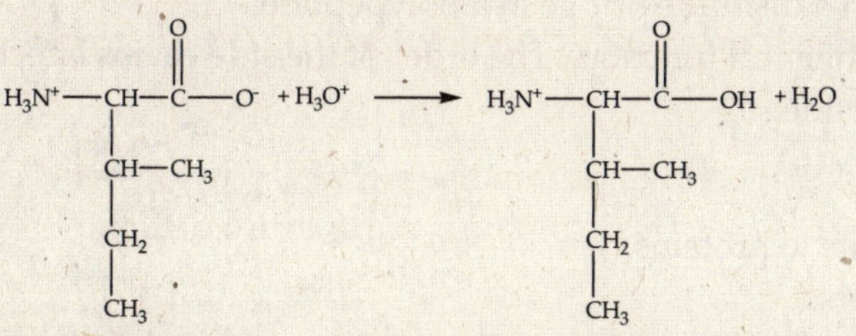

(5)

(6) net charge = +1;

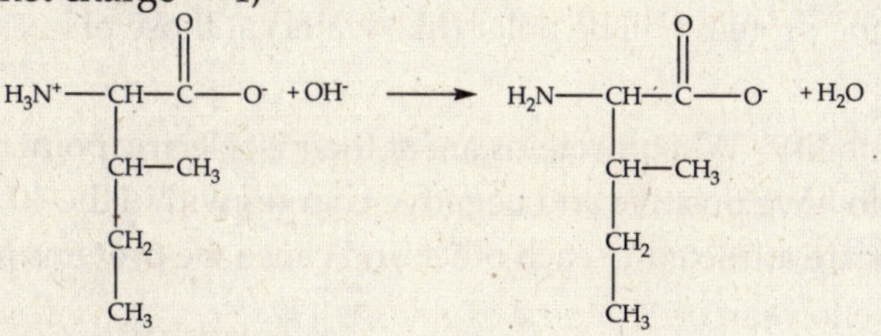

(7)

(8) net charge = -1

(9) isoelectric point

(10) Yes, this amino acid is able to act as a buffer because it has the ability to react with both H_3O^+ and OH^-.

Section 19.3 Review:

(1) No, the amino acid from the Section 19.1 Review cannot produce a disulfide bond with another amino acid because it is not cysteine. Cysteine is the only amino acid that forms disulfide bonds.

(2)

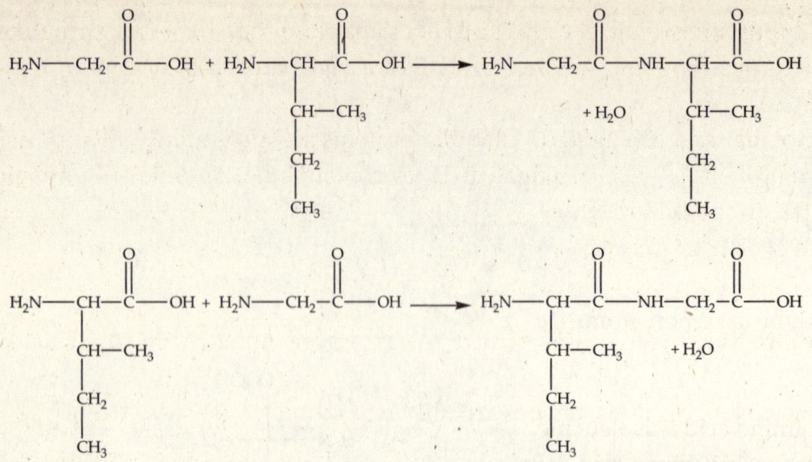

(3) dipeptides; (4) glycylisoleucine, isoleucylglycine; (5) Gly-Ile, Ile-Gly; (6) see below

(7) Peptide linkages are also called peptide bonds or amide linkages.

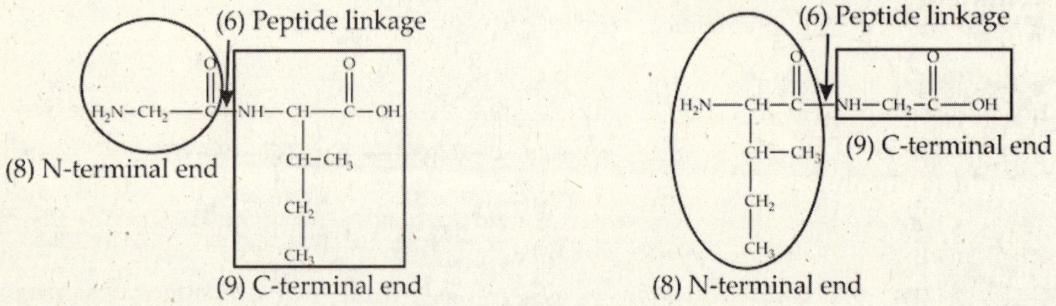

Section 19.4 Review:

(1) A loop forms when two cysteine residues form a disulfide bridge in a polypeptide.

(2) Yes, the order of the amino acids influences biological function. The order of the side chains effects the three dimensional structure of the polypeptide.

Section 19.5 Review:

(1) Polysaccharides have similar molecular weights to proteins.

(2) Both polysaccharides and proteins are polymers.

(3) "Like dissolves like" best explains protein solubility. Proteins will dissolve in water when they are above or below their isoelectric point because the protein will be polar (like water) at those pH values.

"Opposites attract" best explains protein insolubility. When proteins are at their isoelectric point, they do not have a net charge; however, they do have positive and negative charges within the molecule. These positive and negative charges are attracted to each other and cause the proteins to be more insoluble.

(4) a. Catalytic proteins lower the activation energy for reactions. Regulatory proteins affect the rate of body processes which may involve more than one chemical reaction.

b. Structural proteins are responsible for the mechanical strength of skin, bone, hair, and fingernails. Protective proteins are antibodies that counteract viruses, bacteria, and other foreign substances.

c. Storage proteins provide a way to store small molecules or ions for nourishment. Nerve impulse transmission proteins receive small molecules that pass between the synapses of nerve cells.

d. Movement proteins are found in muscles. Transport proteins are responsible for moving small molecules and ions through the body.

(5) Yes, a protein can be both fibrous and simple because fibrous proteins are rod-shaped or stringlike molecules that intertwine to form strong fibers. These proteins are also simple if they only contain amino acid residues.
(6) Yes, a protein can be both globular and conjugated. Globular proteins are more spherical in shape than fibrous proteins. These proteins are also conjugated if they contain other organic or inorganic components in addition to the amino acid residues.
(7) Conjugated proteins contain prosthetic groups.

Section 19.6 Review:
(1) All proteins have a primary structure that contains a backbone of carbon and nitrogen atoms held together by peptide bonds.
(2) Different proteins have different lengths of backbones and different sequences of side chains.

Section 19.7 Review:
(1) Hydrogen bonding is responsible for secondary structure.
(2) The carbonyl oxygen atoms and the amide hydrogen atoms in the backbone of a protein (primary structure) hydrogen bond to each other in order to form the secondary structure.

Section 19.8 Review:
(1) Disulfide bridges form when two cysteine amino acid residues undergo oxidation and form a covalent bond. Salt bridges form when ionized acidic amino acid residue side chains and ionized basic amino acid residue side chains form an ionic bond.
(2) Neutral side chains can be involved in hydrogen bonding, if they are polar and contain oxygen or nitrogen atoms.
(3) Amino acid residues with neutral, nonpolar side chains can be involved in hydrophobic interactions.

Section 19.9 Review:
(1) A subunit is a polypeptide chain. A prosthetic group is a non-amino acid part of a conjugated protein. It can be organic or inorganic.
(2) Quaternary structure is defined as the arrangement of subunits to form a larger protein. If a conjugated protein contains two or more subunits, then it will have quaternary structure. If the conjugated protein only contains one subunit, then it will not have quaternary structure.

Section 19.10 Review:
(1) When simple lipids are hydrolyzed, glycerol or an alcohol is produced as well as fatty acids or salts of fatty acids.
(2) When proteins are hydrolyzed, smaller peptides or amino acids are produced.
(3) Primary structure is changed for a protein that undergoes hydrolysis; consequently, all other levels of structure are affected.
(4) A protein can be denatured by extreme temperatures, extreme pH values, organic solvents, detergents, and heavy-metal ions.
(5) The higher levels of structure (secondary, tertiary, and quaternary) are changed for a protein that undergoes denaturation. Primary structure is not affected.
(6) All four levels of structure are important for a protein to perform its original biological function.

Tying It All Together with a Laboratory Application:

(1) L-

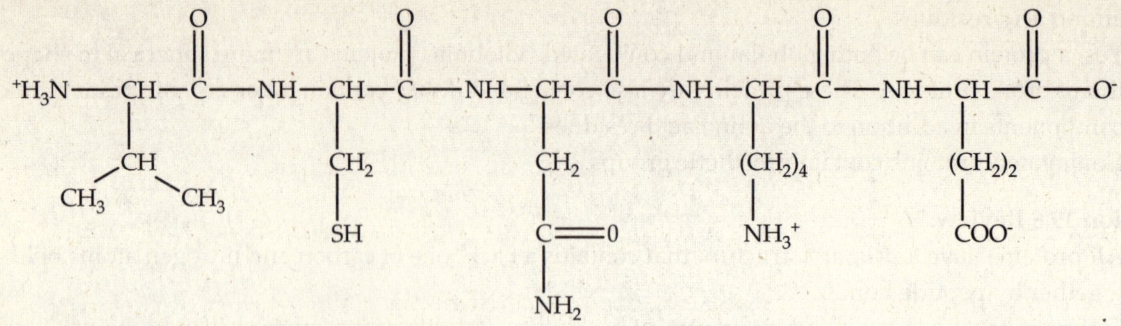

(2)

(3) valine
(4) glutamate
(5) 120
(6) order
(7) primary
(8) secondary
(9) α-helix
(10) tertiary

(11)

Residue	Classification	Interactions
valine	**neutral, nonpolar**	**hydrophobic interactions**
cysteine	**netural, polar**	**disulfide bridges**
asparagine	**neutral, polar**	**hydrogen bonding**
lysine	**basic, polar**	**salt bridges**
glutamate	**acidic, polar**	**salt bridges**

(12) unlikely
(13) shorter
(14) quaternary
(15) it only contains 1 peptide chain
(16) near
(17) 3
(18) 1
(19) 1
(20) denature
(21) native
(22) hydrolysis (or digestion)
(23) primary
(24) hydrolysis
(25) charges
(26) lysine, glutamate
(27) neutral
(28) polarity
(29) Valine
(30) cysteine
(31) oxidizing agent
(32) a source of hydrogen
(33) heavy metal ions
(34) asparagine

SELF-TEST QUESTIONS

Multiple Choice

1. The main distinguishing feature between various amino acids is:
   a. the length of the carbon chain.
   b. the number of amino groups.
   c. the composition of the side chains.
   d. the number of acid groups.

2. The amino acid valine is represented below. Which of the lettered carbon atoms is the alpha carbon atom?

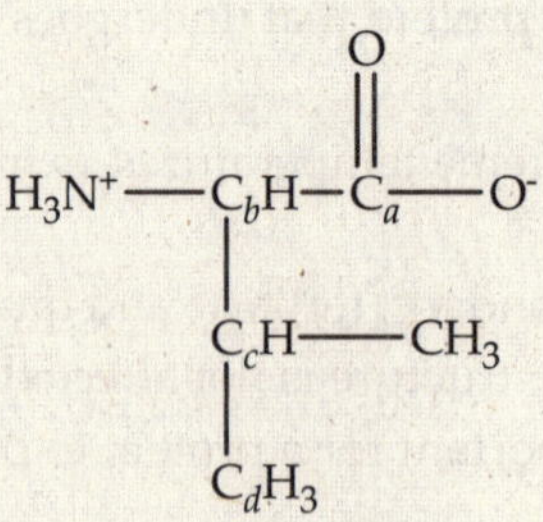

   a. *a*
   b. *b*
   c. *c*
   d. *d*

3. What kind of amino acid is this compound?

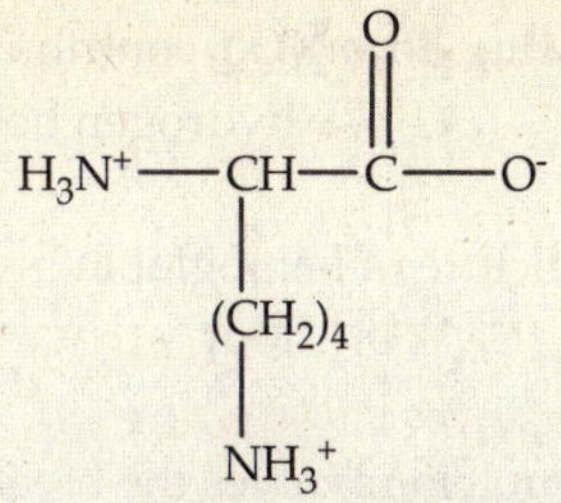

a. acidic
b. basic
c. neutral
d. more than one response is correct

4. Which of the following best represents alanine at its isoelectric point?

a.
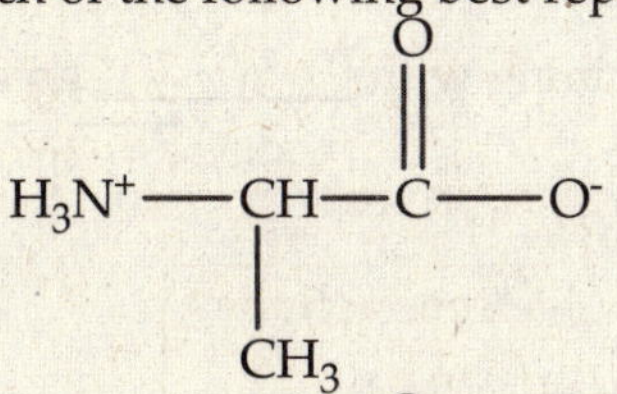

b.
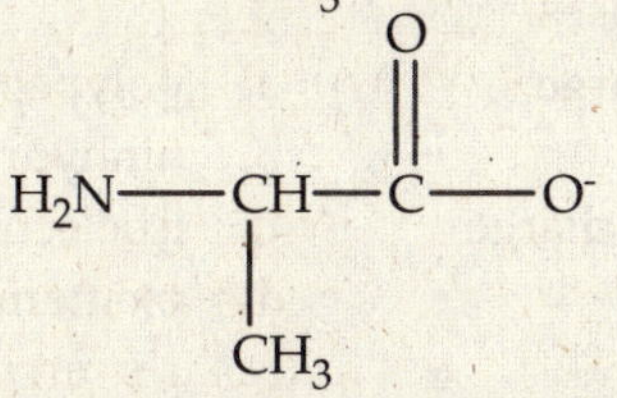

c.
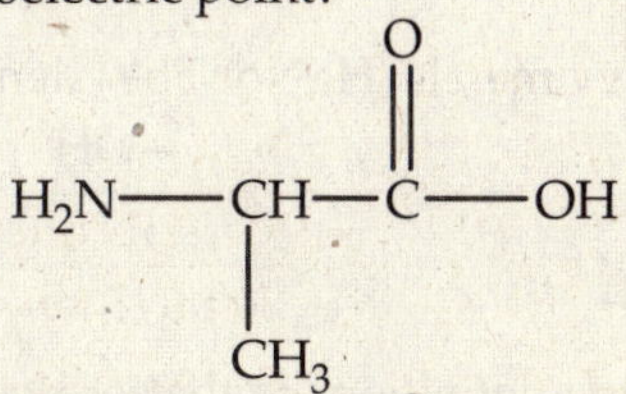

d.
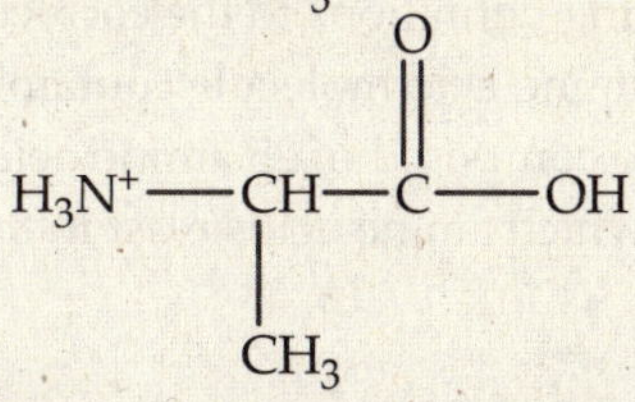

5. Which of the following is a linkage present in all peptides?

a. NH_2-CH_2-

b.
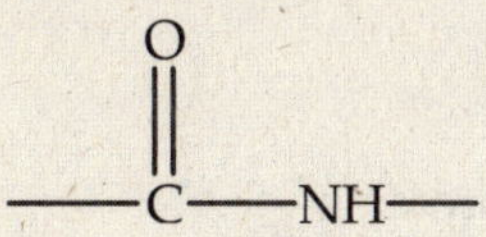

c.
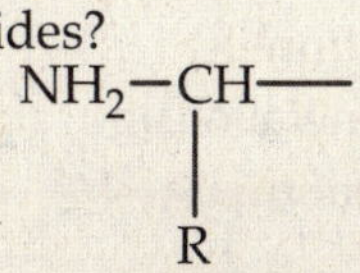

d.
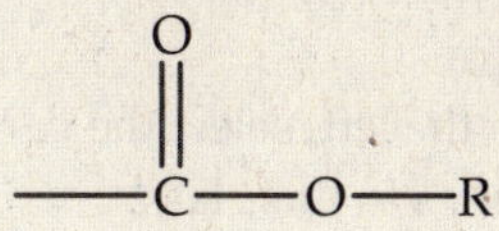

6. Prosthetic groups are found in __________ proteins.
a. all b. simple c. conjugated d. no

7. A protein that is relatively spherical in shape and fairly soluble in water is a __________ protein.
a. simple b. conjugated c. fibrous d. globular

8. Which of the following characteristics of a protein would be classified as a primary structural feature?
a. amino acid sequence
b. pleated-sheet configuration
c. α-helix configuration
d. the shape of the protein molecule

9. Which protein serves as an antibody?
a. myoglobin b. hemoglobin c. collagen d. immunoglobulin

10. Which of the following side-group interactions involves nonpolar groups?
a. salt bridges
b. hydrogen bonds
c. disulfide bonds
d. hydrophobic interactions

11. Which of the following does **NOT** contribute to the tertiary structure of proteins?
    a. salt bridges b. hydrogen bonds c. disulfide bonds d. peptide bonds

12. The quaternary structure of hemoglobin involves __________ subunits.
    a. two b. four c. six d. eight

13. Denaturation of a protein involves the breakdown of:
    a. primary structure.
    b. secondary and tertiary structures.
    c. primary and secondary structures.
    d. primary, secondary, and tertiary structures.

14. Ions of heavy metals (Hg^{2+} or Pb^{2+}) denature proteins by combining with __________ groups.
    a. $-NH_2$ b. $-OH$ c. $-SH$ d. $-\underset{\underset{O}{\|}}{C}-NH-$

Matching

Match the following definitions to the correct words given as responses.

15. a dipolar amino acid molecule containing both a + and a – charge
16. a substance composed of 25 amino acids linked together
17. the pH at which amino acids exist in the form that has no net charge

a. polypeptide
b. ninhydrin
c. isoelectric point
d. zwitterion

Select the correct polypeptide for each description on the left.

18. stimulates milk production
19. controls carbohydrate metabolism
20. decreases urine production

a. prolactin
b. insulin
c. oxytocin
d. vasopressin

For each description on the left, select the correct protein class.

21. hemoglobin belongs to this class
22. proteins that function as enzymes
23. insulin belongs to this class
24. keratin belongs to this class
25. collagen belongs to this class

a. regulatory proteins
b. transport proteins
c. structural proteins
d. catalytic proteins

For each of the interactions given on the left, choose a response that indicates the level of protein structure associated with each type of interaction.

26. hydrogen bonds
27. amide bonds
28. hydrophobic interaction
29. ionic bonds

a. primary structure
b. secondary structure
c. tertiary structure
d. more than one response is correct

True-False

30. Amino acids found in living systems are generally in the D- form.
31. Amino acids are white crystalline solids.
32. Cysteine is the only amino acid that can form disulfide bridges.
33. Six structural isomers could be formed if tripeptides were made from serine, valine, and glycine.
34. Proteins are the building blocks of amino acids.
35. Acids cannot hydrolyze proteins.

Chapter 20: Enzymes

CHAPTER OUTLINE

20.1 General Characteristics of Enzymes
20.2 Enzyme Nomenclature and Classification
20.3 Enzyme Cofactors
20.4 The Mechanism of Enzyme Action
20.5 Enzyme Activity
20.6 Factors Affecting Enzyme Activity
20.7 Enzyme Inhibition
20.8 The Regulation of Enzyme Activity
20.9 Medical Application of Enzymes

LEARNING OBJECTIVES/ASSESSMENT

When you have completed your study of this chapter, you should be able to:

1. Describe the general characteristics of enzymes and explain why enzymes are vital in body chemistry. (Section 20.1; Exercise 20.2)
2. Determine the function and/or substrate of an enzyme on the basis of its name. (Section 20.2; Exercise 20.12)
3. Identify the general function of cofactors. (Section 20.3; Exercise 20.14)
4. Use the lock-and-key theory to explain specificity in enzyme action. (Section 20.4; Exercise 20.20)
5. List two ways of describing enzyme activity. (Section 20.5; Exercise 20.26)
6. Identify the factors that affect enzyme activity. (Section 20.6; Exercise 20.28)
7. Compare the mechanisms of competitive and noncompetitive enzyme inhibition. (Section 20.7; Exercise 20.34)
8. Describe the three methods of cellular control over enzyme activity. (Section 20.8; Exercise 20.38)
9. Discuss the importance of measuring enzyme levels in the diagnosis of disease. (Section 20.9; Exercise 20.46)

SOLUTIONS FOR THE END OF CHAPTER EXERCISES

GENERAL CHARACTERISTICS OF ENZYMES (SECTION 20.1)

☑20.2 Enzyme catalysis of a reaction is superior to normal laboratory conditions because enzymes are specific in the type of reaction they catalyze and enzyme activity can be regulated by the cell.

20.4 Many types of enzymes are needed because enzymes can only catalyze specific reactions and a body requires many reactions to occur in order to maintain life.

20.6 Absolute specificity means that an enzyme can only catalyze the reaction of one substance.

ENZYME NOMENCLATURE AND CLASSIFICATION (SECTION 20.2)

20.8 The ending of EC names and most common names of enzymes is *–ase*.

20.10

	Enzyme	Substrate		Enzyme	Substrate
a.	sucrase	sucrose	d.	maltase	maltose
b.	amylase	amylose	e.	arginase	arginine
c.	lactase	lactose			

☑20.12

	Enzyme	Substrate	Type of Reaction
a.	succinate dehydrogenase	succinate	dehydrogenation
b.	L-amino acid reductase	L-amino acid	redox (specifically reduction of the amino acid)

c.	cytochrome oxidase	cytochrome	redox (specifically oxidation of the cytochrome)
d.	glucose-6-phosphate isomerase	glucose-6-phosphate	isomerization

ENZYME COFACTORS (SECTION 20.3)

☑20.14 When an apoenzyme and a cofactor combine, they form an active enzyme.

20.16 Iron, zinc, and magnesium ions are some typical inorganic ions that serve as cofactors.

THE MECHANISM OF ENZYME ACTION (SECTION 20.4)

20.18 The equation, $E + S \rightleftarrows ES \rightarrow E + P$, shows that the enzyme (E) and substrate (S) establish an equilibrium with the enzyme-substrate complex (ES). This is a reversible reaction. The enzyme-substrate complex can break apart into the enzyme and substrate or react to produce the enzyme and the product (P). The reaction to produce the product is not a reversible reaction.

☑20.20 The lock-and-key theory explains enzyme specificity because only one substrate will fit into the active site of the enzyme. The active site of the enzyme specifies which substrates can bind to the enzyme.

20.22 The induced-fit theory best explains the ability of one enzyme to catalyze the reactions of propanoic acid, butanoic acid, and pentanoic acid because the active site would need to be able to accommodate several similar, but different, substrates.

ENZYME ACTIVITY (SECTION 20.5)

20.24 Enzyme activity in an experiment is observed by any method that allows the rate of product formation or reactant usage to be determined. The disappearance or appearance of a characteristic color is an example.

☑20.26 An enzyme international unit is the quantity of enzyme that catalyzes the conversion of 1 micromole of substrate per minute. This is useful in medical diagnosis because a specific enzyme activity of a patient can be measured and compared to normal activity.

FACTORS AFFECTING ENZYME ACTIVITY (SECTION 20.6)

☑20.28
a. As substrate concentration increases, the enzyme activity increases until a maximum reaction rate is reached when all of the available enzyme is saturated with substrate and the enzyme activity cannot increase unless additional enzyme is added.
b. As enzyme concentration increases, the enzyme activity increases linearly. (Note: The enzyme is the limiting reagent and the substrate is in excess).
c. As pH increases, the enzyme activity increases until the optimum pH value is attained, after which the enzyme activity decreases.
d. As temperature increases, the enzyme activity increases until the optimum temperature is attained, then the enzyme activity decreases.

20.30 V_{max} for an enzyme can be determined by measuring the turnover rate for a specific enzyme concentration with increasing concentrations of substrate. The turnover rates will approach a steady value (V_{max}) when the substrate concentration is high enough for all of the enzyme units to be saturated.

20.32 The pH of enzymes is maintained near 7.0 to prevent the enzymes from denaturing. Enzymes are proteins that can lose their three dimensional structure if the pH is drastically changed.

ENZYME INHIBITION (SECTION 20.7)

☑20.34 In competitive enzyme inhibition, an inhibitor binds to the active site of an enzyme and prevents the substrate from binding to the active site. In noncompetitive enzyme inhibition, an inhibitor binds to the enzyme at a location other than the active site and changes the shape of the active site, thus preventing the substrate from binding to the active site.

20.36
a. One antidote for cyanide poisoning is sodium thiosulfate. This antidote works because the cyanide ion reacts with the thiosulfate ion to produce the thiocyanate ion and the sulfite ion. Unlike cyanide, thiocyanate does not bind to the iron of cytochrome oxidase.
b. One antidote for heavy-metal poisoning is a chelating agent, like ethylenediaminetetraacetic acid. This antidote works because the heavy-metal ions combine with the EDTA and are tightly bound. Consequently, the heavy-metal ions cannot react with proteins in the body.

THE REGULATION OF ENZYME ACTIVITY (SECTION 20.8)

☑20.38 Enzyme activity is controlled by the activation of zymogens, by allosteric regulation, and by genetic control.

20.40 Enzymes that catalyze blood clotting are zymogens because under normal circumstances, blood needs to flow freely through the body. Only when bleeding occurs is clotting action necessary. Consequently, prothrombin flows through the body in the blood stream, but it does not cause the blood to clot until it is converted into thrombin which catalyzes clotting.

20.42 Activators are modulators that increase the activity of an enzyme. Inhibitors are modulators that decrease the activity of an enzyme.

20.44 Genetic control of enzyme activity is the cellular production of additional enzyme molecules in order to increase the production from an enzyme-catalyzed reaction. Enzyme induction is the synthesis of enzymes in response to a temporary need of the cell.

MEDICAL APPLICATION OF ENZYMES (SECTION 20.9)

☑20.46
a. CK — This enzyme assay is useful in the diagnosis of a heart attack.
b. ALP — This enzyme assay is useful in the diagnosis of bone or liver disease.
c. amylase — This enzyme assay is useful in the diagnosis of diseases of the pancreas.

20.48 LDH is a good initial diagnostic test because LDH occurs in multiple forms (isoenzymes) and can be used to diagnose a wide range of diseases (anemias, acute live diseases, congestive heart failure, and muscular diseases).

ADDITIONAL EXERCISES

20.50 Arginine has a basic, polar side chain; therefore, it can form salt bridges. If the amino acid arginine is found in the active site of carboxypeptidase and is responsible for holding the C-terminal end of the polypeptide in place so the cleavage of the peptide bond can occur, the C-terminal end of the polypeptide and the arginine side chain are held together by a salt bridge.

20.52 Propanoic acid chloride reacts with the side chain of a serine residue on an enzyme to cause irreversible inhibition of that enzyme. The mechanism by which this irreversible inhibition occurs is the formation of an ester by reaction of the propanoic acid chloride with the alcohol functional group on the serine side chain. Therefore, the structure of the enzyme has been changed and the enzyme is unable to perform its former function.

20.54 When an enzyme is added to a reaction at equilibrium, the equilibrium will not shift because the enzyme is a catalyst and catalysts speed up both the forward and reverse reactions by lowering the activation energy for both the forward and reverse reactions. Equilibrium would be reached more quickly, since the rates of both reactions will increase.

ALLIED HEALTH EXAM CONNECTION

20.56

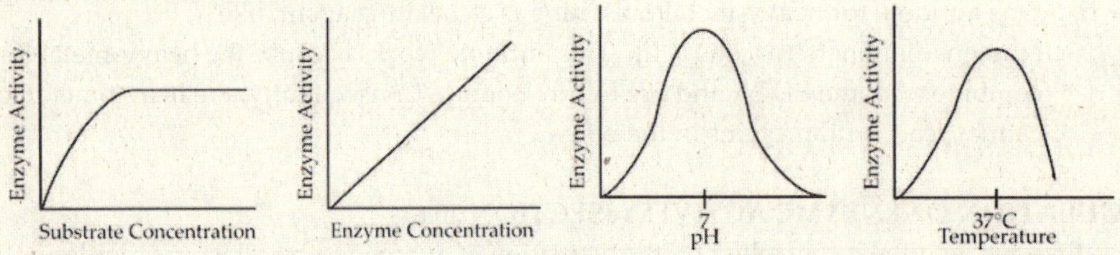

20.58 Saliva contains mucus, water, and (b) amylase, which partially digest polysaccharides.

20.60 The site on an enzyme molecule that does the catalytic work is called the (d) active site.

20.62 (c) Feedback inhibition is the prevention of enzyme activity by a product of a reaction.

CHEMISTRY FOR THOUGHT

20.64 Food preservation by freezing lowers enzyme activity because at such low temperatures enzyme activity virtually ceases.

20.66 Urease can catalyze the hydrolysis of urea, but not the hydrolysis of methyl urea because the active site on urease can accommodate urea, but not methyl urea. Urease has specificity for urea, not methyl urea. The shapes and sizes of these two compounds are quite different.

20.68 Enzymes that are used for laboratory or clinical work are stored in refrigerators because the temperature in the refrigerator is below the optimum temperature for enzyme activity but above the temperature that would result in the denaturing of the enzymes. This preserves the enzymes and allows them to last longer than if they were stored at room temperature.

20.70 The presence of ammonia can be detected using an indicator, like phenolphthalein, since ammonia is a base. Alternatively, its presence can be detected by smell. Ammonia has a distinct smell, which differs from that of urea.

20.72 Blanching corn by placing the corn in boiling water will prevent enzyme activity because the organisms on the corn will have their proteins denatured at this temperature. The blanching also halts the internal enzyme processes that would continue to ripen the corn and cause the flavor and food value to change over time. After blanching, the corn can be frozen, remain safe for human consumption, and maintain its food value for a long period of time.

20.74 An enzyme might be selected oven an inorganic catalyst for a certain industrial process because the enzyme might be more specific to the reaction or allow the reaction to occur at a lower temperature or more neutral pH. Frequently inorganic catalysts are associated with extreme pH values and high temperatures.

ADDITIONAL ACTIVITIES

Section 20.1 Review:

(1) What is catalytic power?

(2) Imagine you have a box of pens and pencils of all different descriptions. If you were to place the pen or pencil you are currently using in the box, how could you describe it to someone else and have them successfully find your current writing implement?

(3) If your friend finds a copy of your writing implement, but not your exact writing implement, is that absolute specificity, relative specificity, or stereochemical specificity?

(4) If your friend finds a writing implement that produces the same color of ink or graphite on the paper as your writing implement, but it is not your writing implement or a copy of your writing implement, is that absolute specificity, relative specificity, or stereochemical specificity?

(5) Are enzymes self-regulating? Explain.

Section 20.2 Review:

(1) Which two endings might indicate a substance is an enzyme?

(2) How are enzymes classified?

(3) What is a substrate?

Section 20.3 Review:

(1) What is the difference between a prosthetic group and a cofactor?

(2) What is the difference between a cofactor and a coenzyme?

(3) What is an apoenzyme?

(4) How do vitamins relate to enzymes?

Section 20.4 Review:

(1) What structural feature on an enzyme is identified in both the lock-and-key theory and the induced-fit theory?

(2) How does the structural feature in (1) differ in the lock-and-key theory and the induced-fit theory?

Section 20.5 Review:

(1) What is the turnover number for an enzyme?

(2) What type of experiments could be used to determine the turnover number of an enzyme?

(3) Who is more likely to determine enzyme activity as a turnover number: a research biochemist or a hospital lab technician?

(4) Who is more likely to determine enzyme activity levels in enzyme international units: a research biochemist or a hospital lab technician?

Section 20.6 Review:

(1) What is typically the limiting factor in the rate of an enzyme-catalyzed reaction?

(2) What does a maximum rate imply about the enzyme molecules?

(3) Temperature and pH have similar trends with respect to enzyme activity. Explain.

Section 20.7 Review:

(1) Inhibitors are divided into two categories. One of the categories is shown on the diagram below. Label the other category.

(2) What types of inhibitors fall into the category you labeled in (1)?

(3) Reversible inhibitors are divided into two categories. One of the categories is shown on the diagram below. Label the other category.

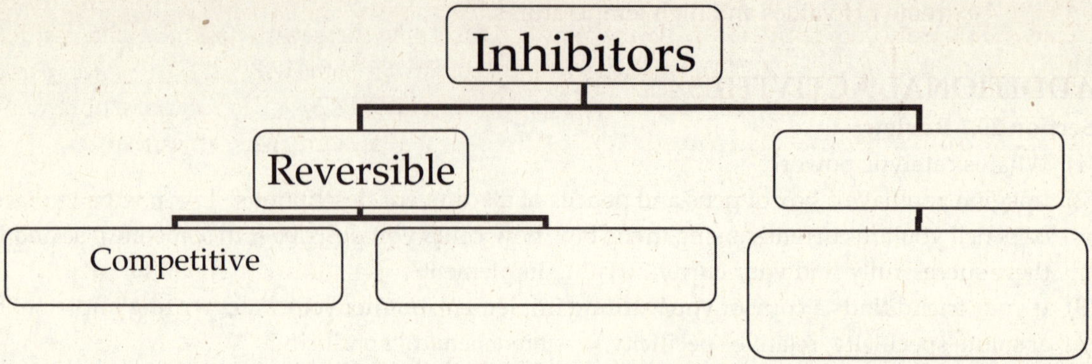

(4) What is the competition for competitive inhibitors?

Section 20.8 Review:

(1) What is another name for zymogens?
(2) How do zymogens differ from apoenzymes?
(3) What are the two categories of modulators?
(4) How does the name "allosteric enzymes" indicate that these enzymes have distinctive binding sites for modulators?
(5) What biomolecules are responsible for genetic control of enzymes?

Section 20.9 Review:

(1) Name a body fluid checked for changes in enzyme concentration that are caused by injury or disease.
(2) What remains constant in all of the LDH isoenzymes?

Tying It All Together with a Laboratory Application:

Our stomachs contain gastric juices with a low pH that help to digest the foods we eat, including proteins. Food that has been cooked will probably contain proteins that have been (1) ______ and no longer maintain their native state. The (2) ______ environment of the stomach favors digestion; however, this can be a time consuming process. The digestive enzyme pepsin (3) ______ the rate of digestion of proteins by (4) ______ the activation energy of the reaction. This enzyme does not affect the digestion of lipids because enzymes have (5) ______. The two theories that help explain this phenomenon both refer to the (6) ______ of the enzyme. The (7) ______ says the (6) is rigid, while (8) ______ says the (6) adapts to the substrate.

Pepsin is stored in the inactive form pepsinogen. Pepsinogen is an example of a (9) ______ or ______ because when pepsin is needed, the gastric glands of the stomach secret pepsinogen into the stomach where the change in pH activates the enzyme by cleaving a bond. If pepsin had required a cofactor, which could have been a(n) (10) ______ or a(n) ______, then the inactive form of pepsin would be called a(n) (11) ______.

The enzyme (12) ______ of pepsin can be measured by adding 0.1 g of lean hamburger meat to two test tubes (A & B) and 0.1 g of lard to two additional test tubes (C & D), then adding 3 mL of 0.01 M HCl to test tubes A and C and 3 mL of a 0.01 M HCl solution containing 10 μg pepsin/mL solution to test tubes B and D. The test tubes are then covered and allowed to react for 1 hour at 25°C. The amount of protein digestion can be compared between test tubes (13) ______ and ______ by checking the (14) ______ of the resulting peptides. Test tube (15) ______ should have more digested protein. The amount of lipid digestion can be compared between test tubes (16) ______ and ______ by checking the amounts of

(17) ______ and ______ that were produced. The contents of (18) ______ test tube should have undergone more digestion than the contents of the other test tube. The test tube with the highest enzyme activity should be (19) ______.

To measure the effect of temperature on the activity of this enzyme, several test tubes like test tube (20) ______ should be prepared. They should be placed in water baths with (21) ______ temperatures for (22) ______ amount(s) of time. The activity of this enzyme should (23) ______ until the optimum temperature is reached and then should (24) ______ as the temperature is raised further.

To measure the effect of substrate concentration on the activity of this enzyme, several test tubes like test tube (25) ______ should be prepared; however, the amount of (26) ______ should be varied. The reaction time(s) and temperature(s) should be (27) ______ for the test tubes. The activity of the enzyme should (28) ______ until (29) ______ is reached. This occurs because the enzyme (30) ______.

To measure the effect of enzyme concentration on the activity of this enzyme, several test tubes like test tube (31) ______ should be prepared; however the amount of (32) ______ should be varied. The reaction time(s) and temperature(s) should be (33) ______ for the test tubes. The activity of the enzyme should increase with (34) ______ enzyme concentration.

To measure the effect of pH on the activity of this enzyme, several test tubes like test tube (35) ______ should be prepared; however, the concentration of (36) ______ should be varied. The reaction time(s) and temperature(s) should be (37) ______ for the test tubes. The activity of the enzyme should increase with (38) ______ pH until the (39) ______ pH is reached. A student performs this reaction and observes that at pH values of 6 or higher the enzyme activity appears to be zero. The student hypothesizes the pepsin is inactivated at higher pH values because pepsin is a (40) ______ that has multiple levels of (41) ______ which are necessary for biological function. At high pH values, pepsin is (42) ______.

The addition of a heavy metal ion to a test tube containing hydrochloric acid, pepsin, and lean hamburger meat would (43) ______ the enzyme activity because heavy metal ions are (44) ______. They are not classified as competitive or noncompetitive because they are (45) ______ (44), not (46) ______ (44).

SOLUTIONS FOR THE ADDITIONAL ACTIVITIES

Section 20.1 Review:

(1) Catalytic power is the ability of a catalyst to speed up a reaction by lowering the activation energy.

(2) You might describe it as a pen or a pencil, the color of the outside, the color of the ink or graphite, the status of the eraser, the appearance of the tip, whether or not it has a cap, whether or not it has a clip, the brand, the length, the width, and any other distinguishing characteristic of the writing implement.

(3) Finding a copy of your writing implement is absolute specificity. The writing implement fits all of the criteria of the original.

(4) Finding a writing implement with the same color of ink or graphite is an example of relative specificity. This new writing implement shares common characteristics with the original; however, it is not the original, nor is it a copy of the original.

(5) Enzymes are not self-regulating. The cell controls the regulating action of enzymes.

Section 20.2 Review:

(1) The endings *–in* and *–ase* indicate a substance is an enzyme.

(2) Enzymes are classified according to the types of reactions they catalyze.

(3) A substrate is the substance upon which an enzyme acts.

Section 20.3 Review:

(1) A prosthetic group is tightly bound to, and forms an integral part of, the enzyme structure. A cofactor is weakly bound to the enzyme and is easily separated from the protein structure.

(2) A cofactor is any nonprotein component of an enzyme that is weakly bound to the enzyme and is easily separated from the protein structure. A coenzyme is an organic cofactor. All coenzymes are cofactors, not all cofactors are coenzymes.

(3) An apoenzyme is the protein component of an enzyme that requires a cofactor.

(4) Vitamins serve as the precursors for coenzymes that are needed for some enzymes to function properly.

Section 20.4 Review:

(1) The active site of an enzyme is the structural feature on an enzyme that is identified in both the lock-and-key theory and the induced-fit theory.

(2) The active site in the lock-and-key theory is fixed or rigid and the active site in the induced-fit theory is flexible and adapts to the incoming substrates.

Section 20.5 Review:

(1) The turnover number for an enzyme is the rate at which one molecule of enzyme converts its substrate into product.

(2) An enzyme assay could be used to determine the turnover number of an enzyme.

(3) A research biochemist is more likely to determine enzyme activity as a turnover number.

(4) A hospital lab technician is more likely to determine enzyme activity levels in enzyme international units.

Section 20.6 Review:

(1) The limiting factor in the rate of an enzyme-catalyzed reaction is the concentration of the enzyme.

(2) A maximum rate implies the enzyme molecules are saturated with substrate and cannot generate product any faster under the imposed conditions.

(3) Extreme temperatures and pH values can denature enzymes. Enzymes have optimum temperatures and pH values.

Section 20.7 Review:

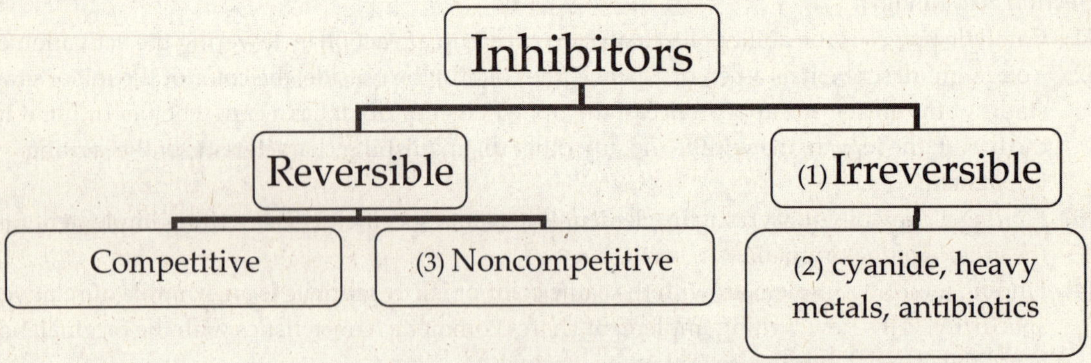

(4) Competitive inhibitors compete with the substrate for the active site of the enzyme.

Section 20.8 Review:

(1) Another name for zymogens is proenzymes.

(2) Zymogens must be released from storage and activated at the location of the reaction. Activation involves the cleavage of one or more peptide bonds of the zymogen. Apoenzymes require a cofactor in order to form an active enzyme.

(3) Modulators are either activators or inhibitors.

(4) The *–steric* part of the name "allosteric enzymes" might indicate they have distinctive binding sites for modulators because it reflects the same root word found in stereochemistry and stereoisomers that indicates a specific three dimensional structure.
(5) Nucleic acids are the biomolecules responsible for genetic control of enzymes.

Section 20.9 Review:
(1) Blood is a body fluid checked for changes in enzyme concentration that are caused by injury or disease.
(2) All of the LDH isoenzymes contain 4 subunits.

Tying It All Together with a Laboratory Application:
(1) denatured
(2) acidic
(3) increases
(4) decreasing
(5) specificity
(6) active site
(7) lock-and-key theory
(8) induced-fit theory
(9) zymogen, proenzyme
(10) coenzyme, inorganic ion
(11) apoenzyme
(12) activity
(13) A, B
(14) size (concentration, amount, molecular weight)
(15) B
(16) C, D
(17) glycerol, fatty acids
(18) neither
(19) B
(20) B
(21) different
(22) the same
(23) increase
(24) decrease
(25) B
(26) lean hamburger
(27) constant
(28) increase
(29) V_{max}
(30) saturates
(31) B
(32) enzyme
(33) constant
(34) increasing
(35) B
(36) hydrochloric acid
(37) constant
(38) increasing
(39) optimum
(40) protein
(41) structure
(42) denaturated
(43) decrease
(44) inhibitors
(45) irreversible
(46) reversible

SELF-TEST QUESTIONS

Multiple Choice

1. Enzymes which act upon only one substrate exhibit:
   a. catalytic specificity.
   b. relative specificity.
   c. binding specificity.
   d. absolute specificity.

2. Which of the following enzyme properties is explained by the lock-and-key model?
   a. specificity
   b. high molecular weight
   c. high turnover rate
   d. susceptibility to denatuartion

3. The induced-fit theory of enzyme action extends the lock-and-key theory in which of the following ways?
   a. assumes enzymes and substrates are rigid
   b. assumes the shape of the substrates changes (conforms) to fit the enzyme
   c. assumes enzymes have no active site
   d. assumes the enzyme shape changes to accommodate the substrate

4. Enzyme turnover numbers are expressed in:
   a. activity/mg.
   b. units/minute.
   c. units/mg.
   d. molecules/minute.

5. Under saturating conditions, an enzyme-catalyzed reaction has a velocity, v. Which of the following would increase the rate of the reaction?
   a. a decrease in the substrate concentration
   b. an increase in the substrate concentration
   c. a decrease in the enzyme concentration
   d. an increase in the enzyme concentration

6. Which of the following has the least effect on the rate of an enzyme-catalyzed reaction?
   a. the pressure of the reaction mixture
   b. the pH of the reaction mixture
   c. the temperature of the reaction mixture
   d. the enzyme concentration in the reaction mixture

7. In noncompetitive inhibition,
   a. substrate and inhibitor bind at separate locations on the enzyme.
   b. the inhibitor forms a covalent bond with the enzyme.
   c. substrate and inhibitor bind at the same location on the enzyme.
   d. the enzyme becomes permanently inactivated.

8. The use of ethanol as a treatment for ethylene glycol poisoning is an excellent example of:
   a. feedback inhibition.
   b. noncompetitive inhibition.
   c. competitive inhibition.
   d. enzyme denaturation.

9. An enzyme is sometimes generated initially in an inactive form called:
   a. a coenzyme.
   b. an activator.
   c. a cofactor.
   d. a zymogen.

10. Enzymes which exist in more than one form are called:
    a. proenzymes.
    b. apoenzymes.
    c. isoenzymes.
    d. allosteric enzymes.

11. Which of the following is an enzyme whose activity is changed by the binding of modulators?
    a. isoenzymes
    b. allosteric enzymes
    c. proenzymes
    d. apoenzymes

Matching

Select an enzyme that matches each description on the left.

12. name is based only on nature of reaction catalyzed
13. name is based on enzyme substrate
14. name is based on both the substrate and the nature of the reaction catalyzed
15. name is an early nonsystematic type

a. pepsin
b. oxidase
c. succinate dehydrogenase
d. sucrase

Match the enzyme components below to the correct term given as a response.

16. the protein portion of an enzyme
17. an organic, but nonprotein portion of an enzyme
18. a nonprotein molecule or ion required by an enzyme

a. cofactor
b. apoenzyme
c. coenzyme
d. proenzyme

Match each description on the left with a response from the right.

19. reacts with cytochrome oxidase
20. stops bacterial synthesis of folic acid
21. inhibits succinate dehydrogenase
22. inhibits bacterial cell-wall construction

a. penicillin
b. cyanide
c. sulfanilamide
d. malonate

Match the correct disease or condition given as a response to the enzyme useful in diagnosing the disease or condition.

23. amylase
24. lysozyme
25. acid phosphatase

a. infectious hepatitis
b. pancreatic diseases
c. prostate cancer
d. monocytic leukemia

True-False

26. The presence of enzymes enables reactions to proceed at lower temperatures.
27. Enzymes are destroyed during chemical reactions in which they participate.
28. Some enzymes are carbohydrates.
29. All enzymes act only on a single substance.
30. Coenzymes are always organic compounds.
31. The optimum pH of most of the enzymes in the human body is near pH 7.
32. A plot of reaction rate (vertical axis) versus temperature of enzyme-catalyzed reactions most frequently yields a straight line.
33. Enzyme induction is an example of genetic control of enzyme activity.
34. Lactate dehydrogenase exists in five different isomeric forms.
35. The enzyme-substrate complex always produces an enzyme and a product.

Chapter 21: Nucleic Acids and Protein Synthesis

CHAPTER OUTLINE

21.1 Components of Nucleic Acids
21.2 The Structure of DNA
21.3 DNA Replication
21.4 Ribonucleic Acid (RNA)
21.5 The Flow of Genetic Information
21.6 Transcription: RNA Synthesis
21.7 The Genetic Code
21.8 Translation and Protein Synthesis
21.9 Mutations
21.10 Recombinant DNA

LEARNING OBJECTIVES/ASSESSMENT

When you have completed your study of this chapter, you should be able to:

1. Identify the components of nucleotides and correctly classify the sugars and bases. (Section 21.1; Exercises 21.2 and 21.4)
2. Describe the structure of DNA. (Section 21.2; Exercise 21.10)
3. Outline the process of DNA replication. (Section 21.3; Exercise 21.20)
4. Contrast the structures of DNA and RNA and list the function of the three types of cellular RNA. (Section 21.4; Exercises 21.26 and 21.28)
5. Describe what is meant by the terms *transcription* and *translation*. (Section 21.5; Exercise 21.32)
6. Describe the process by which RNA is synthesized in cells. (Section 21.6; Exercise 21.34)
7. Explain how the genetic code functions in the flow of genetic information. (Section 21.7; Exercise 21.38)
8. Outline the process by which proteins are synthesized in cells. (Section 21.8; Exercise 21.44)
9. Describe how genetic mutations occur and how they influence organisms. (Section 21.9; Exercise 21.48)
10. Describe the technology used to produce recombinant DNA. (Section 21.10; Exercise 21.52)

SOLUTIONS FOR THE END OF CHAPTER EXERCISES

COMPONENTS OF NUCLEIC ACIDS (SECTION 21.1)

☑21.2 DNA contains 2-deoxyribose. RNA contains ribose.

☑21.4

a.	guanine	purine
b.	thymine	pyrimidine
c.	uracil	pyrimidine
d.	cytosine	pyrimidine
e.	adenine	purine

21.6 thymidine 5′-monophosphate

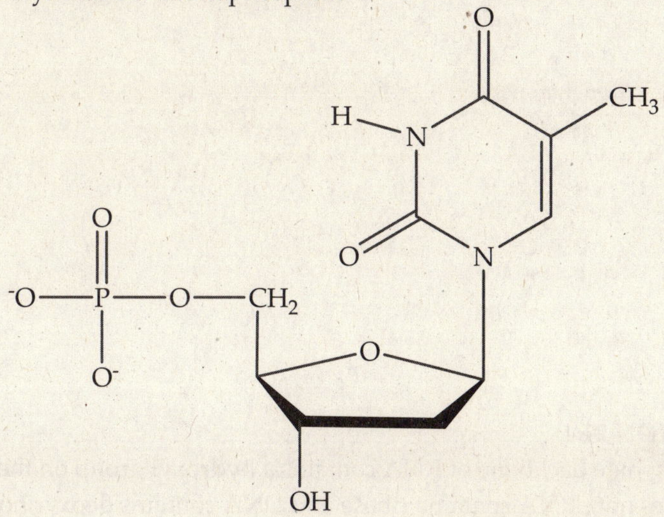

THE STRUCTURE OF DNA (SECTION 21.2)

21.8 The 3′ end of the DNA segment AGTCAT is T.
The 5′ end of the DNA segment AGTCAT is A.

☑21.10 The secondary structure of DNA proposed by Watson and Crick is two complementary polynucleotide chains held together by hydrogen bonds between the A/T and C/G base pairs. The resulting structure is frequently called the double helix..

21.12 a. CAGTAG 15 hydrogen bonds b. TTGACA 14 hydrogen bonds

21.14 5′ ATGCATC 3′ original strand
3′ TACGTAG 5′ complementary strand

5′ GATGCAT 3′ complementary strand read 5′→3′

DNA REPLICATION (SECTION 21.3)

21.16 A chromosome is a structure within a cell that contains one molecule of DNA coiled about small, basic proteins called histones. Each human cell contains 46 chromosomes. The approximate number of genes in a human cell is 25,000.

21.18 A replication fork is the point along the DNA helix at which the DNA helix unwinds for replication to begin.

☑21.20 The steps for DNA replication are: (1) unwinding the double helix, (2) synthesizing DNA segments, and (3) closing the nicks.

21.22 The new DNA strand is formed from the 5′ end to the 3′ end.

21.24

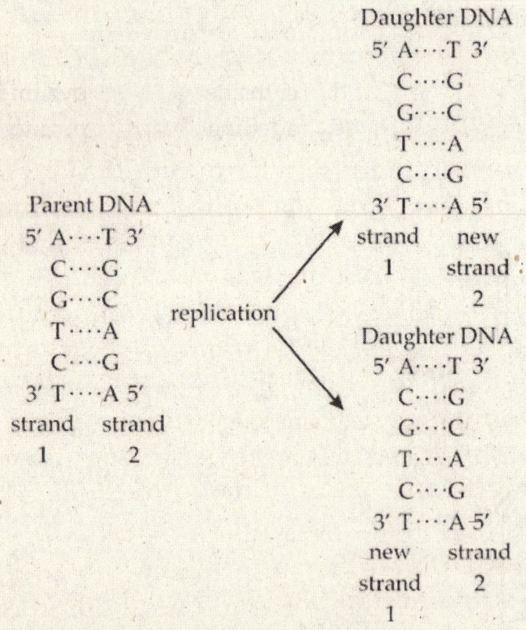

RIBONUCLEIC ACID (RNA) (SECTION 21.4)

☑21.26 The sugar in the sugar phosphate backbone of RNA contains a hydroxyl group on the 2′ carbon, while in DNA it does not. RNA contains ribose and DNA contains deoxyribose.

☑21.28 Messenger RNA functions as a carrier of genetic information from the DNA of the cell nucleus directly to the cytoplasm. It contains 75-3000 nucleotides.
Ribosomal RNA is located in the cytoplasm of the cell in organelles called ribosomes which serve as the sites of protein synthesis. Ribosomal RNA may contain 120, 1700, or 3700 nucleotides.
Transfer RNA delivers amino acids to the site of protein synthesis. Strands of transfer RNA contain 73-93 nucleotides.

21.30 The two important regions of the tRNA molecule are the anticodon that binds to the mRNA and the 3′ end that binds to an amino acid.

THE FLOW OF GENETIC INFORMATION (SECTION 21.5)

☑21.32 Transcription is the transfer of necessary information from a DNA molecule onto a molecule of messenger RNA.
Translation is the decoding of the messenger RNA into an amino acid sequence, resulting in the synthesis of a specific protein.

TRANSCRIPTION: RNA SYNTHESIS (SECTION 21.6)

☑21.34 5′ GCCATATTG 3′ DNA template
3′ CGGUAUAAC5′ mRNA

5′ CAAUAUGGC 3′ mRNA read 5′→3′

21.36 Introns are segments of DNA that do not carry amino acid code. Exons are segments of DNA that carry an amino acid code. hn-RNA is heterogeneous nuclear RNA. It is formed when both introns and exons are transcribed.

THE GENETIC CODE (SECTION 21.7)

☑21.38

a.	UAU	5′ AUA 3′	tyrosine	c.	UCA	5′ UGA 3′	serine
b.	CAU	5′ AUG 3′	histidine	d.	UCU	5′ AGA 3′	serine

21.40
a. False Each codon is composed of ~~four~~ bases. (Correction: not four bases, three bases)
b. True Some amino acids are represented by more than one codon.
c. False ~~All~~ codons represent an amino acid. (Correction: Most, not all; some codons represent stop signals)
d. False Each living species is thought to ~~have its own unique genetic code~~. (Correction: share the same genetic code, not have its own unique genetic code)
e. True The codon AUG at the beginning of a sequence is a signal for protein synthesis to begin at that codon.
f. False ~~It is not known whether or not~~ the code contains stop signals for protein synthesis. (Correction: UAG, UAA, and UGA are the known stop signals for protein synthesis.)

TRANSLATION AND PROTEIN SYNTHESIS (SECTION 21.8)

21.42 A polysome is the configuration of several ribosomes simultaneously translating the same mRNA.

☑21.44 The stages of protein synthesis are: (1) initiation of the polypeptide chain, (2) elongation of the chain, and (3) termination of polypeptide synthesis.

21.46 The A site is the aminoacyl site located on the mRNA-ribosome complex next to the P site where an incoming tRNA carrying the next amino acid will bond. The P site is the peptidyl site and it is the location on the ribosome that the initiating codon must occupy for the initiation process to begin.

MUTATIONS (SECTION 21.9)

☑21.48 A genetic mutation is a change in the genetic code that results from an incorrect sequence of bases on a DNA molecule.

21.50 The result of a genetic mutation that causes the mRNA sequence GCC to be replaced by CCC is that instead of the amino acid alanine, the polypeptide formed will contain the amino acid proline.

RECOMBINANT DNA (SECTION 21.10)

☑21.52 Recombinant DNA contains segments of DNA from two different organisms.

21.54 Substances likely to be produced on a large scale by genetic engineering are: human insulin (needed by diabetics allergic to insulin from pigs or cattle), human growth hormone (treatment of dwarfism), interferon (needed to fight viral infections), hepatitis B vaccine (protection against hepatitis), malaria vaccine (protection against malaria), and tissue plasminogen activator (needed to promote the dissolving of blood clots).

ADDITIONAL EXERCISES

21.56 A hydrolysis reaction could be used to cleave the phosphate ester bonds that join two nucleotides together and excise introns from hnRNA. Esterification of the phosphoric acid on one nucleotide with the alcohol group on the third carbon in another nucleotide could be used to join the exon segments together.

21.58
a. The constituent monosaccharides of sucrose are fructose and glucose.
b. Enzyme Q is controlled by enzyme induction because the production of enzyme Q is controlled by the presence of sucrose.
c. When the sucrose is introduced, the ten molecules of enzyme Q initially present all begin to hydrolyze the sucrose. As the sucrose undergoes hydrolysis, the monosaccharides are separated and can be used to provide energy for the production of additional enzyme Q molecules. The production of enzyme Q may also be an equilibrium process in which more enzyme Q is produced when sucrose is present because, as the enzyme Q molecules form an enzyme-substrate complex with the sucrose molecules, the enzyme Q molecule concentration decreases, and in order to reestablish equilibrium, more enzyme Q molecules are produced. As quickly as the enzyme Q molecules are produced, though, they are "removed" because they form an enzyme-substrate complex with the sucrose; therefore, as long an unhydrolyzed sucrose is present, more enzyme Q molecules are produced.

ALLIED HEALTH EXAM CONNECTION

21.60 The order of information transfer according to the central dogma of molecular biology is DNA, RNA, and proteins. DNA is transcribed into RNA which is translated into protein.

21.62
a. Plasmids are the circular pieces of bacterial DNA into which genes are inserted to form recombinant DNA.

b. Restriction enzymes are the enzymes that cleave a gene and a plasmid so that both pieces have "sticky ends" which can be joined together to form recombinant DNA.

c. DNA ligase is the enzyme that joins the "sticky ends" of the gene and the plasmid together to form recombinant DNA.

21.64 (a) Thymine is found in DNA but not in RNA.

21.66 The site of protein synthesis within a eukaryotic cell is (a) the ribosomes.

CHEMISTRY FOR THOUGHT

21.68 Some benefits of genetic engineering include improved availability of vaccines, enzymes and hormones for humans and improved production of agricultural food products. Some concerns about genetic engineering include the long-term health effects of overusing these enzymes and the effects of the agricultural hormone residues on humans.

21.70 If DNA replication were not semiconservative, another possible structure for the daughter DNA molecules would have the reverse sequence of bases.

21.72 AZT, a drug used to fight HIV, is believed to act as an enzyme inhibitor. The type of enzyme inhibition most likely caused by this drug is competitive inhibition. AZT is similar to the substrate of the virus enzyme and thus most likely would be a competitor for the enzyme.

21.74 Original attempts to modify plants date back to the Egyptians and possibly earlier. Mendel (1865) was the first to formally study the genetics of pea plants.

21.76 If DNA were single stranded, it would fold (like RNA), making it difficult to access specific genetic information; it would not be as chemically stable; it would be much less resistant to mutation; and it could not be replicated using the current mechanism.

ADDITIONAL ACTIVITIES

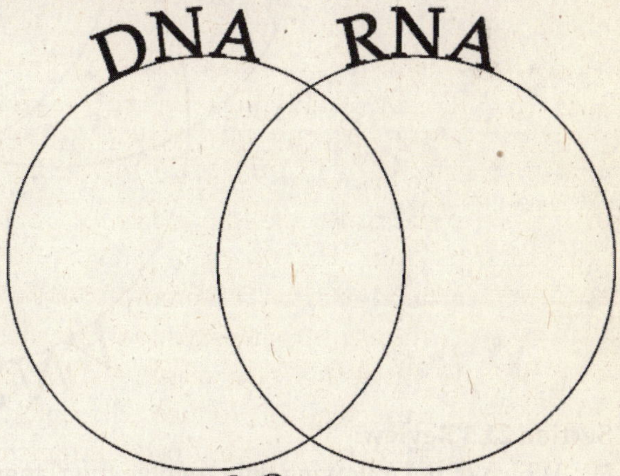

Section 21.1 Review:

(1) Insert the following words into the appropriate area of the diagram on the right: only found in the cell nucleus, in cytoplasm, contains nucleotides, polymer, uracil, thymine, cytosine, adenine, guanine, D-ribose, and D-deoxyribose.

(2) Insert the bases uracil, thymine, cytosine, adenine, and guanine into the appropriate area of the pyrimidine/purine diagram to the right.

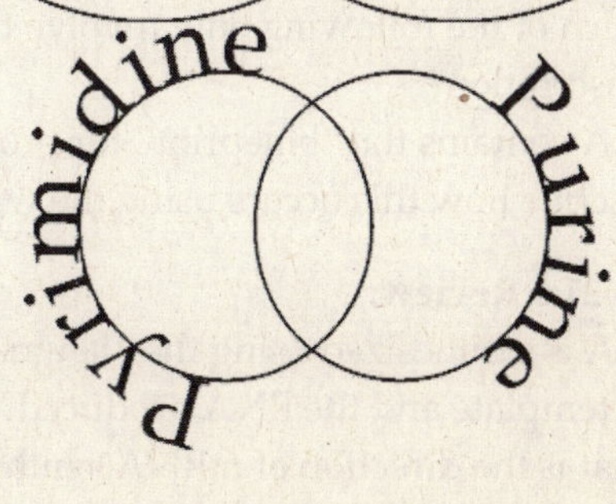

(3) Number the carbon atoms in the structure below.

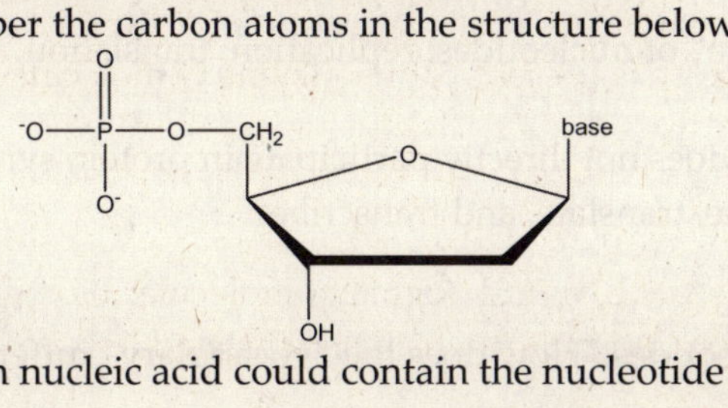

(4) Which nucleic acid could contain the nucleotide in (3)?

Section 21.2 Review:

(1) Why are the linkages between nucleotides called *phosphodiester bonds*?
(2) Why is the sugar-phosphate chain called the *nucleic acid backbone*?
(3) How can the 5′ end of a nucleotide be identified?
(4) How can the 3′ end of a nucleotide be identified?
(5) Why are the percentages of guanine and cytosine in DNA always the same?

Section 21.3 Review:

(1) Rank these components in order increasing size: cell, DNA, chromosome, gene.
(2) What is semiconservative replication?
(3) Why does one DNA daughter strand not have any nicks, but the other one does?
(4) PCR requires heating and cooling DNA repeatedly. How can the human body replicate DNA without these extreme temperatures?

Section 21.4 Review:

Complete the diagram below by adding the following phrases in the appropriate location:

80-85% of RNA in the cell	contains loops	in nucleus	smallest RNA
carries genetic information	at least 20 types	single-stranded	lasts < 1 hr.
present during protein synthesis	has codon	in cytoplasm	has anticodon
component of ribosomes	polymer	deliver amino acids	

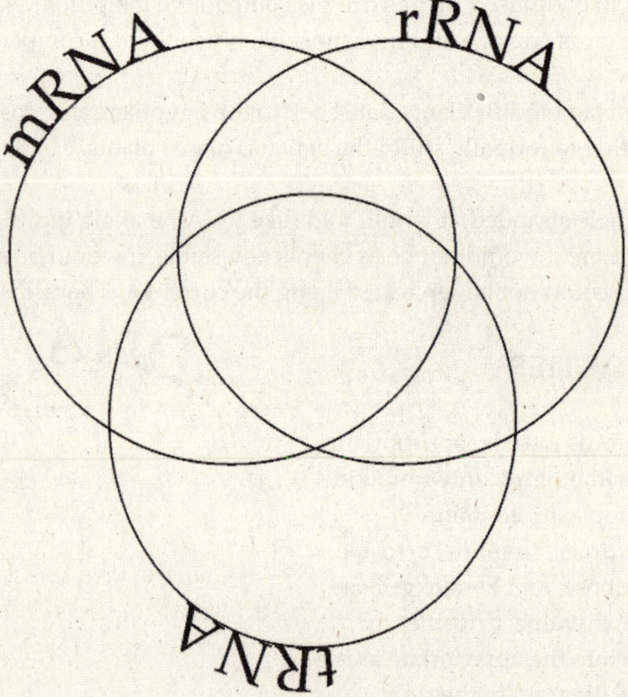

Section 21.5 Review:

(1) Which of the following only involve the "language" of nucleotides: replication, translation, transcription?
(2) DNA contains the "blueprint" for proteins, but it does not directly participate in protein synthesis. Describe how this occurs using the words replicate, translate, and transcribe.

Section 21.6 Review:

(1) RNA is synthesized using the "language" of nucleotides. How does the "vocabulary" differ between the template and the RNA produced?
(2) What is the direction of mRNA synthesis?
(3) How does (2) compare to the direction of daughter DNA strand synthesis?

(4) What is the direction for "reading" a strand of DNA during replication and transciption?
(5) Distinguish between introns and exons.
(6) When hnRNA undergoes enzyme-catalyzed reactions to cut and splice the hnRNA, what type of RNA is produced?

Section 21.7 Review:
Identify each of the descriptions below as degenerate, precise, universal, chain initiation, or chain termination.
(1) The code is almost the same for all living organisms.
(2) Each codon represents only one amino acid or stop.
(3) The codon required for protein synthesis to begin.
(4) One amino acid may be represented by several codons.
(5) One of the codons required for protein synthesis to end.

Section 21.8 Review:
(1) Protein synthesis occurs in three major steps: start, growth, stop. What are the scientific names for these steps?
(2) What is the direction of protein growth?
(3) Which end of the mRNA is closest to the P site of the ribosome during the "start" step of protein synthesis?
(4) What is *translocation*?

Section 21.9 Review:
(1) Will all mutations cause a difference in the synthesized protein? Explain.
(2) What are the causes of mutations?

Section 21.10 Review:
(1) What are the major types of enzymes used in genetic engineering?
(2) What is the difference between a vector and a plasmid?
(3) What are the steps for forming recombinant DNA?

Tying It All Together with a Laboratory Application:
The formation of (1) ______ DNA uses the DNA from two different organisms. The formation of (1) DNA is possible because the genetic code is (2) ______. When (1) DNA undergoes replication, (3) ______ are formed. This is a (4) ______ process because each of the "new" molecules contains (5) ______ and ______.

When (1) DNA undergoes transcription, (6) ______ is formed. (6) can then be used for (7) ______ synthesis. (8) ______ binds to the amino acids and delivers them to (6) which is "complexed" with a(n) (9) ______. The (10) ______ on the (6) only allows the (8) with the correct (11) ______ to leave an amino acid.

In order to synthesize (1) DNA, a(n) (12) ______ must be isolated from one organism using a(n) (13) ______. The (13) is also used on a(n) (14) ______, like a plasmid. Then, (15) ______ is used to join the (16) ______ of the fragments. A technique, like (17) ______, could be used to copy the (1) DNA. In addition to the (1) DNA, DNA polymerase, the cofactor $MgCl_2$, the four (18) ______ building blocks, and primers will be needed. The (18)s contain a phosphate group, (19) ______ (a sugar), and (20) ______, ______, ______, or ______ (nitrogen-containing bases).

SOLUTIONS FOR THE ADDITIONAL ACTIVITIES

Section 21.1 Review:

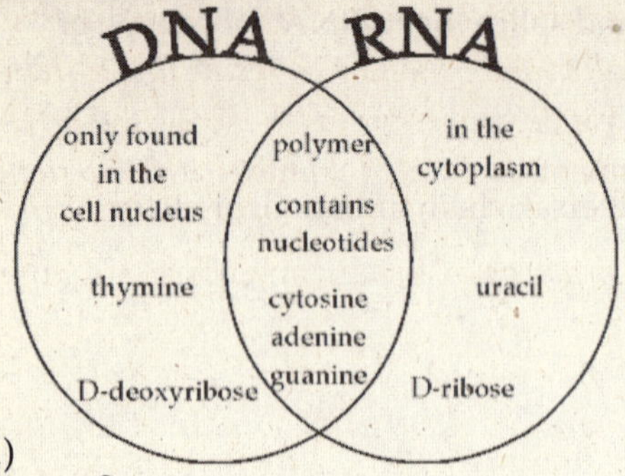

(1)

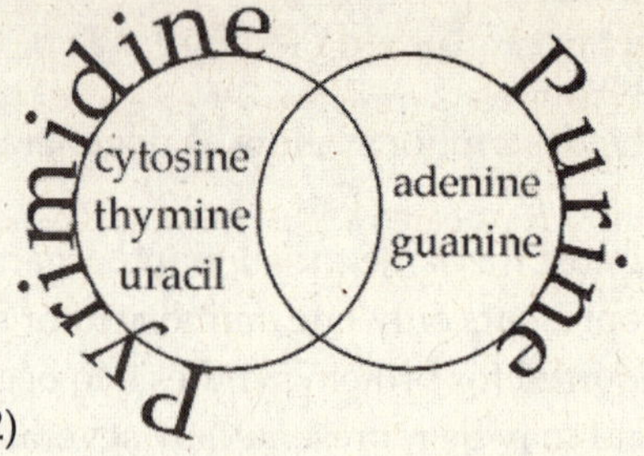

(2)

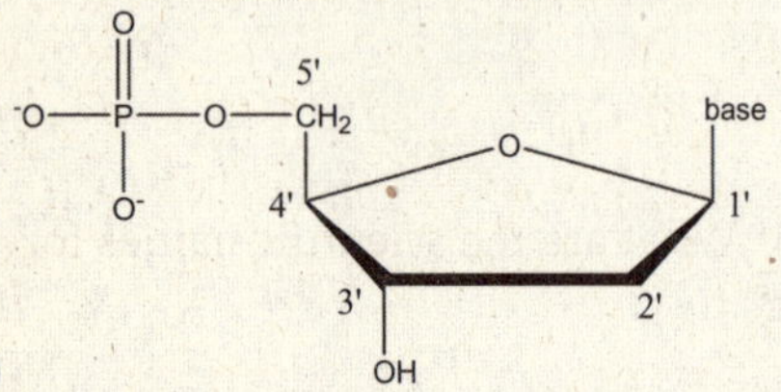

(3)

(4) DNA

Section 21.2 Review:

(1) The linkages between nucleotides are called *phosphodiester bonds* because a phosphate is bonded to two sugars through oxygen atoms; therefore, two phosphate ester bonds are present.

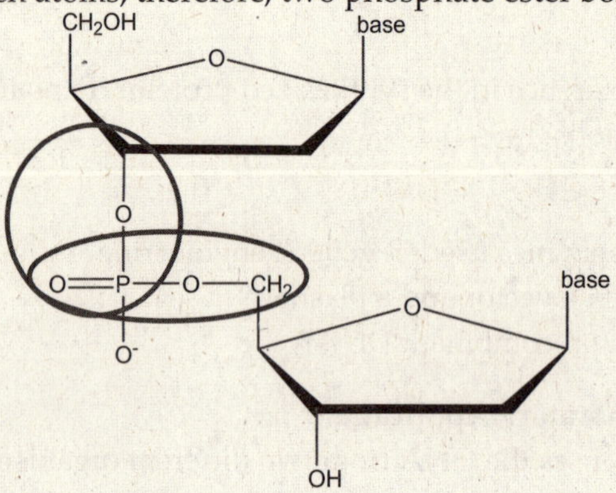

(2) The sugar-phosphate chain is called the *nucleic acid backbone* because it is the repeating unit of the polymer. The bases are the side chains.

(3) The 5′ end of a nucleotide be identified as the end with a phosphate group.

(4) The 3′ end of a nucleotide be identified as the end with a sugar group.

(5) DNA contains two complementary strands hydrogen bonded to form a double helix. The hydrogen bonds are formed between complementary base pairs. Guanine and cytosine are complementary base pairs; therefore, DNA always contains the same percentages of these bases.

Section 21.3 Review:

(1)

gene

DNA

chromosome

cell

(2) Semiconservative replication is the process by which one molecule of DNA is used to produce two "new" molecules of DNA. Each of the "new" DNA molecules

contains 1 new strand (called "daughter") and 1 old strand (called "parent"). The process is considered semiconservative because each new molecule contains half of the old molecule.

(3) As the DNA unwinds at a replication fork, both of the parent strands are read as templates from the 3′ end to the 5′ end. The daughter strands grow from the 5′ end to the 3′ end. The template strands are antiparallel to each other; therefore, one of the daughter strands grows continuously as the DNA parent strands unwind while the other grows in Okazaki fragments separated by nicks as the parent strands unwind.

(4) The human body replicates DNA without these extreme temperatures by using enzymes to aid in the replication process.

Section 21.4 Review:

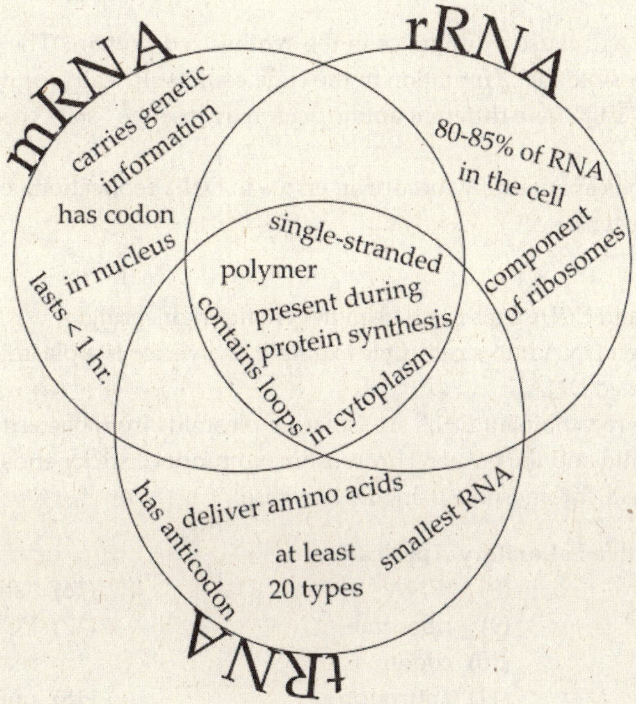

Section 21.5 Review:

(1) Replication and transcription only involve the "language" of nucleotides. Replication is the formation of new DNA molecules from old DNA molecules. Transcription is the process of copying segments of DNA into mRNA that can leave the nucleus.

(2) DNA is replicated within the nucleus. When a protein is needed, the DNA unwinds and is transcribed into mRNA. The mRNA leaves the nucleus and enters the cytoplasm in order to align with a ribosome. The tRNA delivers amino acids to the mRNA-ribosome complex and the mRNA is translated into a protein.

Section 21.6 Review:

(1) The DNA template has a sugar-phosphate backbone that contains D-deoxyribose rather than the D-ribose of the synthesized RNA. Also, the DNA contains thymine and does not contain uracil, while the synthesized RNA contains uracil and does not contain thymine.

(2) A molecule of mRNA is synthesized from 5′ end to 3′ end.

(3) DNA daughter strand synthesis also occurs from the 5′ end to the 3′ end.

(4) A strand of DNA is read from the 3′ end to the 5′ end during replication and transcription.

(5) Introns do not carry genetic code and exons carry genetic code.

(6) mRNA is produced when hnRNA undergoes enzyme-catalyzed reactions to cut and splice the hnRNA.

Section 21.7 Review:

(1) universal; (2) precise; (3) chain initiation; (4) degenerate; (5) chain termination

Section 21.8 Review:

(1) Start = initiation of the polypeptide chain; growth = elongation of the chain; stop = termination of polypeptide synthesis
(2) Proteins are synthesized from the N-terminal end to the C-terminal end.
(3) The 5′ end of the mRNA is closest to the P site of the ribosome during the "start" step of protein synthesis.
(4) *Translocation* is the movement of the ribosome along the mRNA during protein synthesis.

Section 21.9 Review:

(1) No, not all mutations will cause a difference in the synthesized protein. The genetic code is degenerate, so it is possible that a mutation in the code could still "call" for the same amino acid. Some mutations will "call" for a different amino acid, may lose the "start" codon, or may lose the "stop" codon.
(2) The causes of mutations are naturally occurring errors in DNA replication, ionizing radiation, and chemicals called mutagens.

Section 21.10 Review:

(1) Restriction enzymes and DNA ligases are used in genetic engineering.
(2) A vector carries foreign DNA into a cell. One example of a vector is a plasmid which is a circular piece of double-stranded DNA.
(3) The steps for forming recombinant DNA are isolating plasmids from bacterium, using restriction enzymes on the plasmid and the human chromosome to produce sticky ends, and splicing the plasmid and the human chromosome using DNA ligase.

Tying It All Together with a Laboratory Application:

(1) recombinant
(2) universal
(3) 2 molecules of DNA
(4) semiconservative
(5) 1 strand of parent DNA, 1 strand of daughter DNA
(6) mRNA
(7) protein
(8) tRNA
(9) ribosome
(10) codon
(11) anticodon
(12) gene
(13) restriction enzyme
(14) vector
(15) DNA ligase
(16) "sticky ends"
(17) PCR (polymerase chain reaction)
(18) nucleotide
(19) D-deoxyribose
(20) adenine, cytosine, guanine, thymine

SELF-TEST QUESTIONS

Multiple Choice

1. One of the primary differences between DNA and RNA involves:
   a. the phosphate.
   b. the pentose sugar.
   c. the bonding between sugar and base.
   d. the bonding between sugar and phosphate.

2. Which of the following bases is a purine?
   a. uracil
   b. adenine
   c. thymine
   d. cytosine

3. A sample of nucleic acid is to be analyzed. For which of the following would an analysis be useful in deciding whether the sample was DNA or RNA?
   a. guanine
   b. cytosine
   c. adenine
   d. thymine

4. The structural backbone of all nucleic acids consists of alternating molecules of:
   a. a sugar and a base.
   b. a sugar and a phosphate.
   c. a base and a phosphate.
   d. purine and pyrimidine bases.

5. If the nucleotide sequence in one strand of DNA were T-C-G, what would the complementary strand written in 5′-3′ sequence be?
   a. A-C-G
   b. T-G-C
   c. C-G-A
   d. A-G-C

6. A sample of double helical DNA contained 26% of the base thymine (T). What should the percentage of adenine (A) be?
   a. 26%
   b. 24%
   c. 22%
   d. 20%

7. How many hydrogen bonds are between C and G in double-stranded helical DNA?
   a. 0
   b. 1
   c. 2
   d. 3

8. What is mRNA synthesis called?
   a. replication
   b. translocation
   c. translation
   d. transcription

9. Histidine is represented by the codons CAC and CAU. This is an example of:
   a. degeneracy of the genetic code.
   b. punctuation in the genetic code.
   c. a nonsense codon.
   d. the universal nature of the genetic code.

10. The genetic code consists of ______ three-letter codons.
    a. 16
    b. 32
    c. 64
    d. 86

11. What are segments of eukaryotic DNA that code for amino acids called?
    a. exons
    b. introns
    c. hnRNA
    d. codons

12. Recombinant DNA is DNA:
    a. formed by combining portions of DNA from two different sources.
    b. found within plasmids.
    c. with an unusual ability to recombine.
    d. which contains two template strands.

Matching

Characteristics are given below for various nucleic acids. Match each description to the correct type of nucleic acid as a response.

13. represents a large percentage of cellular RNA
14. serves as master template for the formation of all nucleic acids in the body
15. contains base sequences called codons
16. has lowest molecular weight of all nucleic acids
17. replicated during cell division
18. delivers amino acids to site of protein synthesis
19. contains anti-codon
20. helps to serve as a site for protein synthesis
21. carries the directions for protein synthesis to the site of protein synthesis

a. DNA
b. mRNA
c. tRNA
d. rRNA

True-False

22. Purine bases contain both a five- and a six-member ring.
23. DNA and RNA contain identical nucleic acid backbones.
24. During DNA replication, only one DNA strand serves as a template for the formation of a complementary strand.
25. A complementary DNA strand formed during replication is identical to the strand serving as a template.
26. RNA is usually double-stranded.
27. Some amino acids are represented by more than one codon.
28. The genetic code has some code words that don't code for any amino acids.
29. The genetic code is thought to be different for humans and bacteria.
30. Incoming tRNA, carrying an amino acid, attaches at the peptidyl site.
31. N-formylmethionine is always the first amino acid in a growing peptide chain in prokaryotic cells.
32. Protein synthesis generally takes place within the cell nucleus.
33. Some genetic mutations might aid an organism rather than hinder it.
34. Viruses contain both nucleic acids and protein.
35. Restriction enzymes act at sites on DNA called palindromes.

Chapter 22: Nutrition and Energy for Life

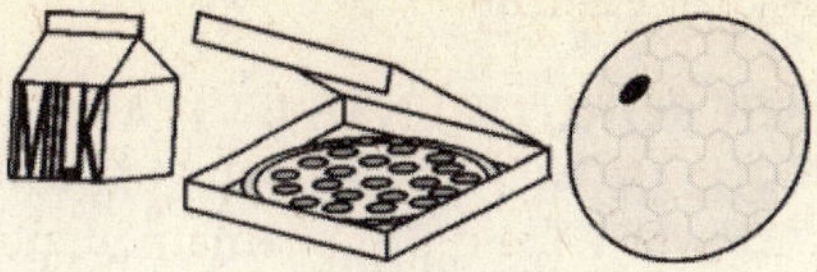

CHAPTER OUTLINE

22.1 Nutritional Requirements
22.2 The Macronutrients
22.3 Micronutrients I: Vitamins
22.4 Micronutrients II: Minerals
22.5 The Flow of Energy in the Biosphere
22.6 Metabolism and an Overview of Energy Production
22.7 ATP: The Primary Energy Carrier
22.8 Important Coenzymes in the Common Catabolic Pathway

LEARNING OBJECTIVES/ASSESSMENT

When you have completed your study of this chapter, you should be able to:

1. Describe the difference between macronutrients and micronutrients in terms of amounts required and their functions in the body. (Section 22.1; Exercise 22.2)
2. Describe the primary functions in the body of each macronutrient. (Section 22.2; Exercise 22.4)
3. Distinguish between and classify vitamins as water-soluble or fat-soluble on the basis of name and behavior in the body. (Section 22.3; Exercise 22.10)
4. List a primary function in the body for each major mineral. (Section 22.4; Exercise 22.16)
5. Describe the major steps in the flow of energy in the biosphere. (Section 22.5; Exercise 22.24)
6. Differentiate among metabolism, anabolism, and catabolism. (Section 22.6; Exercise 22.26)
7. Outline the three stages in the extraction of energy from food. (Section 22.6; Exercise 22.28)
8. Explain how ATP plays a central role in the production and use of cellular energy. (Section 22.7; Exercise 22.34)
9. Explain the role of coenzymes in the common catabolic pathway. (Section 22.8; Exercise 22.46)

SOLUTIONS FOR THE END OF CHAPTER EXERCISES

NUTRITIONAL REQUIREMENTS (SECTION 22.1)

☑22.2 Macronutrients are nutrients needed by the body in gram quantities every day. Micronutrients are nutrients needed by the body in milligram or microgram quantities every day.

THE MACRONUTRIENTS (SECTION 22.2)

☑22.4 Carbohydrates are used for energy and materials for the synthesis of cell and tissue components. Lipids are a concentrated source of energy; they store fat-soluble vitamins and help carry them through the body, and provide essential fatty acids. Lipids are also an essential component of cell membranes. Proteins are used to make new tissues, maintain and repair cells, synthesize enzymes, hormones, and other nitrogen-containing compounds in the body, and provide energy.

22.6

a.	potato chips	carbohydrates, lipids, proteins
b.	buttered toast	carbohydrates, lipids, proteins
c.	plain toast with jam	carbohydrates, proteins
d.	cheese sandwich	carbohydrates, lipids, proteins
e.	a lean steak	proteins, lipids
f.	a fried egg	lipids, proteins, carbohydrates

22.8 The daily values are 65 g of total fat; 20 g of saturated fat, 300 g of total carbohydrates, and 25 g of fiber for a 2000-Calorie diet.

MICRONUTRIENTS I: VITAMINS (SECTION 22.3)

☑22.10 a. tocopherol fat-soluble; c. folic acid water-soluble
b. niacin water-soluble; d. retinol fat-soluble

22.12 Large doses of fat-soluble vitamins are potentially dangerous because excess fat-soluble vitamins will be stored in the body's fat reserves and stay in the body. Fat-soluble vitamins can be harmful in large quantities; however, large doses of water-soluble vitamins will be excreted and any effect due to a higher concentration will be only temporary.

22.14 a. scurvy vitamin C (ascorbic acid); c. pernicious anemia vitamin B_{12} (cobalamin)
b. beriberi vitamin B_1 (thiamin); d. pellagra niacin

MICRONUTRIENTS II: MINERALS (SECTION 22.4)

☑22.16 Calcium is used in building inorganic structural components of bones and teeth.
Phosphorus is used in building inorganic structural components of bones and teeth.
Potassium is found in ionic form, distributed throughout the body's various fluids. It is important for proper function of the nervous system.
Sulfur is used in building inorganic structural components of bones and as a component of proteins.
Sodium is found in ionic form, distributed throughout the body's various fluids. It is important for proper function of the nervous system.
Chloride is found in ionic form, distributed throughout the body's various fluids. It is important for the proper function of the nervous system.
Magnesium is found in ionic form, distributed throughout the body's various fluids. It is important for proper function of the nervous system and is a cofactor for some proteins.

22.18 Trace minerals are components of vitamins, enzymes, hormones, and specialized proteins.

THE FLOW OF ENERGY IN THE BIOSPHERE (SECTION 22.5)

22.20 Hydrogen is the fuel for the fusion process occurring on the sun.

22.22 $6\ CO_2 + 6\ H_2O + \text{energy} \rightarrow C_6H_{12}O_6 + 6\ O_2$

☑22.24 Plants use the energy from the sun to produce glucose. Animals eat the plants that contain glucose. During cellular respiration, glucose and other storage forms of energy are converted into energy. Some of that energy is captured within the cells in the form of ATP.

METABOLISM AND AN OVERVIEW OF ENERGY PRODUCTION (SECTION 22.6)

☑22.26 Catabolism includes all breakdown processes. Anabolism includes all synthesis processes.

☑22.28 a. Formation of ATP Stage III; c. Consumption of O_2 Stage III
b. Digestion of fuel molecules Stage I; d. Generation of acetyl CoA Stage II

22.30 The main purpose of the catabolic pathway is to produce ATP.

ATP: THE PRIMARY ENERGY CARRIER (SECTION 22.7)

22.32 ATP is involved in both anabolic and catabolic processes.

☑22.34 ATP is called the primary energy carrier because the catabolism of food results in the production of ATP which stores energy until it is delivered to cells for their use.

22.36 P*i* stands for inorganic phosphate, or HPO_4^{2-}. PP*i* stands for pyrophosphate, or $P_2O_7^{4-}$.

22.38 Phosphoenolpyruvate is called a high-energy compound because it releases 14.8 kcal per mole. Glycerol 3-phosphate only releases 2.2 kcal per mole and is, therefore, not a high-energy compound.

22.40 The triphosphate end of ATP is the high-energy portion of the ATP molecule. The energy is stored in the bonds within this portion of the ATP molecule.

22.42 Mitochondria serve as the site for most ATP synthesis in the cells.

22.44 Mitochondria have an outer membrane, an intermembrane space, and an inner membrane that forms cristae and the matrix. The enzymes needed for energy production are located within the inner membrane and inside the matrix.

IMPORTANT COENZYMES IN THE COMMON CATABOLIC PATHWAY (SECTION 22.8)

☑22.46
a. coenzyme A — This coenzyme is part of acetyl coenzyme A which is formed from all foods as they pass through Stage II of catabolism. Coenzyme A is, therefore, an acetyl-group carrier or acyl transfer compound.
b. FAD — This coenzyme forms coenzyme $FADH_2$ by being reduced and carries hydrogen atoms from the citric acid cycle to the electron transport chain. FAD is, therefore, an oxidizing agent and it facilitates electron transfer.
c. NAD^+ — This coenzyme forms coenzyme NADH by being reduced and carries hydrogen atoms from the citric acid cycle to the electron transport chain. NAD^+ is an oxidizing agent that facilitates electron transfer.

22.48 $^-OOC\text{-}CH_2\text{-}CH_2\text{-}COO^-$ is oxidized. FAD is reduced.

22.50
a. $HO\text{-}CH_2\text{-}COOH + NAD^+ \rightarrow O{=}CH\text{-}COOH + NADH + H^+$
b. $HO\text{-}CH_2CH_2\text{-}OH + FAD \rightarrow HO\text{-}CH{=}CH\text{-}OH + FADH_2$

22.52

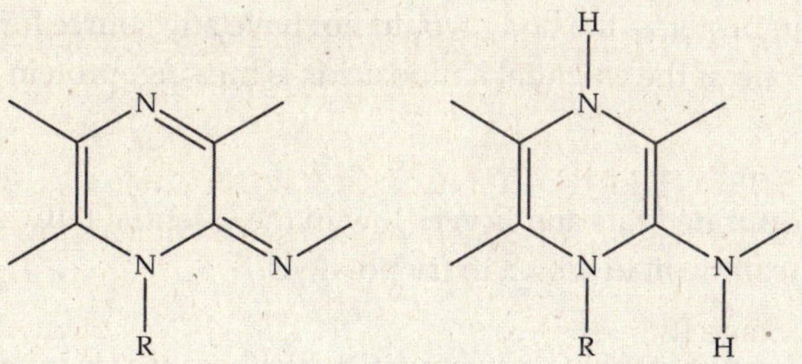

FAD FADH$_2$

ADDITIONAL EXERCISES

22.54 $CH_4 + O_2 + 2\ NAD^+ \rightarrow CO_2 + 2\ NADH + 2\ H^+$

Each NAD^+ that is reduced to NADH accepts two electrons and one proton (H^+); therefore, in this reaction, four electrons are transferred in total to NADH.

22.56 Proteins are needed for the maintenance and repair of cells and for the production of enzymes, hormones, and other important nitrogen-containing compounds of the body. Protein synthesis is the main process in the human body that would be interrupted by a deficiency of an essential amino acid. The elongation of a protein would be impossible if the tRNA could not find the amino acid needed to lengthen the protein chain.

ALLIED HEALTH EXAM CONNECTION

22.58
a. The fat-soluble vitamins include A, D, E, and K.
b. Thiamin is vitamin B_1.
c. Vitamin D helps prevent rickets.
d. Vitamin K helps clot blood.
e. Vitamin B_{12} is useful in combating pernicious anemia.

22.60 Oxygen (a) is not a requirement for photosynthesis, it is produced during photosynthesis. (b) Carbon dioxide, (c) sunlight, and (d) chlorophyll are all required for photosynthesis to occur.

22.62 Fat produces the greatest amount of energy (9 Calories) from 1 gram of macronutrient. Carbohydrates and proteins both produce about the same amount of energy (4 Calories) per gram.

22.64 Amino acids that cannot be manufactured by the body are called (a) essential amino acids.

22.66 Water soluble vitamins include (b) vitamin C.

22.68 $6\ CO_2$ + light energy + $6\ H_2O \rightarrow C_6H_{12}O_6 + 6\ O_2$ represents (c) photosynthesis.

CHEMISTRY FOR THOUGHT

22.70
a. If a diet were extremely low in carbohydrates, the body would have to use stored compounds for fuel, or use proteins in the muscles as a fuel source. This might be beneficial if an individual needed to lose weight, but in the long-term, such a diet can lead to the break-down of important body structures and tissues.
b. If a diet were extremely low in lipids, the body would have no source for the essential fatty acid linoleic acid.
c. If a diet were extremely low in proteins, the body would not have any source for the essential amino acids. If any one of the essential amino acids is missing, protein synthesis cannot occur.

22.72 Palm oil has a high percentage of saturated fats and is very low in the essential fatty acid, linoleic acid; therefore, it is an unhealthy oil in which to fry potatoes.

22.74 An orange might be better for a person that taking a vitamin C tablet because (1) vitamin tablets are not as highly regulated as medications and may not be what they claim to be, (2) the orange contains many other materials that may be of great benefit in addition to the vitamin C, and (3) the orange tastes better.

22.76 Since excess protein cannot be stored in the body, we must include protein in our diet every day, especially proteins containing the essential amino acids.

ADDITIONAL ACTIVITIES

Section 22.1 Review:

(1) What do the prefixes *macro-* and *micro-* mean?
(2) What are the subclasses of macronutrients?
(3) What are the subclasses of micronutrients?
(4) What are the four general categories of materials the human body must receive from diet?
(5) How are Reference Daily Intakes, Daily Reference Values, and Daily Values related?

Section 22.2 Review:

(1) Carbohydrates are the main dietary source of __________.
(2) What is the difference between simple and complex carbohydrates?
(3) Why must lipids be included in a healthy diet?
(4) What type of fat is considered the "wrong type" for healthy eating?
(5) Will a food that is a "complete protein" for a child always be a "complete protein" for an adult? Explain.

Section 22.3 Review:

(1) What is the main role of water-soluble vitamins in the body?
(2) Which water-soluble vitamin is an exception to (1)?
(3) Do the fat-soluble vitamins have the same role in the body as water-soluble vitamins? Explain.
(4) Can a person overdose on vitamins? Explain.

Section 22.4 Review:

(1) What is the difference between major minerals and trace minerals in terms of dietary requirements?
(2) What is the difference between major minerals and trace minerals in terms of functions in the body?
(3) If a patient is told to limit their sodium intake, what seasoning should they avoid?
(4) Does (3) contain elemental sodium? Explain.

Section 22.5 Review:

(1) Classify the following processes as endothermic or exothermic.
    (a) nuclear fusion
    (b) photosynthesis
    (c) cellular respiration
(2) Identify carbon dioxide as a reactant, product, or neither for the following processes.
    (a) nuclear fusion
    (b) photosynthesis
    (c) cellular respiration
(3) Repeat (2) for oxygen.
(4) Repeat (2) for water.

Section 22.6 Review:

(1) Metabolism can be divided into two subgroups as shown below. Label the subgroups with their scientific names.

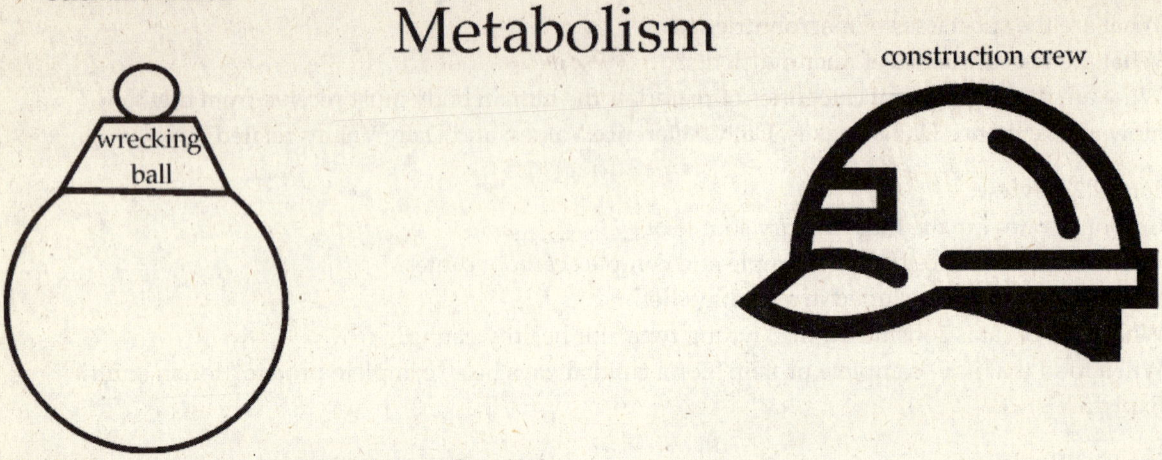

(2) Which of the processes in (1) is exothermic?

(3) Which of the processes in (1) is endothermic?

(4) Using the construction/wrecking ball analogy, describe the three stages of the catabolism of food.

Section 22.7 Review:

(1) Where is the energy stored that is released during the oxidation of macronutrients?

(2) What are the components of ATP?

(3) How does ATP differ from the nucleotide containing adenine?

(4) What is the sign on ΔG for exothermic reactions?

(5) Complete the following diagram for the hydrolysis of ATP.

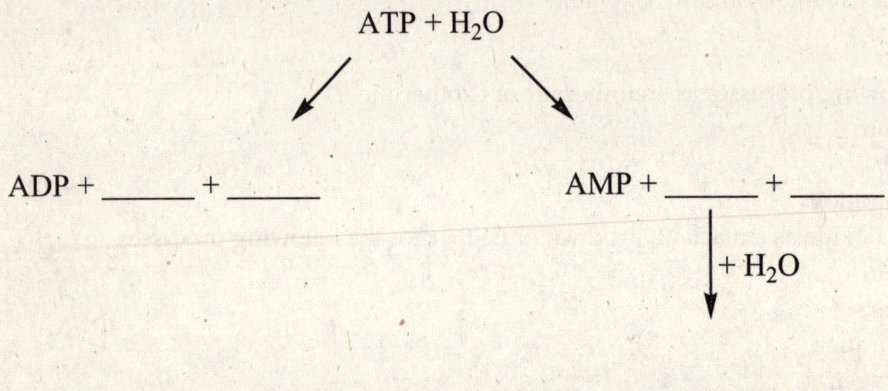

(6) Do the charges balance in each of the hydrolysis processes shown in (5)? Explain.

(7) Which of the processes in (5) produces more energy?

(8) What is one of the important processes that occurs in the mitochondria of a cell?

Section 22.8 Review:

(1) Label the following symbol for Coenzyme A using the information from this section.

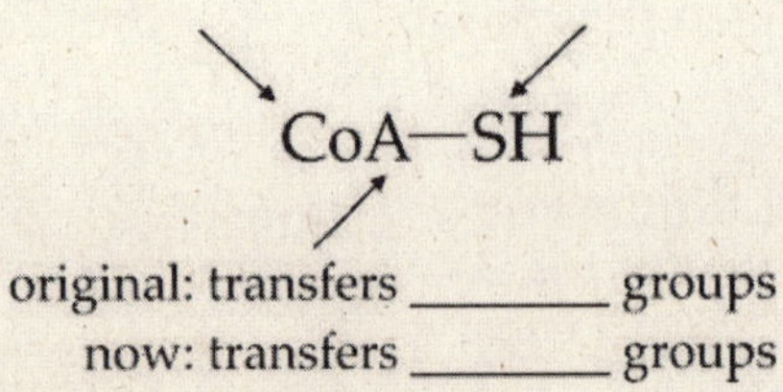

(2) Write a balanced equation for the reduction of oxygen to water by the addition of electrons and hydrogen ions.
(3) Write the equation for the decomposition of NADH into NAD^+, electrons, and hydrogen ions.
(4) Write the equation for the decomposition of $FADH_2$ into FAD, electrons, and hydrogen ions.
(5) Write the balanced equation for the reduction of oxygen to water by the addition of NADH. Include any additional hydrogen ions needed to balance the reaction.
(6) Write the balanced equation for the reduction of oxygen to water by the addition of $FADH_2$. Include any additional hydrogen ions needed to balance the reaction.
(7) If coenzymes are chemical group shuttles, what do the different coenzymes in this section shuttle?

Tying It All Together with a Laboratory Application:

The Calories in one serving of reduced-fat peanut butter, based on the food label on the next page, are (1) ______. Calories come from (2) ______, not from (3) ______, which include vitamins and minerals. The percentage of Calories from fat is (4) ______, carbohydrates is (5) ______, and protein is (6) ______. Reduced-fat peanut butter (7) ______ the recommended percentage of fat in the diet, which is no more than (8) ______. Reduced-fat peanut butter (9) ______ the recommended percentage of carbohydrates in the diet, which is about (10) ______. Reduced-fat peanut butter can still be included as part of a healthy diet, though.

Calories	?
Total Fat	12 g
Saturated Fat	2.5 g
Cholesterol	0 mg
Sodium	210 mg
Total Carbohydrate	15 g
Dietary Fiber	2 g
Sugars	4g
Protein	8 g
Vitamin A	0%
Vitamin C	0%
Calcium	0%
Iron	4%
Niacin	25%
Vitamin B_6	6%
Folic Acid	6%
Magnesium	15%
Zinc	6%
Copper	10%

Reduced-fat peanut butter contains (11) ______ g of complex carbohydrates, not including the fiber. It also contains (12) ______ g of unsaturated fats. If the ratio of unsaturated to saturated fats is higher for reduced-fat peanut butter, then the consistency of reduced-fat peanut butter may be more (13) ______ than regular peanut butter. The daily value percentage can be listed for all of the nutritional categories except sugars and proteins. The Daily Value for protein is called (14) ______, but it cannot be turned into a universal percentage because (15) ______.

Reduced-fat peanut butter might contain (16) ______ amino acids, which can only be obtained through diet; however, the labeling laws do not require that (16) amino acids be identified on product packaging. Reduced-fat peanut butter could be considered a source of (17) ______ protein for some people, if it contains all the (16) amino acids in the proper proportions needed by the body.

Vitamin A is a (18) ______-soluble vitamin, unlike Vitamin C which is a (19) ______-soluble vitamin. Neither vitamin A or vitamin C is contained in reduced-fat peanut butter. The magnesium and copper in reduced-fat peanut butter (20) ______ in metallic form, which is common for (21) ______. Magnesium is a (22) ______ mineral, while copper is a (23) ______ mineral.

The energy stored in peanut-butter originally came from (24) ______, where nuclear fusion occurs. The sunlight allows carbon dioxide and water to be converted through (25) ______ into glucose and oxygen. The glucose is then converted into other (26) ______ (storage or mechanical) forms of energy.

When a person eats reduced-fat peanut butter, the body digests the peanut butter through (27) ______ processes that result in (28) ______ at the end of the common (27) pathway. This pathway utilizes a(n) (29) ______ called CoA to (30) ______ acyl groups.

SOLUTIONS FOR THE ADDITIONAL ACTIVITIES

Section 22.1 Review:

(1) *Macro-* means large. *Micro-* means very small.
(2) The subclasses of macronutrients are carbohydrates, proteins, and lipids.
(3) The subclasses of micronutrients are vitamins and minerals.
(4) The four general categories of materials the human body must receive from diet are macronutrients, micronutrients, water, and fiber.
(5) Reference Daily Intakes are the Daily Values for protein as well as 19 vitamins and minerals. Daily Reference Values are the Daily Values for other nutrients of public health importance.

Section 22.2 Review:

(1) energy
(2) Simple carbohydrates are monosaccharides and disaccharides. Complex carbohydrates are polysaccharides.
(3) Lipids are a necessary part of the diet as a source of energy, source and transportation for fat-soluble vitamins, and source of essential fatty acids.
(4) Saturated fat is considered the "wrong type" for healthy eating.
(5) A food that is a "complete protein" for a child will not always be a "complete protein" for an adult because a complete protein contains all of the essential amino acids in the proper proportions needed by the body. The daily needs of each essential amino acid will vary by individual.

Section 22.3 Review:

(1) Water-soluble vitamins serve as coenzymes in the body.
(2) Vitamin C does not function as a coenzyme. It participates in the synthesis of collagen for connective tissue.
(3) The fat-soluble vitamins do not have the same role in the body as water-soluble vitamins. They serve many different roles: participation in the synthesis of visual pigments, regulation of calcium and phosphorus metabolism, prevention of oxidation of vitamin A and fatty acids, synthesis of blood-clotting substances.
(4) A person can overdose on vitamins, especially fat-soluble vitamins. The water-soluble vitamins can be removed from the body by the kidneys; however, the fat-soluble vitamins are stored in fat reserves which are not filtered by the body.

Section 22.4 Review:

(1) Major minerals are found in the body in quantities of greater than 5 g. Minor minerals are found in the body in quantities smaller than 5 g. In general, a healthy diet includes more major minerals than trace minerals.

(2) Major minerals serve structural functions as well as being distributed throughout the body's fluids. Trace minerals are components of vitamins, enzymes, hormones, and specialized proteins.
(3) If a patient is told to limit their sodium intake, the patient should avoid table salt (sodium chloride).
(4) Table salt does not contain elemental sodium. It contains sodium ions. Mineral names do not reflect that the element is part of a compound; however, most minerals are inorganic compounds.

Section 22.5 Review:

(1) (a) exothermic (b) endothermic (c) exothermic
(2) (a) neither (b) reactant (c) product
(3) (a) neither (b) product (c) reactant
(4) (a) neither (b) reactant (c) product

Section 22.6 Review:

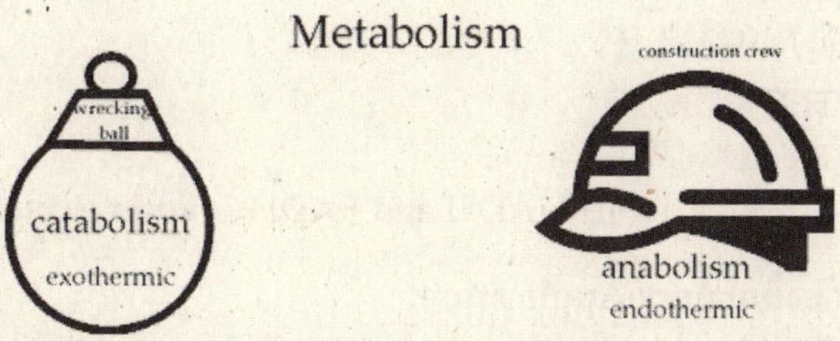

(4) Stage I of the catabolism of food is the "wrecking ball" phase. Large molecules are broken into smaller molecules (proteins → amino acids, carbohydrates → monosaccharides, lipids → fatty acids and glycerol). Stage II of the catabolism of food involves further breakdown of the small molecules into even simpler units, like the acetyl portion of acetyl CoA. These "refining" processes are unique for each type of material; however, the major product of these processes is the same. Stage III of the catabolism of food takes the refined product and "recycles" it through the citric acid cycle, the electron transport chain, and oxidative phosphorylation to produce high energy ATP.

Section 22.7 Review:

(1) The energy released during the oxidation of macronutrients is stored in some of the bonds of ATP.
(2) The components of ATP are ribose, adenine, and triphosphate.
(3) ATP differs from the nucleotide containing adenine only in the size of the phosphate group attached to the ribose. ATP contains a triphosphate group, but the nucleotide contains a monophosphate group.
(4) The sign on ΔG for exothermic reactions is negative.
(5)

$ATP + H_2O$

↙ $ADP + P_i + H^+$

↘ $AMP + PP_i + 2\ H^+$ → ($+ H_2O$) → $2\ P_i$

(6) Yes, see calculations below.

$ATP + H_2O = 4- + 0 = 4-$

$ADP + P_i + H^+ = 3- + 2- + 1+ = 4-$

$AMP + PP_i + 2\ H^+ = 2- + 4- + 2(1+) = 4-$

$PP_i + H_2O = 4- + 0 = 4-$

$2\ P_i = 2(2-) = 4-$

(7) The conversion of ATP to AMP produces more energy than the conversion of ATP to ADP, even if the pyrophosphate is not hydrolyzed.
(8) ATP synthesis occurs in the mitochondria of a cell.

Section 22.8 Review:

(1)
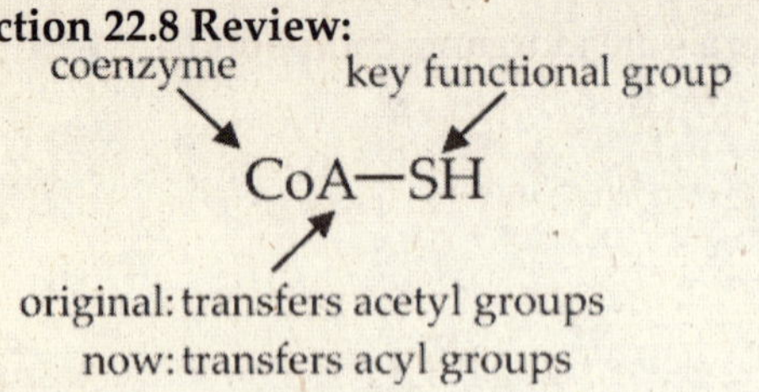

(2) $O_2 + 4\ e^- + 4\ H^+ \rightarrow 2\ H_2O$
(3) $NADH \rightarrow NAD^+ + 2\ e^- + H^+$
(4) $FADH_2 \rightarrow FAD + 2\ e^- + 2\ H^+$
(5) $O_2 + 2\ NADH + 2\ H^+ \rightarrow 2\ H_2O + 2\ NAD^+$
(6) $O_2 + 2\ FADH_2 \rightarrow 2\ H_2O + 2\ FAD$
(7) Coenzyme A shuttles acyl groups, while NADH and $FADH_2$ shuttle electrons and hydrogen ions.

Tying It All Together with a Laboratory Application:

(1) 192 Calories (fiber = Calorie-free)
(2) macronutrients
(3) micronutrients
(4) 56%
(5) 27%
(6) 17%
(7) exceeds
(8) 30%
(9) has less than
(10) 58%
(11) 9
(12) 9.5
(13) liquid
(14) Reference Daily Intake
(15) RDI varies based on age and gender
(16) essential
(17) complete
(18) fat
(19) water
(20) are not
(21) minerals
(22) major
(23) trace
(24) the sun
(25) photosynthesis
(26) storage
(27) catabolic
(28) ATP
(29) coenzyme
(30) transfer

SELF-TEST QUESTIONS

Multiple Choice

1. Nutrients required by the body in relatively large amounts are:
   a. fiber. b. macronutrients. c. micronutrients. d. vitamins.

2. The "D" in DV stands for:
   a. daily. b. determined. c. dietary. d. deficiency.

3. Which of the following is an example of a complex carbohydrate nutrient for humans?
   a. starch b. cellulose c. glucose d. lactose

4. Which of the following has an established DV?
   a. carbohydrates b. lipids c. proteins d. water

5. Which of the following is a water-soluble vitamin?
   a. vitamin A b. vitamin B_2 c. vitamin E d. vitamin D

6. Which of the following is a trace mineral?
   a. phosphorus b. sulfur c. potassium d. copper

7. It is recommended that no more than ______ percent of our calories be obtained from fats.
   a. 10 b. 30 c. 40 d. 50

8. Mitochondria are called "power stations of the cell" because:
   a. most of the cellular energy is consumed there.
   b. mitochondria are rich in fat molecules.
   c. mitochondria exist at higher temperatures than other cell components.
   d. mitochondria are the major sites of ATP synthesis.

9. During the energy production process, what product of the second stage is passed on to the third stage?
   a. CO_2 b. H atoms c. acetyl CoA d. H_2O

10. FAD often accompanies the enzyme-catalyzed:
    a. oxidation of $-CH_2CH_2-$ to $-CH{=}CH-$.
    b. oxidation of aldehydes.
    c. oxidation of alcohol groups.
    d. transfer of acetyl groups.

Matching

Select the stage in the extraction of energy from food where the following molecules are produced.

11. majority of ATP
12. amino acids
13. acetyl CoA
14. carbon dioxide

a. stage I
b. stage II
c. stage III
d. more than one response is correct

Select the mineral that best matches each description on the left.

15. a major mineral found in the body in largest amounts
16. a vitamin component
17. a hormone component

a. cobalt
b. iodine
c. calcium
d. zinc

Select the coenzyme that best matches each description on the left.

18. serves as an electron acceptor in the oxidation of an alcohol
19. a derivative of the vitamin riboflavin
20. contains a sulfhydryl group (-SH)

a. coenzyme Q
b. coenzyme A
c. FAD
d. NAD^+

Select the process that best matches each description on the left.

21. energy-releasing solar process
22. conversion of CO_2 and H_2O to carbohydrates
23. oxidation of glucose to CO_2 and H_2O

a. cellular respiration
b. photosynthesis
c. nuclear fusion
d. cellular anabolism

True-False

24. Carbon dioxide is a source of carbon for the earth's organic compounds.
25. ΔG for the hydrolysis of ATP is a positive value.
26. Nine amino acids are listed as being essential.
27. The base present in ATP is dexoyribose.
28. An ATP molecule at pH 7.4 (body pH) has a -3 charge.
29. On a mass basis, water is the most abundant compound found in the human body.

30. Sucrose, a disaccharide, is correctly classified as a complex carbohydrate.
31. Most nutritional studies indicate that typical diets in the U.S. contain too high a percentage of complex carbohydrates.
32. Lipids are digested faster than carbohydrates or proteins.
33. Concern about vitamin overdoses focuses primarily on water-soluble vitamins.
34. Most fat-soluble vitamins are known to function in the body as coenzymes.
35. Vitamin K is important in the process of blood clotting.

Chapter 23: Carbohydrate Metabolism

CHAPTER OUTLINE

23.1 The Digestion of Carbohydrates
23.2 Blood Glucose
23.3 Glycolysis
23.4 The Fates of Pyruvate
23.5 The Citric Acid Cycle
23.6 The Electron Transport Chain
23.7 Oxidative Phosphorylation
23.8 The Complete Oxidation of Glucose
23.9 Glycogen Metabolism
23.10 Gluconeogenesis
23.11 The Hormonal Control of Carbohydrate Metabolism

LEARNING OBJECTIVES/ASSESSMENT

When you have completed your study of this chapter, you should be able to:

1. Identify the products of carbohydrate digestion. (Section 23.1; Exercise 23.2)
2. Explain the importance to the body of maintaining proper blood sugar levels. (Section 23.2; Exercise 23.6)
3. List the starting material and products of the glycolysis pathway. (Section 23.3; Exercise 23.10)
4. Describe how the glycolysis pathway is regulated in response to cellular needs. (Section 23.3; Exercise 23.14)
5. Name the three fates of pyruvate. (Section 23.4; Exercise 23.20)
6. Identify the two major functions of the citric acid cycle. (Section 23.5; Exercise 23.24)
7. Describe how the citric acid cycle is regulated in response to cellular energy needs. (Section 23.5; Exercise 23.32)
8. Explain the function of the electron transport chain and describe how electrons move down the chain. (Section 23.6; Exercises 23.34 and 23.36)
9. List the major features of the chemiosmotic hypothesis. (Section 23.7; Exercise 23.46)
10. Calculate the amount of ATP produced by the complete oxidation of a mole of glucose. (Section 23.8; Exercise 23.48)
11. Explain the importance of the processes of glycogenesis and glycogenolysis. (Section 23.9; Exercise 23.50)
12. Describe gluconeogenesis and the operation of the Cori cycle. (Section 23.10; Exercise 23.58)
13. Describe how hormones regulate carbohydrate metabolism. (Section 23.11; Exercise 23.62)

SOLUTIONS FOR THE END OF CHAPTER EXERCISES

THE DIGESTION OF CARBOHYDRATES (SECTION 23.1)

☑23.2
a. starch — glucose
b. lactose — glucose, galactose
c. sucrose — fructose, glucose
d. maltose — glucose

BLOOD GLUCOSE (SECTION 23.2)

23.4 Blood sugar level is the concentration of glucose in the blood.
Normal fasting level is the concentration of glucose in the blood after an 8-12 hour fast.

☑23.6
a. hypoglycemia — blood sugar level is below the normal fasting level
b. hyperglycemia — blood sugar level is above the normal fasting level
c. renal threshold — blood sugar level is above 108 mg/100 mL and the sugar is not completely reabsorbed from the urine by the kidneys
d. glucosuria — blood sugar level is above the renal threshold and glucose appears in the urine

23.8 Severe hypoglycemia can be very serious because it can cause convulsions and shock.

GLYCOLYSIS (SECTION 23.3)

☑23.10 The starting material for glycolysis is glucose. The major products of glycolysis are pyruvate and ATP.

23.12 Two steps in glycolysis require ATP and two steps produce ATP. The glycolysis of one molecule of glucose results in the net production of 2 molecules of ATP.

☑23.14 The third control point in glycolysis is the conversion of phosphoenolpyruvate to pyruvate catalyzed by the enzyme pyruvate kinase. This process is responsive to cellular needs because this enzyme is inhibited by higher concentrations of ATP.

23.16 A high concentration of glucose 6-phosphate inhibits glycolysis because hexokinase, the enzyme that catalyzes the conversion of glucose to glucose 6-phosphate, is inhibited by glucose 6-phosphate. This feedback inhibition decreases the rate of glycolysis when a high concentration of glucose 6-phosphate is present.

THE FATES OF PYRUVATE (SECTION 23.4)

23.18 Aerobic describes a process that requires oxygen to take place. Anaerobic describes a process that does not require oxygen to take place.

☑23.20 a. Under aerobic conditions in the body, the fate of pyruvate is conversion to acetyl CoA.
b. Under anaerobic conditions in the body, the fate of pyruvate is conversion to lactate.

23.22 During fermentation, the pyruvate molecule's carboxylate group is converted to carbon dioxide. Ethanol formation allows yeast to survive with limited oxygen supplies because alcoholic fermentation is an anaerobic process.

THE CITRIC ACID CYCLE (SECTION 23.5)

☑23.24 The primary function of the citric acid cycle in ATP production is to meet the cellular needs for ATP by generating the reduced coenzymes NADH and $FADH_2$. The other vital role served by the citric acid cycle is the production of intermediates for biosynthesis.

23.26

a.	The fuel needed by the cycle	acetyl CoA
b.	The form in which carbon atoms leave the cycle	CO_2
c.	The form in which hydrogen atoms and electrons leave the cycle	NADH, $FADH_2$

23.28 a. NADH 3 b. $FADH_2$ 1 c. GTP 1 d. CO_2 2

23.30 succinate + FAD $\rightarrow$ $FADH_2$ + fumarate

☑23.32 a. The enzymes at the three control points of the citric acid cycle are citrate synthetase (step 1), isocitrate dehydrogenase (step 3), and the α-ketoglutarate dehydrogenase complex (step 4).

b. The citric acid cycle produces both NADH and ATP. If the cell has high levels of NADH and ATP, then in all likelihood, the energy needs of the cell are or will be adequately met. Therefore, when the NADH levels are high, NADH serves as an inhibitor to slow the production of NADH and $FADH_2$. When ATP levels are high, the entry of acetyl CoA into the citric acid cycle is reduced. Energy production is lowered because the requirements of the cell are already being met.

If the cell has low levels of NADH and ATP, the cell's energy requirements are probably not being met. If levels of ATP are low, levels of ADP and usually high. High ADP levels activate the cycle and stimulate the production of more ATP. Similarly, if the levels of NADH are low, this compound is not available to inhibit the control point enzymes and the cycle is stimulated to produce more NADH and ATP.

c. Two of the control point enzymes are activated by ADP. This benefits the cell because a high level of ADP indicates a low level of ATP, so the cell can respond immediately to low ATP levels. If the cycle were simply stimulated by a feedback mechanism involving only high or low levels of ATP, the cellular response to low levels of ATP (and low energy) would be slower. Since ADP stimulates the cycle, the cell can respond immediately to its presence rather than waiting for the level of ATP to fall and its activity as an inhibitor to be reversed. A second benefit occurs because ADP is converted directly into ATP, the substance which stimulates the pathway is the direct substrate of that pathway, and conversion between the two compounds is, therefore, quick.

THE ELECTRON TRANSPORT CHAIN (SECTION 23.6)

☑23.34 NADH and $FADH_2$ bring electrons to the electron transport chain.

☑23.36 The cytochromes in the electron transport chain function to pass electrons along the chain to oxygen atoms and to provide H^+ ions to the oxygen, resulting in the formation of water.

OXIDATIVE PHOSPHORYLATION (SECTION 23.7)

23.38 NADH and $FADH_2$ are oxidized during oxidative phosphorylation.
ADP is phosphorylated during oxidative phosphorylation.

23.40 1.5 ATP molecules are synthesized when one $FADH_2$ passes electrons to molecular oxygen.

23.42 When 10 acetyl CoA molecules enter the common catabolic pathway, 100 ATP molecules are formed by oxidation of acetyl CoA.

23.44 $$\frac{686\text{ kcal}}{1\text{ mole glu cos e}}\left(\frac{1\text{ mol ATP}}{7.3\text{ kcal}}\right) = 94\text{ mol ATP}$$

☑23.46 ATP synthesis results from protons crossing the inner mitochondrial membrane.

THE COMPLETE OXIDATION OF GLYCOSE (SECTION 23.8)

☑23.48 a. Glycolysis produces 2 moles of ATP.
b. The citric acid cycle produces 2 moles of ATP.
c. The electron transport chain and oxidative phosphorylation together produces 28 moles of ATP.

GLYCOGEN METABOLISM (SECTION 23.9)

☑23.50 Glycogenesis is the process by which glucose is converted to glycogen. Glycogenolysis is the process by which glycogen is broken back down into glucose. These are reverse processes.

23.52 The high-energy compound involved in the conversion of glucose to glycogen is uridine triphosphate (UTP).

23.54 Liver glycogen can be used to raise blood sugar levels because glycogenolysis can occur in the liver, but not in muscle tissue. The enzyme glucose-6-phosphatase essential for this process is found in the liver, but not in muscle tissue.

GLUCONEOGENESIS (SECTION 23.10)

23.56 The principal site for gluconeogenesis is the liver.

☑23.58 Lactate, glycerol, and certain amino acids undergo gluconeogenesis. Amino acids are derived from proteins, glycerol is derived from lipids, and lactate has a variety of sources, including pyruvate, which is produced from carbohydrates.

23.60 The Cori cycle would not operate if the body's muscles were completely oxidizing the glucose supplied because the body would not produce lactate (required for the Cori cycle).

THE HORMONAL CONTROL OF CARBOHYDRATE METABOLISM (SECTION 23.11)

☑23.62
a. Glucagon increases the blood sugar level by enhancing the breakdown of glycogen in the liver, while insulin reduces the blood sugar level by enhancing the absorption of glucose from the blood into cells of active tissue.
b. Glucagon inhibits glycogen formation, while insulin increases the rate of glycogen formation.

ADDITIONAL EXERCISES

23.64

$$10.0\text{ g glucose}\left(\frac{1\text{ mole glucose}}{180\text{ g glucose}}\right)\left(\frac{32\text{ moles ATP}}{1\text{ mole glucose}}\right)\left(\frac{6.02\times10^{23}\text{ molecules ATP}}{1\text{ mole ATP}}\right)$$

$$=1.1\times10^{24}\text{ molecules ATP}$$

23.66 $\text{pyruvate} + \text{NADH} + \text{H}^+ \rightleftarrows \text{lactate} + \text{NAD}^+$

Under anaerobic conditions, the cellular supply of oxygen is not adequate for the oxidation of NADH to NAD^+; therefore, cells reduce pyruvate to lactate in order to regenerate NAD^+. In order to continue generating NAD^+ from this reaction, the lactate produced must then be removed from the muscle cells. By removing the lactate, the equilibrium shifts to the right and continues to produce the NAD^+ needed by the cell. Under anaerobic conditions, the lactate does not accumulate in the liver because gluconeogenesis will occur converting the lactate back to pyruvate (and then into glucose).

23.68

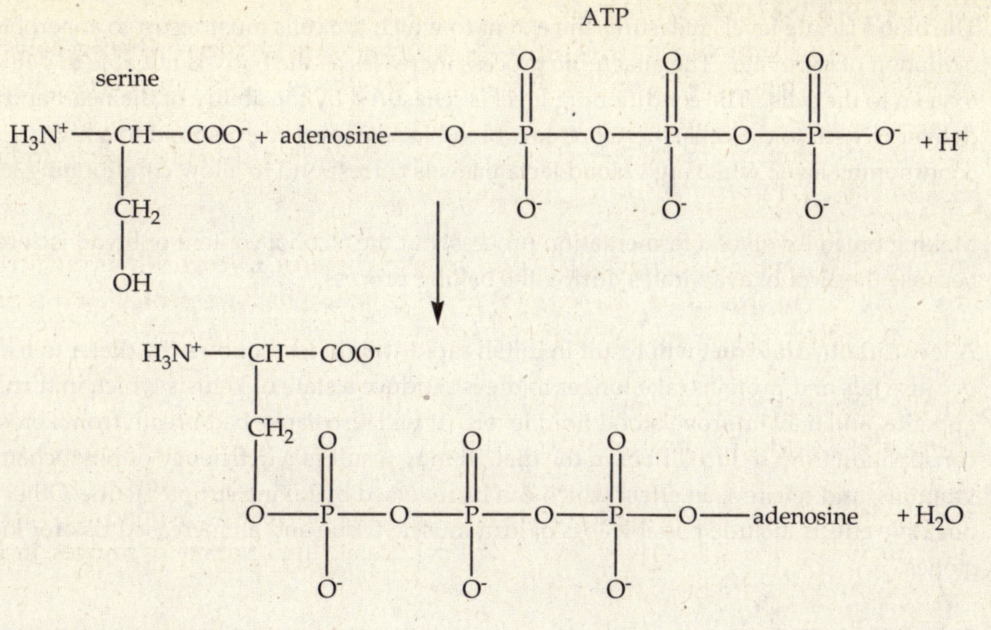

When the specific serine residue in the pyruvate dehydrogenase enzyme complex is phosphorylated by ATP, a phosphate ester bond forms as does a molecule of water.

ALLIED HEALTH EXAM CONNECTION

23.70 Glycolysis (a) requires 1 ATP molecule to activate a molecule of glucose and (c) does not require oxygen. Glycolysis occurs in the cellular cytoplasm (not the mitochondria).

23.72 The citric acid cycle (b) occurs in the mitochondria, (c) is the first stage of aerobic cellular respiration, and (d) is also called the Krebs cycle. The citric acid cycle, the third stage in the oxidation of fuel molecules, produces carbon dioxide and NADH as well as $FADH_2$, GTP, CoA-SH, and hydrogen ions. The reduced coenzymes NADH and $FADH_2$ combine with oxygen to form water in the electron transport chain.

23.74 Most soluble food substances enter the bloodstream through the (a) small intestine.

23.76 Insulin is produced in the (c) pancreas.

23.78 For the aerobic respiration pathway, electron transport systems are located in the (d) mitochondrion.

CHEMISTRY FOR THOUGHT

23.80 The citric acid cycle does not utilize oxygen and for a time it can continue without oxygen. The final product of the citric acid cycle will build up, if oxygen is not present because oxygen is the final acceptor of electrons in the electron transport chain, which uses the final product of the citric acid cycle. The build up of the final product from the citric acid cycle will inhibit the citric acid cycle; therefore, the citric acid cycle cannot continue without an adequate supply of oxygen to facilitate the removal of these electrons through the electron transport chain.

23.82 The bottle exploded because the fermentation process produces CO_2 gas, which pressurized the bottle and caused it to explode.

23.84 The blood lactate levels measures the extent to which the cells must resort to anaerobic oxidation of pyruvate. The anaerobic process increases as the body is not able to deliver the oxygen to the cells. The conditioning level is measured by the ability of the heart and lungs to deliver oxygen to the cells; therefore, low blood lactate levels correspond to a high conditioning level, while high blood lactate levels correspond to a low conditioning level.

23.86 Making bread involves a fermentation process, but the alcohol content of bread is zero because the alcohol evaporates during the baking process.

23.88 A low carbohydrate diet will result in initial rapid weight loss, allows the dieter to feel fuller because fats and proteins take longer to digest, induce a state of ketosis which in turn reduces appetite, and may improve blood lipid levels. It will also cause bad breath from ketosis and constipation from reduced fiber in the diet. It may result in a deficiency of phytochemicals, vitamins, and calcium, an effect which can be reversed by taking supplements. Other negative effects include possible loss of lean muscle tissue and an increased risk for kidney stones.

23.90 If drinking milk always seems to make you slightly nauseous, you might be lactose intolerant and need a lactase supplement.

ADDITIONAL ACTIVITIES

Section 23.1 Review:

(1) Describe the digestion of carbohydrates using the words monosaccharides, disaccharides, and polysaccharides.
(2) How do digested carbohydrates enter the bloodstream?

Section 23.2 Review:

Complete the following diagram with the vocabulary from this section.

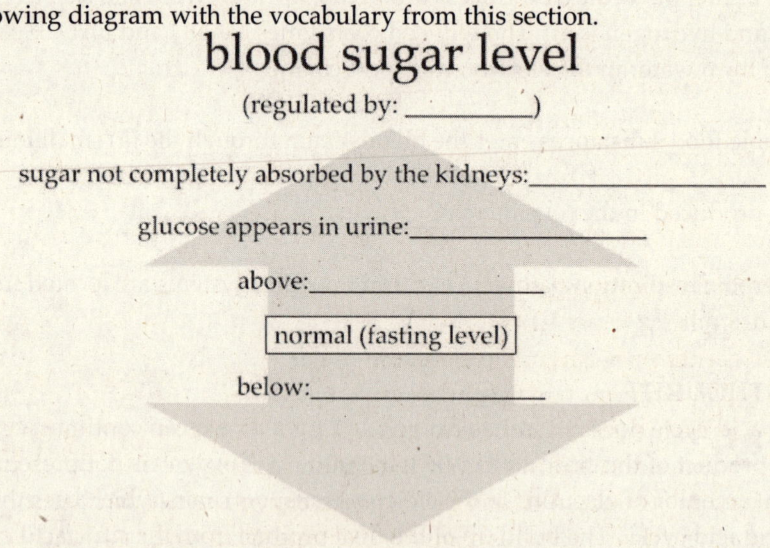

Section 23.3 Review: Complete the following diagram for glycolysis.

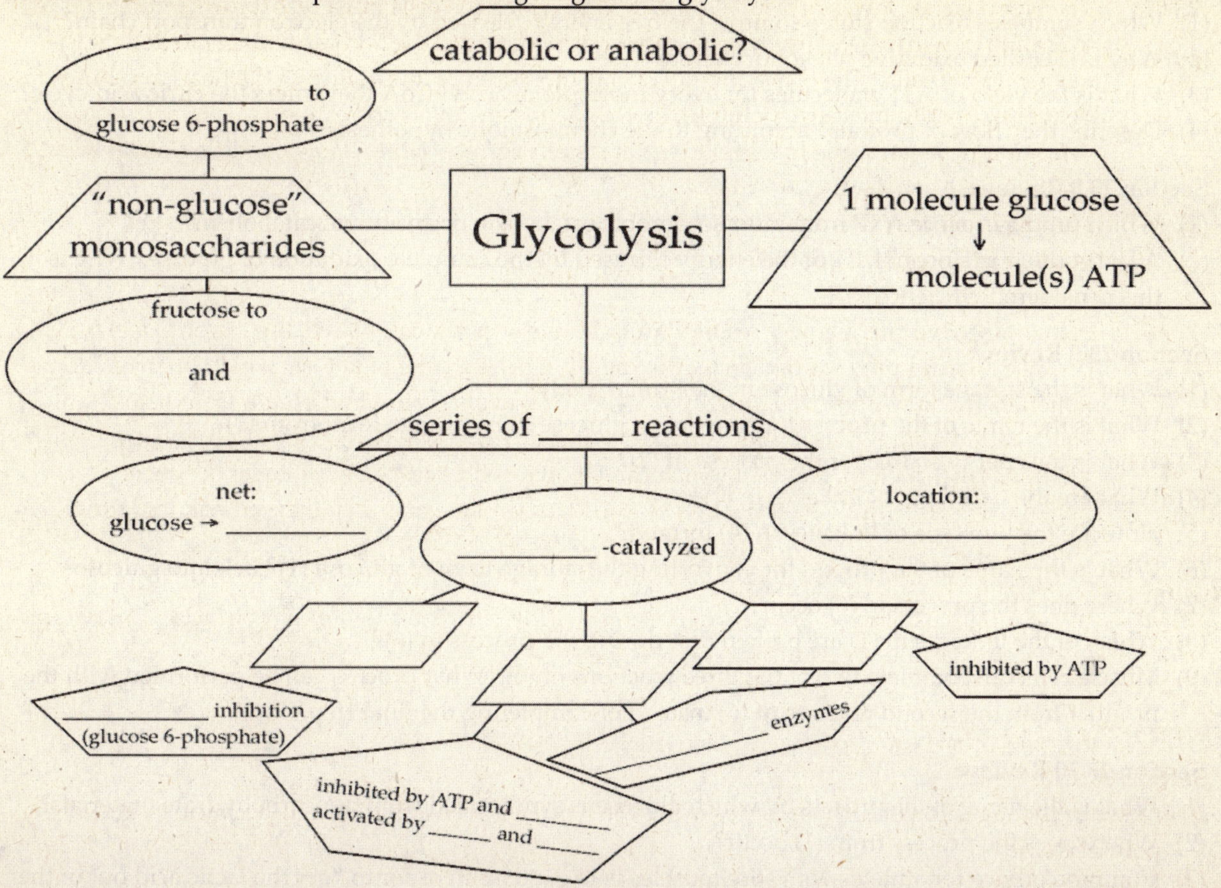

Section 23.4 Review:

(1) What is meant by the term *fates* in the expression, "The Fates of Pyruvate"?
(2) What is common to all three fates of pyruvate from a coenzyme perspective?
(3) What is the fate of pyruvate when sufficient oxygen is present?
(4) What are the "fates" of the product from (3)?
(5) What are the fates of pyruvate when insufficient oxygen is present?
(6) Which of the processes in (5) are possible in the human body?

Section 23.5 Review:

(1) What are the other names for the citric acid cycle?
(2) Where does the citric acid cycle occur within the cell?
(3) How many main reactions are part of the citric acid cycle?
(4) Which coenzymes play an important role in the citric acid cycle?
(5) Which compound is the fuel for the citric acid cycle?
(6) Which high energy compound is produced by the citric acid cycle?
(7) What is the relationship between ATP and the allosteric enzyme citrate synthetase?
(8) Which other regulating enzyme behaves similarly to citrate synthetase?
(9) What inhibits the α-ketoglutarate dehydrogenase complex?

Section 23.6 Review:

(1) What compound is produced as NADH and $FADH_2$ participate in the electron transport chain?
(2) Why are the electron carriers in the electron transport chain in order of increasing electron affinity?
(3) What are the similarities among the cytochromes?

Section 23.7 Review:

(1) Which synthesis reaction stores some of the free energy released by the electron transport chain?
(2) Why is (1) called oxidative phosphorylation?
(3) What is the yield of ATP molecules for every molecule of acetyl CoA that enters the citric acid cycle?
(4) Describe the "flow of protons" according to the chemiosmotic hypothesis.

Section 23.8 Review:

(1) Which produces more ATP from glucose: aerobic catabolism or anaerobic catabolism?
(2) ATP production stores 34.1% of the energy released by the complete oxidation of glucose. Why is this considered efficient?

Section 23.9 Review:

(1) What is the storage form of glucose in the human body?
(2) What is the name of the process for converting glucose to its storage form from (1)?
(3) What is the energy source for the process in (2)?
(4) What are the two types of linkages in (1)?
(5) How are the two type of linkages in (4) formed?
(6) What is the name of the process for converting the storage form of glucose (1) back into glucose?
(7) Where does the process in (6) occur?
(8) Which of the linkages in (1) are broken first during the process in (6)?
(9) Muscle cells can complete two of the three reactions of (6). What process can be performed with the product from the second reaction of (6) instead of completing the final step of (6)?

Section 23.10 Review:

(1) What is the name of the process by which glucose is synthesized from noncarbohydrate materials?
(2) Where does the process from (1) occur?
(3) Common advice for athletes with sore muscles is to exercise in order to "get the lactic acid out of the muscles." Explain why exercise helps relieve sore muscles.

Section 23.11 Review:

(1) Which hormones are produced in the pancreas?
(2) Which cells in the pancreas produce each of the hormones in (1)?
(3) Which of the "glucose reactions" does each of the hormones in (1) stimulate?
(4) Epinephrine is also known as the "fight or flight" hormone. Explain how this hormone provides an opportunity to "fight or flight."

Tying It All Together with a Laboratory Application:

After fasting overnight, a sample of blood is taken from three students and checked for their blood sugar (or (1)_______) levels. The results of the test are:

Student	Blood Sugar Level (mg/dL)
A	65
B	115
C	72

Student A could be classified as (2) ______ because these results indicate a blood sugar level that is (3) ______ compared to the normal fasting level. Student B could be classified as (4) ______ because these results indicate a blood sugar level that is (5) ______ compared to the normal fasting level. Student C could be classified as (6) ______ because these results indicate a blood sugar level that is (7) ______ compared to the normal fasting level.

Student B also had a sample of urine tested for glucose. The sample came back positive for (8) ______ (glucosuria or insulin shock), which indicates the blood sugar level is above the (9) ______. Student B may have (10) ______.

The students then participated in an oral glucose tolerance test. They were given "glucola" (a solution with 75 g of glucose) and their blood was tested after 30 minutes, 1 hour, 2 hours, and 3 hours. After 30 minutes and 1 hour, the blood sugar level was elevated in all three students; however, after 2 hours, the blood sugar levels in Students A and C were significantly lower then after the 1 hour reading. The blood sugar level for Student B was still elevated at 2 hours. At the 3 hour mark, Students A and C have "normal" blood sugar level, but Student B does not.

The hormone (11) ______ is responsible for the lowering of the blood sugar level in Students A and C. It causes glucose to breakdown into (12) ______ in a process called (13) ______. Under aerobic conditions (12) is converted into (14) ______. Under anaerobic conditions (12) is converted into (15) ______ which is stored in the (16) ______.

(14) is the fuel for the (17) ______. The coenzymes (18) ______ and ______ undergo (19) ______ during the (17). After the citric acid cycle, (20) ______ and ______ participate in the (21) ______ to produce water and regenerate (18). The (21) releases a significant amount of energy that can be used to synthesize (22) ______.

Excess glucose can be stored in the form of (23) ______ through a process called (24) ______. When the glucose is needed, (23) can be broken down through a process called (25) ______. The hormone (26) ______ activates (25) in the liver. If the first two stages of (25) occur in the muscles rather than the liver, glucose cannot be formed because the muscles lack (27) ______; however, the intermediate (28) ______ can enter the (29) ______ pathway to produce energy. This process can also be stimulated in the muscles by the hormone (30) ______.

Student A's overnight fast lasted about 19 hours. Consequently, this student may have produced glucose from noncarbohydrate materials in a process called (31) ______. If (31) involved lactate from muscle tissue, this is one of the steps in the (32) ______ cycle.

SOLUTIONS FOR THE ADDITIONAL ACTIVITIES

Section 23.1 Review:

(1) Polysaccharides and disaccharides are hydrolyzed into monosaccharides during the digestion of carbohydrates.
(2) Digested carbohydrates enter the bloodstream through the lining of the small intestine.

Section 23.2 Review:

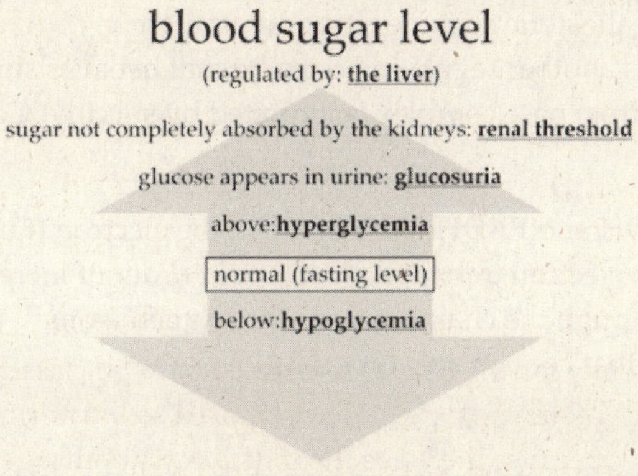

Section 23.3 Review:

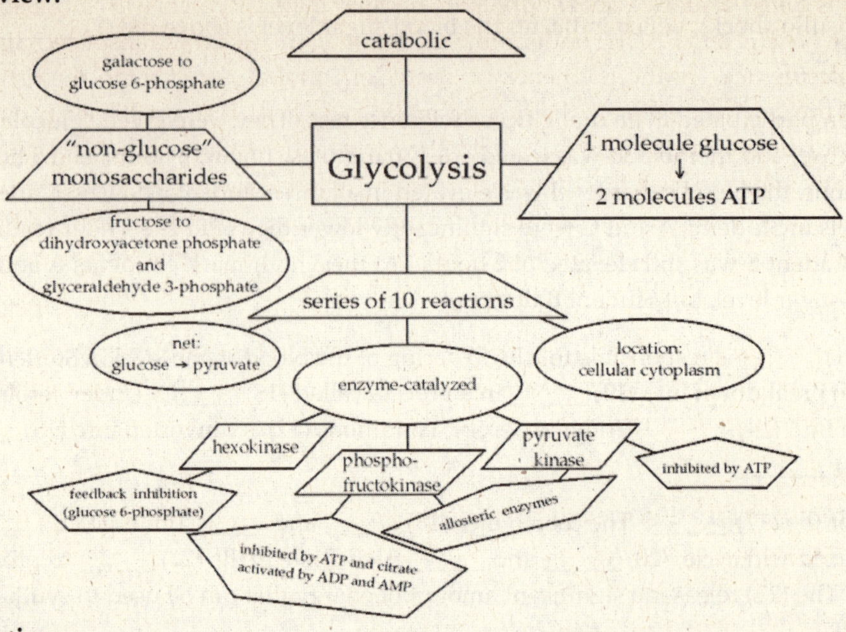

Section 23.4 Review:

(1) The term *fates* in the expression, "The Fates of Pyruvate" means what happens to the pyruvate molecule.
(2) All three fates of pyruvate regenerate NAD^+ so that glycolysis can continue.
(3) When sufficient oxygen is present, pyruvate is converted to acetyl CoA.
(4) The acetyl CoA either enters the citric acid cycle to produce carbon dioxide or undergoes biosynthesis to produce compounds like the fatty acid components of triglycerides.
(5) Lactate fermentation or alcoholic fermentation are the fates of pyruvate when insufficient oxygen is present.
(6) Alcohol fermentation cannot occur in the human body; however, lactate fermentation can occur in the human body.

Section 23.5 Review:

(1) The citric acid cycle is also known as the tricarboxylic acid cycle and the Krebs cycle.
(2) The citric acid cycle occurs within the mitochondria.
(3) Eight main reactions are part of the citric acid cycle.
(4) NAD^+ and FAD are coenzymes that play an important role in the citric acid cycle.
(5) Acetyl CoA is the fuel for the citric acid cycle.
(6) GTP is a high energy compound produced by the citric acid cycle.
(7) ATP is an inhibitor for the allosteric enzyme citrate synthetase.
(8) Isocitrate dehydrogenase is another regulating enzyme that behaves similarly to citrate synthetase.
(9) The α-ketoglutarate dehydrogenase complex is inhibited by succinyl CoA, NADH, and ATP.

Section 23.6 Review:

(1) Water is produced as NADH and $FADH_2$ participate in the electron transport chain.
(2) The electron carriers in the electron transport chain are in order of increasing electron affinity because the electrons must flow through the chain towards the oxygen atom.
(3) Cytochromes are proteins that contain an iron group.

Section 23.7 Review:

(1) The synthesis of ATP from ADP and P_i stores some of the free energy released by the electron transport chain.

(2) This process is called oxidative phosphorylation because the ADP reacts with inorganic phosphate as a result of the NADH and $FADH_2$ undergoing oxidation in the electron transport chain.
(3) Ten ATP molecules are produced for every molecule of acetyl CoA that enters the citric acid cycle.
(4) The "flow of protons" according to the chemiosmotic hypothesis is across the inner mitochondrial membrane into the intermembrane space. Simultaneously the electrons flow through the electron transport chain, and then back through the inner mitochondrial membrane through a channel formed by the enzyme F_1-ATPase as a result of proton concentration and electric potential differences on either side of the inner mitochondrial membrane.

Section 23.8 Review:
(1) Aerobic catabolism of glucose produces more ATP than anaerobic catabolism of glucose.
(2) Compared to other systems that use/store chemical energy, 34.1% is a high percentage; therefore, the process is considered efficient.

Section 23.9 Review:
(1) The storage form of glucose in the human body is glycogen.
(2) The process for converting glucose to glycogen is called glycogenesis.
(3) The energy source for glycogenesis is UTP.
(4) The two types of linkages in glycogen are $\alpha(1\rightarrow4)$ and $\alpha(1\rightarrow6)$.
(5) The $\alpha(1\rightarrow4)$ linkages are formed as each glucose unit lengthens the chain. The $\alpha(1\rightarrow6)$ linkages are formed by a branching enzyme.
(6) The process for converting glycogen back into glucose is glycogenolysis.
(7) Glycogenolysis occurs in the liver, kidneys, and intestinal cells.
(8) The $\alpha(1\rightarrow4)$ linkages in glycogen are broken first during glycogenolysis.
(9) Muscle cells can complete two of the three reactions of glycogenolysis. Glucose 6-phosphate is the product of the second reaction of glycogenolysis. This compound can then enter the glycolysis pathway and release energy.

Section 23.10 Review:
(1) Gluconeogenesis is the process by which glucose is synthesized from noncarbohydrate materials. The name can be divided into three parts "gluco-" refers to glucose, "neo-" refers to in a new, different, or modified way, and "-genesis" refers to production. This could be combined to define gluconeogenesis as the production of glucose in a modified way (not from carbohydrates).
(2) Gluconeogenesis occurs primarily in the liver.
(3) During exercise, blood goes to the muscles to deliver oxygen and retrieve waste products. Some of the lactate in the muscles is taken by the blood to the liver where the lactate can be converted back into pyruvate and then into glucose by gluconeogenesis. The glucose then reenters the bloodstream. If the exercise is aerobic and enough oxygen is available to the cells, then the glucose can undergo aerobic glycolysis to produce acetyl CoA.

Section 23.11 Review:
(1) Insulin and glucagon are produced by the pancreas.
(2) Insulin is produced by the β-cells of the pancreas. Glucagon is produced by the α-cells of the pancreas.
(3) Insulin stimulates glycolysis. Glucagon stimulates glycogenolysis.
(4) Epinephrine stimulates glycogen breakdown in the muscles and the liver. When glycogen is broken down in the muscles, glucose 6-phosphate is produced which undergoes glycolysis to produce energy. This energy is produced quickly and enables a person (or animal) with raised epinephrine levels to either "fight or flight."

Tying It All Together with a Laboratory Application:

(1) glucose
(2) hypoglycemic
(3) low
(4) hyperglycemic
(5) high
(6) normal
(7) average
(8) glucosuria
(9) renal threshold
(10) diabetes
(11) insulin
(12) pyruvate
(13) glycolysis
(14) acetyl CoA
(15) lactate
(16) muscles
(17) citric acid cycle
(18) NAD^+, FAD
(19) reduction
(20) NADH, $FADH_2$
(21) electron transport chain
(22) ATP
(23) glycogen
(24) glycogenesis
(25) glycogenolysis
(26) glucagon
(27) glucose 6-phosphatase
(28) glucose 6-phosphate
(29) glycolysis
(30) epinephrine
(31) gluconeogenesis
(32) Cori

SELF-TEST QUESTIONS

Multiple Choice

1. The digestion of carbohydrates produces glucose, fructose, and:
   a. starch. b. amylose. c. lactose. d. galactose.

2. The central compound in carbohydrate metabolism is:
   a. glycogen. b. glucose. c. fructose. d. galactose.

3. The glycolysis pathway is located within the ______ of cells.
   a. cytoplasm b. nucleus c. mitochondria d. ribosomes

4. The net production of ATP in glycolysis from one molecule of glucose is ______ ATP molecules.
   a. 0 b. 1 c. 2 d. 4

5. The net production of NADH in glycolysis from one molecule of glucose is ______ NADH molecules.
   a. 0 b. 1 c. 2 d. 4

6. Enzymes regulate the glycolysis pathway at ______ control points.
   a. 2 b. 3 c. 4 d. 5

7. The muscle pain which follows prolonged and vigorous contraction of skeletal muscles is the result of:
   a. lactate. b. citrate. c. NADH. d. pyruvate.

8. How many molecules of pyruvate are produced from the glycolysis of one molecule of glucose?
   a. 0 b. 1 c. 2 d. 3

9. In the liver, how many net molecules of ATP can be formed in the electron transport chain by oxidative phosphorylation from each molecule of cytoplasmic NADH produced during glycolysis?
   a. 1.0 b. 1.5 c. 2.0 d. 2.5

10. The complete oxidation of glucose in the liver results in the formation of ______ molecules of ATP.
    a. 18 b. 32 c. 34 d. 38

11. Which of the following exerts an effect on blood sugar levels opposite to that of insulin?
    a. cholesterol b. glucagon c. vasopressin d. aldosterone

12. What disease is commonly associated with glucosuria?
    a. diabetes mellitus b. hepatitis c. hypoglycemia d. hyperinsulinism

13. The first step of the citric acid cycle involves the reaction of oxaloacetate with ______ to form citrate.
    a. coenzyme A b. acetyl CoA c. CO_2 d. malate

14. The high-energy compound formed in the citric acid cycle is:
    a. CTP. b. ATP. c. UTP. d. GTP.

15. One turn of the citric acid cycle produces ______ molecules of NADH.
    a. 1 b. 2 c. 3 d. 4

16. The citric acid cycle is inhibited by:
    a. ATP. b. CO_2. c. NAD^+. d. FAD.

17. One product of the electron transport chain is:
    a. O_2. b. H_2O. c. CO_2. d. NADH.

18. How many sites in the electron transport chain can support the synthesis of ATP?
    a. 1 b. 2 c. 3 d. 4

Matching

Match each description on the left with the condition or disease on the right.

19.	a high glucose level in the blood	a.	glucosuria
20.	a low glucose level in the blood	b.	hypoglycemia
21.	a high glucose level in the urine	c.	hyperglycemia
		d.	galactosemia

Match each description on the left with a product on the right.

22.	the product of alcoholic fermentation	a.	acetaldehyde
23.	produced by carbohydrate oxidation under aerobic conditions	b.	lactate
24.	produced by glycolysis under anaerobic conditions in the human body	c.	acetyl CoA
		d.	ethanol

Match each characteristic on the left with the cycle or process on the right.

25.	glucose is converted to glycogen	a.	glycogenesis
26.	synthesis of glucose from non-carbohydrate sources	b.	gluconeogenesis
27.	breakdown of glycogen to glucose	c.	glycogenolysis
		d.	glycolysis

True-False

28. Fructose and galactose are not metabolized by humans.
29. The blood sugar level in a hypoglycemic individual is higher than the normal fasting level.
30. The glycolysis pathway is inhibited by ATP.
31. Part of the Cori cycle involves the conversion of pyruvate to glucose.
32. About 90% of gluconeogenesis takes place in the liver.
33. Oxidative phosphorylation is a process coupled with the electron transport chain.
34. The chemiosmotic hypothesis pertains to the flow of cytoplasmic NADH across the mitochondrial membrane.
35. The net production of ATP from one mole of glucose in brain cells is greater than that in liver cells.

Chapter 24: Lipid and Amino Acid Metabolism

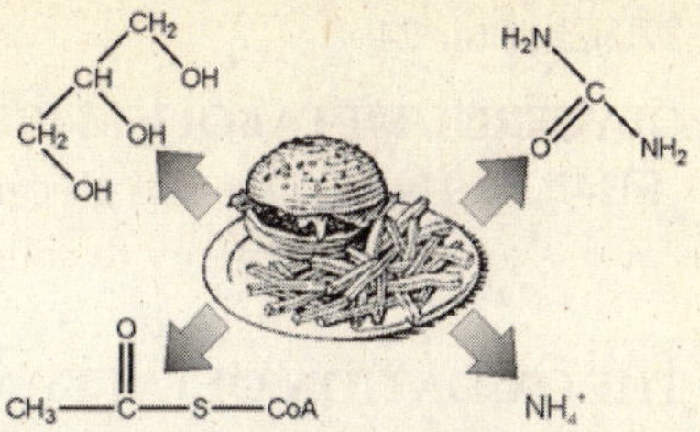

CHAPTER OUTLINE

24.1 Blood Lipids
24.2 Fat Mobilization
24.3 Glycerol Metabolism
24.4 The Oxidation of Fatty Acids
24.5 The Energy from Fatty Acids
24.6 Ketone Bodies
24.7 Fatty Acid Synthesis
24.8 Amino Acid Metabolism
24.9 Amino Acid Catabolism: The Fate of the Nitrogen Atoms
24.10 Amino Acid Catabolism: The Fate of the Carbon Skeleton
24.11 Amino Acid Biosynthesis

LEARNING OBJECTIVES/ASSESSMENT

When you have completed your study of this chapter, you should be able to:

1. Describe the digestion, absorption, and distribution of lipids in the body. (Section 24.1; Exercises 24.2 and 24.4)
2. Explain what happens during fat mobilization. (Section 24.2; Exercise 24.8)
3. Identify the metabolic pathway by which glycerol is catabolized. (Section 24.3; Exercise 24.12)
4. Outline the steps of β-oxidation for fatty acids. (Section 24.4; Exercise 24.22)
5. Determine the amount of ATP produced by the complete catabolism of a fatty acid. (Section 24.5; Exercise 24.26)
6. Name the three ketone bodies and list the conditions that cause their overproduction. (Section 24.6; Exercise 24.28)
7. Describe the pathway for fatty acid synthesis. (Section 24.7; Exercise 24.34)
8. Describe the source and function of the body's amino acid pool. (Section 24.8; Exercise 24.38)
9. Write equations for transamination and deamination reactions. (Section 24.9; Exercise 24.44)
10. Explain the overall results of the urea cycle. (Section 24.9; Exercise 24.54)
11. Describe how amino acids can be used for energy production, the synthesis of triglycerides, and gluconeogenesis. (Section 24.10; Exercise 24.58)
12. State the relationship between intermediates of carbohydrate metabolism and the synthesis of nonessential amino acids. (Section 24.11; Exercise 24.64)

SOLUTIONS FOR THE END OF CHAPTER EXERCISES

BLOOD LIPIDS (SECTION 24.1)

☑24.2 The products of triglyceride digestion are glycerol, fatty acids, and monoglycerides.

☑24.4 Insoluble lipids form complexes with proteins to form lipoprotein aggregates (chylomicrons). These complexes pass into the lymph system and then into the bloodstream. They are modified by the liver into smaller lipoprotein particles, which allows most lipids to be transported to various parts of the body by the bloodstream.

24.6 The major types of lipoproteins are chylomicrons, very low-density lipoproteins, low-density lipoproteins, and high-density lipoproteins.

FAT MOBILIZATION (SECTION 24.2)

☑24.8 Fat mobilization is the process whereby epinephrine stimulates the hydrolysis of triglycerides (in adipose tissue) into fatty acids and glycerol, which enter the bloodstream.

24.10 Resting muscles and liver cells use fatty acids rather than glucose to satisfy their energy needs.

GLYCEROL METABOLISM (SECTION 24.3)

☑24.12 The two fates of glycerol after it has been converted to an intermediate of glycolysis are to provide energy to cells or be converted to glucose.

THE OXIDATION OF FATTY ACIDS (SECTION 24.4)

24.14 A fatty acid is prepared for catabolism by a conversion into fatty acyl CoA. This activation occurs outside the mitochondria in the cytoplasm of the cell.

24.16 FAD and NAD^+ are hydrogen-transporting coenzymes that play important roles in the oxidation of fatty acids.

24.18

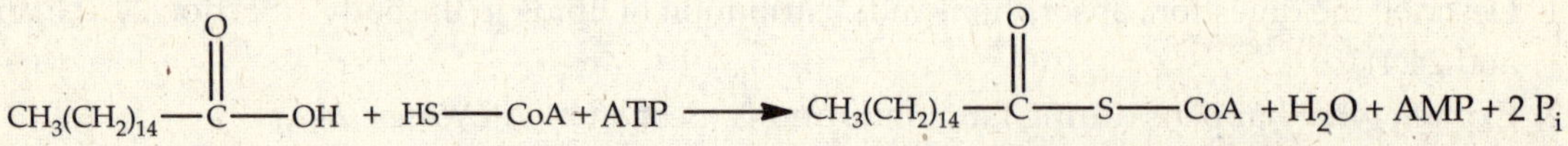

24.20 The oxidation of fatty acids is referred to as the fatty acid spiral rather than the fatty acid cycle because the process results in the production of acetyl CoA only. The fatty acid is not regenerated as it would be in a cycle.

☑24.22 a.

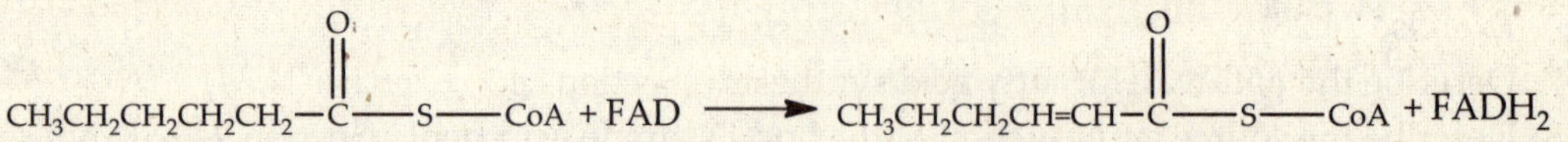

b.

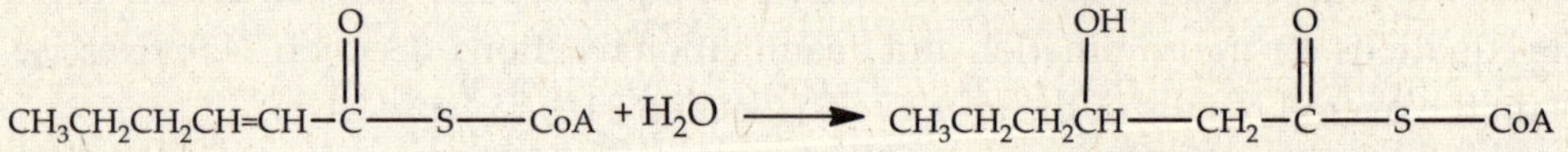

c.

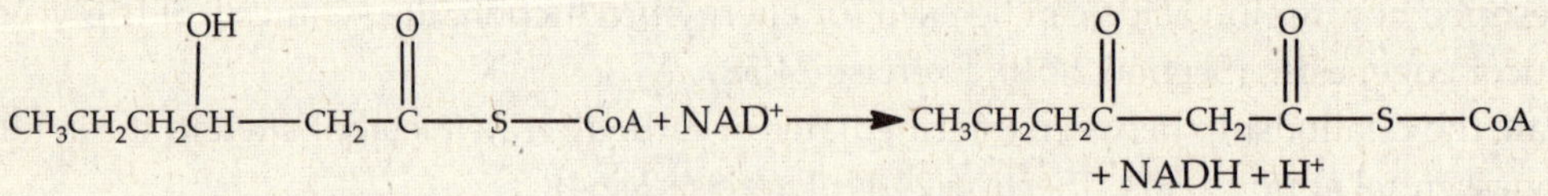

d.

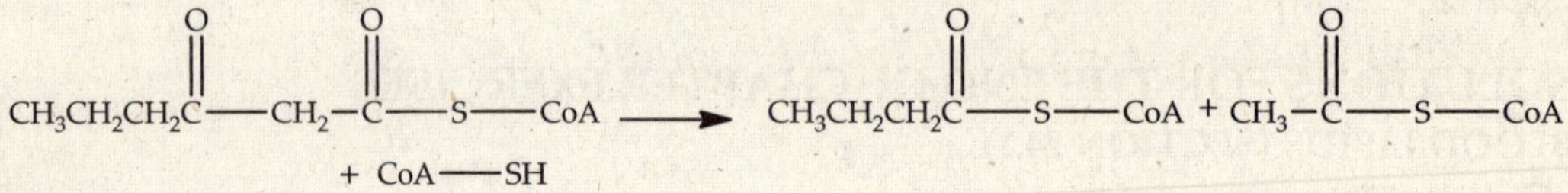

THE ENERGY FROM FATTY ACIDS (SECTION 24.5)

24.24 For every molecule of acetyl CoA that enters the citric acid cycle, 10 molecules of ATP are produced.

☑24.26 As a 10-carbon fatty acid passes through the fatty acid spiral, 5 acetyl CoA, 4 $FADH_2$, and 4 NADH molecules are produced. Consequently, 64 molecules of ATP can be produced.

Oxidation product or step	ATP/unit	Total ATP
Activation step		-2
5 acetyl CoA	10	50
4 $FADH_2$	1.5	6
4 (NADH + H^+)	2.5	10
Total		64

KETONE BODIES (SECTION 24.6)

☑24.28 Ketone bodies are the compounds synthesized within the liver from excess acetyl CoA. They are acetoacetate, β-hydroxybutyrate, and acetone.

24.30 Ketone bodies are formed in the liver. The brain, heart, and skeletal muscles can use these compounds to meet energy needs.

24.32 Ketosis is frequently accompanied by acidosis because two of the ketone bodies are acids. As the concentration of the ketone bodies rises in the blood, the pH of the blood decreases.

FATTY ACID SYNTHESIS (SECTION 24.7)

☑24.34 Acetyl CoA supplies the carbon atoms for fatty acid synthesis.

24.36 The liver can convert glucose to fatty acids via acetyl CoA; however, it does not have the enzyme that can catalyze the conversion of acetyl CoA to pyruvate which is needed for gluconeogenesis.

AMINO ACID METABOLISM (SECTION 24.8)

☑24.38 The amino acid pool is the body's store of available amino acids for maintenance. These amino acids arise from proteins eaten (and hydrolyzed), the body's degraded tissue, and the synthesis of amino acids in the liver.

24.40 Protein turnover is the dynamic process in which body proteins are continuously hydrolyzed and resynthesized.

24.42 Other than proteins synthesis, amino acids are used for the production of purine and pyrimidine bases of nucleic acids, heme structures for hemoglobin and myoglobin, choline and ethanolamine for phospholipids, and neurotransmitters such as acetylcholine and dopamine.

AMINO ACID CATABOLISM: THE FATE OF THE NITROGEN ATOMS (SECTION 24.9)

☑24.44
a. An amino group is removed from an amino acid and donated to an α-keto acid. *transamination*
b. Ammonium ion is produced. *deamination*
c. New amino acids are synthesized from other amino acids. *transamination*
d. A keto acid is produced from an amino acid. *deamination and transamination*

24.46

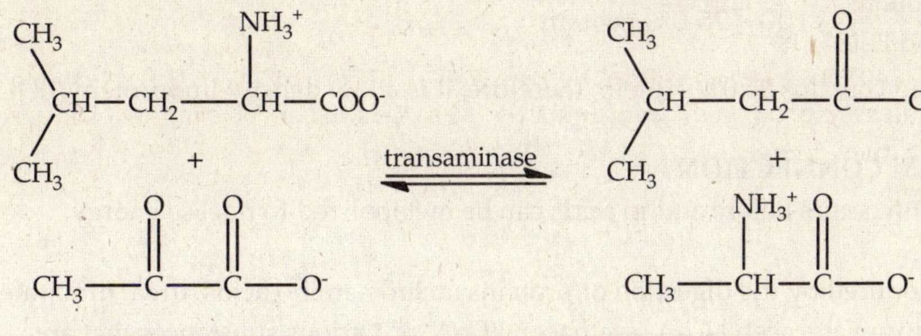

24.48

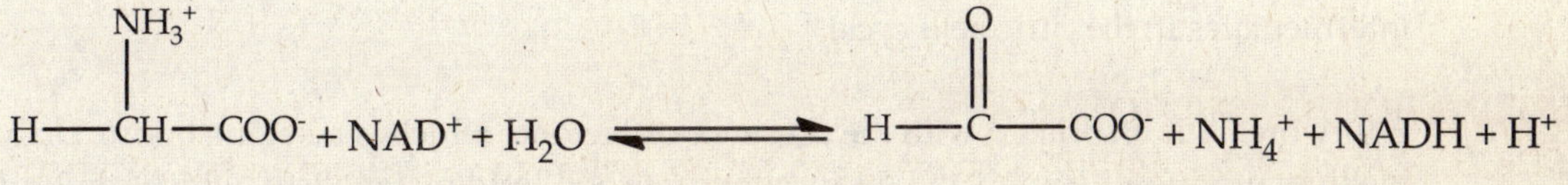

24.50 Urea is synthesized in the liver.

24.52 Carbamoyl phosphate enters the urea cycle by combining with ornithine.

☑24.54

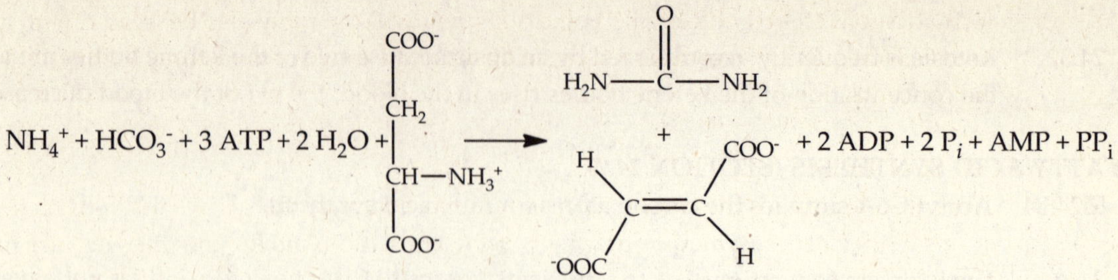

AMINO ACID CATABOLISM: THE FATE OF THE CARBON SKELETON (SECTION 24.10)

24.56 The amino acids isoleucine, phenylalanine, tryptophan, and tyrosine can be both glucogenic and ketogenic.

☑24.58 a. aspartate → oxaloacetate glucogenic
b. leucine → acetyl CoA ketogenic
c. tyrosine → acetoacetyl CoA ketogenic

AMINO ACID BIOSYNTHESIS (SECTION 24.11)

24.60 Essential amino acids are amino acids which must be obtained from the diet because, unlike nonessential amino acids, they cannot be synthesized within the liver in the amounts needed by the body.

24.62 Transaminases participate in the biosynthesis of nonessential amino acids and they transfer the nitrogen atoms of all amino acids to α-keto acids when the disposal of nitrogen is necessary. These functions are essential to the survival of the body.

☑24.64 Two general sources of intermediates for the biosynthesis of amino acids are the glycolysis pathway and citric acid cycle.

ADDITIONAL EXERCISES

24.66 Enzyme concentrations may be changed and regulated by allosteric regulation and by enzyme induction.

24.68 $$\frac{1.87 \text{ mg protein}}{7.48 \text{ mg blood lipid}} \times 100 = 25.0\% \text{ protein}$$

This blood lipid contains 25.0% protein; therefore, it is a low density lipoprotein (LDL).

ALLIED HEALTH EXAM CONNECTION

24.70 (b) Deamination occurs before amino acids can be metabolized to release energy.

24.72 Substances produced by the digestion of proteins include amino acids, urea, fumarate, ADP, P_i, AMP, PP_i, pyruvate, acetyl CoA, acetoacetyl CoA, or various substances that are intermediates in the citric acid cycle.

24.74 Fruity breath is associated with the presence of acetone, which is a ketone body. Diabetes mellitus produces an imbalance in carbohydrate and lipid metabolism, which can lead to the excessive production of acetyl CoA and a substantial increase in the level of ketone bodies in the blood of untreated diabetics. This patient is probably suffering from ketoacidosis and should be given insulin to restore normal glucose metabolism and reduce the rate of formation of ketone bodies. The hot, dry patient will probably also be treated with an intravenous solution containing sodium bicarbonate in order to restore his fluid and acid-base balance.

CHEMISTRY FOR THOUGHT

24.76 The half-lives of enzymes are comparatively short because this allows the body to recycle the atoms in the enzyme for more productive purposes and removes enzymes that are no longer participating in reactions.

24.78 For women, low HDL levels are the strongest predictor of heart disease. If women as a group have fewer heart attacks than do men as a group, then the HDL levels of women must be higher than the HDL levels for men.

24.80 It might be undesirable to reduce blood cholesterol to very low levels because the HDL or good cholesterol level needs to be high enough to help maintain health. Cholesterol is an essential component of cell membranes, which in turn are essential for health and life.

ADDITIONAL ACTIVITIES

Section 24.1 Review:

(1) Why are triglycerides and phosphoglycerides digested into component substances before they are absorbed into cells of the intestinal mucosa, if they are resynthesized once they are absorbed?
(2) What does the term *lipoprotein* imply?
(3) How do the densities of proteins and lipids compare?

Section 24.2 Review:

(1) What part of the cell is responsible for fatty acid oxidation?
(2) Fatty acids cannot cross the blood-brain barrier; however, fatty acids can be used to supply energy to the brain during a fast. Explain.
(3) Which horomone stimulates the hydrolysis of triglycerides?
(4) What is formed when fatty acids combine with serum albumin?
(5) How is glycerol transported in the body?

Section 24.3 Review:

(1) Glycerol can be converted to dihydroxyacetone phosphate. What is the chemical process by which dihydroxyacetone phosphate can be converted into pyruvate?
(2) What could the pyruvate from (1) be used for in the body?

Section 24.4 Review:

(1) Is the process for converting a fatty acid to a fatty acyl CoA endothermic or exothermic? Explain.
(2) What part of the cell can a fatty acyl CoA enter that a fatty acid molecule cannot?
(3) What process can occur once the fatty acyl CoA enters the part of the cell identified in (2)?
(4) What is the name of the process when one fatty acyl CoA repeats the process in (3) until all of the carbon atoms from the original molecule are in molecules of acetyl CoA?

Section 24.5 Review:

(1) As food is digested and turned into ATP, is the food oxidized or reduced?
(2) If excess energy is stored in fat, why are low-fat diets encouraged?

Section 24.6 Review:

(1) What happens if more acetyl CoA is produced by fatty acid oxidation than can be processed through the citric acid cycle?
(2) How does the body process ketone bodies at a low concentration in the blood?
(3) How does the body process ketone bodies at a high concentration in the blood?
(4) What are the consequences of prolonged ketoacidosis?

Section 24.7 Review:

(1) Does the phrase, "Use it or lose it," apply to the energy provided by excess fat or carbohydrates in the diet?
(2) Where does biosynthesis of fatty acids occur in the cell?
(3) Acetyl CoA is produced during the fatty acid spiral in the mitochondria. How is it involved in the biosynthesis of fatty acids?
(4) Which type of fatty acids cannot be synthesized in the body?
(5) What would a nutritionist call the fatty acids in (4) that are needed by the body?
(6) Acetyl CoA is one of the "fates of pyruvate." Is pyruvate one of the "fates of acetyl CoA" in humans?

Section 24.8 Review:

(1) Does the phrase, "Use it or lose it," apply to proteins?
(2) What are the sources for the amino acid pool?
(3) What are the uses of the amino acid pool?

Section 24.9 Review:

(1) What processes do transaminases catalyze?
(2) What undergoes oxidation during oxidative deamination?
(3) On the figure below, identify and label the Stage 1 reaction(s).
(4) On the figure below, identify and label the Stage 2 reaction(s).
(5) On the figure below, identify and label the Stage 3 reaction(s).

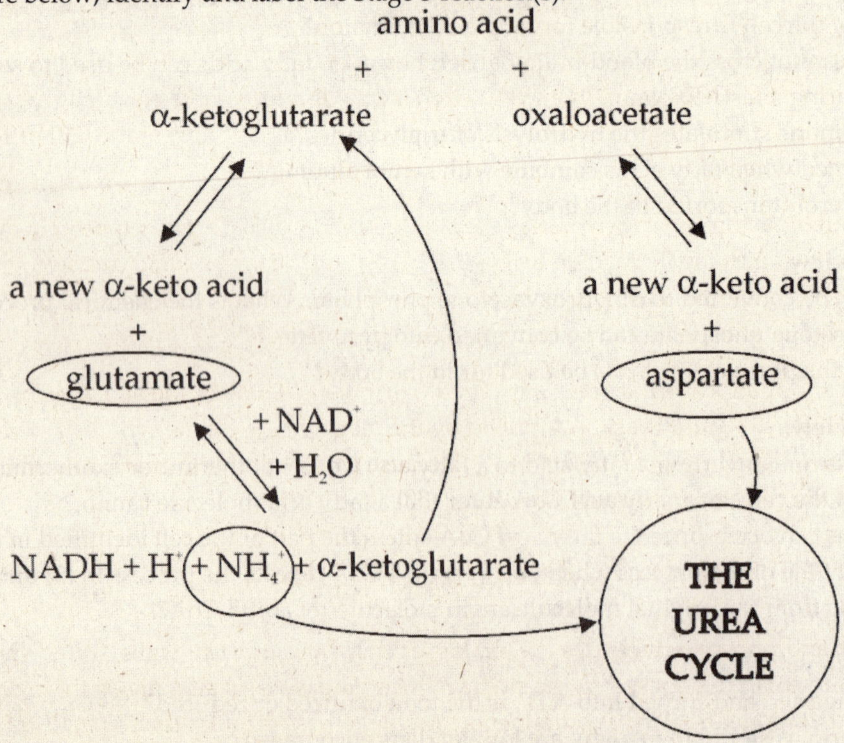

Section 24.10 Review:

(1) Can amino acids be used as a source of energy in the body?
(2) Can amino acids be converted into pyruvate?
(3) Can amino acids be converted into source material for gluconeogenesis?
(4) Can amino acids be converted into source material for ketone bodies and fatty acids?
(5) Give one reason why a person suffering from anorexia nervosa might have heart trouble.

Section 24.11 Review:

(1) Why is cirrhosis of the liver dangerous for the health of the entire body?
(2) What is the relationship between essential and nonessential amino acids?
(3) Which enzymes are a part of the relationship in (2)?

Tying It All Together with a Laboratory Application:

When a patient has (1) ______ breath, it suggests that (2) ______ is the macronutrient serving as a major energy source for the body. The cause of this condition could be a number of reasons, including that the patient might have (3) ______ or be fasting.

Urinalysis that reveals glucose and (4) ______ indicates that (3) is likely because both of these materials are above their respective (5) ______. The presence of glucose in the urine is called (6) ______ and the presence of (4) in the urine is called (7) ______. The blood condition in which (4) is revealed is called (8) ______. A patient who suffers from (1) breath, (7), and (8) is said to have (9) ______. This condition can lead to (10) ______ blood pH (or (11) ______) because two of the compounds associated with (4) are (12) ______.

(13) ______ can be used as a treatment for patients with (3) in order to use the glucose from their diet and retard the (14) ______ mechanism that was activated in order to meet the body's energy needs. The sequence of reactions for producing energy from lipids is called the (15) ______. Lipids store (16) ______ energy than carbohydrates of similar mass. Consequently, when excess carbohydrates or lipids are consumed, they undergo (17) ______ to produce fatty acids which can be converted to (18) ______ and stored for later use.

Blood work done within 6 hours of eating a lipid-rich meal will likely show elevated (19) ______ levels (compared to fasting). These (19)s are distinguished by (20) ______ because of the (20) differences between lipids and proteins.

Excess proteins are not stored as (18); consequently, if the (21) ______ (from diet, protein degradation, and biosynthesis) becomes too large, then more (22) ______ will be formed and excreted in the urine. (23) ______ and ______ are the amino acids produced by transamination and (24) ______ that can be used to form (22) by participating in or supplying intermediates to the (25) ______ cycle.

The degradation and rebuilding of proteins is called (26) ______. Some of the amino acids required for this process are (27) ______ and must be supplied by the diet. Others can be synthesized in the body. Protein can also be used as a source of (28) ______ because the (29) ______ of amino acids, specifically the carbon skeletons, produces intermediates of the (30) ______.

SOLUTIONS FOR THE ADDITIONAL ACTIVITIES

Section 24.1 Review:

(1) Triglycerides and phosphoglycerides are too large to pass through cell walls. When these compounds are eaten they pass into the stomach where lipases and acid digest these compounds into their smaller components. These components are able to pass through the cell walls of the intestinal mucosa where they are use to resynthesize triglycerides and phosphoglycerides.

(2) The term *lipoprotein* implies the compound contains both lipid and protein components.
(3) Proteins are more dense than lipids.

Section 24.2 Review:

(1) The mitochondria are responsible for fatty acid oxidation.
(2) Glucose and other compounds can cross the blood-brain barrier; therefore, the fatty acids must be broken down and used to form compounds that can cross the barrier in order to supply energy to the brain during a fast.
(3) Epinephrine stimulates the hydrolysis of triglycerides.
(4) A lipoprotein forms when fatty acids combine with serum albumin.
(5) Glycerol is water soluble and it dissolves in the blood for transport in the body.

Section 24.3 Review:

(1) Glycolysis is the chemical process by which dihydroxyacetone phosphate can be converted into pyruvate.
(2) Pyruvate can be used for energy production or gluconeogenesis.

Section 24.4 Review:

(1) The process for converting a fatty acid to a fatty acyl CoA is endothermic because a molecule of ATP must be broken down into AMP and PP_i to release the energy required for this reaction.
(2) A fatty acyl CoA can enter the mitochondria inside the cell, but a fatty acid molecule cannot.
(3) Once the fatty acyl CoA enters the mitochondria, β-oxidation can occur.
(4) The fatty acid spiral is the process by which one fatty acyl CoA molecule undergoes repeated β-oxidation until all of the carbon atoms from the original molecule are in molecules of acetyl CoA.

Section 24.5 Review:

(1) Food is oxidized as it is digested and turned into ATP.
(2) Low-fat diets are encouraged for many reasons. The body will use energy from carbohydrate sources before it uses energy from fat sources. If excess carbohydrates are consumed, they can be stored in the form of fat. The American diet typically includes too much dietary fat. This fat is excessive because the body does not need as much energy as the fat provides; therefore, this fat is stored in the body. The body does need some fat-soluble vitamins; however, these can be obtained in adequate amounts from a low-fat diet.

Section 24.6 Review:

(1) Ketone bodies form in the liver, if more acetyl CoA is produced by fatty acid oxidation than can be processed through the citric acid cycle.
(2) Ketone bodies at a low concentration in the blood can be oxidized to meet energy needs of cells.
(3) Ketone bodies at a high concentration in the blood result in a condition called ketonemia (ketones in the blood) as well as ketonuria (ketones in the urine), if the renal threshold for ketone bodies is exceeded. If the level is high enough, acetone will be expelled through the lungs and "acetone breath" results. At this level, the condition is called ketosis.
(4) The consequences of prolonged ketoacidosis are debilitation, coma, and death.

Section 24.7 Review:

(1) No, the phrase, "Use it or lose it," does not apply to the energy provided by excess fat or carbohydrates in the diet. They are stored as fat.
(2) The biosynthesis of fatty acids occurs in the cytoplasm of the cell.
(3) Acetyl CoA reacts with oxaloacetate and water in the mitochondria to produce citrate that can be transported into the cytoplasm. In the cytoplasm, the citrate is converted back into acetyl CoA and oxaloacetate. The acetyl CoA is then used to synthesize fatty acids.

(4) Polyunsaturated fatty acids cannot be synthesized in the body.
(5) A nutritionist would call those polyunsaturated fatty acids essential fatty acids.
(6) No, pyruvate is not one of the "fates of acetyl CoA" in humans.

Section 24.8 Review:

(1) Yes, the phrase, "Use it or lose it," applies to proteins. They cannot be stored for later use.
(2) The sources for the amino acid pool are dietary proteins, biosynthesis in the liver, and degradation of body tissues.
(3) The uses of the amino acid pool include rebuilding body tissues, biosynthesis of nitrogen-containing compounds, and catabolism for energy and excretion of excess nitrogen.

Section 24.9 Review:

(1) Transaminases catalyze reactions that involve the transfer of an amino group from amino acids to other amino acids or α-keto acids.
(2) During oxidative deamination, glutamate is oxidized.

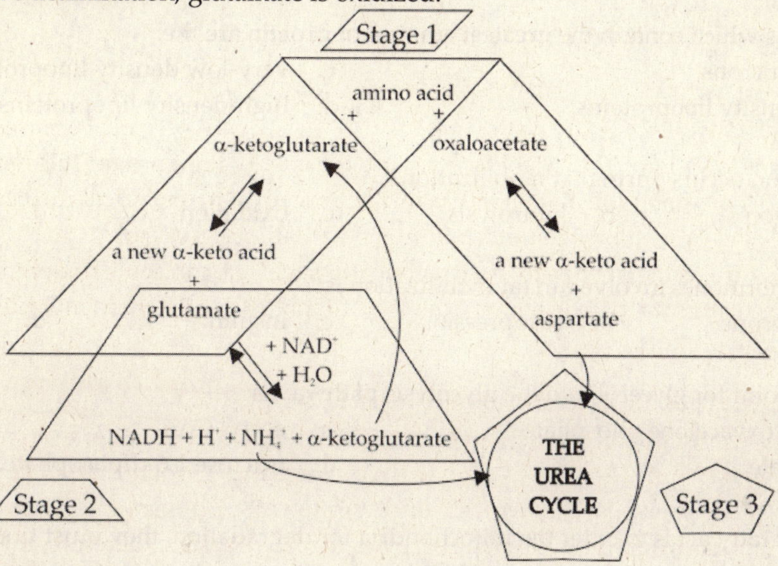

Section 24.10 Review:

(1) Yes, amino acids can be used as a source of energy in the body.
(2) Yes, amino acids can be converted into pyruvate.
(3) Yes, amino acids can be converted into source material for gluconeogenesis.
(4) Yes, amino acids can be converted into source material for ketone bodies and fatty acids.
(5) A person suffering from anorexia nervosa might have heart trouble because the heart is a muscle and the body will hydrolyze proteins from body tissues, including muscles, in order to supply amino acids for gluconeogenesis during starvation. The loss of protein from the heart may weaken it.

Section 24.11 Review:

(1) The liver is extremely active biochemically. It is a major site of biosynthesis. Liver damage, including cirrhosis, will decrease the ability of the liver to function.
(2) Essential amino acids or other compounds can be used to synthesize nonessential amino acids in the body.
(3) Transaminases are needed for the synthesis of nonessential amino acids.

Tying It All Together with a Laboratory Application:

(1) acetone	(4) ketone bodies	(7) ketonuria	(10) low
(2) fat	(5) renal thresholds	(8) ketonemia	(11) ketoacidosis
(3) diabetes	(6) glucosuria	(9) ketosis	(12) acids

(13) Insulin
(14) fat mobilization
(15) fatty acid spiral
(16) more
(17) biosynthesis
(18) body fat
(19) lipoprotein
(20) density
(21) amino acid pool
(22) urea
(23) Glutamate, aspartate
(24) deamination
(25) urea
(26) protein turnover
(27) essential
(28) energy
(29) catabolism
(30) citric acid cycle

SELF-TEST QUESTIONS

Multiple Choice

1. What reaction occurs during the digestion of triglycerides?
   a. hydrolysis b. hydrogenation c. hydration d. oxidation

2. The products of triglyceride digestion are fatty acids, some monoglycerides, and:
   a. triglycerides. b. diglycerides. c. glucose. d. glycerol.

3. Lipoproteins which contain the greatest amount of protein are the:
   a. chylomicrons.
   b. low-density lipoproteins.
   c. very-low density lipoproteins.
   d. high-density lipoproteins.

4. What reaction occurs during fat mobilization?
   a. reduction b. hydrolysis c. oxidation d. hydration

5. One of the hormones involved in fat mobilization is:
   a. aldosterone. b. vasopressin. c. insulin. d. epinephrine.

6. The entry point for glycerol into the glycolysis pathway is:
   a. dihydroxyacetone phosphate.
   b. pyruvate.
   c. acetyl CoA.
   d. fructose 1,6-diphsophate.

7. In order for fatty acids to enter the mitochondria for degradation, they must first be converted to:
   a. acetyl CoA. b. fatty acyl CoA. c. pyruvate. d. malonate.

8. Which of the following processes takes place during the fatty acid spiral?
   a. addition of water to a double bond
   b. addition of H atoms to a double bond
   c. oxidation of an –OH group to a ketone
   d. more than one response is correct

9. During one run through the fatty acid spiral, which bond of the following fatty acid would be broken?

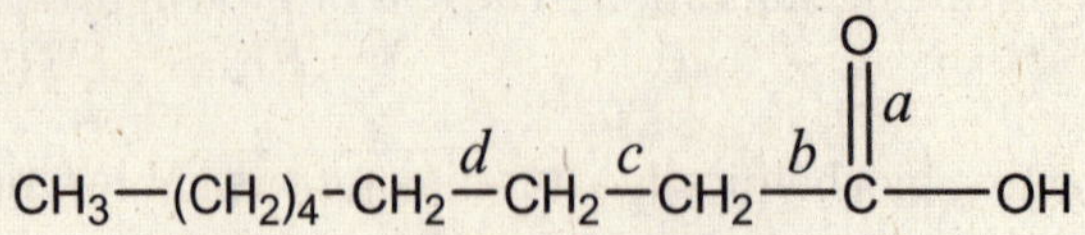

   a. bond a b. bond b c. bond c d. bond d

10. Which of the following is a product of the fatty acid spiral?
    a. pyruvate
    b. $CO_2 + H_2O$
    c. acetyl CoA
    d. more than one response is correct

11. How many runs through the fatty acid spiral would be required to completely break down one molecule of a 12-carbon fatty acid?
    a. 1 b. 5 c. 6 d. 12

12. How many $FADH_2$ molecules are produced from one turn of the fatty acid spiral?
    a. 1 b. 2 c. 3 d. 4

13. How many ATP molecules ultimately result from one turn of the fatty acid spiral?
    a. 8 b. 12 c. 14 d. 17

14. The concentration of ketone bodies builds up when increased amounts of ______ are oxidized.
    a. amino acids b. fatty acids c. glucose d. glycerol

15. Which of the following types of proteins tends to have the shortest half-life in the body?
    a. enzymes
    b. connective tissue proteins
    c. plasma proteins
    d. muscle proteins

16. The primary function of the urea cycle in the body is to:
    a. produce ATP from ADP.
    b. convert keto acids into amino acids.
    c. convert amino acids into keto acids.
    d. convert ammonia ions into urea.

17. Which of the following substances is a product of a transamination reaction?
    a. amino acid
    b. keto acid
    c. ammonia
    d. more than one response is correct

18. The nitrogen of amino acids appears in the urine of mammals primarily as:
    a. uric acid. b. urea. c. ammonia. d. N_2.

19. Which of the following substances enters into the urea cycle?
    a. malonyl CoA
    b. acetyl CoA
    c. phosphocreatine
    d. carbamoyl phosphate

20. Which one of the following is an intermediate formed in the urea cycle?
    a. CO_2 b. acetyl CoA c. pyruvate d. aspartate

21. Deamination of an amino acid produces:
    a. ammonium ions.
    b. CO_2.
    c. an α-keto acid.
    d. more than one response is correct

22. The carbon skeletons of amino acids are ultimately catabolized through:
    a. glycolysis.
    b. the citric acid cycle.
    c. glycogenolysis.
    d. the urea cycle.

23. A number of amino acids can be synthesized in the body from intermediates of:
    a. the citric acid cycle.
    b. oxidative phosphorylation.
    c. the urea cycle.
    d. the fatty acid spiral.

True-False

24. Fat has a caloric value more than twice that of glycogen and starch.
25. Upon complete oxidation to CO_2 and H_2O, fatty acids produce more net energy than a carbohydrate containing the same number of carbon atoms.
26. The glycerol resulting from triglyceride hydrolysis can be converted to pyruvate.
27. The glycerol resulting from triglyceride hydrolysis can be converted to glucose.

28. Starvation can lead to ketosis.
29. Amino acids in excess of immediate body requirements are stored for later use.
30. Amino acids can be catabolized for energy production.
31. Urea formation is an energy-yielding process.
32. The carbon atoms of some amino acids can be used in fatty acid synthesis.
33. The carbon atoms of some amino acids can be used in glucose synthesis.
34. Biosynthesis of fatty acids occurs in the mitochondria of the cell.
35. The liver is the most important organ in fatty acid synthesis.

Chapter 25: Body Fluids

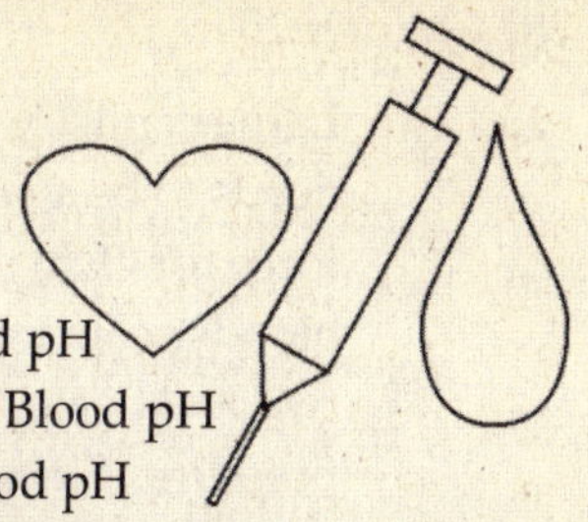

CHAPTER OUTLINE

25.1 A Comparison of Body Fluids
25.2 Oxygen and Carbon Dioxide Transport
25.3 Chemical Transport to the Cells
25.4 The Constituents of Urine
25.5 Fluid and Electrolyte Balance
25.6 Acid-Base Balance
25.7 Buffer Control of Blood pH
25.8 Respiratory Control of Blood pH
25.9 Urinary Control of Blood pH
25.10 Acidosis and Alkalosis

LEARNING OBJECTIVES/ASSESSMENT

When you have completed your study of this chapter, you should be able to:

1. Compare the chemical compositions of plasma, interstitial fluid, and intracellular fluid. (Section 25.1; Exercise 25.2)
2. Explain how oxygen and carbon dioxide are transported within the bloodstream. (Section 25.2; Exercise 25.10)
3. Explain how materials move from the blood into the body cells and from the body cells into the blood. (Section 25.3; Exercise 25.14)
4. List the normal and abnormal constituents of urine. (Section 25.4; Exercise 25.18)
5. Discuss how proper fluid and electrolyte balance is maintained in the body. (Section 25.5; Exercise 25.22)
6. Explain how acid-base balance is maintained in the body. (Section 25.6; Exercise 25.26)
7. Explain how buffers work to control blood pH. (Section 25.7; Exercise 25.28)
8. Describe respiratory control of blood pH. (Section 25.8; Exercise 25.30)
9. Describe urinary control of blood pH. (Section 25.9; Exercise 25.34)
10. List the causes of acidosis and alkalosis. (Section 25.10; Exercise 25.38)

SOLUTIONS FOR THE END OF CHAPTER EXERCISES

A COMPARISON OF BODY FLUIDS (SECTION 25.1)

☑25.2 Plasma and interstitial fluid have nearly the same chemical composition.

25.4
a. protein — intracellular fluid
b. K^+ — intracellular fluid
c. Na^+ — plasma, interstitial fluid
d. HPO_4^{2-} — intracellular fluid
e. HCO_3^- — plasma, interstitial fluid
f. Mg^{2+} — intracellular fluid
g. Cl^- — plasma, interstitial fluid

25.6
a. The principal anion in blood plasma is chloride, Cl^-.
b. The principal anion in interstitial fluid is chloride, Cl^-.
c. The principal anion in intracellular fluid is hydrogen phosphate, HPO_4^{2-}.

OXYGEN AND CARBON DIOXIDE TRANSPORT (SECTION 25.2)

25.8
a. $HHb + O_2 \rightleftharpoons HbO_2^- + H^+$
b. $HHb + CO_2 \rightleftharpoons HHbCO_2$

☑25.10
a. O_2 as dissolved gas in plasma — 2%
b. O_2 as oxyhemoglobin — 98%
c. CO_2 as carbaminohemoglobin — 25%
d. CO_2 as dissolved gas in plasma — 5%
e. CO_2 as bicarbonate ions — 70%

25.12 Large capillary surface areas are critical to O_2 and CO_2 transport processes because the body needs to be able to diffuse large amounts of O_2 and CO_2 through the capillaries. The driving force for O_2 to move from the lung alveoli into the bloodstream and CO_2 from the bloodstream into the lung alveoli is the difference in the partial pressures of these gases in the two locations. Gases will move from a region of higher pressure to a region of lower pressure. Since the partial pressure of oxygen is higher in the lungs than in the blood oxygen moves from the lungs into the blood. The movement of carbon dioxide occurs in the opposite direction.

CHEMICAL TRANSPORT TO THE CELLS (SECTION 25.3)

☑25.14 a. At the venous end of a capillary osmotic pressure has the greater influence on the direction of fluid movement through capillary walls. This is because venous blood pressure is much lower than arterial blood pressure. The blood moves more slowly and the concentration of dissolved substances become more important in determining the direction of fluid movement.

b. At the arterial end of a capillary blood pressure has the greater influence on the direction of fluid movement through capillary walls. Since arterial blood pressure is high this factor outweighs osmotic pressure in determining the direction of fluid movement.

THE CONSTITUENTS OF URINE (SECTION 25.4)

25.16 The organic material normally excreted in the largest amount in urine is urea.

☑25.18 Abnormal constituents of urine include large amounts of glucose, protein, ketone bodies, pus, hemoglobin, red blood cells, and large amounts of bile pigments.

FLUID AND ELECTROLYTE BALANCE (SECTION 25.5)

25.20 Water leaves the body via the kidneys (urine), the lungs (water vapor in expired air), the skin (diffusion and perspiration), and the intestines (feces).

☑25.22 Fluid balance is primarily maintained in the body by urine output.

25.24 As the body loses water, salivary secretions decrease, a dry feeling develops in the mouth, thirst is recognized, and water is drunk to relieve the condition.

ACID-BASE BALANCE (SECTION 25.6)

☑25.26 Three systems that cooperate to maintain blood pH in an appropriate narrow range are blood buffer, respiratory, and urinary. The blood buffer system provides the most rapid response to changes in blood pH.

BUFFER CONTROL OF BLOOD pH (SECTION 25.7)

☑25.28 Three major buffer systems of the blood are the bicarbonate system, the phosphate system, and the plasma protein system.

RESPIRATORY CONTROL OF BLOOD pH (SECTION 25.8)

☑25.30 The respiratory mechanism for controlling blood pH is as follows. As CO_2 and H_2O are exhaled, more carbonic acid is removed from the blood to form CO_2 in order to restore the equilibrium: $H_2CO_3 \rightleftarrows H_2O + CO_2$. The blood pH rises because the equilibrium is also linked to the production of H^+ ions. $H_2CO_3 \rightleftarrows H^+ + HCO_3^-$

If carbonic acid is removed from the system, the above equilibrium will shift to the left, lowering the concentration of H^+ and raising the blood pH. A reduced rate of breathing will maintain a higher concentration of CO_2 and lower the blood pH.

The reverse response will occur, if the rate of breathing is too low. If insufficient CO_2 and H_2O are exhaled, carbonic acid builds up and causes the second equilibrium to shift to the right. This increases the concentration of H^+ and lowers the blood pH. An increased rate of breathing will remove the CO_2 more quickly and raise the blood pH.

25.32 Hypoventilation is a symptom of high pH in the body (alkalosis).

URINARY CONTROL OF BLOOD pH (SECTION 25.9)

☑25.34 When kidneys function to control blood pH, the concentration of H^+ in the blood decreases. As a result, the CO_2 in the blood is converted to bicarbonate ions in the blood, in order to maintain the equilibria: $H_2O + CO_2 \rightleftarrows H_2CO_3 \rightleftarrows H^+ + HCO_3^-$. Removal of H^+ ions causes more carbonic acid to form HCO_3^-. The source of carbonic acid is CO_2. Overall CO_2 is converted to HCO_3^-.

25.36 As hydrogen ions are removed from the blood by the kidneys, sodium ions enter the bloodstream. This ionic shift maintains electron charge balance when the kidneys function to increase the pH of blood.

ACIDOSIS AND ALKALOSIS (SECTION 25.10)

☑25.38 Respiratory acid-base imbalances are caused by breathing too quickly and deeply or too slowly and shallowly, whereas metabolic acid-base imbalances are caused by various body processes that either produce or use acid.

25.40 An individual who takes an overdose of a depressant narcotic will probably breathe more slowly and shallowly. This will likely lead to respiratory acidosis. Diabetes mellitus can cause metabolic acidosis.

ADDITIONAL EXERCISES

25.42

$$25.0 \text{ g urea}\left(\frac{1 \text{ mole urea}}{60.0 \text{ g urea}}\right)\left(\frac{2 \text{ moles N}}{1 \text{ mole urea}}\right)\left(\frac{1 \text{ mole a.a.}}{1 \text{ mole N}}\right)\left(\frac{6.02\times10^{23} \text{ molecules a.a.}}{1 \text{ mole a.a.}}\right)$$

$$= 5.02\times10^{23} \text{ amino acid molecules}$$

25.44 When aldosterone is secreted from the adrenal gland, sodium ions are reabsorbed from the renal tubes into the blood. Chloride ions follow the sodium ions to maintain electrical neutrality. This causes the solute concentration in the blood to increase and the water follows the sodium chloride because water is the solvent and the osmotic pressure difference causes the water solvent to flow into the blood in an attempt to dilute the ionic solution in the blood.

25.46 Heavy exercise can cause metabolic acidosis because when the body has used all of its glucose and glycogen supplies, it mobilizes fat to provide energy. This in turn produces ketone bodies which can lead to ketoacidosis.

ALLIED HEALTH EXAM CONNECTION

25.48 Movement of respiratory gases into and out of blood is primarily dependent upon the presence of a partial pressure gradient.

25.50 The normal constituents of urine include: (a) creatinine, (b) urea, (c) water, and (d) ammonia. (e) Sugar, (f) hormones, and (g) blood are not normal constituents of urine.

25.52 The liquid portion of the blood is called (c) plasma.

25.54 Water reabsorption in the collecting duct of the kidneys is controlled by (b) ADH from the posterior pituitary.

25.56 The presence of protein in the urine is called (c) albuminuria.

CHEMISTRY FOR THOUGHT

25.58 A hysterical individual might develop alkalosis because he or she is hyperventilating which causes too much carbon dioxide to be exhaled, which in turn disturbs the blood's normal carbon dioxide/carbonic acid ratio. In order to restore equilibrium, the concentration of carbonic acid will decrease and the blood pH will consequently increase, according to the related equilibria: $H_2O + CO_2 \rightleftarrows H_2CO_3 \rightleftarrows H^+ + HCO_3^-$. This can result in respiratory alkalosis.

25.60 Ordinary water may be the most effective "sports drink" because the body needs to be hydrated in order to maintain fluid and electrolyte balance in the body. Water comprises a large percentage of the composition of the human body and drinking water can help restore natural balance within the body during and after exercise.

25.62 If the concentration of H^+ ions in the urine exceeds the urine's buffering capacity, the pH of the urine will drop below pH 6 to produce very acidic urine.

25.64 Meat and grains can make urine acidic, while fruits and vegetables can make urine basic.

ADDITIONAL ACTIVITIES

Section 25.1 Review:

(1) Rank intracellular fluid, plasma, and interstitial fluid in order of increasing volumes in the body.
(2) What class of fluids contains both plasma and interstitial fluid?
(3) What are the major differences between the cations and anions inside intracellular fluid and the category from (2)?
(4) Rank intracellular fluid, plasma, and interstitial fluid in order of increasing protein content.

Section 25.2 Review:

(1) How is most oxygen transported through the bloodstream?
(2) How is most carbon dioxide transported through the bloodstream?
(3) How does oxygen enter the bloodstream?
(4) Why is chloride shift physiologically important?
(5) What enzyme is responsible for breaking carbonic acid into carbon dioxide and water?
(6) What enzyme is responsible for the formation of carbonic acid from carbon dioxide and water?

Section 25.3 Review:

(1) What are the various mechanisms of chemical transport in the body?
(2) Do capillary walls act more like osmotic membranes or dialyzing membranes? Explain.
(3) What factors govern the movement of water and dissolved materials through the capillary walls?

(4) How does the importance of the factors in (3) change as blood is pumped from the heart through the body and back to the heart?

Section 25.4 Review:

(1) Explain why urinalysis may be the first test performed on a patient entering the hospital.
(2) What is the major component of urine?
(3) Approximately how many grams of (2) should be excreted by an adult, if they also excrete 40 g of dissolved solids?
(4) If the entire volume of urine was due to the major component of (2), how many milliliters of urine should be excreted?
(5) Draw a pH scale for the "normal" urine pH range. Mark the neutral pH, the average pH for urine, the pH range for fruit/vegetable-rich diets, and the pH range for protein-rich diets.
(6) Is the average pH for urine acidic, basic, or neutral?

Section 25.5 Review:

(1) Is the relationship between water content and fat content in the body generally direct or inverse?
(2) Why are marathon runners generally given sports drinks at aid stations rather than just water?
(3) Fluid enters the body through food and drink. How does fluid leave the body?
(4) Which hormones influence urine output?

Section 25.6 Review:

(1) What is the $[H^+]$ range for blood in the normal pH range?
(2) Is normal blood pH acidic, basic, or neutral?
(3) What are the three systems to maintain normal blood pH?

Section 25.7 Review:

(1) Two of the three ketone bodies are acids. Explain how the blood is able to maintain a healthy pH, even when a low concentration of ketone bodies are present.
(2) What happens to blood pH when a high concentration of ketone bodies are present in the blood?
(3) What is buffering according to biochemists?

Section 25.8 Review:

(1) During exercise, which is more likely to occur: hyperventilation or hypoventilation?
(2) During meditation, which is more likely to occur: hyperventilation or hypoventilation?

Section 25.9 Review:

(1) Which buffer systems are involved in the urinary system?
(2) Which buffer system is most responsible for keeping urine near pH 6?

Section 25.10 Review:

(1) When blood pH is normal and acid-base balance exists, which is in higher concentration: bicarbonate ions or carbonic acid?
(2) During respiratory alkalosis, which is in higher concentration: bicarbonate ions or carbonic acid?
(3) What is the cause of respiratory alkalosis?
(4) During respiratory acidosis, which is in higher concentration: bicarbonate ions or carbonic acid?
(5) What is the cause of respiratory acidosis?
(6) During metabolic acidosis, which is in higher concentration: bicarbonate ions or carbonic acid?
(7) What is the cause of metabolic acidosis?
(8) During metabolic alkalosis, which is in higher concentration: bicarbonate ions or carbonic acid?
(9) What is the cause of metabolic alkalosis?

Tying It All Together with a Laboratory Application:

(1) ______ and urine are body fluids examined in a hospital setting on a regular basis. The pH of (1) should be in the range of (2) ______, while the pH of urine should be in the range of (3) ______. The pH of (1) is regulated by (4) ______, respiratory, and urinary systems. The pH of urine is regulated by the (5) ______ buffer.

One of the main pathways of chemical transport within the body is the (6) ______. The main cation found in the (1) plasma is (7) ______ and the main anion is (8) ______. Nutrients are delivered to the cells because of (9) ______ and waste products are reclaimed from the cells because of (10) ______. The body disposes of these waste products in (11) ______, ______, ______, ______, and ______. These processes also involve (12) ______ and electrolytes. When the (12) level in the body is low, (13) ______ helps maintain electrolyte balance.

Hyperventilation, prolonged vomiting, and ingesting excess antacid can lead to (14) ______, which is a condition in which the pH of the blood is (15) ______. Hypoventilation, uncontrolled diabetes mellitus, severe diarrhea, and aspirin overdose can lead to (16) ______, which is a condition in which the pH of the blood is (17) ______. These conditions are dangerous because death can occur if the pH of (1) is below (18) ______ or above (19) ______. Respiratory causes of (14) and (16) change the concentration of (20) ______ in the (1).

SOLUTIONS FOR THE ADDITIONAL ACTIVITIES

Section 25.1 Review:

(1) volume in the body: plasma < interstitial fluid < intracellular fluid
(2) Extracellular fluids contain both plasma and interstitial fluid.
(3) Intracellular fluids contain mostly potassium cations and hydrogen phosphate anions. Extracellular fluids contain mostly sodium cations and chloride anions.
(4) protein content: interstitial fluid < plasma < intracellular fluid

Section 25.2 Review:

(1) Most oxygen is transported through the bloodstream bound to hemoglobin in the form of oxyhemoglobin.
(2) Most carbon dioxide is transported through the bloodstream as bicarbonate ions.
(3) Oxygen enters the bloodstream by diffusing from the alveoli of the lungs through the capillary walls into the red blood cells.
(4) The chloride shift is physiologically important because it maintains electrolyte balance and osmotic pressure relationships between red blood cells and plasma.
(5) Carbonic anhydrase is responsible for breaking carbonic acid into carbon dioxide and water.
(6) Carbonic anhydrase is responsible for the formation of carbonic acid from carbon dioxide and water.

Section 25.3 Review:

(1) Chemical transport in the body requires the substances to become part of the moving bloodstream. They may dissolve in the water-based plasma, chemically bond to cellular components, or form a suspension in the plasma.
(2) Capillary walls act more like dialyzing membranes than osmotic membranes because they allow materials other than the solvent (water) to pass through the walls.
(3) Blood pressure and osmotic pressure govern the movement of water and dissolved materials through the capillary walls.
(4) Blood pressure exceeds osmotic pressure as blood is pumped away from the heart; however, osmotic pressure exceeds blood pressure as blood is returned to the heart.

Section 25.4 Review:

(1) Urine is an excellent indicator of health. The components of urine can be used to diagnosis illness.

(2) The major component of urine is water.

(3) 960 g of water

$$\frac{4\%}{100\%} = \frac{40\text{ g}}{x}$$

$$4\%\,x = (100\%)(40\text{ g})$$

$$x = 1000\text{ g urine}$$

$$\text{water} = 96\%\text{ of urine} = (96\%)(1000\text{ g}) = 960\text{ g water}$$

(4) 960 mL

$$960\text{ g water}\left(\frac{1\text{ mL}}{1\text{g}}\right) = 960\text{ mL}$$

(5)

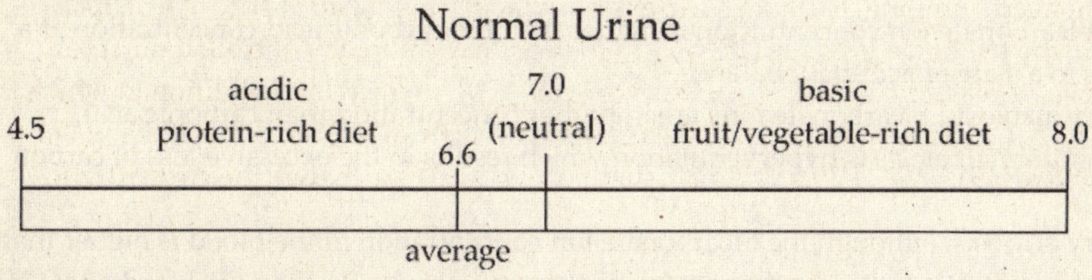

(6) The average pH for urine is acidic.

Section 25.5 Review:

(1) The relationship between water content and fat content in the body is inverse. High fat content corresponds to low water content and vice versa.

(2) Marathon runners generally given sports drinks at aid stations rather than just water because they are sweating profusely and need to replenish not only the fluids lost, but also the electrolyte balance in the body.

(3) Fluids leave the body through urine, feces, perspiration, diffusion through the skin, and exhaled air.

(4) Vasopressin helps reabsorb water in the kidneys. Aldosterone helps reabsorb sodium ions from the kidneys.

Section 25.6 Review:

(1) The $[H^+]$ range for blood in the normal pH range is 3.5×10^{-8} M - 4.5×10^{-8} M.

$$[H^+] = 10^{-pH}$$

$$10^{-7.35} = 4.5 \times 10^{-8}\text{ M}$$

$$10^{-7.45} = 3.5 \times 10^{-8}\text{ M}$$

(2) Normal blood pH is basic.

(3) The three systems to maintain normal blood pH are the buffer system, respiratory system, and urinary system.

Section 25.7 Review:

(1) The blood has three buffer systems that respond to the presence of ketone bodies in the blood and attempt to maintain a "normal" pH.

(2) If the concentration of ketone bodies in the blood is too high, then the buffer capacity of the blood could be overwhelmed and the pH of the blood will change.

(3) According to biochemists, buffering is the replacement of a stronger acid with a weaker acid.

Section 25.8 Review:

(1) During exercise, hyperventilation is more likely to occur than hypoventilation because the rate of breathing is increased. Some athletes develop patterns for breathing to ensure the oxygen in the air breathed into the lungs is exchanged for the carbon dioxide waste products. These patterns help athletes guard against hyperventilation.

(2) During meditation, hypoventilation is more likely to occur than hyperventilation because the rate of breathing is slowed. Most "meditation" instructors remind their students to breathe as they become increasingly relaxed. This generally prevents hypoventilation.

Section 25.9 Review:

(1) The bicarbonate buffer system and the phosphate buffer system are involved in the urinary system.

(2) The phosphate buffer system is most responsible for keeping urine near pH 6.

Section 25.10 Review:

(1) In the blood, the bicarbonate ion concentration is higher than the carbonic acid concentration at a normal pH that is in a state of acid-base balance.

(2) During respiratory alkalosis, bicarbonate ions are in higher concentration than carbonic acid.

(3) The cause of respiratory alkalosis is hyperventilation which results in the excessive loss of carbon dioxide.

(4) During respiratory acidosis, although the bicarbonate ion concentration in the blood is higher than the carbonic acid concentration, the bicarbonate ion concentration is lower than the bicarbonate ion concentration in blood at normal pH.

(5) The cause of respiratory acidosis is hypoventilation which results in the accumulation of carbon dioxide.

(6) During metabolic acidosis, the bicarbonate ion concentration is higher than the carbonic acid concentration; however, the blood contains a lower bicarbonate ion concentration than blood at normal pH.

(7) Metabolic acidosis is caused by excessive H^+ ions in the blood.

(8) During metabolic alkalosis, the bicarbonate ion concentration is greater than the carbonic acid concentration. The bicarbonate ion concentration is also higher than in blood at a normal pH.

(9) Metabolic alkalosis is caused by a loss of acid because of vomiting or ingestion of alkaline substances.

Tying It All Together with a Laboratory Application:

(1) Blood
(2) 7.35-7.45
(3) 4.5-8.0
(4) buffer
(5) phosphate
(6) bloodstream
(7) Na^+
(8) Cl^-
(9) blood pressure
(10) osmotic pressure
(11) urine, feces, perspiration, diffusion from the skin, water vapor
(12) water
(13) aldosterone
(14) alkalosis
(15) raised
(16) acidosis
(17) lowered
(18) 6.8
(19) 7.8
(20) CO_2

SELF-TEST QUESTIONS

Multiple Choice

1. Which of the following fluids contain similar concentrations of Na^+ and Cl^- ions?
   a. blood plasma and interstitial fluid
   b. interstitial fluid and intracellular fluid
   c. blood plasma and intracellular fluid
   d. blood plasma, interstitial fluid, and intracellular fluid

2. A body fluid is analyzed and found to contain a low concentration of Na^+ and a high concentration of K^+. The fluid is most likely:
   a. blood plasma.
   b. interstitial fluid.
   c. intracellular fluid.
   d. all three have about the same concentrations

3. Which of the following reactions takes place in red blood cells at the lungs?
   a. $H_2CO_3 \rightarrow H_2O + CO_2$
   b. $H^+ + HbO_2^- \rightarrow HHb + O_2$
   c. $CO_2 + H_2O \rightarrow H_2CO_3$
   d. $H_2CO_3 \rightarrow H^+ + HCO_3^-$

4. Most oxygen is carried to various parts of the body, via the bloodstream, in the form of:
   a. a dissolved gas.
   b. bicarbonate ion.
   c. oxyhemoglobin.
   d. carbon dioxide.

5. During respiration reactions at the lungs and cells, the function of the chloride shift is to:
   a. eliminate toxic chlorine from the body.
   b. maintain charge balance in red blood cells.
   c. maintain pH balance in red blood cells.
   d. activate the carbonic anhydrase enzyme in red blood cells.

6. Which of the following would be considered to be an abnormal constituent in urine?
   a. ammonium ion
   b. protein
   c. creatinine
   d. bicarbonate salts

7. The vasopressin mechanism and the aldosterone mechanism both tend to regulate the:
   a. fluid and electrolyte levels in the body.
   b. rate of hemoglobin production in the body.
   c. rate of glucose oxidation in the body.
   d. CO_2 levels in the cells.

8. Which of the following maintains a constant pH for the blood?
   a. respiration reactions associated with breathing
   b. kidney activity
   c. formation and excretion of perspiration
   d. more than one response is correct

9. Which buffer system is regulated in part by the kidneys and by the respiratory system?
   a. bicarbonate
   b. protein
   c. phosphate
   d. ammonium

10. The three major buffer systems of the blood are the plasma proteins, the bicarbonate buffer, and the:
   a. succinate buffer.
   b. lactate buffer.
   c. ammonium buffer.
   d. phosphate buffer.

11. Exhaling CO_2 and H_2O:
    a. raises blood pH.
    b. increases the blood concentration of H_2CO_3.
    c. lowers blood pH.
    d. has no effect on H_2CO_3 concentration.

12. What is the normal range for blood pH?
    a. 7.15-7.25 b. 7.25-7.35 c. 7.35-7.45 d. 7.45-7.55

Matching

For each cause of blood acid-base imbalance listed on the left, select the resulting condition from the responses.

13. prolonged diarrhea	a. respiratory acidosis
14. excessive intake of baking soda	b. respiratory alkalosis
15. hypoventilation	c. metabolic acidosis
16. hyperventilation	d. metabolic alkalosis

For each condition listed on the left, identify the resulting abnormal urine constituent on the right.

17. hepatitis	a. ketone bodies
18. starvation	b. glucose (in large amounts)
19. diabetes mellitus	c. bile pigments
	d. more than one constituent results

Select a correct name for each formula on the left.

20. HbO_2^-	a. carbaminohemoglobin
21. HHb	b. oxyhemoglobin
22. $HHbCO_2$	c. deoxyhemoglobin
	d. carboxyhemoglobin

True-False

23. Plasma is classified as an interstitial fluid.
24. Osmotic pressure differences between plasma and interstitial fluid always tend to move fluid into the blood.
25. Osmotic pressure exceeds heart pressure at the venous end of the circulatory system.
26. An increased rate of breathing tends to lower blood pH.
27. The excretion of H^+ ions decreases urine pH.
28. The pH of blood is increased by the excretion of H^+ in urine.
29. Vomiting may give rise to metabolic acidosis.
30. The concentration of protein in plasma is much higher than the protein concentration in the interstitial fluid.
31. The body is 82-90% water.
32. Aldosterone is also called the antidiuretic hormone (ADH).
33. Death can result if blood pH falls below 6.8.
34. H_2CO_3 is a moderately strong acid.
35. Urine is buffered by the phosphate buffer.

ANSWERS TO CHAPTER SELF-TESTS

CHAPTER 1

1. C	8. B	15. A	22. E	29. F
2. C	9. B	16. D	23. F	30. T
3. A	10. C	17. B	24. T	31. T
4. A	11. D	18. A	25. T	32. F
5. D	12. A	19. C	26. F	33. F
6. B	13. C	20. B	27. T	34. F
7. B	14. C	21. C	28. F	35. T

CHAPTER 2

1. D	6. C	11. E	16. A	21. T
2. B	7. B	12. G	17. B	22. T
3. A	8. A	13. F	18. B	23. T
4. C	9. A	14. D	19. F	24. T
5. D	10. E	15. C	20. F	25. F

CHAPTER 3

1. A	8. B	15. D	22. T	29. A
2. D	9. B	16. C	23. B	30. D
3. A	10. C	17. D	24. A	31. A
4. D	11. D	18. T	25. C	32. C
5. C	12. C	19. T	26. B	33. E
6. A	13. C	20. T	27. B	34. A
7. B	14. A	21. T	28. C	35. B

CHAPTER 4

1. A	8. B	15. D	22. F	29. E
2. D	9. B	16. F	23. D	30. A
3. A	10. B	17. F	24. B	31. D
4. D	11. C	18. T	25. B	32. A
5. A	12. B	19. F	26. B	33. C
6. C	13. B	20. F	27. C	34. D
7. D	14. B	21. F	28. A	35. B

CHAPTER 5

1. B	8. D	15. F	22. D	29. C
2. A	9. B	16. T	23. E	30. A
3. D	10. B	17. T	24. C	31. A; Fe
4. D	11. F	18. F	25. A	32. A; H_2
5. B	12. T	19. A	26. D	33. B
6. C	13. T	20. E	27. B	34. B
7. B	14. F	21. B	28. A	35. A; Ag_2O

CHAPTER 6

1. A
2. A
3. B
4. B
5. D
6. A
7. A
8. B
9. B
10. A
11. C
12. D
13. A
14. C
15. F
16. T
17. F
18. T
19. T
20. F
21. T
22. T
23. F
24. F
25. T
26. F
27. F
28. D
29. C
30. B
31. D
32. D
33. C
34. B
35. A

CHAPTER 7

1. B
2. A
3. D
4. B
5. A
6. B
7. A
8. D
9. B
10. C
11. D
12. B
13. D
14. T
15. F
16. F
17. T
18. T
19. T
20. T
21. F
22. T
23. F
24. F
25. T
26. T
27. F
28. F
29. T
30. F
31. C
32. B
33. B
34. A
35. D

CHAPTER 8

1. D
2. A
3. D
4. C
5. A
6. B
7. C
8. B
9. B
10. B
11. T
12. T
13. F
14. T
15. F
16. F
17. T
18. T
19. F
20. F
21. C
22. A
23. B
24. C
25. C
26. A
27. B
28. C
29. A
30. D
31. B
32. A
33. A
34. B
35. C

CHAPTER 9

1. B
2. D
3. D
4. C
5. C
6. D
7. A
8. C
9. D
10. B
11. F
12. T
13. T
14. T
15. F
16. T
17. F
18. F
19. T
20. T
21. F
22. C
23. D
24. A
25. B
26. C
27. B
28. A
29. A
30. B
31. D
32. A
33. C
34. D
35. B
36. A
37. A
38. C
39. B
40. A

CHAPTER 10

1. D
2. A
3. C
4. D
5. D
6. C
7. D
8. A
9. B
10. D
11. T
12. F
13. F
14. T
15. T
16. F
17. F
18. C
19. D
20. B

CHAPTER 11

1. D	9. C	17. A	25. T	33. F
2. A	10. A	18. C	26. F	34. T
3. C	11. C	19. C	27. T	35. F
4. A	12. D	20. B	28. T	36. T
5. B	13. D	21. A	29. F	37. F
6. B	14. A	22. B	30. T	38. T
7. B	15. B	23. F	31. F	39. F
8. A	16. D	24. F	32. T	40. T

CHAPTER 12

1. B	7. B	13. C	19. B	25. T
2. C	8. C	14. A	20. T	26. F
3. D	9. D	15. D	21. T	27. T
4. D	10. C	16. B	22. F	28. T
5. C	11. C	17. C	23. T	29. F
6. C	12. D	18. A	24. F	30. T

CHAPTER 13

1. C	7. B	13. C	19. C	25. T
2. C	8. A	14. D	20. A	26. T
3. A	9. C	15. B	21. F	27. F
4. B	10. A	16. C	22. T	28. T
5. A	11. B	17. A	23. F	29. F
6. B	12. A	18. B	24. T	30. F

CHAPTER 14

1. A	6. A	11. C	16. D	21. F
2. B	7. B	12. D	17. A	22. T
3. B	8. B	13. B	18. B	23. T
4. B	9. A	14. D	19. T	24. F
5. A	10. B	15. C	20. F	25. T

CHAPTER 15

1. B	7. C	13. C	19. A	25. F
2. B	8. C	14. A	20. B	26. T
3. D	9. D	15. B	21. F	27. T
4. B	10. B	16. D	22. F	28. T
5. D	11. B	17. C	23. T	29. F
6. D	12. B	18. D	24. T	30. F

CHAPTER 16

1. A	7. C	13. A	19. F	25. T
2. D	8. D	14. D	20. F	26. T
3. B	9. A	15. B	21. F	27. T
4. B	10. B	16. C	22. T	28. T
5. A	11. D	17. A	23. F	29. T
6. C	12. C	18. T	24. T	30. F

CHAPTER 17

1. B
2. A
3. B
4. C
5. C
6. D
7. B
8. C
9. C
10. A
11. B
12. D
13. B
14. D
15. C
16. B
17. C
18. A
19. C
20. A
21. D
22. T
23. F
24. F
25. T
26. T
27. T
28. T
29. F
30. T

CHAPTER 18

1. B
2. A
3. C
4. B
5. A
6. C
7. C
8. A
9. C
10. A
11. C
12. D
13. B
14. D
15. D
16. C
17. A
18. C
19. A
20. D
21. B
22. F
23. T
24. T
25. T
26. T
27. F
28. F
29. F
30. T

CHAPTER 19

1. C
2. B
3. B
4. A
5. B
6. C
7. D
8. A
9. D
10. D
11. D
12. B
13. B
14. C
15. D
16. A
17. C
18. A
19. B
20. D
21. B
22. D
23. A
24. C
25. C
26. D
27. A
28. C
29. C
30. F
31. T
32. T
33. T
34. F
35. F

CHAPTER 20

1. D
2. A
3. D
4. D
5. D
6. A
7. A
8. C
9. D
10. C
11. B
12. B
13. D
14. C
15. A
16. B
17. C
18. A
19. B
20. C
21. D
22. A
23. B
24. D
25. C
26. T
27. F
28. F
29. F
30. T
31. T
32. F
33. T
34. T
35. F

CHAPTER 21

1. B
2. B
3. D
4. B
5. C
6. A
7. D
8. D
9. A
10. C
11. A
12. A
13. D
14. A
15. B
16. C
17. A
18. C
19. C
20. D
21. B
22. T
23. F
24. F
25. F
26. F
27. T
28. T
29. F
30. F
31. T
32. F
33. T
34. T
35. T

CHAPTER 22

1. B
2. A
3. A
4. C
5. B
6. D
7. B
8. D
9. C
10. A
11. C
12. A
13. B
14. C
15. C
16. A
17. B
18. D
19. C
20. B
21. C
22. B
23. A
24. T
25. F
26. T
27. F
28. F
29. T
30. F
31. F
32. F
33. F
34. F
35. T

CHAPTER 23

1. D
2. B
3. A
4. C
5. C
6. B
7. A
8. C
9. D
10. B
11. B
12. A
13. B
14. D
15. C
16. A
17. B
18. C
19. C
20. B
21. A
22. D
23. C
24. B
25. A
26. B
27. C
28. F
29. F
30. T
31. T
32. T
33. T
34. F
35. F

CHAPTER 24

1. A
2. D
3. D
4. B
5. D
6. A
7. B
8. D
9. C
10. C
11. B
12. A
13. C
14. B
15. A
16. D
17. D
18. B
19. D
20. D
21. D
22. B
23. A
24. T
25. T
26. T
27. T
28. T
29. F
30. T
31. F
32. T
33. T
34. F
35. T

CHAPTER 25

1. A
2. C
3. A
4. C
5. B
6. B
7. A
8. D
9. A
10. D
11. A
12. C
13. C
14. D
15. A
16. B
17. C
18. A
19. D
20. B
21. C
22. A
23. F
24. T
25. T
26. F
27. T
28. T
29. F
30. T
31. F
32. F
33. T
34. F
35. T